GABA in Plants

GABA in Plants

Biosynthesis, Plant Development, and Food Security

Edited by

Dr. Samiksha Singh
Department of Botany
S.N. Sen B.V. Post Graduate College
Chhatrapati Shahu Ji Maharaj University
Kanpur, Uttar Pradesh, India

Dr. Durgesh Kumar Tripathi
Amity Institute of Organic Agriculture (AIOA)
Amity University, Uttar Pradesh, India

Dr. Vijay Pratap Singh
Department of Botany
C.M.P. Degree College
University of Allahabad
Prayagraj, Uttar Pradesh, India

WILEY

This edition first published 2025

Registered Offices
John Wiley & Sons, Inc., 111 River Street, Hoboken, NJ 07030, USA
John Wiley & Sons Ltd, New Era House, 8 Oldlands Way, Bognor Regis, West Sussex, PO22 9NQ, UK

For details of our global editorial offices, customer services, and more information about Wiley products visit us at www.wiley.com.

Wiley also publishes its books in a variety of electronic formats and by print-on-demand. Some content that appears in standard print versions of this book may not be available in other formats.

Library of Congress Cataloging-in-Publication Data applied for:

ISBN 9781394217755 (Cloth)

Cover Design: Wiley
Cover Image: © Hsing Chin/Getty Images

Set in 9.5/12.5pt STIXTwoText by Straive, Pondicherry, India
Printed and bound by CPI Group (UK) Ltd, Croydon, CR0 4YY
C9781394217755_060125

The manufacturer's authorized representative according to the EU General Product Safety Regulation is Wiley-VCH GmbH, Boschstr. 12, 69469 Weinheim, Germany, e-mail: Product_Safety@wiley.com.

Contents

Contributors

Nader Adamipour
Department of Horticultural Science
Faculty of Agriculture
University of Kurdistan
Sanandaj, Iran

Fazilet Özlem Albayrak
Department of Biology
Faculty of Science
Mersin University
Mersin, Turkey

Nimisha Amist
Department of Botany
University of Allahabad
Prayagraj, Uttar Pradesh, India

and

Department of Botany
Ewing Christian College
Prayagraj, Uttar Pradesh, India

Koravantakamparambil Sulaiman Anjitha
Plant Physiology and Biochemistry
Division
Department of Botany
University of Calicut
Malappuram, Kerala, India

Zeba Azim
Department of Botany
Ewing Christian College
University of Allahabad
Prayagraj, Uttar Pradesh, India

and

Plant Physiology Laboratory
Department of Botany
University of Allahabad
Prayagraj, Uttar Pradesh, India

Melike Bor
Department of Biology
Faculty of Science
Ege University
Izmir, Turkey

İrem Çetinkaya
Department of Biology
Faculty of Science
Ege University
Izmir, Turkey

V B Chandana Kumari
Department of Biotechnology
and Bioinformatics
JSS Academy of Higher Education
and Research
Mysuru, Karnataka, India

Petronia Carillo
Department of Environmental, Biological
and Pharmaceutical Sciences
and Technologies
University of Campania "Luigi
Vanvitelli"
Caserta, Italy

Andrea Carra
Council for Agricultural Research and Economics
Research Centre for Forestry and Wood
Casale Monferrato, Italy

Manjunath Dammalli
Department of Biotechnology
Siddaganga Institute of Technology
Tumkur, Karnataka, India

Neeraj Kumar Dubey
Department of Botany
Rashtriya Snatkottar Mahavidyalaya
Jaunpur, Uttar Pradesh, India

Parammal Faseela
Department of Botany
Kahm Unity Women's College
Manjeri, Kerala, India

Nagma Firdose
Department of Pharmacology
JSS Medical College
JSS Academy of Higher Education & Research
Mysuru, Karnataka, India

Bhavya Somaplara Gangadharappa
Department of Biotechnology
M. S. Ramaiah Institute of Technology
Bangalore, Karnataka, India

Srijita Ghosh
Post Graduate Department of Botany
Scottish Church College
Kolkata, West Bengal, India

Nihal Gören-Sağlam
Department of Biology
Faculty of Science
Istanbul University
Istanbul, Turkey

Gubbi Vani Ishika
Department of Biotechnology and Bioinformatics
JSS Academy of Higher Education and Research
Mysuru, Karnataka, India

Lakshmi Jayaram
Department of Biotechnology and Bioinformatics
JSS Academy of Higher Education and Research
Mysuru, Karnataka, India

Kolothodi Chandran Jisha
Department of Botany
MES Asmabi College
Thrissur, Kerala, India

Joy Mulakkal Joel
Plant Physiology and Biochemistry Division
Department of Botany
University of Calicut
Malappuram, Kerala, India

Riya Johnson
Plant Physiology and Biochemistry Division
Department of Botany
University of Calicut
Malappuram, Kerala, India

Shubhra Khare
Department of Applied Sciences and Humanities
Invertis University
Bareilly, Uttar Pradesh, India

Emad Hamdy Khedr
Department of Pomology
Faculty of Agriculture
Cairo University
Giza, Egypt

Rangoli Krishna
Department of Applied Sciences
and Humanities
Invertis University
Bareilly, Uttar Pradesh, India

Kuldeep Lahry
Institut Régional du Cancer de
Montpellier (ICM)
Université de Montpellier
Montpellier, France

Farzad Nazari
Department of Horticultural Science
Faculty of Agriculture
University of Kurdistan
Sanandaj, Iran

Km Niharika
Department of Botany
University of Lucknow
Lucknow, Uttar Pradesh, India

Louis Noble
Plant Physiology and Biochemistry Division
Department of Botany
University of Calicut
Malappuram, Kerala, India

Tülay Öztürk
Department of Biology
Faculty of Science
Ege University
Izmir, Turkey

Şerife Palabıyık
Department of Biology
Faculty of Science
Ege University
Izmir, Turkey

Akhilesh Kumar Pandey
Department of Biotechnology
Faculty of Bioscience and Biotechnology
Invertis University
Bareilly, Uttar Pradesh, India

Jogeswar Panigrahi
Department of Biotechnology
Berhmpur University
Bhanja, Bihar, Odisha, India

Shashank M. Patil
Department of Biotechnology
and Bioinformatics
JSS Academy of Higher Education
and Research
Mysuru, Karnataka, India

V. H. Pushpa
Department of Pharmacology
JSS Medical College
JSS Academy of Higher Education &
Research
Mysuru, Karnataka, India

Jos Thomas Puthur
Plant Physiology and Biochemistry
Division
Department of Botany
University of Calicut
Malappuram, Kerala, India

Deepthi Puttegowda
Department of Biotechnology
and Bioinformatics
JSS Academy of Higher Education
and Research
Mysuru, Karnataka, India

Ranjith Raj
Department of Pharmacology
JSS Medical College
JSS Academy of Higher Education and
Research
Mysuru, Karnataka, India

Kakkuzhiyulla Parambath Raj Aswathi
Plant Physiology and Biochemistry
Division
Department of Botany
University of Calicut
Malappuram, Kerala, India

Ramith Ramu
Department of Biotechnology
and Bioinformatics
JSS Academy of Higher Education
and Research
Mysuru, Karnataka, India

Somayeh Rastegar
Department of Horticultural Sciences
Faculty of Agriculture and Natural
Resources
University of Hormozgan
Bandar Abbas, Iran

Aryadeep Roychoudhury
Discipline of Life Sciences
School of Sciences
Indira Gandhi National Open University
New Delhi, India

Nair Gopalakrishnan Sarath
Department of Botany
Mar Athanasius College
Kothamangalam, Kerala, India

Pegah Sayyad-Amin
Department of Horticultural Science
and Landscape
Ferdowsi University of Mashhad
Mashhad, Iran

Akhila Sen
Department of Botany
Mar Athanasius College
Kothamangalam, Kerala, India

Ajey Singh
Department of Botany
University of Lucknow
Lucknow, Uttar Pradesh, India

Amarjeet Singh
Department of Botany
Post Graduate College
Ghazipur, Uttar Pradesh, India

Kunwar Deelip Singh
Department of Zoology
Rashtriya Snatkottar Mahavidyalaya
Jaunpur, Uttar Pradesh, India

Narsingh Bahadur Singh
Department of Botany
Ewing Christian College
University of Allahabad
Uttar Pradesh, India

and

Plant Physiology Laboratory
Department of Botany
University of Allahabad
Prayagraj, Uttar Pradesh, India

Ran Vijay Singh
Department of Zoology
Rashtriya Snatkottar Mahavidyalaya
Jaunpur, Uttar Pradesh, India

Sudhir Singh
Centre for Molecular Biology
Central University of Jammu
Bagla, Jammu and Kashmir, India

Navya Sreepathi
Department of Biotechnology
and Bioinformatics and Department of
Pharmacology
JSS Academy of Higher Education
and Research
Mysuru, Karnataka, India

Nishtha Srivastava
Department of Applied Sciences &
Humanities
Faculty of Applied Sciences
Invertis University
Bareilly, Uttar Pradesh, India

Jaime A. Teixeira da Silva
Independent Researcher
Ikenobe, Kagawa-ken, Japan

Filiz Vardar
Department of Biology
Faculty of Science
Marmara University
Istanbul, Turkey

Mathew Veena
Plant Physiology and Biochemistry Division
Department of Botany
University of Calicut
Malappuram, Kerala, India

Ravi Kumar Yadav
Department of Botany
Kashi Naresh Government P. G. College
Bhadohi, Uttar Pradesh, India

Satyendra Kumar Yadav
Department of Zoology
Rashtriya Snatkottar Mahavidyalaya
Jaunpur, Uttar Pradesh, India

Vijay Bahadur Yadav
Department of Botany
Rashtriya Snatkottar Mahavidyalaya
Jaunpur, Uttar Pradesh, India

Preface

A four-carbon non-proteinogenic amino acid, gamma-aminobutyric acid (GABA), initially discovered as a metabolite in plants during the mid-20th century in potato tubers. But, the focus shifted to animals after the revelation of its function in neurotransmission especially due to its role in calming anxiety. Recently, this was accompanied by the release of CRISPR-edited and GABA enriched tomatoes in the market. Eventually, the research has also been performed for elucidating function of GABA in plants after the finding of its active role in combating various biotic and abiotic stresses. Because of its unique cellular working, recently, GABA has also been identified as a signalling molecule regulating plant development and stress acclimation. Therefore, it would'nt be unexpected that this four carbon signalling molecule may play a very pivotal role in sustaining crop production in the present scenario of climate change.

The current global population is near 8 billion which is expected to increase to 10 billion by 2050. So, we will be in need of 30% extra crop yields. But current global crop production is severely hampered by the unexpected episode of climate change. Therefore, researchers should seek some strategies for increasing crop production for feeding the global population. Since GABA plays very promising roles in plant growth and development, and stress acclimation, this four carbon metabolite/signaling molecule could be a very promising agent in boosting crop productivity under changing environmental conditions.

In the present book we have tried to compile potential roles of GABA in plant biology from developmental and stress acclimation points of view. In this edited book, attempts has been made to put together plant GABA related research from leading laboratories around the globe in one place, in order to make it easily accessible to researchers, students, academicians, etc. In this book, 18 book chapters have been compiled. These book chapters encompass knowledge about discovery and biosynthetic mechanisms of GABA, GABA crosstalk with other signaling molecules during plant development, and biotic and abiotic stress tolerance, and GABA and plant derived therapeutics. We believe that this book will be useful for the readers from academic and non-academic backgrounds.

Samiksha Singh
Durgesh Kumar Tripathi
Vijay Pratap Singh

1

Discovery and Background of GABA in Plants

Gubbi Vani Ishika[1*], *Deepthi Puttegowda*[1*], *Ranjith Raj*[2], *Manjunath Dammalli*[3], *and Ramith Ramu*[1]

[1] *Department of Biotechnology and Bioinformatics, JSS Academy of Higher Education and Research, Mysuru, Karnataka, 570015, India*
[2] *Department of Pharmacology, JSS Medical College, JSS Academy of Higher Education and Research, Mysuru, Karnataka, 570015, India*
[3] *Department of Biotechnology, Siddaganga Institute of Technology, Tumkur, Karnataka, 572103, India*

Abbreviations

ALMT	Aluminum-activated malate transporters
GABA	Gamma-aminobutyric acid
GAD	Glutamic acid decarboxylase
GHB	Gamma hydroxybutyrate
MDA	Malondialdehyde
ROS	Reactive oxygen species
SlGAD	*Solanum lycopersicum* glutamic acid decarboxylase
SSA	Succinic semialdehyde
Ssdh	Succinic semialdehyde dehydrogenase
TCAC	Tricarboxylic acid cycle

Introduction

Gamma-aminobutyric acid (GABA) is a chemical that helps plants adapt to different growing conditions. GABA is a four-carbon nonproteinogenic amino acid that is present in all plants and plant parts. It has been researched in both eukaryotes and prokaryotes. It was first identified in plants over seven decades ago in potato (*Solanum tuberosum*) tubers. Since then, extensive research has been conducted on its physiological significance, and it has been established that it functions as a signal molecule in plants in addition to being a metabolite. Food security and crop productivity are seriously at risk from the possible effects of climate change on plant development (Li et al. 2021). GABA is a four-C,

* Equal contribution

GABA in Plants: Biosynthesis, Plant Development, and Food Security, First Edition. Edited by Samiksha Singh, Durgesh Kumar Tripathi, and Vijay Pratap Singh.

nonprotein amino acid that makes up 75% of the pool of free amino acids. Prokaryotic and eukaryotic species include GABA. Glutamic acid decarboxylase (GAD), an enzyme found in the cytosol, is the primary source of endogenous GABA. The so-called GABA shunt, which is involved in a variety of physiological processes, including carbon flow into the tricarboxylic acid cycle (TCAC), cytosolic pH regulation, osmoregulation, signaling, and energy production, is how it is metabolized in the mitochondrial matrix (Shelp et al. 2017, 2021).

Its physiological function has been the subject of several research. Furthermore, it has been demonstrated that in plants, it serves as a signaling molecule as well as a metabolite. It has the ability to control plant growth and respond to both biotic and abiotic stresses, among other things (Seifikalhor et al. 2019). GABA functions as a signal for root growth, fruit ripening, pollen tube elongation (to enter the ovule), and seed germination in the context of *Agrobacterium tumefaciens*-mediated plant gene transformation.

In response to drought stress, it emerges as a result of plant reactions that activate antioxidant enzymes and regulate stomatal opening; a high-GABA concentration makes a plant more resilient to stress (Li et al. 2021). GABA's function in both stress-adaptation (salinity, hypoxia/anoxia, drought, temperature, heavy metals, plant–insect interaction, and ROS-related responses) and nonstress-related biological pathways (e.g., plant–microbe interaction, contribution to the carbon and nitrogen metabolism, and regulation of signal transduction pathways) has been specifically highlighted (Seifikalhor et al. 2019 and Kaspal et al. 2021).

History

It was initially found that GABA was a naturally occurring compound in plants in the early to mid-twentieth century, although the credit for its identification in plants goes to a number of researchers who studied different species of plants. The identification of GABA in plants is typically associated with the isolation and characterization of the compound in various plant tissues. These are some of the main figures who helped uncover GABA in plants. The year 1949 saw the discovery of it in plants (potato tuber) (Steward et al. 1949). However, later studies on animals showed that it was also significant in the brains of mammals as a neurotransmitter.

Research on plants gained additional speed when it was discovered that abiotic stress causes a rise in GABA concentration. The fact is that higher plants have been found to contain GABA, several animals, and bacteria as well as invertebrates. GABA was first identified in 1883 as a metabolic byproduct in microorganisms and plants. After the occurrence of GABA in potato tubers was confirmed in 1949, other reports of its presence in the brain emerged. Three distinct investigations on GABA in the mammalian central nervous system were given by Roberts and Frankel (1950) and Roberts (2007) in the same year in the same issue of Biological Chemistry.

Global experts are still working to establish GABA's status as an inhibitory neurotransmitter years later. Ernst Florey extracted a substance from horse brains in 1953 and administered it exogenously to cats and crayfish. They noticed inhibition in a few of their receptors in both studies and aversions. This substance was subsequently separated from the cow brain and recognized as GABA. It was also shown that artificially produced GABA inhibited the same crayfish receptors, confirming Florey's hypothesis that GABA functions in the brain as an inhibitory neurotransmitter. GABA was later proposed to serve as a metabolite and not be engaged in signaling because it is over 1000 times more abundant in the vertebrate brain than other neurotransmitters and participates in the Krebs cycle (GABA shunt). The lack of considerable GABA levels in invertebrates led to the conclusion that GABA is merely a metabolite and fails to meet the criteria for being a neurotransmitter (Florey and McLennan 1955).

Later studies on nematodes, however, demonstrated that GABA is, in fact, a potential neurotransmitter. The absence of it in invertebrates has been disproved by the discovery that it is also detected in crayfish and in larger quantities in Ascaris motor neurons. (**Deo** Rashmi et al. 2018). Ernest H. Wood and Albert F. Haight were among the first to identify and characterize GABA in plants in 1950. They described the presence of GABA in various plant tissues. Arthur M. Andrews and Ray B. Morrison in 1960 contributed to an understanding of GABA in plant tissues. The plasma membrane and organelle membranes are two examples of the membranes that GABA can pass through in a cell. Intercellular and intracellular GABA transport are also involved in this mechanism. It was not until 1999 that GABA transporters were found in plants, having first been identified in animals. According to Ramesh et al. (2018), *Arabidopsis thaliana* is one plant species that can effectively develop when given GABA as its main source of nitrogen (Ramesh et al. 2015). This suggests that plants do indeed have GABA transporters.

Background

The amino acid GABA, not part of the standard proteinogenic amino acids, was initially discovered in 1949 in potato tubers before its identification in animal brain extracts. Growing data in the 1950s and 1960s revealed that GABA may have an inhibitory neurotransmitter role in animals. It has been shown to inhibit crayfish stretch receptor neurons' impulses. However, GABA in mammalian nerve terminals was not identified until the study of Bloom and Iversen in 1971. About 10 years later, the activation of GABAA (ionotropic) and GABAB (metabotropic) receptors was found to be the primary mechanism responsible for GABA's inhibitory neurotransmitter effect. Mammals respond to excitatory neurotransmitters by activating GABAA receptors, which causes membrane hyperpolarization and a modest impact (Bloom and Iversen 1969; Watanabe and Fukuda 2015).

The regulation of brain function and development in mammals is greatly influenced by GABA receptors. Still, in addition to neural cells, these receptors have also been identified in human organs and other tissues. The role of GABA as an animal signaling molecule has been extensively studied after 50 years. In contrast, its comprehension in plants has primarily revolved (Bouch and Fromm 2004). Since the 1990s, there has been growing data that supports GABA's classification as a carbon–nitrogen metabolite and suggests that it may really serve as a signaling drug in plants. The evidence for this statement encompasses several factors, such as the presence of GABA concentration gradients, compartmentalized GABA metabolism, the fluctuating concentration of GABA in plant tissue, and its quick increase in response to different stimuli (Yue et al. 2014).

The control of the plant anion channel family ALMTs by GABA, for example, has led to new insights into the functions of GABA in plants. These insights suggest that changing anion flux across cell membranes is a means of translation of GABA metabolism into membrane signaling. The sequence similarity between animal GABAA receptors and ALMTs is minimal, yet GABA increases GABAA channel activity in mammals while inhibiting ALMT activity in plants. Nonetheless, this regulation leads to a relatively hyperpolarized state in both plant and animal cells Li et al.2016; Nonaka et al. 2017). The alteration in membrane potential due to GABA in plants, influencing tissue growth, strongly indicates its signaling role in plants. GABA's role in plants was initially confined to its definition as a carbon–nitrogen metabolite (Palacios et al. 1981). Nevertheless, since the 1990s, a growing body of evidence has pointed toward GABA's potential function as a signaling molecule in plants. Notable observations include the variability in GABA concentration in plant tissues, its rapid escalation in response to diverse stresses, the existence of concentration gradients, and the impact of GABA on plant growth and development Stress. Moreover, recent discoveries, such as the regulation of plant anion channels (ALMTs) by GABA, have challenged previous notions. While these ALMTs bear little resemblance to animal GABAA receptors, they exhibit sensitivity to GABA in a manner opposite to that observed in animals. Even though adult animal neurons and plants have different equilibrium potentials for chloride, this differential response causes cells to become hyperpolarized (Palacios et al. 1981; Ramesh et al. 2017).

Previously, arguments against GABA acting as a signal in both animals and plants were rooted in its ubiquitous presence and lack of localized gradients. However, analogs to the discoveries in animal biology, localized gradients, and the identification of specific receptors in plants challenge these assumptions, underscoring the potential signaling role of GABA in plant biology. In the area of plant research, there's research to describe the intricacies of GABA-regulated ion channels and their connectivity to physiological processes. This entails exploring the dissimilarities and commonalities between ALMTs and animal GABAA receptors while illuminating GABA's unique impacts on plants and its potential intersections with other signaling molecules. The continual examination and exploration of GABA's multifaceted role in both animal and plant systems offer an intriguing avenue for further research, promising novel insights into the realm of biological signaling and cellular regulation (Ramesh et al. 2017).

The GABA Metabolic Pathway in Plants

In plants, cytosolic GAD converts glutamate into GABA irreversibly. GABA may be produced via oxidative stress-induced breakdown of polyamines or a nonenzymatic mechanism starting from proline. As shown in Figure 1.1, the expression of five GAD-encoding genes in Arabidopsis fluctuates according to the activities and growth conditions of the plant in different plant organs (Bouche and Fromm 2004; Shelp et al. 2012c): GAD is a Ca^{2+}-dependent calmodulin-binding protein that is induced by changes in the neutral pH and cytosolic Ca^{2+} levels in intact plants.

Research on the development of pollen tubes revealed that GAD is positively regulated on Ca^{2+} channels that depend on extracellular GABA signals, i.e., by administering GABA.

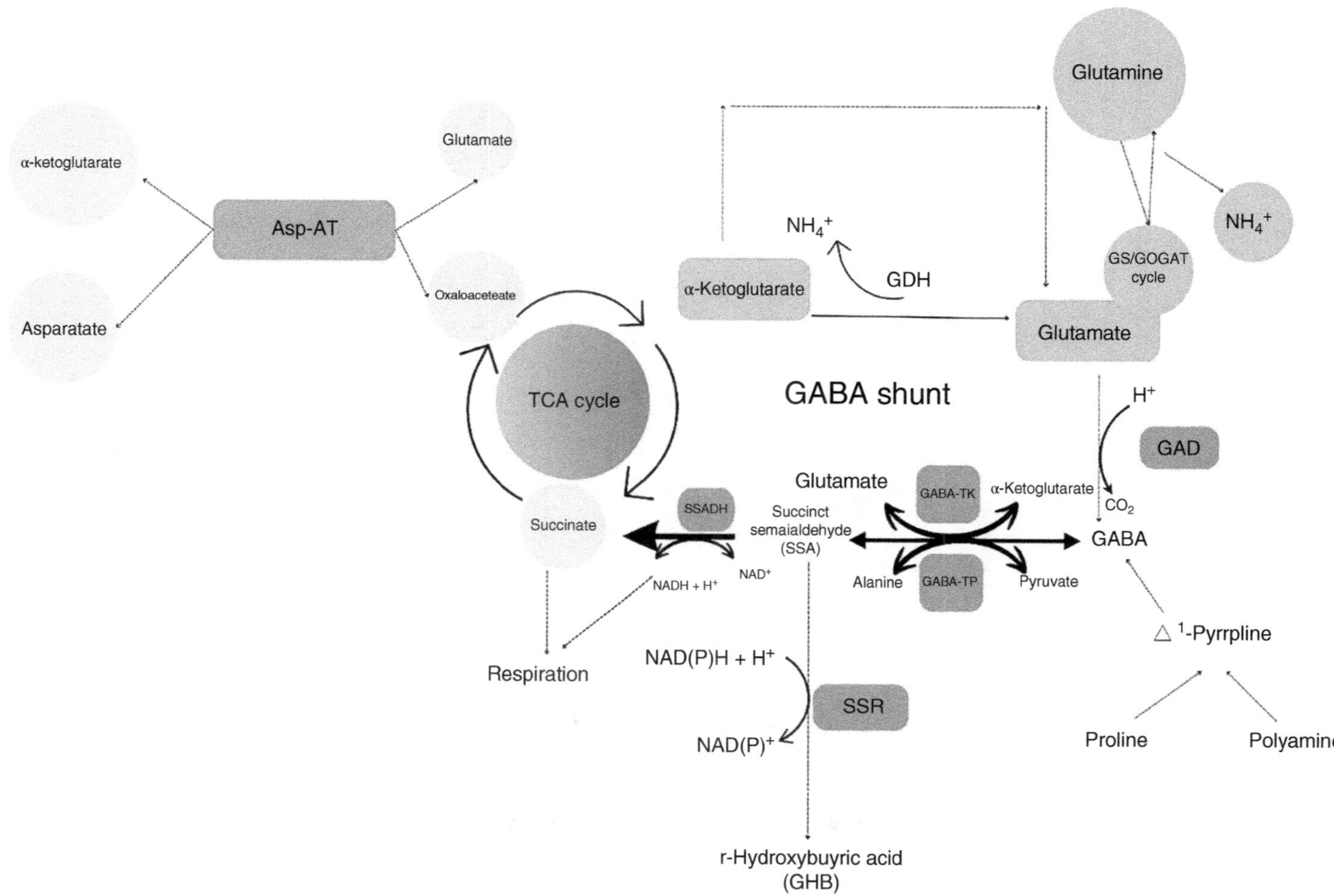

Figure 1.1 Metabolic synthesis of GABA.

As a result, variations in intracellular GABA levels stimulated the production of genes linked to metabolism and signaling. Through cytosolic acidification following plant cell wounding and vacuolar content release, GAD activation was demonstrated to be independent of Ca^{2+} (Carroll et al. 1994).

In the mitochondrial membrane, GABA permease (GABA-P) establishes a connection between the GABA shunt and the TCA cycle. This allows cytosolic GABA to be taken up into the mitochondria, where the TCA cycle ultimately provides energy and carbon skeletons. The GABA shunt, which skips two TCA cycle steps from α-ketoglutarate to succinate to succinyl-CoA, is used to break down GABA (Long et al. 2020). The GABA transaminase in the mitotic matrix changes GABA into succinic semialdehyde (SSA) (GABA-T). According to Ling et al. (2013) and Ramos-Ruiz et al. (2019), there are two types of GABA-T that use either pyruvate or α-ketoglutarate (GABA-TK). As amino acceptors, GABA-TP produces either glutamate or alanine as the final product. The TCA cycle components succinate and NADH is produced when SSA is broken down by SSA dehydrogenase (SSADH). The ultimate result of both chemicals is ATP, which is supplied by electrons to the electron transport chain in the mitochondria. The modulation of SSADH's negative feedback by ATP and NADH. A possible reason for the connection between GABA and pathways other than the TCAC, such as signaling routes. SSA reductase, on the other hand, can convert SSA to γ-hydroxybutyrate (GHB). This enzyme is most likely present in the cytosol or in chloroplasts. Stress from the environment causes GHB to accumulate, and it has been shown that ssadh mutants of Arabidopsis that accumulate GHB are less active in lower chlorophyll concentrations. Fewer blooms and less substance than the wild variety. It's unclear how GABA and GHB are related at this time. By inhibiting the correct TCA cycle enzymes and then compensating with an enhanced flow through the GABA shunt, the GABA shunt's circumvention of the TCA cycle was shown (Ramos-Ruiz et al. 2019). On the other hand, Arabidopsis mutants deficient in the mitochondrial GABA transporter displayed elevated TCA cycle activity and decreased absorption of GABA into the mitochondria. Plants include iron-activated malate transporter proteins or ALMT (Meyer et al. 2006). According to theory, these represent GABA receptors that are often triggered by anions and inhibited by GABA, which is obtained from plant tissue. It has been suggested that the control of ALMT by GABA represents a signaling mechanism, perhaps by altering membrane potentials, which can then initiate physiological alterations throughout the whole plant. It is still unclear how ALMT is distributed and characterized in plant cells and across plant tissue (Ramos-Ruiz et al. 2019; Bouché et al. 2003).

Structure and Conformation of GABA

GABA is primarily found as a zwitterion when the amino group is protonated, and the carboxyl group is deprotonated (Figure 1.2). The environment has an impact on its conformance. According to (Thacker and Popelier 2018), the electrostatic attraction between the two functional groups in the gas phase greatly favors a highly folded conformation. Quantum chemistry simulations indicate that the stabilization is around 50 kcal/mol. An extended shape is seen in the solid state, with a *trans* conformation at the

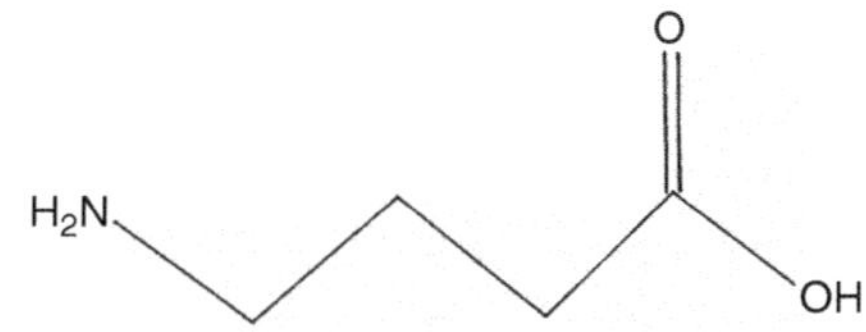

Figure 1.2 Structure of GABA.

amino end and a gauche conformation at the carboxyl end. This is caused by the packing interactions with the molecules in the vicinity. Solvation effects result in the identification of five distinct conformations in solution, some of which are folded and some of which are stretched. Since GABA has been shown to bind to many receptors with various conformations, its conformational flexibility is crucial to its biological function. To improve binding control, several GABA analogs used in pharmaceutical applications have more rigid structures (Majumdar and Guha 1988).

Roles and Functions of GABA

GABA's several roles in plants have been thoroughly outlined in Figure 1.3. GABA is an intrinsic signaling molecule that regulates plant growth and development in addition to its role in metabolism (Renault et al. 2011; Carillo 2018). In addition, it is essential for controlling the conversion of carbon to nitrogen. There is strong evidence that GABA contributes to the development of tolerance to a range of environmental challenges, including salt, low light, drought, nitrogen deficiency, and temperature changes (Kinnersley and Turano 2000). It may also help to prolong and improve the quality of crops while they are in storage, in addition to promoting plant development and stress reduction through the strengthening of antioxidant defense systems.

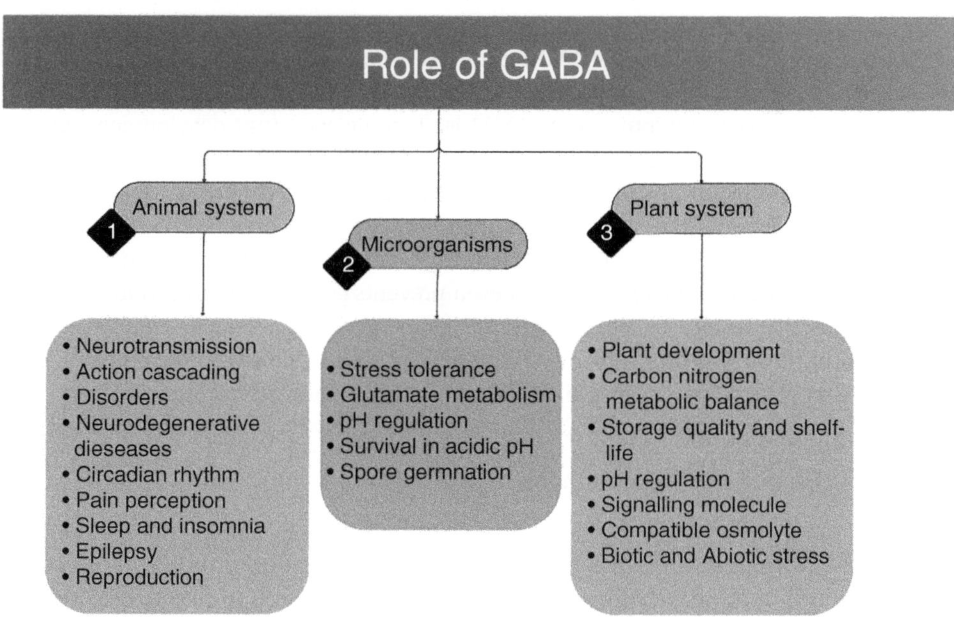

Figure 1.3 Role of GABA.

It has been established that the recycling and redistribution of nitrogen during abiotic stress-induced leaf aging is related to GABA metabolism (Jalil et al. 2017). In addition, studies have demonstrated that it plays a part in protecting plants from biotic stresses such as necrotrophic fungi and insects (Seifi et al. 2013; Bown and Shelp 2016). Michaeli and Fromm (2015) suggest that the various functions of GABA in plants are probably interrelated and difficult to distinguish apart.

Plant Development

Sexual reproduction in flowering plants is crucial for their life cycle. Fertilization cannot take place until the pollen grains rest on the stigma; instead, the pollen tube must grow through the pistil and toward the ovule. The amount of GABA present in the bloom influences the pollen tube's orientation and elongation. GABA creates a gradient along the pollen tube's path, reaching its highest levels near the micropyle, the entry point for the pollen tube into the ovule (Takayama et al. 2017). Studies on *Arabidopsis* mutants lacking POP2, a GABA-T encoding gene, showed that excessive GABA hindered pollen tube growth and disrupted the gradient, leading to misdirected tubes that could not enter the ovule. Experiments on *Picea wilsonii* revealed that both low and high concentrations of GABA influenced pollen tube elongation, where low concentrations promoted growth while excessive amounts inhibited it. High levels of GABA can also impact pollen tube growth by interfering with calcium ion influx and GAD functions, which are crucial in calcium feedback control (Matsuyama et al. 2009).

GABA is involved in the ripening of fruit as well. Under normal circumstances, GABA levels in fruits peak during the fully grown green stage along with the immature seeds, then quickly decline as the fruits ripen and the tomato seeds become mature. Modifying *Solanum lycopersicum* GAD (SlGAD) genes in tomatoes influences GABA levels, affecting plant growth and development. Upregulation of SlGAD2 results in stunted growth and delayed flowering, while overexpressing SlGAD3 causes high GABA levels without causing morphological abnormalities. However, manipulating SlGAD3 leads to delayed fruit development due to altered ethylene production and sensitivity (Xu et al. 2010; Takayama et al. 2017).

In many different types of plants, GABA also influences seed germination and root growth. It influences root growth and facilitates the germination of many seeds. For instance, it activates gene expression related to seed starch degradation during germination. Excessive GABA inhibits primary root elongation and even prevents germination in certain seeds by altering the carbon–nitrogen balance. Furthermore, an abundance of GABA inhibits adventitious root growth in plants like poplar through various mechanisms affecting cell wall metabolism and phytohormone signaling (Zhao et al. 2017; Ramos-Ruiz et al. 2019).

Carbon and Nitrogen Metabolic Balance

For plants to grow and produce, carbon and nitrogen are essential components. For the best possible plant development, these elements must be absorbed efficiently. GABA can serve as a nitrogen store in plants, and through the GABA shunt, carbon structures are integrated into the TCA cycle (Chen et al. 2020). For instance, *A. thaliana* exhibits robust growth when supplied with GABA as the primary nitrogen source. *Medicago truncatula* has a nitrogen deficit, which causes the GABA content in phloem exudates to nearly treble when 50% of the nodules are removed. Supplementing with GABA significantly boosts GABA concentration in nodules, decreases glutamate concentration in exudates, and restores N_2 fixation postexcision after four to five days (Sulieman 2011).

GABA first increases during the "maturation-drying" stage of Arabidopsis seeds, and it subsequently diminishes after germination. Changes in GAD affect the nitrogen-to-carbon ratio in *Arabidopsis* seeds, as shown by the accumulation of GABA in dry seeds in genetically modified plants, a decrease in the number of sugars and organic acids, and a noticeable increase in total protein and different amino acids (Scholz et al. 2015). Significant changes in the amounts of sucrose and starch result from interference with the GABA shunt, which impacts the carbon metabolism of the cell wall. Therefore, it is acknowledged that GABA is the essential component that connects the nitrogen and carbon metabolisms in plants (Ramos-Ruiz et al. 2019; Li et al. 2021).

In poplar seedling stems cultivated in low-nitrogen conditions, the application of exogenous GABA increases TCA intermediates and nonstructural carbs. In addition, in poplar, GABA significantly suppresses the low nitrogen-induced rise in leaf antioxidant enzymes, suggesting that it regulates the C : N ratio for growth by conserving energy in nitrogen-scarce environments (Sulieman and Schulze 2010; Žárský 2015).

Enhancement of Storage Quality and Shelf Life

The utilization of GABA was found to be an effective method in preserving postharvest quality and enhancing storage performance in various fruits like bananas, citrus, cucumber, peaches, and pears. Studies by different authors have shown that GABA treatment leads to increased antioxidant enzyme activity, reduction in peel browning, decreased levels of OSs and malondialdehyde, improved gene expression, heightened activities of antioxidant enzymes, sustained mitochondrial structure, and elevated endogenous GABA concentration in treated fruits (Malekzadeh et al. 2017). Notably, fruits treated with GABA during postharvest periods had superior quality in terms of reduced chilling injury index and weight loss. This was especially true for zucchini and cut flowers. The significant role that GABA components play in avoiding damage caused by cold stress at the cellular level was indicated by increased GABA content and the activity of associated enzymes (GABA-T and GAD) during storage. Moreover, pre- and postharvest GABA treatments were observed to improve cut flower indices, with positive effects on chilling injury and browning when applied at different concentrations and stages, maintaining membrane integrity, enhancing antioxidant capacity, and increasing PAL enzyme activity. Therefore, it was found that using GABA could potentially lessen the effects of chilling harm in cut Anthurium flowers (Palma et al. 2019; Ramos-Ruiz et al. 2019).

pH Regulation

Since GAD consumes H^+ in an acidic environment, GAD activity increases, which raises the possibility that GABA, which is produced during stress, plays a role in regulating pH. Successful demonstrations of GABA build-up in response to acidic conditions in the cytosol have also been made. Weak acid treatment of *Asparagus sprengeri* cells resulted in GABA accumulation and cytosolic pH alterations; GABA concentration increased to 300% in just a few seconds (Crawford et al. 1994). In addition, after ammonium absorption, GAD activity increased in cultivated carrot cells along with a pH decrease and an increase in GABA concentrations. A subsequent rise in pH resulted in a reduction in GAD activity, illustrating GABA's function in reducing cytosolic acidification (Snedden, 2018).

Compatible Osmolyte

Plants can be stressed by a variety of biotic and abiotic factors due to their immobility. One strategy to deal with stressful situations is to produce osmoprotectants or suitable chemicals like proline, GABA, glycine betaine, and carbs (Wang et al. 2020). GABA is widely distributed throughout all plants and plant sections in addition to its build-up, especially under the role it plays as an efficient osmoprotectant molecule, which can be explained under stressful situations. Over time, a number of investigations have demonstrated the build-up of GABA in response to diverse biotic and abiotic stressors. Numerous data indicate that GABA enhances a variety of metabolic processes, photosynthetic activities, relative water levels, osmolyte accumulation, and leaf turgor (Shelp et al. 2012a). Black pepper plants exposed to PEG stress responded well to a pretreatment with GABA, which produced a priming effect that increased the plants' ability to withstand osmotic stress. GABA treatment enhanced the ability of maize plants to photosynthesize, exchange gases, and maintain an antioxidant defense mechanism against oxidative stress. Research into GABA's function in plants is still under progress despite the fact that its role as a neurotransmitter in mammals is well documented. The role of GABA gradient in pollen tube guidance provides evidence in favor of the notion. Certain glutamate receptor-like (GLRs) proteins have been found in plants, and it is anticipated that these proteins will be involved in the control of growth, wound healing, and plant defense signals. Certain GABA receptor agonists and antagonists, like picrotoxin and bicuculline, cause phenotypic changes in plants. Plant cell membranes have been found to include several GABA-binding sites (Shelp et al. 2012b; Vijayakumari and Puthur 2016).

Biotic and Abiotic Stress

The plant's ability to withstand abiotic stress was enhanced by positive regulation of the GABA shunt and related pathways, while endogenous GABA levels were raised by exogenous GABA. It was discovered that applying exogenous GABA to a range of crops, such as lentils, melon, rice, and wheat, was beneficial in lowering germination or growth inhibition. When it was applied in response to unfavorable environmental conditions like extreme heat, drought, water, salt, light, or oxygen stress (Liu et al. 2011).

It has been demonstrated that elevated GAD expression and GABA accumulation are correlated with wheat cultivars' resistance to salt and osmotic stress, which showed a correlation between partially higher photosynthetic capability and antioxidant enzyme activity and improved wheat seedling development under salt stress. The restoration of growth in heat-stressed rice seedlings was also followed by an increase in several antioxidant enzymes. Drought-tolerant barley seedlings showed increased expression of GABA receptor genes. Under stress, the GABA shunt in plants that is controlled by Ca^{2+} through GAD was triggered. In plants under stress, GABA treatment resulted in increased GABA shunt activity, enhanced photosynthetic capacity, increased levels of endogenous GABA, and increased levels of antioxidant enzymes (Aghdam et al. 2015).

ROS increased in tandem with a reduction in malondialdehyde (MDA), a marker of oxidative stress. Aghdam et al. (2015) concluded that the treated plants' membrane integrity was preserved. Potentially, the enzymes that the GABA shunt avoids are impacted by oxidative stress. Thus, reduced efficiency of the TCAC could arise from stressed plants' ability to compensate for this sensitivity through the GABA shunt. To prevent ROS from rising, it appears that a functional GABA shunt is required.

Degradation of proteins and lipids is commonly reported in combination with abiotic stress-induced leaf senescence. Products of degradation are sent to the metabolic process and then dispersed to budding plant sections or ripening fruits. GABA is produced when amino acids are partially converted to glutamate during the breakdown of proteins. Through increased activity of the GABA shunt, specifically GABA-T, and additional downstream pathways, the plant's metabolic system is once again able to access the nitrogen and carbon skeletons. In these studies, *Arabidopsis thaliana* GABA-T mutants were employed (Aghdam et al. 2015; Aghdam et al. 2016).

Temperature Stress

When plants are exposed to extreme temperatures, whether it's cold (low temperatures) or heat (high temperatures), GABA can accumulate in plant tissues. This accumulation helps plants cope with temperature-induced stress by regulating cellular homeostasis and minimizing damage caused by temperature extremes. GABA stabilizes and preserves the integrity of proteins and cell membranes. When plants are stressed by high temperatures, they may receive some protection from the nonprotein amino acid GABA. This study proposes the hypothesis that a decrease in GABA concentrations within the cells of mung bean plants experiencing heat stress leads to increased susceptibility to heat-related damage in their reproductive functions. There are two types of temperature such as low temperature and high temperature (Li et al. 2021).

Low Temperature

Cold temperatures are a major factor that can greatly limit plant productivity. When plants experience cold stress, they often show increased GABA levels. For example, when barley or wheat seedlings are exposed to freezing or low temperatures, GABA buildup increases noticeably, and genes involved in the GABA shunt pathway are prompted to express (Mazzucotelli et al. 2006). In the perennial grass *Brachypodium sylvaticum*, GABA levels significantly rise in response to freezing stress.

High Temperature

Elevated temperatures can be a critical factor limiting plant growth and development. Researchers have examined the relationship between heat stress and GABA in numerous plant species. The application of GABA appears to play a protective role for plants exposed to heat stress. When GABA was externally applied to four-day-old rice seedlings under heat stress conditions, it notably enhanced their growth and survival rates. This improvement was attributed to the enhancement of leaf turgor, the upregulation of osmoprotectants, and antioxidants (Li et al. 2016).

Similar to this, when GABA was administered exogenously, creeping bent grass showed enhanced heat tolerance. Numerous processes were regulated by GABA, including the TCAC, osmotic potential, photosynthesis, and metabolic balance. The application of GABA to heat-stressed plants enhanced carbon fixation, assimilation, and leaf water status by promoting the synthesis of osmolytes and subsequently reducing oxidative damage (Priya et al. 2019).

Drought

Drought, much like the stresses mentioned earlier, acts as a significant impediment to crop development and production. Plants often store more GABA in response to drought stress. Many plants, including turnip, bean, soybean, and sesame, have detached leaves that exhibit this rise in GABA levels. There have also been reports of GABA accumulation brought on by drought in tomatoes, *Phyllanthus* species, and creeping bent grass. According to Ayenew et al. (2015), these results imply that the build-up of GABA during drought is a particular reaction to this stress and helps control the stomata's opening to minimize water loss.

Also, white clover's ability to withstand drought is enhanced when the endogenous GABA content is raised externally. This improvement is attributable to upregulating the GABA shunt, proline metabolism, and polyamines. Improved nitrogen recycling, photosystem II protection, a reduction in drought-induced cell elongation, fatty acid desaturation, wax biosynthesis, and a delay in leaf senescence in creeping bent grass are all linked to GABA's role in improving drought tolerance.

Heavy Metals

Heavy metals are significant soil pollutants that can lead to food contamination by accumulating in the edible parts of crop plants. Heavy metal stress frequently leads to the build-up of ROS in addition to ion toxicity. The amount of GABA in rice roots increases when heavy metals such as chromium (Cr) are present, according to metabolome analysis. Similarly, higher levels of GABA have been seen in soybean plants under stress from zinc (Zn) and copper (Cu) (Kang et al. 2015).

GABA concentrations were lower in plants treated with higher (100 μM) Zn concentrations than in *Nicotiana tabacum* plants treated with moderate (10 μM) Zn values. GABA treatment boosted the resistance of rice seedlings to arsenic (As [III]) stress by encouraging the expression of genes associated with the GABA shunt, triggering the antioxidant enzyme system, and successfully lowering accumulation (Li et al. 2021).

It's interesting to note that long-term GABA accumulation – as opposed to short-term, potentially hazardous GABA elevation – is more effective in promoting As (III) tolerance (Kumar et al. 2017). Conversely, in the presence of cadmium (Cd) stress, endogenous GABA content appeared to decrease in duckweed. Under Cd stress conditions, the addition of exogenous GABA promoted the shedding of root-like structures (rhizoid abscission), while the addition of glutamate (Glu) facilitated rhizoid abscission. In addition, it's worth noting that having a high GABA content does not always guarantee enhanced plant tolerance to metal stress (Ludewig et al. 2008).

ROS

Abiotic stress conditions can lead to the accumulation of ROS in plants. In response to abiotic stress, the GABA shunt is initiated, which essentially limits the generation of ROS in plant tissues (Bouché et al. 2003). The final stage of the GABA shunt is catalyzed by an enzyme known as succinic-semialdehyde dehydrogenase (SSADH). Studies have shown that mutants with T-DNA insertions in the SSADH gene, also known as ssadh SSADH mutants, display specific traits such as reduced leaf area, necrotic lesions, bleached leaves, shorter hypocotyls, and fewer flowers. These mutants, according to Ludwig et al. (2008), show increased susceptibility to heat and UV stress, which leads to the build-up of significant amounts of ROS, which eventually induces cell death in stressed tissues.

Interestingly, it was found that GHB, a consequence of SSA, was five times more numerous in ssadh mutants than in wild-type Arabidopsis. In ssadh mutants, treatment with γ-vinyl-γ-aminobutyrate, a specific GABA transaminase inhibitor (GABA-T/POP2), or a mutation in the POP2 gene reduced cell death, increased proliferation, and inhibited ROS production. Boa et al. (2015) reported that the symptoms of tomato mutants with silenced SlSSADH, which were obtained by the virus-induced gene silencing (VIGS) system, were comparable to those of *Arabidopsis* ssadh mutants. Among these were twisted leaves, slowed growth, and an overabundance of ROS. GHB dehydrogenase can convert GHB into SSA, while SSA reductase can convert SSA into GHB in both plants and animals. In ssadh mutants, SSA and/or GHB build up because SSADH is unable to convert SSA to succinate. Furthermore, compared to wild-type plants, pop2, and pop2ssadh mutant plants have much higher GABA levels. However, the phenotypes of the double pop2ssadh mutants revert to those of the wild type, suggesting that the high-GABA level is not responsible for the phenotype of the ssadh mutants. Instead, increased SSA and/or GHB levels are the cause of the characteristics linked to ssadh mutants (Ludewig et al. 2008).

Salt

Salinity in the soil is a major environmental stressor that affects agricultural productivity worldwide (Snedden, 2018). There have been three main hypothesized cellular responses for salt tolerance in plants: tissue tolerance to the build-up of Na^+ ions, the capacity to reject Na^+ ions, and tolerance to osmotic stress. Numerous researches have shown the build-up of defense compounds in plants and have illuminated the processes behind plants' reactions to salinity (Wang et al. 2018). As an illustration, it has been discovered that the pop2-1 mutant, which is GABA-T defective, is susceptible to salt stress but not osmotic stress. However, salt tolerance rather than sensitivity was shown by the pop2-5 mutant, which accumulates high quantities of GABA in its roots (Liu et al. 2011).

The variations in the reactions of these distinct pop2 mutants could be related to the GABA levels in each of them, as excessive accumulation of GABA above a particular threshold can be harmful to plants. Research employing mutants with decreased capacity to produce GABA, such as pop2-5 and gad1,2, has demonstrated that GABA activates H^+-ATPase, diminishes Na^+ absorption, restricts H_2O_2-induced K^+ efflux, and decreases ROS levels (Su et al. 2019). In addition, GAD activity was not consistently linked with GABA content, despite the fact that GABA accumulation was observed in *Nicotiana sylvestris* and cytoplasmic male sterile (CMS) II plants during both short- and long-term salt stress. Conversely, on the seventh day of the salt treatment, the GABA content of *N. tabacum* plants treated with 500 mM NaCl increased, possibly due to the high-saline level and differences in plant developmental phases (Ludewig et al. 2008). The GABA content of the plants originally dropped on the first and third day of salt treatment.

Additional studies have shown that in saline conditions, the levels of GABA and GAD mRNA in poplar and other wheat cultivars increase significantly. Important metabolic enzymes required for the tricarboxylic acid (TCA) cycle were physicochemically hindered by salt in wheat leaves; however, the elevated activity of the GABA shunt offered an alternative carbon source that allowed the TCA cycle to continue operating within mitochondria. By avoiding salt-sensitive enzymes, this enabled wheat plants to have increased leaf respiration (Che-Othman et al. 2020). When applied to plants that are stressed by salt, exogenous GABA has also been shown to increase the content of endogenous GABA, activate enzymatic antioxidant activity, lessen plant damage from salt, and improve salt tolerance in maize, white clover, muskmelon, germinated hull-less barley, and tomato. Hormones and other stimuli are linked to the build-up of GABA in response to salt. Research has shown that durum wheat exposed to high nitrogen or high light treatments in combination with salt accumulates GABA. GABA serves as a temporary nitrogen storage area as well. It has been shown that giving plants exposed to NaCl GABA has an impact on how much ethylene, H_2O_2, and ABA (abscisic acid) are produced. Exogenous GABA has been demonstrated in studies to modulate several genes related to ABA and ethylene (Vijayakumari and Puthur 2016; Li et al. 2021).

Conclusion

GABA has a different purpose in plants than it does as an animal neurotransmitter. GABA is largely used by plants as a nonprotein amino acid and a signaling molecule in a variety of physiological processes. When it comes to reacting to environmental stresses like salt, drought, and disease attacks, GABA is a key player. These stressors cause plants to store GABA, which aids in ion balance regulation, oxidative stress management, and cellular homeostasis maintenance. The importance of GABA in plant biology can be attributed to its capacity to improve stress tolerance, provide damage protection, and facilitate general environmental adaptation. Understanding GABA's function can help create plant varieties that are more resilient and tolerant of stress, which makes it an important target for research in crop improvement and agriculture.

What People Will Learn by Reading This Chapter

After reading this chapter, readers will have a thorough understanding of the many functions that GABA performs in the field of plant biology. They will learn about the entirely distinct role that GABA, primarily recognized as an animal neurotransmitter, plays in plants. In particular, in response to environmental challenges, including salt, drought, and disease attacks, the chapter explores the many ways in which plants use GABA as a nonprotein amino acid and a signaling molecule. Readers will learn how crucial GABA is in assisting plants in adjusting to challenges as they read on. The chapter emphasizes the role that GABA plays in controlling ion balance, mitigating oxidative stress, and preserving cellular homeostasis. Through understanding these mechanisms, readers will see how GABA basically plays a major role in improving plants' ability to withstand stress, shielding them from harm, and promoting general environmental adaptation. As the chapter highlights the possibilities for utilizing GABA discoveries in crop improvement and agriculture, the practical ramifications of this information become clear. The importance of GABA as a research target will be evident to readers, who may imagine a time when this knowledge will enable the creation of plant varieties that are more resilient and stress-tolerant. In summary, the chapter provides an in-depth examination of the complex function of GABA in plant biology, highlighting its potential to improve agricultural methods, and crop sustainability.

How This Chapter Helps People Due to its Collective Content

Offering a thorough grasp of GABA actions in the particular setting of plant biology, this chapter acts as a knowledge center. Readers will obtain knowledge about the various physiological functions that GABA is used in plants by exploring the roles of GABA as a signaling molecule and nonprotein amino acid.

This chapter's collective content empowers readers by emphasizing the role that GABA plays in improving stress tolerance, offering defense against harm, and promoting general environmental adaptation in plants. In the field of agriculture and agricultural improvement, the knowledge offered becomes an invaluable instrument. Understanding the role of GABA will help researchers and practitioners develop novel approaches to produce plant types that are more stress-tolerant and resilient, which will ultimately aid in the advancement of sustainable farming techniques.

References

Aghdam, M.S., Naderi, R., Sarcheshmeh, M.A., and Babalar, M. (2015). Amelioration of postharvest chilling injury in anthurium cut flowers by γ-aminobutyric acid (GABA) treatments. *Postharvest Biology and Technology* 110: 70–76.

Ayenew, B., Degu, A., Manela, N. et al. (2015). Metabolite profiling and transcript analysis reveal specificities in the response of a berry derived cell culture to abiotic stresses. *Frontiers in Plant Science* 6: 728.

Bao, H., Chen, X., Lv, S. et al. (2015). Virus-induced gene silencing reveals control of reactive oxygen species accumulation and salt tolerance in tomato by γ-aminobutyric acid metabolic pathway. *Plant, Cell & Environment* 38 (3): 600–613.

Bloom, F. E., & Iversen, L. L. (1969). Localizing 3H-GABA in nerve terminals of rat cerebral cortex by electron microscopic autoradiography. *Nature*, 221 (5184), 865–867.

Bloom, F.E. and Iversen, L.L. (1971). Localizing 3H-GABA in nerve terminals of rat cerebral cortex by electron microscopic autoradiography. *Nature* 229 (5287): 628–630.

Bouche, N. and Fromm, H. (2004). GABA in plants: just a metabolite? *Trends in Plant Science* 9 (3): 110–115.

Bouché, N., Fait, A., Bouchez, D. et al. (2003). Mitochondrial succinic-semialdehyde dehydrogenase of the γ-aminobutyrate shunt is required to restrict levels of reactive oxygen intermediates in plants. *Proceedings of the National Academy of Sciences* 100 (11): 6843–6848.

Bown, A.W. and Shelp, B.J. (2016). Plant GABA: not just a metabolite. *Trends in Plant Science* 21 (10): 811–813.

Bown, A.W. and Shelp, B.J. (2020). Does the GABA shunt regulate cytosolic GABA? *Trends in Plant Science* 25 (5): 422–424.

Carillo, P. (2018). GABA shunt in durum wheat. *Frontiers in Plant Science* 9: 100.

Carroll, A.D., Fox, G.G., Laurie, S. et al. (1994). Ammonium assimilation and the role of [gamma]-aminobutyric acid in pH homeostasis in carrot cell suspensions. *Plant Physiology* 106 (2): 513–520.

Chen, W., Meng, C., Ji, J. et al. (2020). Exogenous GABA promotes adaptation and growth by altering the carbon and nitrogen metabolic flux in poplar seedlings under low nitrogen conditions. *Tree Physiology* 40 (12): 1744–1761.

Crawford, L.A., Bown, A.W., Breitkreuz, K.E., and Guinel, F.C. (1994). The synthesis of [gamma]-aminobutyric acid in response to treatments reducing cytosolic pH. *Plant Physiology* 104 (3): 865–871.

Florey, E.R. and McLennan, H. (1955). Effects of an inhibitory factor (factor I) from brain on central synaptic transmission. *The Journal of Physiology* 130 (2): 446.

Che-Othman, M.H., Jacoby, R.P., Millar, A.H., and Taylor, N.L. (2020). Wheat mitochondrial respiration shifts from the tricarboxylic acid cycle to the GABA shunt under salt stress. *New Phytologist* 225 (3): 1166–1180.

Jalil, S.U., Ahmad, I., and Ansari, M.I. (2017). Functional loss of GABA transaminase (GABA-T) expressed early leaf senescence under various stress conditions in Arabidopsis thaliana. *Current Plant Biology* 9: 11–22.

Kang, S.M., Radhakrishnan, R., You, Y.H. et al. (2015). *Enterobacter asburiae* KE17 association regulates physiological changes and mitigates the toxic effects of heavy metals in soybean. *Plant Biology* 17: 1013–1022.

Kaspal, M., Kanapaddalagamage, M.H., and Ramesh, S.A. (2021). Emerging roles of γ aminobutyric acid (GABA) gated channels in plant stress tolerance. *Plants* 10 (10): 2178.

Kinnersley, A.M. and Turano, F.J. (2000). Gamma aminobutyric acid (GABA) and plant responses to stress. *Critical Reviews in Plant Sciences* 19 (6): 479–509.

Kumar, N., Dubey, A.K., Upadhyay, A.K. et al. (2017). GABA accretion reduces Lsi-1 and Lsi-2 gene expressions and modulates physiological responses in *Oryza sativa* to provide tolerance towards arsenic. *Scientific Reports* 7 (1): 8786.

Li, Z., Yu, J., Peng, Y., and Huang, B. (2016). Metabolic pathways regulated by γ-aminobutyric acid (GABA) contributing to heat tolerance in creeping bentgrass (*Agrostis stolonifera*). *Scientific Reports* 6 (1): 30338.

Li, L., Dou, N., Zhang, H., and Wu, C. (2021). The versatile GABA in plants. *Plant Signaling & Behavior* 16 (3): 1862565.

Liu, C., Zhao, L., & Yu, G. (2011). The dominant glutamic acid metabolic flux to produce γ-amino butyric acid over proline in Nicotiana tabacum leaves under water stress relates to its significant role in antioxidant activity. *Journal of Integrative Plant Biology*, 53 (8), 608–618.

Ling, Y., Chen, T., Jing, Y. et al. (2013). γ-Aminobutyric acid (GABA) homeostasis regulates pollen germination and polarized growth in *Picea wilsonii*. *Planta* 238: 831–843.

Long, Y., Tyerman, S.D., and Gilliham, M. (2020). Cytosolic GABA inhibits anion transport by wheat ALMT1. *The New Phytologist* 225 (2): 671–678.

Ludewig, F., Hüser, A., Fromm, H. et al. (2008). Mutants of GABA transaminase (POP2) suppress the severe phenotype of succinic semialdehyde dehydrogenase (ssadh) mutants in *Arabidopsis*. *PLoS One* 3 (10): e3383.

Majumdar, D. and Guha, S. (1988). Conformation, electrostatic potential and pharmacophoric pattern of GABA (gamma-aminobutyric acid) and several GABA inhibitors. *Journal of Molecular Structure: THEOCHEM* 180: 125–140.

Malekzadeh, P., Khosravi-Nejad, F., Hatamnia, A.A., and Sheikhakbari, M.R. (2017). Impact of postharvest exogenous γ-aminobutyric acid treatment on cucumber fruit in response to chilling tolerance. *Physiology and Molecular Biology of Plants* 23: 827–836.

Matsuyama, A., Yoshimura, K., Shimizu, C. et al. (2009). Characterization of glutamate decarboxylase mediating γ-amino butyric acid increase in the early germination stage of soybean (*Glycine max* [L.] Merr). *Journal of Bioscience and Bioengineering* 107 (5): 538–543.

Mazzucotelli, E., Tartari, A., Cattivelli, L., and Forlani, G. (2006). Metabolism of γ-aminobutyric acid during cold acclimation and freezing and its relationship to frost tolerance in barley and wheat. *Journal of Experimental Botany* 57 (14): 3755–3766.

Meyer, A., Eskandari, S., Grallath, S., and Rentsch, D. (2006). AtGAT1, a high affinity transporter for γ-aminobutyric acid in *Arabidopsis thaliana*. *The Journal of Biological Chemistry* 281 (11): 7197–7204.

Michaeli, S. and Fromm, H. (2015). Closing the loop on the GABA shunt in plants: are GABA metabolism and signaling entwined? *Frontiers in Plant Science* 6: 419.

Nonaka, S., Someya, T., Zhou, S., Takayama, M., Nakamura, K., & Ezura, H. (2017). An Agrobacterium tumefaciens strain with gamma-aminobutyric acid transaminase activity shows an enhanced genetic transformation ability in plants. *Scientific reports*, 7 (1), 42649.

Palacios, J.M., Wamsley, J.K., and Kuhar, M.J. (1981). High affinity GABA receptors – autoradiographic localization. *Brain Research* 222 (2): 285–307.

Palma, F., Carvajal, F., Jiménez-Muñoz, R. et al. (2019). Exogenous γ-aminobutyric acid treatment improves the cold tolerance of zucchini fruit during postharvest storage. *Plant Physiology and Biochemistry* 136: 188–195.

Priya, M., Sharma, L., Kaur, R. et al. (2019). GABA (γ-aminobutyric acid), as a thermo-protectant, to improve the reproductive function of heat-stressed mungbean plants. *Scientific Reports* 9 (1): 7788.

Ramesh, S.A., Tyerman, S.D., Xu, B. et al. (2015). GABA signalling modulates plant growth by directly regulating the activity of plant-specific anion transporters. *Nature Communications* 6 (1): 7879.

Ramesh, S.A., Tyerman, S.D., Gilliham, M., and Xu, B. (2017). γ-Aminobutyric acid (GABA) signalling in plants. *Cellular and Molecular Life Sciences* 74: 1577–1603.

Ramesh, S.A., Kamran, M., Sullivan, W. et al. (2018). Aluminum-activated malate transporters can facilitate GABA transport. *The Plant Cell* 30 (5): 1147–1164.

Ramos-Ruiz, R., Martinez, F., and Knauf-Beiter, G. (2019). The effects of GABA in plants. *Cogent Food & Agriculture* 5 (1): 1670553.

Rashmi, D., Zanan, R., John, S. et al. (2018). γ-aminobutyric acid (GABA): Biosynthesis, role, commercial production, and applications. *Studies in Natural Products Chemistry* 57: 413–452.

Renault, H., El Amrani, A., Palanivelu, R. et al. (2011). GABA accumulation causes cell elongation defects and a decrease in expression of genes encoding secreted and cell wall-related proteins in*Arabidopsis thaliana*. *Plant and Cell Physiology* 52: 894–908. https://doi.org/10.1093/pcp/pcr041.

Roberts, M.R. (2007). Does GABA act as a signal in plants? Hints from molecular studies: hints from molecular studies. *Plant Signaling & Behavior* 2 (5): 408–409.

Roberts, E. and Frankel, S. (1950). γ-Aminobutyric acid in brain: its formation from glutamic acid. *The Journal of Biological Chemistry* 187: 55–63.

Scholz, S.S., Reichelt, M., Mekonnen, D.W. et al. (2015). Insect herbivory-elicited GABA accumulation in plants is a wound-induced, direct, systemic, and jasmonate-independent defense response. *Frontiers in Plant Science* 6: 1128.

Seifi, H.S., Curvers, K., De Vleesschauwer, D. et al. (2013). Concurrent overactivation of the cytosolic glutamine synthetase and the GABA shunt in the ABA-deficient sitiens mutant of tomato leads to resistance against *Botrytis cinerea*. *The New Phytologist* 199 (2): 490–504.

Seifikalhor, M., Aliniaeifard, S., Hassani, B. et al. (2019). Diverse role of γ-aminobutyric acid in dynamic plant cell responses. *Plant Cell Reports* 38: 847–867.

Shelp, B.J., Bozzo, G.G., Trobacher, C.P. et al. (2012a). Hypothesis/review: contribution of putrescine to 4-aminobutyrate (GABA) production in response to abiotic stress. *Plant Science* 193: 130–135.

Shelp, B.J., Mullen, R.T., and Waller, J.C. (2012b). Compartmentation of GABA metabolism raises intriguing questions. *Trends in Plant Science* 17 (2): 57–59.

Shelp, B.J., Bozzo, G.G., Zarei, A. et al. (2012c). Strategies and tools for studying the metabolism and function of γ-aminobutyrate in plants. II. Integrated analysis. *Botany* 90 (9): 781–793.

Shelp, B.J., Bown, A.W., and Zarei, A. (2017). 4-Aminobutyrate (GABA): a metabolite and signal with practical significance. *Botany* 95: 1015–1032.

Shelp, B.J., Aghdam, M.S., and Flaherty, E.J. (2021). γ-Aminobutyrate (GABA) regulated plant defense: mechanisms and opportunities. *Plants* 10 (9): 1939.

Snedden, W.A. (2018). Regulation of the γ-aminobutyrate-synthesizing enzyme, glutamate decarboxylase, by calcium-calmodulin: a mechanism for rapid activation in response to stress. In: *Plant Responses to Environmental Stresses* (ed. H.R. Lerner), 549–574. Routledge.

Steward, F.C., Thompson, J.F., and Dent, C.E. (1949). γ-Aminobutyric acid: a constituent of potato tubers? *Science* 110: 439–440.

Su, N., Wu, Q., Chen, J. et al. (2019). GABA operates upstream of H+-ATPase and improves salinity tolerance in Arabidopsis by enabling cytosolic K+ retention and Na+ exclusion. *Journal of Experimental Botany* 70 (21): 6349–6361.

Sulieman, S. (2011). Does GABA increase the efficiency of symbiotic N2 fixation in legumes? *Plant Signaling & Behavior* 6 (1): 32–36.

Sulieman S, Schulze J. Phloem-derived γ-aminobutyric acid (GABA) is involved in upregulating nodule N_2 fixation efficiency in the model legume *Medicago truncatula*. *Plant, Cell and Environment* 2010;33(12):2162–2172.

Seifikalhor, M., Aliniaeifard, S., Hassani, B., Niknam, V., & Lastochkina, O. (2019). Diverse role of γ-aminobutyric acid in dynamic plant cell responses. *Plant cell reports*, 38, 847–867.

Takayama, M., Matsukura, C., Ariizumi, T., and Ezura, H. (2017). Activating glutamate decarboxylase activity by removing the autoinhibitory domain leads to hyper γ-aminobutyric acid (GABA) accumulation in tomato fruit. *Plant Cell Reports* 36: 103–116.

Thacker, J.C. and Popelier, P.L. (2018). Fluorine gauche effect explained by electrostatic polarization instead of hyperconjugation: An interacting quantum atoms (IQA) and relative energy gradient (REG) study. *The Journal of Physical Chemistry A* 122 (5): 1439–1450.

Vijayakumari, K. and Puthur, J.T. (2016). γ-Aminobutyric acid (GABA) priming enhances the osmotic stress tolerance in *Piper nigrum* Linn. plants subjected to PEG-induced stress. *Plant Growth Regulation* 78: 57–67.

Wang, Q., Chen, D., Wu, M. et al. (2018). MFS transporters and GABA metabolism are involved in the self-defense against DON in *Fusarium graminearum*. *Frontiers in Plant Science* 9: 438.

Wang, Y., Xiong, F., Nong, S. et al. (2020). Effects of nitric oxide on the GABA, polyamines, and proline in tea (*Camellia sinensis*) roots under cold stress. *Scientific Reports* 10 (1): 12240.

Watanabe, M. and Fukuda, A. (2015). Development and regulation of chloride homeostasis in the central nervous system. *Frontiers in Cellular Neuroscience* 9: 371.

Xu, J.G., Hu, Q.P., Duan, J.L., and Tian, C.R. (2010). Dynamic changes in γ-aminobutyric acid and glutamate decarboxylase activity in oats (*Avena nuda* L.) during steeping and germination. *Journal of Agricultural and Food Chemistry* 58 (17): 9759–9763.

Yue, X., Gao, X.Q., Wang, F. et al. (2014). Transcriptional evidence for inferred pattern of pollen tube-stigma metabolic coupling during pollination. *PLoS One* 9 (9): e107046.

Žárský, V. (2015). Signal transduction: GABA receptor found in plants. *Nature Plants* 1 (8): 1–2.

Zhao, G.C., Xie, M.X., Wang, Y.C., and Li, J.Y. (2017). Molecular mechanisms underlying γ-aminobutyric acid (GABA) accumulation in giant embryo rice seeds. *Journal of Agricultural and Food Chemistry* 65 (24): 4883–4889.

2

GABA Biosynthesis Pathways and its Signaling in Plants

Nader Adamipour[1], Farzad Nazari[1], and Jaime A. Teixeira da Silva[2]

[1] Department of Horticultural Science, Faculty of Agriculture, University of Kurdistan, Sanandaj, 66177-15175, Iran
[2] Independent Researcher, Ikenobe, Kagawa-ken, 7610799, Japan

GABA Production and Degradation Pathway

γ-Aminobutyric acid (GABA) is a four-carbon amino acid (nonprotein) that has been conserved widely from lower to higher organisms and was first discovered in 1949 in potato (*Solanum tuberosum* L.) tubers (Bouche and Fromm 2004; Yuan et al. 2023). GABA is distributed in the cytoplasm and organelles of plant cells (e.g., mitochondria, peroxisomes, and vacuoles) (Yuan et al. 2023). In plants, GABA plays multiple roles, including in physiological processes associated with growth and development, amino acid biosynthesis, energy production for downstream biosynthetic pathways, and signaling associated with biotic and abiotic stresses (Ramos-Ruiz et al. 2019). In plants, the metabolism of GABA takes place primarily via the GABA shunt, bypassing the tricarboxylic acid (TCA) cycle and involving three enzymes (Bouche and Fromm 2004; Sita and Kumar 2020). To synthesize and maintain optimal levels of GABA, the optimal functioning of the GABA shunt is necessary, which includes three main reactions that are catalyzed by a cytosolic enzyme, glutamate decarboxylase (GAD), and two mitochondrial enzymes, GABA-transaminase (GABA-T) and succinic semialdehyde dehydrogenase (SSADH) (Akihiro et al. 2008). In the cytosol, GAD irreversibly catalyzes the α-decarboxylation of glutamate to GABA, serving as the rate-limiting step of GABA biosynthesis (Figure 2.1) (Michaeli et al. 2011; Podlešáková et al. 2019; Sita and Kumar 2020; Yuan et al. 2023).

In this reaction, one proton is used, and this may reduce cytosolic pH and release a molecule of CO_2 (Shelp et al. 2017; Khan et al. 2021). GAD, a Ca^{2+}-dependent calmodulin (CaM)-binding protein that is stimulated by changes in cytosolic Ca^{2+} levels and has a neutral pH in intact plants (Bouché et al. 2004; Ramos-Ruiz et al. 2019), is not activated by Ca^{2+}/CaM at its acidic pH optimum (Shelp et al. 2017). The number of genes encoding GAD differs among plants. For example, GAD is coded by 5, 3, and 3 genes in *Arabidopsis thaliana* (L.) Heynh. (Ramos-Ruiz et al. 2019), apple (*Malus domestica* Borkh.) (Trobacher

GABA in Plants: Biosynthesis, Plant Development, and Food Security, First Edition. Edited by Samiksha Singh, Durgesh Kumar Tripathi, and Vijay Pratap Singh.

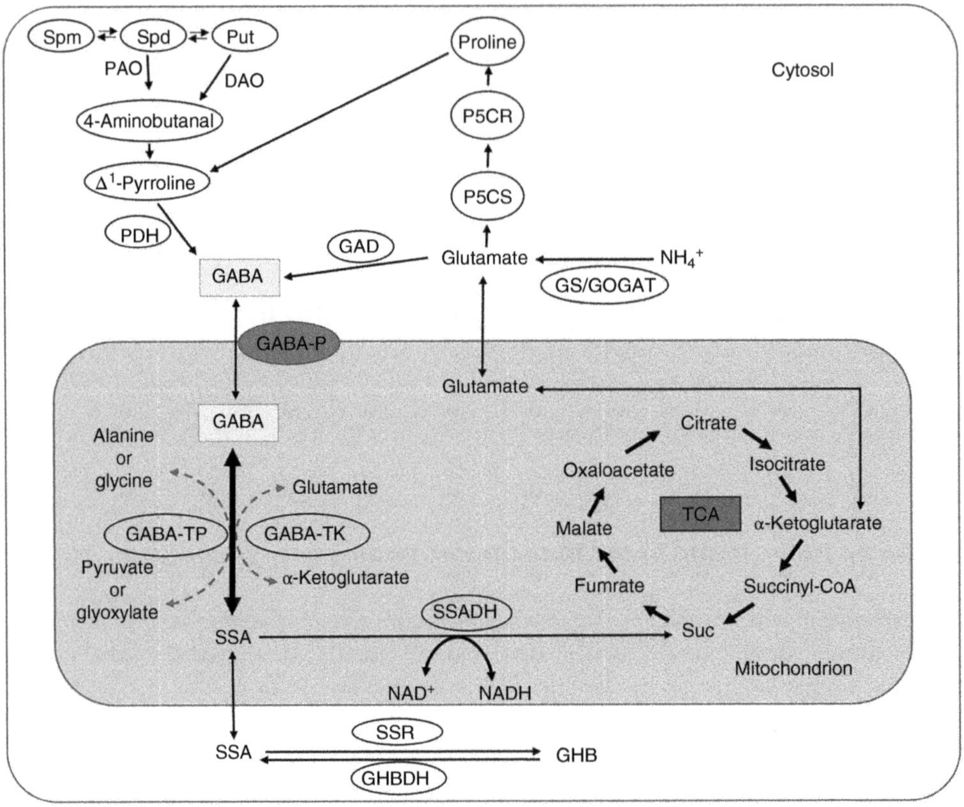

Figure 2.1 GABA biosynthesis and catabolism pathways in plants. Abbreviations: glutamine synthetase/glutamate synthetase (GS/GOGAT), glutamate decarboxylase (GAD), succinate reductase (SSR), succinic semialdehyde (SSA), γ-hydroxybutyric acid (GHB), nicotinamide adenine dinucleotide (NAD^+), GABA-transaminase (GABA-T), NAD^+-dependent succinic semialdehyde dehydrogenase (SSADH), GHB dehydrogenase (GHBDH), NAD^+-dependent succinic semialdehyde dehydrogenase (SSADH), GABA permease (GABA-P), Put, Spd, Spm, diamine oxidases (DAO), polyamine oxidase (PAO), Δ^1-pyrroline dehydrogenase (PDH), P5C synthase (P5CS), and P5C reductase (P5CR) (based on Podlešáková et al. 2019; Yuan et al. 2023).

et al. 2013), and tea (*Camellia sinensis* (L.) Kuntze) (Mei et al. 2016), respectively. The production of glutamate as a precursor of GABA occurs in different locations within the cell (Yuan et al. 2023). Nitrogen is assimilated in the cytoplasm via the glutamine synthetase/glutamate synthetase (GS/GOGAT) cycle, while glutamine and α-ketoglutarate also produce glutamate in plastids (Figure 2.1) (Yuan et al. 2023). Mitochondrial membrane-localized GABA permease (GABA-P) links the GABA shunt and TCA cycle, thus making cytosolic GABA available in mitochondria, with energy and carbon skeletons provided by the TCA cycle (Figure 2.1) (Michaeli et al. 2011). In the mitochondrial matrix, two types of GABA-Ts convert GABA to succinic semialdehyde (SSA) using either α-ketoglutarate (GABA-TK) or pyruvate as amino acid acceptors to produce glutamate or alanine (Figure 2.1) (Bouche and Fromm 2004; Ramos-Ruiz et al. 2019). SSA is then converted by SSADH to succinate and NADH, serving as electron donors in the mitochondrial electron transport chain, resulting in the production of ATP (Figure 2.1) (Ramos-Ruiz et al. 2019; Hasan et al. 2021). In the GABA shunt, GABA bypasses the TCA cycle in two steps: GABA-TK → succinate-CoA → succinate (Figure 2.1) (Ramos-Ruiz et al. 2019; Yuan et al. 2023). In a reversible reaction, SSA is converted to γ-hydroxybutyric acid (GHB), another product of GABA catabolism, by succinate reductase (SSR) (Figure 2.1) (Yuan et al. 2023). GHB also generates SSA under the action of GHB dehydrogenase (GHBDH) (Figure 2.1) (Sita and Kumar 2020; Hasan et al. 2021). Depending on the type of SSR, this reaction can occur in the cytoplasm, mitochondria, or plastids, and like the GABA shunt, GABA's catabolism to GHB is impacted by abiotic or biotic stress, and the detoxification of SSA builds resistance to stress by reducing SSA to GHB (Figure 2.1) (Yuan et al. 2023).

GAD activity was reduced in the CMSII mutant of woodland tobacco (*Nicotiana sylvestris*), but GABA content more than doubled in salinity stress, indicating that the accumulation of GABA can be mediated (Akçay et al. 2012). Therefore, it has been suggested that, in addition to the GABA shunt pathway, GABA can be synthesized from alternative pathways such as those associated with polyamines (PAs) and proline degradation under stress conditions (Wang et al. 2014a; Signorelli et al. 2015; Yuan et al. 2023). In the GABA biosynthetic pathway via proline, a plastid-located nonenzymatic pathway has been proposed in which proline is decarboxylated to form Δ^1-pyrroline, and finally GABA, driven by pyrroline dehydrogenase (Yuan et al. 2023). Another GABA biosynthetic pathway involves the degradation of PAs (putrescine (Put), spermidine (Spd), and spermine (Spm)), explained in detail in Section 4.6. (Zarei et al. 2016; Liao et al. 2017; Yuan et al. 2023).

Is GABA Only a Metabolite?

GABA is not only a primary metabolite but also serves as a signaling molecule in plants (Bouche and Fromm 2004; Fromm 2020). In animals, GABA research has focused mainly on its role, as GABA receptors, as signaling molecules in neurotransmission, while its role was also shown in bacteria, suggesting its role in plants as well (Fromm 2020). Plant-related research to test the function of GABA had focused on the improvement of morphological and physiological characteristics, such as increases in the activity of antioxidant enzymes and relative water content or a reduction of ion leakage, malondialdehyde (MDA), and lipid peroxidation, when GABA was applied to mitigate abiotic stresses (drought, salinity, and heat), in perennial ryegrass (*Lolium perenne* L.) (Krishnan et al. 2013; Ahmadi et al. 2023), wheat (*Triticum aestivum* L.) (Li et al. 2016a; Farooq et al. 2017), sunflower (*Helianthus annuus* L.) (Abdel Razik et al. 2021), strawberry (*Fragaria × ananassa* Duch.) (Golnari et al. 2021) and white clover (*Trifolium repens* sensu Baker f.) (Cheng et al. 2018). None of those studies provided direct evidence of GABA's role in stress-related signaling, whereas the interaction of GABA with reactive oxygen species (ROS), Ca^{2+}, nitric oxide (NO), PAs, amino acids (e.g., proline), hormones, and its role in biotic and abiotic stresses confirmed its role in a signaling network (Khan et al. 2021). Moreover, when the effective role of GABA in *A. thaliana* pollen tube growth and the impact of disturbed GABA catabolism on fertilization were determined, the role of GABA in signaling was suggested (Kaspal et al. 2021). This functionality was fortified by the identification of GABA's role in negatively regulating anion flux through plant-specific aluminum-activated malate transporters (ALMTs) in *A. thaliana*, altering the plasma membrane potential and leading to downstream physiological responses (e.g., stomatal movement and cell turgor) (Xu et al. 2021a, 2021b). In addition, the identification of specific GABA receptors, the possibility of direct action of GABA on ALMTs, the facilitating role of ALMTs in GABA transport, the revelation of the mechanism of GABA's effect in guard cells through ALMTs, the involvement of ALMTs in GABA's responses in pollen, and the possibility of involvement of 14-3-3 proteins, H^+-ATPases, and potassium (K^+) channels in the GABA signaling network are additional evidence that GABA is a signaling molecule in plants (Xu et al. 2021a). In the following sections, we examine in greater detail studies that provided evidence of the role of GABA as a signaling molecule.

Role of GABA in Signaling between Eukaryotes and Bacteria

The rapid increase of GABA in the apoplasm following pathogen infection suggests its role in defense responses and possibly signaling (Ramesh et al. 2017). External application of GABA (5 mM) increased tomato (*Lycopersicon esculentum* L.) resistance to *Botrytis cinerea* (Seifi et al. 2013). The impaired degradation of GABA in the triple mutant strain-*ΔgabT2/T3/T1* was caused by the deletion of three *GABA-T* genes in *Pseudomonas syringae* DC3000. This strain showed a 2.5-fold greater accumulation of GABA than the wild-type strain and attenuated its infection on *A. thaliana* leaves, further reducing colonization in *A. thaliana pop2* mutants, in which GABA-T production was disrupted (Park et al. 2010). These studies support the defensive role of GABA against pathogen infection.

Separately, and as a unique plant-pathosystem, the ability of *Agrobacterium tumefaciens* to transfer DNA via the Ti plasmid, as a strategy to transform plant cells, has made it a useful tool for plant genetic engineering (Li et al. 2021). Following infection, bacterial T-DNA is integrated into the plant genome, and transformed plant cells produce opines, which regulate the quorum sensing (QS) system, and causing crown gall disease (Li et al. 2021). *N*-(3-oxooctanoyl) hemoserine lactone (OC8-HSL), which is degraded by *A. tumefaciens* lactonase AttM, is a QS signal molecule that increases Ti plasmid conjugation and amplification, and the severity of tumor symptoms (Li et al. 2021). GABA increases rapidly during biological stress, such as in wounded tissues (Bor et al. 2009; Ramesh et al. 2017). Tie et al. (2012) noted that japonica rice (*Oryza sativa* L.) showed a higher frequency of transformation than indica rice because, compared to japonica rice, the frequency of T-DNA transfer by *A. tumefaciens* was very low in indica rice. T-DNA transfer was higher in a GABA-deficient tomato line than in the control, indicating that GABA inhibits T-DNA transfer (Nonaka et al. 2017). A new strain of *A. tumefaciens* with amplified GABA transaminase activity could degrade GABA and displayed higher T-DNA transfer in two tomato cultivars and sweetcane (*Erianthus arundinaceus* (Retz.) Jeswiet) (Nonaka et al. 2017). In the *Agrobacterium*–plant interaction, GABA accumulation inactivates OC8-HSL, and high GABA concentrations in tumors suppress Ti plasmid conjugation, reducing the susceptibility to infection (Li et al. 2021). Transgenic tobacco (*Nicotiana tabacum* L.) plants with high GABA content were less susceptible to *A. tumefaciens* C58 infection than wild-type plants (Chevrot et al. 2006). Similarly, plants inoculated with the *atu2422* mutant of *A. tumefaciens* showed more severe crown gall symptoms than the wild type because they were unable to take up GABA (Haudecoeur et al. 2009). This collection of evidence suggests that GABA acts in defense against bacterial pathogens, specifically as a signal between eukaryotes and bacteria.

The Role of GABA Signaling in Regulating Pollen Tube Growth

During angiosperm reproduction, after pollen grains make contact with a compatible stigma, they form a tube that travels through the style to the micropyle, allowing for the transport of sperm cells to the egg, a process whose regulation involves several signals (Palanivelu et al. 2003; Sauter 2009). The pollen–pistil interaction (*POP*) gene mediates the direction and growth of pollen tubes in *A. thaliana* pistils (Wilhelmi and Preuss 1996). In vivo, the growth rate of the pollen tube increases in response to low concentrations of GABA but is delayed when the concentration exceeds 1 mM, showing a biphasic effect of GABA on pollen tube growth (Yue et al. 2014). GABA agonists and antagonists are used to investigate the role of GABA in pollen tube germination. Muscimol and bicuculline can detect GABA receptors in mammals (Ramesh et al. 2015). *A. thaliana* and grapevine (*Vitis vinifera* L.) pollen tube growth was inhibited by muscimol, a GABA receptor agonist, but was antagonized by bicuculline, a competitive GABA antagonist (Ramesh et al. 2017). The *POP2* gene encodes a pyruvate-dependent GABA-T that participates in GABA catabolism and contributes to the formation of a gradient of GABA leading to the micropyle (Palanivelu et al. 2003; Bouche and Fromm 2004). The GABA gradient is disrupted in *A. thaliana pop2* mutant plants because they accumulate GABA (22 μM) in flowers, causing concentration-dependent defects in pollen tube development, guidance, and fertility (Palanivelu et al. 2003; Yu et al. 2014). New evidence emerged of the need to maintain appropriate levels of GABA in pollen tube growth and its function as a signal in the regulation of pollen tube development, as is noted next.

Another important mediator in GABA signaling in pollen tube growth is cytosolic Ca^{2+} (Yu et al. 2014; Michaeli and Fromm 2015). The application of 1 mM GABA to tobacco pollen protoplasts in vitro increased cytosolic Ca^{2+} levels by activating Ca^{2+}-permeable channels in those protoplasts and growing pollen tubes, with GABA coupling the outward K^+ current by activating Ca^{2+}-permeable channels in the plasma membrane of pollen tubes (Yu et al. 2006, 2014). An autoinhibitory domain in the C-terminal part of plant GADs inhibits GAD activity under normal development, but this inhibitory function is abolished when the Ca^{2+}/CaM complex binds to this domain (Chen et al. 1994; Snedden et al. 1996). Ca^{2+} plays an important role in the regulation of pollen tube growth: as cytosolic Ca^{2+} increases, Ca^{2+}/CaM binding releases GAD inhibition and leads to the rapid synthesis of GABA (Franklin-Tong 1999). GAD thus plays an important role in regulating pollen tube growth by acting as a downstream regulator of Ca^{2+}/CaM and feedback control of Ca^{2+}-permeable channels (Yu et al. 2014).

Thus far, the most credible model of the role of GABA as a signal in pollen tube growth was proposed by Xu et al. (2021a). In that model, GABA regulates pollen tube development via its effects on anion efflux and pH, which may occur through a Ca^{2+}-mediated pathway, facilitating the growth of the pollen tube into the ovules during sexual reproduction.

The Role of GABA Signaling in Regulating Stomatal Aperture

Plants respond to environmental stresses by regulating stomatal opening to reduce water loss and absorb more carbon (Kim et al. 2010; Murata et al. 2015). GABA reduces water loss by closing stomata in response to environmental stresses (Silva et al. 2017; Liu et al. 2021; Kumari et al. 2023). ALMTs and outward rectifying potassium channel (GORK) facilitate stomatal function. During stomatal opening, AtALMT6 and AtALMT9 (in the tonoplast) regulate fluxes of malate and Cl^- to the guard cell vacuole while during stomatal closure, AtALMT12 and AtGORK (in the plasma membrane) regulate the efflux of malate and K^+, and different Ca^{2+}-permeable channels contribute to the increase of cytosolic Ca^{2+} during drought stress (Hasan et al. 2021; Kaspal et al. 2021; Shelp et al. 2021). Xu et al. (2021a) observed that during drought, GABA synthesized in leaves reduced AtALMT9-mediated vacuolar anion uptake into guard cells. Mekonnen et al. (2016) noted that GABA downregulation in *A. thaliana gad1/gad2* mutants resulted in stomata remaining open under drought stress. Drought-induced accumulation of GABA in *gad2* and *almt9* mutants suppressed light-induced stomatal opening, which was unaffected under constant light, and stomatal opening in the *almt12* mutant showed wild-type sensitivity to GABA, whereas dark-induced stomatal closure was insensitive to GABA (Xu et al. 2021b). From these studies, it can be concluded that the accumulation of GABA in the cytosol of the guard cell reduces the opening of stomata and, thus, water loss (Shelp et al. 2021). In addition, GABA plays a role in controlling intracellular pH, identified by regulating membrane transport of H^+ through H^+-ATPase activity (Guo et al. 2023). Roots of the *A. thaliana pop-5* mutant with higher GABA content had a more pronounced H^+ current than *gad1* and *gad2*, and H^+-ATPase was involved in this process to improve plant salt tolerance (Su et al. 2019). The general role of GABA in stomatal movement is as follows. The accumulation of GABA during drought disables the influx of anions into the vacuole by binding to ALMT proteins, especially ALMT6. Moreover, 14-3-3 regulatory proteins increase the activity of phototropin 2 (PHOT2) blue light receptor, potassium channel KAT1, and plasma membrane H^+-ATPase, leading to stomatal opening, whereas GABA suppresses the expression of 14-3-3 proteins. In *gad1/2* mutants, stomatal opening during drought due to their inability to produce GABA leads to the accumulation of 3-14 proteins that bind to many H^+-ATPase proteins, thereby stimulating proton pumping. The plasma membrane is hyperpolarized by additional H^+-ATPase activation and leads to a greater import of K^+ (Mekonnen et al. 2016).

Role of the GABA Shunt in Plants

GABA plays key roles in plant growth and development, regulation of carbon : nitrogen (C : N) metabolism, mediating tolerance to biotic and abiotic stresses, improving shelf life and product quality during storage, and increasing the performance of antioxidant defense systems, removal of ROS, and cytosolic pH homeostasis (Ramos-Ruiz et al. 2019).

The Effect of GABA on Plant Growth

GABA plays an important role in mediating plant growth and development (Du et al. 2020). After external application, GABA is absorbed by plant cells and acts like endogenous GABA, causing beneficial changes in morphological and physiological characteristics (Ramos-Ruiz et al. 2019). External application of 10 mM GABA on day 12 during in vitro culture, overexpression of *GAD2*, and the exogenous application of GABA-T inhibitor increased endogenous GABA accumulation and led to a delay and reduction of adventitious root growth in hybrid poplar (*Populus alba* × *P. glandulosa* cv. "84K") (Xie et al. 2020). The excessive accumulation of GABA can result in an imbalance of intracellular amino acid content and lead to aberrant phenotypes (Khan et al. 2021). Aberrant phenotypes such as dwarfism, etiolated leaves, and sterility were observed in transgenic tobacco plants in which the *GAD2ΔC* gene was overexpressed (Akama and Takaiwa 2007). Kinnersley and Lin (2000a) reported an increase in duckweed (*Lemna minor* L.) dry weight following treatment with 10 mM GABA, finding that the promotion of growth by GABA was associated with an increase in the mineral content (such as N, P, K, Ca, Mg, Zn, and Mn) of the treated plants. A foliar spray of 50 mg/l GABA to three varieties of maize (*Zea mays* L.) improved seedling length, fresh and dry weight of shoots and roots, photosynthetic characteristics such as net photosynthesis, stomatal conductance, intercellular CO_2 concentration and transpiration rate, chlorophyll and proline content, and the activity of antioxidant enzymes (superoxide dismutase (SOD) and peroxidase (POD)) (Li et al. 2016b). In addition, the activity of enzymes involved in nitrogen metabolism, such as nitrate reductase and glutamine synthetase, increased (Li et al. 2016b). A foliar spray of 1 mg/l GABA to white gourd (*Benincasa hispida* (Thunb.) Cogn.) 60 days after planting increased characteristics related to leaves, flowers, seeds, fruit, and yield (Ali et al. 2010). In grapes, GABA is present in the leaves, bunches, and branches in equal amounts, and only in tendrils was the content of GABA as much as five times higher, while the treatment of isolated tendrils with GABA (50 μM) produced a transient and reversible positive effect on tendril coiling (Malabarba et al. 2019). Soaking pak-choi (*Brassica campestris* ssp. chinensis Makino) seeds with 5 mg/l GABA improved the growth and yield and increased the activities of nitrate reductase, nitrite reductase, GAD, and protein in the leaves (Li et al. 2016c).

The Effect of GABA on the Regulation of C : N Metabolism

The optimal absorption of carbon (C) and nitrogen (N), the most important factors in maintaining plant growth, development, and metabolism, is essential for plant growth and productivity. The coordination of C and N pools is essential, and a balance of the C/N ratio has a regulatory role in plant growth and development (Stitt 1999; Coruzzi and Zhou 2001; Martin et al. 2002). Plants regulate intracellular C : N balance by partitioning C and N sources and by modulating C and N metabolic activity (Aoyama et al. 2014). The interaction of C and N is controlled by a complex network of signals derived from nitrate, ammonium, and N-containing metabolites (glutamate, glutamine, and aspartate). Signals resulting from C metabolism through CO_2 and C skeletons from the TCA cycle are required for NH^{4+} absorption (Khan et al. 2021). Carbon skeletons enter the TCA cycle through the GABA shunt, whereby GABA can act as an N-storage metabolite in plants (Li et al. 2021). Normally, GABA is biosynthesized from glutamate and is closely related to the TCA cycle in plants (Khan et al. 2021). Suppression of succinyl-CoA ligase (SCoAL) in transgenic tomato plants showed that GABA bypasses a major part of the TCA cycle that converts 2-oxoglutarate to succinate, showing only relatively minor effects on respiration, photosynthesis, and growth. When SCoAL activity was reduced beyond a certain threshold, the GABA shunt was upregulated, and this was mediated at the transcriptional level, allowing normal respiration rate and cellular functions to be maintained in leaves (Studart-Guimarães et al. 2007). The inhibition of 2-oxoglutarate dehydrogenase in the TCA cycle increased the regulation of the GABA shunt in potato tubers, possibly explaining the increased levels of succinate, while the increased activity of the GABA shunt compensates for the lost activity of the TCA cycle enzymes (Araujo et al. 2008; Michaeli and Fromm 2015). Thus, GABA metabolism serves as a moderator in maintaining C/N balance (Fait et al. 2008). *A. thaliana* grew well in culture medium containing 20 mM GABA as a N source (Breitkreuz et al. 1999). Nitrogen deficiency caused by removing 50% of barrel medic (*Medicago truncatula* Gaertn.) nodules almost tripled the concentration of GABA in phloem exudates (Sulieman and Schulze 2010). After artificially exposing petioles to 15 mM GABA, its concentration (1.86 μmol/g nodule FW) in nodules increased, glutamate concentration in phloem secretions decreased, and N_2 fixation improved 4–5 days after harvest (Sulieman and Schulze 2010). The activity of enzymes involved in N and C metabolism such as nitrate reductase, glutamine synthetase, glutamate synthase, and phosphonol pyruvate carboxylase increased in *A. thaliana* root tissue after culture on in vitro medium supplemented with 50 mM GABA, suggesting the possible role of GABA as a signaling molecule and its important role in maintaining C : N pools (Barbosa et al. 2000). The identification of GABA transporters can contribute to understanding the metabolic and signaling roles of GABA in plants. GABA-mediated C metabolism and respiration depend on carrier protein-regulated transport of GABA from the cytosol to the mitochondria (Khan et al. 2021). AtGABP, an APC family GABA transporter, incorporates GABA into the TCA cycle in *A. thaliana* (Michaeli et al. 2011). The GABA transport function of AtGABP became evident as reduced GABA-mediated CO_2 release in *gabp1-4* and *gabp1-5* mutants and from their isolated mitochondria compared to wild-type mitochondria, which was associated with increased TCA cycle metabolism, indicating a tight functional link between the GABA shunt and the TCA cycle that operates in both directions (Michaeli et al. 2011). The 14-3-3 protein also has a regulatory role in C : N metabolism in plants, showing phosphorylation-dependent interactions with other target proteins to perform their functions. In addition, GABA, together with calcium, plays a role in the regulation of the 14-3-3 gene, which is relevant to C/N balance because the 14-3-3 protein affects several key enzymes in C and N metabolism such as glyceraldehyde-3-PDH, starch synthase III, glutamine synthetase, and nitrate reductase (Khan et al. 2021).

The Role of GABA in Improving Shelf Life and the Storage Quality of Products

GABA treatment significantly affects postharvest quality and improves the storage performance of select horticultural products (Ramos-Ruiz et al. 2019). Various studies recommended the use of GABA as a protective approach to reduce the harmful effects of long-term storage. An increase in the activity of GABA-T and GAD during cold storage after harvesting shows the importance of the GABA shunt in providing metabolites for energy production, reducing power, and helping the fruit to cope with long-term cold stress (Yang et al. 2011a; Palma et al. 2019). Nazoori et al. (2020) reported that a coating treatment of 5 and 10 mM GABA (the best recommended concentrations in this study) at

4 ± 1 °C preserved pomegranate (*Punica granatum* L.) for three months. Moreover, GABA effectively improved aril (edible part of seeds) color, fruit firmness, scent, taste, texture, and freshness, and antioxidant activity and content of phenolic compounds in the peel (skin) and aril reducing the symptoms of chilling injury and MDA formation and ion leakage (Nazoori et al. 2020). In a similar study, improvements in pomegranate fruit firmness indicators, titratable acidity, red color of the skin and aril, and higher levels of total phenol and total anthocyanin were reported when 100 mM GABA was used for 60 days of storage at 10 °C (Lorente-Mento et al. 2023). Spraying 10 mM GABA for three minutes on fresh pistachio (*Pistacia vera* L.) fruit inside the hull after 45 days reduced weight loss and total color change significantly more than the control. In addition, the polyphenol oxidase activity of the hull decreased, the antioxidant activity, anthocyanin content of the pistachio skin (integument), and phenols were maintained, while the rate of respiration and microbial activity of fruit decreased at the end of the storage period (Saeedi et al. 2022). A foliar application of GABA (10, 50, and 100 mM) effectively prevented fruit weight loss, softening, total phenolic content, and improved antioxidant activity after 28 days of storage at 2 and 10 °C in lemon (*Citrus limon* (L.) Burm. f.), although 50 mM was the most effective preharvest treatment (Badiche-El Hilali et al. 2023). The treatment of banana (*Musa* spp., AAA group cv. "Brazil") with 20 mM GABA for 20 minutes inhibited chilling injury and maintained fruit quality during 20 days of storage at 7 °C, with GABA decreasing electrolyte leakage, ROS, and MDA content, and increasing total phenolic and proline content (Wang et al. 2014b). In addition, GABA effectively led to the accumulation of proline by increasing the expression of proline biosynthetic genes and suppressing genes involved in its catabolism while enhancing the antioxidant defense system, serving as an approach to inhibit chilling injury and maintain banana quality (Wang et al. 2014b). In similar studies, 5 mM GABA effectively reduced chilling injury in peach (*Prunus persica* L.) by increasing the expression of genes involved in proline synthesis and suppressing genes involved in its catabolism (Shang et al. 2011). The GABA shunt played a key role in the metabolism of organic acids during the postharvest storage of citrus (*Citrus sinensis* L.) fruits (Sheng et al. 2017). The external application of 0.5 mM GABA increased citrate content in fruits by inhibiting the expression of GAD, which, together with an increase in the activity of POS, SOD, and catalase, decreased fruit decay and increased the accumulation of amino acids, including glutamate, alanine, serine, aspartate, and proline (Sheng et al. 2017). Malekzadeh et al. (2017) attributed the effective role of 5 mM GABA in increasing the postharvest lifetime of *Cucumis sativus* L. fruits to the reduced activity of phospholipase C, lipoxygenases, and PLD, enzymes that are involved in the signaling pathway and membrane deterioration during chilling, attributed to an increase in the activity of SOD. GABA at 5 mM regulated the mitochondrial oxidative defense system and maintained mitochondrial structure, effectively reducing the browning of pear (*Pyrus ussuriensis* Maxim. ex Rupr.) fruits in long-term cold storage (0 ± 0.5 °C) after 180 days (Li et al. 2019). Aghdam et al. (2015) applied GABA as a preharvest spray or as a postharvest stem-end dip of anthurium (*Anthurium andraeanum* L.) cut flowers, noting a dose-dependent positive effect on chilling injury and spathe browning, sustained membrane integrity, and increased total antioxidant capacity and phenylalanine ammonia-lyase activity compared to untreated fruit, thus using GABA to reduce the impact of postharvest chilling injury.

Role of GABA in Cytosolic pH Homeostasis

One function of GAD in *Escherichia coli* is to control cellular pH, and GAD activity is stimulated by low pH with maximum activity at pH 3.8 (Kinnersley and Turano 2000b). Initially, it was believed that plant GADs act in a similar way as bacterial GADs by reducing cellular acidosis for three reasons: (i) the maximum activity of all tested plant GADs lay in the acidic range (approximate pH of 5.8); (ii) increased GAD activity was detected in tissues with low cytoplasmic pH; and (iii) GABA biosynthesis consumes a proton and increases pH, leading to the conclusion that GABA metabolism regulates cytosolic pH in plant tissues exposed to stress-induced acidosis (Kinnersley and Turano 2000b). GABA plays an important role in maintaining cytosolic pH homeostasis (Roberts 2007). Decreased cytosolic pH accelerates GABA accumulation in low oxygen conditions such as flooding, further causing mechanical damage and the rupture of vacuolar membranes, leading to the discharge of organic acids into the cytosol and resulting in a decrease in pH and an increase in cytosol acidification (Khan et al. 2021). When cytoplasmic pH decreases, cytosolic Ca^{2+} accumulates at the same time, and this accumulation as a signal causes the decarboxylation of glutamate by GAD (Khan et al. 2021). A proton is consumed in the carboxylation of glutamate by GAD (Carroll et al. 1994), facilitating the accumulation of GABA, which acts as a sink for excess cytosolic protons and maintains pH homeostasis (Khan et al. 2021).

The Impact of GABA on ROS

The accumulation of ROS occurs in unfavorable environmental conditions. ROS can act as signaling molecules and activate signal transduction pathways in response to stress, or their high accumulation can cause oxidative stress and cause serious damage to plants: lipid peroxidation, protein oxidation, inactivation of enzymes, or DNA damage (You and Chan 2015; Farooq et al. 2019; Ansari et al. 2021; Sachdev et al. 2021). GABA helps to reduce ROS in plants. GABA transporters activated by GABA binding facilitate high Ca^{2+} content in the cytosol. Ca^{2+} binds to the CaM-binding domain, activating GAD, which oversees the conversion of glutamate into GABA (Ansari et al. 2021). The accumulation of GABA shunt metabolites in oxidative stress indicates that the GABA shunt pathway, the accumulation of GABA metabolites, and Ca^{+2}/CaM-mediating signaling mechanisms are key components of the antioxidant machinery related to ROS scavenging, hydrogen peroxide (H_2O_2) balance, maintaining a balance of the cellular redox state, and the acquisition of stress tolerance (Al-Quraan 2015). Other components of the GABA shunt show ROS inhibitory activity under stress (Ansari et al. 2021). Fait et al. (2005) reported that reactive oxygen intermediates (ROI) accumulate in *A. thaliana ssadh* mutants, and γ-vinyl-γ-aminobutyrate (a specific γ-aminobutyrate (GABA)-transaminase inhibitor) treatment prevented ROI accumulation in *ssadh* mutants, inhibited cell death and improved growth. Ludewig et al. (2008) reported H_2O_2 accumulation in *ssadh* mutants, as did Bouché et al. (2003), and this was associated with increased cell death, emphasizing the role of the GABA shunt in preventing the accumulation of ROS. To determine the role of specific CaM in plants' tolerance to oxidative stress, ultraviolet (UVA and UVB) treatments were used to induce T-DNA insertion mutations of seven CaM genes in *A. thaliana*, resulting in the accumulation of ROS and induction of oxidative stress in *cam* mutants, especially *cam1*, *cam4*, *cam5-4*, *cam6-1*, and *cam7-1* (Al-Quraan 2015). In tomatoes, silencing *SlGAD* and *SlGABA-T* involved in GABA biosynthesis and catabolism, respectively, increased the accumulation of ROS in the presence of 200 mM NaCl (i.e., salt stress) (Bao et al. 2015). In addition, GABA-T acts as a key factor under stress in plants by reducing oxidative stress via the removal of ROS (Ansari et al. 2021).

Crosstalk of GABA with Other Signaling Molecules

Abscisic Acid

The role of abscisic acid (ABA) signaling in stomatal movement is evident by regulating the concentration of Ca^{2+} in the cytosol both under nonstressed and stressed conditions (Mori and Schroeder 2004). The concentration of Ca^{2+} in the cytosol increases directly or indirectly under stress (Singh et al. 2017). Intracellular increases in Ca^{2+} levels lead to dimerization of the C-terminal domains by Ca^{2+}-CaM, leading to activation of GADs that convert Glu to GABA (Virdi et al. 2015). GABA production in guard cells is necessary to reduce stomatal opening during the negative regulation of local tonoplasts of *A. thaliana* via ALMT9, which is an anion channel that serves as a major pathway for mediating anion (malate and Cl^-) uptake into the vacuole during stomatal opening (Kovermann et al. 2007; Xu et al. 2021b). Defective stomatal closure in *A. thaliana gad1/2* mutants was due to the lack of GABA accumulation (Mekonnen et al. 2016) while high stomatal conductance and drought sensitivity of *gad2* plants were complemented by the loss of ALMT9 (Xu et al. 2021b). In *A. thaliana*, the loss of ALMT9 impaired light-induced stomatal opening and resulted in greater drought tolerance of *almt9* mutants (De Angeli et al. 2013). Depletion of ALMT9 also reduced GABA's ability to inhibit stomatal opening, which was restored by complementation of native ALMT9, suggesting that GABA inhibits stomatal opening by affecting ALMT9 (Xu et al. 2021b). Xu et al. (2021b) reported that 2 mM GABA prevented ABA-induced stomatal closure and that ABA restored stomatal closure to wild-type levels in mutants (*gad2-1* and *almt9*) that were unable to induce GABA and ALMT9 synthesis (Xu et al. 2021b). Collectively, these findings suggest a crosstalk between GABA and ABA-regulated ALMT9 in mediating stomatal closure under drought stress (Kaspal et al. 2021).

Ethylene

Collectively, the positive correlation between successive increases in Ca^{2+}, GABA, and ethylene concentrations (Kinnersley and Turano 2000b), GABA-mediated stress reduction with ethylene, ABA, and H_2O_2 (Ji et al. 2018), and faster accumulation of GABA than ethylene is evidence showing the role of GABA in ethylene signaling (Podlešáková et al. 2019; Suhel et al. 2023). Stimulation of ethylene biosynthesis through increased expression of 1-aminocyclopropane-1-carboxylic acid (ACC) synthase (ACS) and ACC oxidase (ACO) signaling genes by external application of 250 μM and 200 mM GABA, respectively, was reported in long-stalk starwort (*Stellaria longipes* Goldie) and sunflower (Kathiresan et al. 1997, 1998) and also the reduction of ethylene in the roots of *Caragana intermedia* under NaCl (300 mM) stress (Shi et al. 2010). Ethylene inhibitors modulate GABA concentration or *ALMT* expression to maximize malate efflux, and crosstalk between GABA and ethylene reduces malate efflux by regulating TaALMT1 activity (Ramesh et al. 2017). Treatment of root tips of wheat lines with ethylene inhibitors aminoethoxyvinylglycine and amino isobutyric acid promoted Al^{3+}-activated malate efflux from the root tips, while an ethylene donor (Ethrel) inhibited Al^{3+}-induced efflux from tobacco cells upon *TaALMT1* expression (Tian et al. 2014). On the other hand, in wheat, Al^{3+} stress reduced endogenous GABA production, thus increasing malate flow (Ramesh et al. 2015). In addition, the interaction between GABA and ethylene can regulate fruit taste by modulating tonoplast-localized ALMT-mediated malate storage in fruit during ripening and postharvest storage (Xu et al. 2021a). Apple fruits treated with 10 mM GABA increased GABA and malate content and decreased ethylene production after 70 days of storage (Han et al. 2018). Malate storage in tomato and apple is associated with *SlALMT9* and *MdALMT9/MdMa1*, respectively (Ye et al. 2017; Li et al. 2020). *MdALMT9/MdMa1*, the ortholog of *A. thaliana* ALMT9, encodes a tonoplast-localized channel that catalyzes malate uptake into vacuoles and facilitates malate accumulation in apple (Ramesh et al. 2017; Li et al. 2020; Xu et al. 2021a).

Auxins

Crosstalk between auxin and other plant hormones interferes with GABA signaling in various plant processes involved in plant stress tolerance, root growth, nutrient uptake, stress-related ROS response, and metabolism (Podlešáková et al. 2019). In plants, indole-3-acetic acid (IAA) is the most active auxin, and its internal levels change under stress (Podlešáková et al. 2019). In drought, *LEA* overexpression in the halophyte saltwater cress (*Thellungiella salsuginea* (Pall.) O.E. Schulz) led to the accumulation of GABA (Lugan et al. 2010). TLD1/OsGH3.13 encodes IAA-amido synthetase, which increases *LEA* gene expression in response to stress (Angelovici et al. 2010; Min et al. 2012). GABA modulation regulates ALMT activity, and some ALMT proteins (R-type anion channels) are involved in the inhibition of hypocotyl growth by auxin (Thomine et al. 1997).

Cytokinins

The positive role of cytokinin (CK) in increasing stress tolerance through plants transformed with an isopentenyl transferase (*ipt*) gene encoding a CK biosynthetic enzyme was demonstrated in rice and cassava (*Manihot esculenta* Crantz) (Podlešáková et al. 2019). Barley (*Hordeum vulgare* L.) plants expressing the cytokinin dehydrogenase 1 gene from *A. thaliana* (*AtCKX1*) overexpressed *GAD* and *ALMT* genes related to GABA synthesis in roots, showing a clear relationship between GABA and CK (Pospíšilová et al. 2016). The endogenous content of CK in transgenic creeping bentgrass (*Agrostis stolonifera* L.) plants increased after transformation with the *ipt* gene, leading to 38.5% higher GABA accumulation than the wild type in drought stress, and increased tolerance of these transgenic plants was attributed to the higher accumulation of GABA and other metabolites (Merewitz et al. 2012). Transferring the *ipt* gene to tobacco plants to accumulate more CK resulted in the increased accumulation of GABA when exposed to high zinc levels in soil (Pavlíková et al. 2014).

Gibberellins

Overexpression of two DELLA subfamily members – gibberellin (GA) insensitive gene (GAI) – and GA-like repressor 1 (RGL1) in *Populus* seedlings increased GABA levels threefold (Busov et al. 2006). In grapes, GABA levels increased more than twofold by increasing the expression of *VvGAD1* and *VvGABA-T2* genes before pollination during external treatment of 100 mg/l GA (Jung et al. 2017). GABA content increased 42 and 24 hours after germination when the seeds of two cultivars of indica and japonica rice were soaked with 0.25 mg/l GA_3 (Zhang et al. 2014). GABA (50 mM) significantly increased the concentration of IAA, indole butyric acid (IBA), GA_1, and GA_4 in Italian senna (*Cassia italica*, syn. *Senna italica* Mill.) under salt stress (Alqarawi et al. 2016).

Interaction of GABA, PAs, NO, and H_2O_2

Put, Spd, and Spm are among the most important PAs in plants, and their synthesis and catabolism are strongly affected by stresses (Adamipour et al. 2023). Enzymes involved in the oxidative deamination of PAs are amine oxidases (AOs): diamine oxidase (DAO), copper-containing AOs (CuAOs), and flavin adenine dinucleotide-dependent AOs (Adamipour et al. 2020a, 2023). PAs can be converted to GABA under stress, and the pathway is explained next (Shelp et al. 2012). CuAOs convert 1,3-diaminopropane to 3-aminopropanal or Put to 4-aminobutanal (ABAL), H_2O_2, and NH_3. Polyamine oxidase (PAO) converts Spd and Spm to ABAL and *N*-(3-aminopropyl)-4-aminobutanal, respectively, in addition to 1,3-diaminopropane (DAP) and H_2O_2. ABAL is then spontaneously cyclized to form Δ^1-pyrroline, which is converted to GABA by pyrroline dehydrogenase (Podlešáková et al. 2019). Eight putative DAOs have been identified in *A. thaliana*, and their expression is strongly influenced by stress, while their behavior is unlike that of GADs, confirming the mutual relationship between PAs and GABA (Shelp et al. 2012). Xing et al. (2007) reported a 39% greater accumulation of GABA due to the degradation of PAs under salt stress in soybean (*Glycine max* (L.) Merr.) and suggested that PAs might exert their function by forming GABA under stress. About 25% of the GABA formed in tea leaves in anoxic conditions was obtained from the PA degradation pathway (Liao et al. 2017). In addition, in broad bean (*Vicia faba* L.), 30% accumulation of GABA was caused by the degradation of PAs by DAO (Yang et al. 2011b). A number of downstream ABA signaling mediators such as H_2O_2 and NO are involved in stomatal closure. NADPH oxidase is a source of H_2O_2 in *A. thaliana* guard cells, while the degradation of PAs is also a source of H_2O_2 under drought stress (Adamipour et al. 2020a). H_2O_2 from the breakdown of PAs is a signal for stomatal closure under stress (Adamipour et al. 2020b). GABA acts as a secondary messenger by increasing the expression of NADPH oxidase, leading to the production of H_2O_2 (Suhel et al. 2023). In addition, NO is a mediator signaling molecule in response to stresses (Adamipour et al. 2020b). PAs stimulated NO signaling, thereby enhancing cold tolerance in tobacco (Guo et al. 2014). GABA (5 mM) increased NO content by 25% in soybean plants under NaCl stress while cGMP/PKG, produced as a second messenger by NO in soybean, activated the expression of the mitogen-activated protein kinase (MAPK) gene (Xie et al. 2021). In addition, NO-activated MAPK in tobacco through the transient influx of calcium, then MAPK activated two enzymes, DAO and AMADH (NAD^+-dependent amino aldehyde dehydrogenase) (Yin et al. 2014). DAO converts Put to γ-aminobutyraldehyde, which is then converted to GABA via AMADH (Yin et al. 2014). In response to stresses and high ROS levels, Ca^{2+} channels open in the second step by producing NO, which increases Ca^{2+} level and cytosolic Ca^{2+} concentration by regulating the activity of calcium-dependent proteins, playing a role in plants' adaptation to stress (Bor and Turkan 2019; Podlešáková et al. 2019). GABA signaling increased cytosolic Ca^{2+}, which combines with calmodulin to activate GAD (Suhel et al. 2023).

GABA and Proline Interaction

In stress conditions, the role of proline as an osmotic regulator and in the regulation of cellular homeostasis, including redox balance and energy status, is well known. In addition, this amino acid is considered as a signaling molecule (Podlešáková et al. 2019). Proline is a precursor of GABA biosynthesis through a nonenzymatic reaction. Spontaneous decarboxylation of proline leads to the formation of pyrrolidin-1-yl, which can easily be converted to Δ^1-pyrroline, the substrate of Δ^1-pyrroline dehydrogenase, which produces GABA in peroxisomes (Podlešáková et al. 2019). The relationship between GABA and proline in the regulation of stress tolerance is also well researched. For example, in creeping bentgrass plants treated with 5 mmol GABA twice at 2-day intervals had 11–43% higher proline content than control plants under heat stress (plants exposed to 35/30 °C [day/night]) (Li et al. 2016d). A foliar spray of 15 mmol of GABA at low temperatures (2 hours at 7 °C; 6 hours at 15 °C; 4 hours at 12 °C; 2 hours at 7 °C; and 10 hours at 3 °C) increased proline content from 11.25 to 43.28% on days 14 and 35 compared to the control in ruthenia medic (*Medicago ruthenica* (L.) Trautv.) (Li et al. 2022). External application of GABA (5 mM for 10 minutes) increased internal proline content in peach fruits during storage at 1 °C for up to five weeks by increasing the expression of genes involved in the proline biosynthetic pathway (Shang et al. 2011).

Conclusion and Future Prospects

In this chapter, we provided an overview of the biosynthetic pathways of GABA and its role in signaling and physiological processes as a signaling molecule and metabolite, respectively. In addition, GABA serves as an interface that connects several primary and secondary pathways in plants, providing a carbon skeleton and energy, regulating cytosolic pH, regulating nitrogen metabolism, removing ROS, and helping to increase tolerance to biotic and abiotic stresses. The external application of GABA increases the tolerance of plants to environmental stresses, showing its protective role. Additional research is needed to fully understand the Molecular mechanisms of the protective effects of GABA in response to stresses, so the production of transgenic plants with the ability to accumulate GABA will be useful for inducing stress-tolerant plants. Pollen tube growth, regulation of stomatal opening, and defense response against bacterial contamination are strong lines of evidence of the role of GABA in signaling, but research that provides additional details of this pathway is needed. Some challenges and unknowns remain: How is the balance between high levels of GABA and plant growth maintained under stress? How do plant hormones interact with GABA? Do other metabolites such as amino acids and compounds related to GABA metabolism play a role in the regulation of ALMT channels? What is the relationship between GABA and ROS? Perhaps ongoing and future research will provide answers to these questions. Future studies should also aim to identify the role of GABA in regulating proteins or transport channels (such as aquaporins), how GABA binds and interacts with transporters and ion channels, investigate the functional description of GABA in GORK channels, how to regulate the distribution of GABA in tissues, and the existence of alternative GABA transporters.

References

Abdel Razik, E.S., Alharbi, B.M., Pirzadah, T.B. et al. (2021). γ-Aminobutyric acid (GABA) mitigates drought and heat stress in sunflower (*Helianthus annuus* L.) by regulating its physiological, biochemical and Mol pathways. *Physiologia Plantarum* 172 (2): 505–527.

Adamipour, N., Khosh-Khui, M., Salehi, H. et al. (2020a). Regulation of stomatal aperture in response to drought stress mediating with polyamines, nitric oxide synthase and hydrogen peroxide in *Rosa canina* L. *Plant Signaling & Behavior* 15 (9): 1790844.

Adamipour, N., Khosh-Khui, M., Salehi, H. et al. (2020b). Role of genes and metabolites involved in polyamines synthesis pathways and nitric oxide synthase in stomatal closure on *Rosa damascena* Mill. under drought stress. *Plant Physiology and Biochemistry* 148: 53–61.

Adamipour, N., Nazari, F., Vafaee, Y., and Teixeira da Silva, J.A. (2023). Arbuscular mycorrhizal fungus affects genes and metabolites involved in the polyamine biosynthetic pathway of pot marigold under drought stress. *Journal of Plant Growth Regulation* 43: 1–15.

Aghdam, M.S., Naderi, R., Sarcheshmeh, M.A.A., and Babalar, M. (2015). Amelioration of postharvest chilling injury in anthurium cut flowers by γ-aminobutyric acid (GABA) treatments. *Postharvest Biology and Technology* 110: 70–76.

Ahmadi, F., Nazari, F., Ghaderi, N., and Teixeira da Silva, J.A. (2023). Assessment of morpho-physiological and biochemical responses of perennial ryegrass to gamma-aminobutyric acid (GABA) application under salinity stress using multivariate analyses techniques. *Journal of Plant Growth Regulation* 42 (1): 168–182.

Akama, K. and Takaiwa, F. (2007). C-terminal extension of rice glutamate decarboxylase (OsGAD2) functions as an autoinhibitory domain and overexpression of a truncated mutant results in the accumulation of extremely high levels of GABA in plant cells. *Journal of Experimental Botany* 58 (10): 2699–2707.

Akçay, N., Bor, M., Karabudak, T. et al. (2012). Contribution of gamma amino butyric acid (GABA) to salt stress responses of *Nicotiana sylvestris* CMSII mutant and wild type plants. *Journal of Plant Physiology* 169 (5): 452–458.

Akihiro, T., Koike, S., Tani, R. et al. (2008). Biochemical mechanism on GABA accumulation during fruit development in tomato. *Plant & Cell Physiology* 49 (9): 1378–1389.

Ali, M.M., Ashrafuzzaman, M.U.P.M., Ismail, M.R. et al. (2010). Influence of foliar applied GABA on growth and yield contributing characters of white gourd (*Benincasa hispida*). *International Journal of Agriculture and Biology* 12 (3): 373–376.

Alqarawi, A.A., Hashem, A., Elsayed, F.A. et al. (2016). Protective role of gamma amminobutyric acid on *Cassia italica* Mill under salt stress. *Legume Research* 39 (3): 396–404.

Al-Quraan, N.A. (2015). GABA shunt deficiencies and accumulation of reactive oxygen species under UV treatments: insight from *Arabidopsis thaliana* calmodulin mutants. *Acta Physiologia Plantarum* 37: 1–11.

Angelovici, R., Galili, G., Fernie, A.R., and Fait, A. (2010). Seed desiccation: a bridge between maturation and germination. *Trends in Plant Science* 15 (4): 211–218.

Ansari, M.I., Jalil, S.U., Ansari, S.A., and Hasanuzzaman, M. (2021). GABA shunt: a key-player in mitigation of ROS during stress. *Plant Growth Regulation* 94: 131–149.

Aoyama, S., Huarancca Reyes, T., Guglielminetti, L. et al. (2014). Ubiquitin ligase ATL31 functions in leaf senescence in response to the balance between atmospheric CO_2 and nitrogen availability in *Arabidopsis. Plant & Cell Physiology* 55 (2): 293–305.

Araujo, W.L., Nunes-Nesi, A., Trenkamp, S. et al. (2008). Inhibition of 2-oxoglutarate dehydrogenase in potato tuber suggests the enzyme is limiting for respiration and confirms its importance in nitrogen assimilation. *Plant Physiology* 148 (4): 1782–1796.

Badiche-El Hilali, F., Valverde, J.M., Díaz-Mula, H. et al. (2023). Potential preharvest application of γ-aminobutyric acid (GABA) on improving quality of 'Verna' lemon at harvest and during storage. *Agriculture* 13 (7): 1397.

Bao, H., Chen, X., Lv, S. et al. (2015). Virus-induced gene silencing reveals control of reactive oxygen species accumulation and salt tolerance in tomato by γ-aminobutyric acid metabolic pathway. *Plant, Cell and Environment* 38 (3): 600–613.

Barbosa, J.M., Locy, R.D., Barger, T.W. et al. (2000). GABA increases the rate of nitrate uptake and utilization in *Arabidopsis* roots. In: *Plant Tolerance to Abiotic Stresses in Agriculture: Role of Genetic Engineering,*. NATO Science Series, vol. 83 (ed. J.H. Cherry, R.D. Locy, and A. Rychter). Dordrecht: Springer.

Bor, M., Seckin, B., Ozgur, R. et al. (2009). Comparative effects of drought, salt, heavy metal and heat stresses on gamma-aminobutryric acid levels of sesame (*Sesamum indicum* L.). *Acta Physiologia Plantarum* 31: 655–659.

Bor, M. and Turkan, I. (2019). Is there a room for GABA in ROS and RNS signalling? *Environmental and Experimental Botany* 161: 67–73.

Bouche, N. and Fromm, H. (2004). GABA in plants: just a metabolite? *Trends in Plant Science* 9 (3): 110–115.

Bouché, N., Fait, A., Bouchez, D. et al. (2003). Mitochondrial succinic-semialdehyde dehydrogenase of the γ-aminobutyrate shunt is required to restrict levels of reactive oxygen intermediates in plants. *Proceedings of the National Academy of Science of the United States of America* 100 (11): 6843–6848.

Bouché, N., Fait, A., Zik, M., and Fromm, H. (2004). The root-specific glutamate decarboxylase (GAD1) is essential for sustaining GABA levels in *Arabidopsis. Plant Molecular Biology* 55: 315–325.

Breitkreuz, K.E., Shelp, B.J., Fischer, W.N. et al. (1999). Identification and characterization of GABA, proline and quaternary ammonium compound transporters from *Arabidopsis thaliana. FEBS Letters* 450 (3): 280–284.

Busov, V., Meilan, R., Pearce, D.W. et al. (2006). Transgenic modification of gai or rgl1 causes dwarfing and alters gibberellins, root growth, and metabolite profiles in *Populus. Planta* 224: 288–299.

Carroll, A.D., Fox, G.G., Laurie, S. et al. (1994). Ammonium assimilation and the role of [gamma]-aminobutyric acid in pH homeostasis in carrot cell suspensions. *Plant Physiology* 106 (2): 513–520.

Chen, Y., Baum, G., and Fromm, H. (1994). The 58-kilodalton calmodulin-binding glutamate decarboxylase is a ubiquitous protein in petunia organs and its expression is developmentally regulated. *Plant Physiology* 106 (4): 1381–1387.

Cheng, B., Li, Z., Liang, L. et al. (2018). The γ-aminobutyric acid (GABA) alleviates salt stress damage during seeds germination of white clover associated with Na^+/K^+ transportation, dehydrins accumulation, and stress-related genes expression in white clover. *International Journal of Molecular Sciences* 19 (9): 2520.

Chevrot, R., Rosen, R., Haudecoeur, E. et al. (2006). GABA controls the level of quorum-sensing signal in *Agrobacterium tumefaciens. Proceedings of the National Academy of Science of the United States of America* 103 (19): 7460–7464.

Coruzzi, G.M. and Zhou, L. (2001). Carbon and nitrogen sensing and signaling in plants: emerging "matrix effects". *Current Opinion in Plant Biology* 4 (3): 247–253.

De Angeli, A., Zhang, J., Meyer, S., and Martinoia, E. (2013). AtALMT9 is a malate-activated vacuolar chloride channel required for stomatal opening in *Arabidopsis. Nature Communications* 4 (1): 1804.

Du, C., Chen, W., Wu, Y. et al. (2020). Effects of GABA and vigabatrin on the germination of Chinese chestnut recalcitrant seeds and its implications for seed dormancy and storage. *Plants* 9 (4): 449.

Fait, A., Yellin, A., and Fromm, H. (2005). GABA shunt deficiencies and accumulation of reactive oxygen intermediates: insight from *Arabidopsis* mutants. *FEBS Letters* 579 (2): 415–420.

Fait, A., Fromm, H., Walter, D. et al. (2008). Highway or byway: the metabolic role of the GABA shunt in plants. *Trends in Plant Science* 13 (1): 14–19.

Farooq, M., Nawaz, A., Chaudhry, M.A.M. et al. (2017). Improving resistance against terminal drought in bread wheat by exogenous application of proline and gamma-aminobutyric acid. *Journal of Agronomy and Crop Science* 203 (6): 464–472.

Farooq, M.A., Niazi, A.K., Akhtar, J. et al. (2019). Acquiring control: the evolution of ROS-induced oxidative stress and redox signaling pathways in plant stress responses. *Plant Physiology and Biochemistry* 141: 353–369.

Franklin-Tong, V.E. (1999). Signaling in pollination. *Current Opinion in Plant Biology* 2 (6): 490–495.

Fromm, H. (2020). GABA signaling in plants: targeting the missing pieces of the puzzle. *Journal of Experimental Botany* 71 (20): 6238–6245.

Golnari, S., Vafaee, Y., Nazari, F., and Ghaderi, N. (2021). Gamma-aminobutyric acid (GABA) and salinity impacts antioxidative response and expression of stress-related genes in strawberry cv. Aromas. Brazilian Journal of Botany 44: 639–651.

Guo, Z., Gong, J., Luo, S. et al. (2023). Role of gamma-aminobutyric acid in plant defense response. *Metabolites* 13 (6): 741.

Guo, Z., Tan, J., Zhuo, C. et al. (2014). Abscisic acid, H_2O_2 and nitric oxide interactions mediated cold-induced S-adenosylmethionine synthetase in *Medicago sativa* subsp. *falcata* that confers cold tolerance through up-regulating polyamine oxidation. *Plant Biotechnology Journal* 12 (5): 601–612.

Han, S., Nan, Y., Qu, W. et al. (2018). Exogenous γ-aminobutyric acid treatment that contributes to regulation of malate metabolism and ethylene synthesis in apple fruit during storage. *Journal of Agricultural and Food Chemistry* 66 (51): 13473–13482.

Hasan, M.M., Alabdallah, N.M., Alharbi, B.M. et al. (2021). GABA: a key player in drought stress resistance in plants. *International Journal of Molecular Sciences* 22 (18): 10136.

Haudecoeur, E., Planamente, S., Cirou, A. et al. (2009). Proline antagonizes GABA-induced quenching of quorum-sensing in *Agrobacterium tumefaciens*. *Proceedings of the National Academy of Sciences of the United States of America* 106 (34): 14587–14592.

Ji, J., Yue, J., Xie, T. et al. (2018). Roles of γ-aminobutyric acid on salinity-responsive genes at transcriptomic level in poplar: Involving in abscisic acid and ethylene-signalling pathways. *Planta* 248: 675–690.

Jung, C.J., Hur, Y.Y., Moon, J.S., and Jung, S.M. (2017). Pre-bloom application of gibberellin in "Tamnara" grape increases γ-aminobutyric acid (GABA) production at full bloom. *Horticulture, Environment and Biotechnology* 58: 568–575.

Kaspal, M., Kanapaddalagamage, M.H., and Ramesh, S.A. (2021). Emerging roles of γ aminobutyric acid (GABA) gated channels in plant stress tolerance. *Plants* 10 (10): 2178.

Kathiresan, A., Tung, P., Chinnappa, C.C., and Reid, D.M. (1997). [Gamma]-Aminobutyric acid stimulates ethylene biosynthesis in sunflower. *Plant Physiology* 115 (1): 129–135.

Kathiresan, A., Miranda, J., Chinnappa, C.C., and Reid, D.M. (1998). Gamma-aminobutyric acid promotes stem elongation in *Stellaria longipes*: the role of ethylene. *Plant Growth Regulation* 26 (2): 131–137.

Khan, M.I.R., Jalil, S.U., Chopra, P. et al. (2021). Role of GABA in plant growth, development and senescence. *Plant Gene* 26: 100283.

Kim, T.H., Böhmer, M., Hu, H. et al. (2010). Guard cell signal transduction network: advances in understanding abscisic acid, CO_2, and Ca^{2+} signaling. *Annual Review of Plant Biology* 61: 561–591.

Kinnersley, A.M. and Lin, F. (2000a). Receptor modifiers indicate that 4-aminobutyric acid (GABA) is a potential modulator of ion transport in plants. *Plant Growth Regulation* 32 (1): 65–76.

Kinnersley, A.M. and Turano, F.J. (2000b). Gamma aminobutyric acid (GABA) and plant responses to stress. *Critical Reviews in Plant Sciences* 19 (6): 479–509.

Kovermann, P., Meyer, S., Hörtensteiner, S. et al. (2007). The *Arabidopsis* vacuolar malate channel is a member of the ALMT family. *The Plant Journal* 52 (6): 1169–1180.

Krishnan, S., Laskowski, K., Shukla, V., and Merewitz, E.B. (2013). Mitigation of drought stress damage by exogenous application of a non-protein amino acid γ-aminobutyric acid on perennial ryegrass. *Journal of the American Society for Horticultural Science* 138 (5): 358–366.

Kumari, S., Nazir, F., Jain, K., and Khan, M.I.R. (2023). GABA and potassium modulates defence systems, assimilation of nitrogen and carbon, and yield traits under salt stress in wheat. *Journal of Plant Growth Regulation* 42: 6721–6740.

Li, M.F., Guo, S.J., Yang, X.H. et al. (2016a). Exogenous gamma-aminobutyric acid increases salt tolerance of wheat by improving photosynthesis and enhancing activities of antioxidant enzymes. *Biologia Plantarum* 60 (1): 123–131.

Li, W., Liu, J., Ashraf, U. et al. (2016b). Exogenous γ-aminobutyric acid (GABA) application improved early growth, net photosynthesis, and associated physio-biochemical events in maize. *Frontiers in Plant Science* 7: 919.

Li, J.R., Tian, Z., Wu, X.L. et al. (2016c). Regulation of γ-aminobutyric acid on growth and nitrate metabolism of pak-choi treated with high nitrogen application. *Acta Horticulturae Sinica* 43 (11): 2182–2192.

Li, Z., Peng, Y., and Huang, B. (2016d). Physiological effects of γ-aminobutyric acid application on improving heat and drought tolerance in creeping bentgrass. *Journal of the American Society for Horticultural Science* 141 (1): 76–84.

Li, J., Zhou, X., Wei, B. et al. (2019). GABA application improves the mitochondrial antioxidant system and reduces peel browning in "Nanguo" pears after removal from cold storage. *Food Chemistry* 297: 124903.

Li, C., Dougherty, L., Coluccio, A.E. et al. (2020). Apple ALMT9 requires a conserved C-terminal domain for malate transport underlying fruit acidity. *Plant Physiology* 182 (2): 992–1006.

Li, L., Dou, N., Zhang, H., and Wu, C. (2021). The versatile GABA in plants. *Plant Signaling & Behavior* 16 (3): 1862565.

Li, Y., Yu, X., and Ma, K. (2022). Physiological effects of γ-aminobutyric acid application on cold tolerance in *Medicago ruthenica*. *Frontiers in Plant Science* 13: 958029.

Liao, J., Wu, X., Xing, Z. et al. (2017). γ-Aminobutyric acid (GABA) accumulation in tea (*Camellia sinensis* L.) through the GABA shunt and polyamine degradation pathways under anoxia. *Journal of Agricultural and Food Chemistry* 65 (14): 3013–3018.

Liu, C., Wang, H., Zhang, X. et al. (2021). Activation of the ABA signal pathway mediated by GABA improves the drought resistance of apple seedlings. *International Journal of Molecular Sciences* 22 (23): 12676.

Lorente-Mento, J.M., Valero, D., Martínez-Romero, D. et al. (2023). Preharvest multiple applications of GABA improve quality traits and antioxidant compounds of pomegranate fruit during storage. *Horticulturae* 9 (5): 534.

Ludewig, F., Hüser, A., Fromm, H. et al. (2008). Mutants of GABA transaminase (POP2) suppress the severe phenotype of succinic semialdehyde dehydrogenase (*ssadh*) mutants in *Arabidopsis*. *PLoS One* 3 (10): e3383.

Lugan, R., Niogret, M.F., Leport, L. et al. (2010). Metabolome and water homeostasis analysis of *Thellungiella salsuginea* suggests that dehydration tolerance is a key response to osmotic stress in this halophyte. *The Plant Journal* 64 (2): 215–229.

Malabarba, J., Reichelt, M., Pasquali, G., and Mithöfer, A. (2019). Tendril coiling in grapevine: jasmonates and a new role for GABA? *Journal of Plant Growth Regulation* 38: 39–45.

Malekzadeh, P., Khosravi-Nejad, F., Hatamnia, A.A., and Sheikhakbari Mehr, R. (2017). Impact of postharvest exogenous γ-aminobutyric acid treatment on cucumber fruit in response to chilling tolerance. *Physiology and Molecular Biology of Plants* 23: 827–836.

Martin, T., Oswald, O., and Graham, I.A. (2002). *Arabidopsis* seedling growth, storage lipid mobilization, and photosynthetic gene expression are regulated by carbon: nitrogen availability. *Plant Physiology* 128 (2): 472–481.

Mei, X., Chen, Y., Zhang, L. et al. (2016). Dual mechanisms regulating glutamate decarboxylases and accumulation of gamma-aminobutyric acid in tea (*Camellia sinensis*) leaves exposed to multiple stresses. *Scientific Reports* 6 (1): 23685.

Mekonnen, D.W., Flügge, U.I., and Ludewig, F. (2016). Gamma-aminobutyric acid depletion affects stomata closure and drought tolerance of *Arabidopsis thaliana*. *Plant Science* 245: 25–34.

Merewitz, E.B., Du, H., Yu, W. et al. (2012). Elevated cytokinin content in ipt transgenic creeping bentgrass promotes drought tolerance through regulating metabolite accumulation. *Journal of Experimental Botany* 63 (3): 1315–1328.

Michaeli, S. and Fromm, H. (2015). Closing the loop on the GABA shunt in plants: are GABA metabolism and signaling entwined? *Frontiers in Plant Science* 6: 419.

Michaeli, S., Fait, A., Lagor, K. et al. (2011). A mitochondrial GABA permease connects the GABA shunt and the TCA cycle, and is essential for normal carbon metabolism. *The Plant Journal* 67 (3): 485–498.

Min, D.H., Zhang, X.H., Xu, Z.S. et al. (2012). Induction kinetics of a novel stress-related LEA gene in wheat. *Plant Molecular Biology Reporter* 30: 1313–1321.

Mori, I.C. and Schroeder, J.I. (2004). Reactive oxygen species activation of plant Ca^{2+} channels. A signaling mechanism in polar growth, hormone transduction, stress signaling, and hypothetically mechanotransduction. *Plant Physiology* 135 (2): 702–708.

Murata, Y., Mori, I.C., and Munemasa, S. (2015). Diverse stomatal signaling and the signal integration mechanism. *Annual Review of Plant Biology* 66: 369–392.

Nazoori, F., ZamaniBahramabadi, E., Mirdehghan, S.H., and Rafie, A. (2020). Extending the shelf life of pomegranate (*Punica granatum* L.) by GABA coating application. *Journal of Food Measurement and Characterization* 14 (5): 2760–2772.

Nonaka, S., Someya, T., Zhou, S. et al. (2017). An *Agrobacterium tumefaciens* strain with gamma-aminobutyric acid transaminase activity shows an enhanced genetic transformation ability in plants. *Scientific Reports* 7 (1): 42649.

Palanivelu, R., Brass, L., Edlund, A.F., and Preuss, D. (2003). Pollen tube growth and guidance is regulated by *POP2*, an *Arabidopsis* gene that controls GABA levels. *Cell* 114 (1): 47–59.

Palma, F., Carvajal, F., Jiménez-Muñoz, R. et al. (2019). Exogenous γ-aminobutyric acid treatment improves the cold tolerance of zucchini fruit during postharvest storage. *Plant Physiology and Biochemistry* 136: 188–195.

Park, D.H., Mirabella, R., Bronstein, P.A. et al. (2010). Mutations in γ-aminobutyric acid (GABA) transaminase genes in plants or *Pseudomonas syringae* reduce bacterial virulence. *The Plant Journal* 64 (2): 318–330.

Pavlíková, D., Zemanová, V., Procházková, D. et al. (2014). The long-term effect of zinc soil contamination on selected free amino acids playing an important role in plant adaptation to stress and senescence. *Ecotoxicology and Environmental Safety* 100: 166–170.

Podlešáková, K., Ugena, L., Spíchal, L. et al. (2019). Phytohormones and polyamines regulate plant stress responses by altering GABA pathway. *New Biotechnology* 48: 53–65.

Pospíšilová, H., Jiskrova, E., Vojta, P. et al. (2016). Transgenic barley overexpressing a cytokinin dehydrogenase gene shows greater tolerance to drought stress. *New Biotechnology* 33 (5): 692–705.

Ramesh, S.A., Tyerman, S.D., Xu, B. et al. (2015). GABA signalling modulates plant growth by directly regulating the activity of plant-specific anion transporters. *Nature Communications* 6 (1): 7879.

Ramesh, S.A., Tyerman, S.D., Gilliham, M., and Xu, B. (2017). γ-Aminobutyric acid (GABA) signalling in plants. *Cellular and Molecular Life Sciences* 74: 1577–1603.

Ramos-Ruiz, R., Martinez, F., and Knauf-Beiter, G. (2019). The effects of GABA in plants. *Cogent Food & Agriculture* 5 (1): 1670553.

Roberts, M.R. (2007). Does GABA act as a signal in plants? Hints from molecular studies. *Plant Signaling & Behavior* 2 (5): 408–409.

Sachdev, S., Ansari, S.A., Ansari, M.I. et al. (2021). Abiotic stress and reactive oxygen species: generation, signaling, and defense mechanisms. *Antioxidants* 10 (2): 277.

Saeedi, M., Mirdehghan, S.H., Nazoori, F. et al. (2022). Impact of calcium and γ-aminobutyric acid (GABA) on qualitative attributes and shelf life characteristics of fresh in-hull pistachio during cold storage. *Postharvest Biology and Technology* 187: 111863.

Sauter, M. (2009). A guided tour: pollen tube orientation in flowering plants. *Chinese Science Bulletin* 54: 2376–2382.

Seifi, H.S., Curvers, K., De Vleesschauwer, D. et al. (2013). Concurrent overactivation of the cytosolic glutamine synthetase and the GABA shunt in the ABA-deficient sitiens mutant of tomato leads to resistance against *Botrytis cinerea*. *The New Phytologist* 199 (2): 490–504.

Shang, H., Cao, S., Yang, Z. et al. (2011). Effect of exogenous γ-aminobutyric acid treatment on proline accumulation and chilling injury in peach fruit after long-term cold storage. *Journal of Agricultural and Food Chemistry* 59 (4): 1264–1268.

Shelp, B.J., Bozzo, G.G., Trobacher, C.P. et al. (2012). Contribution of putrescine to 4-aminobutyrate (GABA) production in response to abiotic stress. *Plant Science* 193: 130–135.

Shelp, B.J., Bown, A.W., and Zarei, A. (2017). 4-Aminobutyrate (GABA): a metabolite and signal with practical significance. *Botany* 95 (11): 1015–1032.

Shelp, B.J., Aghdam, M.S., and Flaherty, E.J. (2021). γ-Aminobutyrate (GABA) regulated plant defense: mechanisms and opportunities. *Plants* 10 (9): 1939.

Sheng, L., Shen, D., Luo, Y. et al. (2017). Exogenous γ-aminobutyric acid treatment affects citrate and amino acid accumulation to improve fruit quality and storage performance of postharvest citrus fruit. *Food Chemistry* 216: 138–145.

Shi, S.Q., Shi, Z., Jiang, Z.P. et al. (2010). Effects of exogenous GABA on gene expression of *Caragana intermedia* roots under NaCl stress: regulatory roles for H_2O_2 and ethylene production. *Plant, Cell and Environment* 33 (2): 149–162.

Signorelli, S., Dans, P.D., Coitiño, E.L. et al. (2015). Connecting proline and γ-aminobutyric acid in stressed plants through non-enzymatic reactions. *PLoS One* 10 (3): e0115349.

Silva, M., Rubilar, R., Espinoza, J. et al. (2017). Gas-exchange response and survival of young *Eucalyptus* spp commercial genotypes under water stress. *Revista Bosque* 38 (1): 79–87.

Singh, R., Parihar, P., Singh, S. et al. (2017). Reactive oxygen species signaling and stomatal movement: current updates and future perspectives. *Redox Biology* 11: 213–218.

Sita, K. and Kumar, V. (2020). Role of gamma amino butyric acid (GABA) against abiotic stress tolerance in legumes: a review. *Plant Physiology Reports* 25: 654–663.

Snedden, W.A., Koutsia, N., Baum, G., and Fromm, H. (1996). Activation of a recombinant petunia glutamate decarboxylase by calcium/calmodulin or by a monoclonal antibody which recognizes the calmodulin binding domain. *The Journal of Biological Chemistry* 271 (8): 4148–4153.

Stitt, M. (1999). Nitrate regulation of metabolism and growth. *Current Opinion in Plant Biology* 2 (3): 178–186.

Studart-Guimarães, C., Fait, A., Nunes-Nesi, A. et al. (2007). Reduced expression of succinyl-coenzyme A ligase can be compensated for by up-regulation of the γ-aminobutyrate shunt in illuminated tomato leaves. *Plant Physiology* 145 (3): 626–639.

Su, N., Wu, Q., Chen, J. et al. (2019). GABA operates upstream of H^+-ATPase and improves salinity tolerance in Arabidopsis by enabling cytosolic K^+ retention and Na^+ exclusion. *Journal of Experimental Botany* 70 (21): 6349–6361.

Suhel, M., Husain, T., Pandey, A. et al. (2023). An appraisal of ancient molecule GABA in abiotic stress tolerance in plants, and its crosstalk with other signaling molecules. *Journal of Plant Growth Regulation* 42 (2): 614–629.

Sulieman, S. and Schulze, J. (2010). Phloem-derived γ-aminobutyric acid (GABA) is involved in upregulating nodule N_2 fixation efficiency in the model legume *Medicago truncatula*. *Plant, Cell & Environment* 33 (12): 2162–2172.

Thomine, S., Lelièvre, F., Boufflet, M. et al. (1997). Anion-channel blockers interfere with auxin responses in dark-grown *Arabidopsis* hypocotyls. *Plant Physiology* 115 (2): 533–542.

Tian, Q., Zhang, X., Ramesh, S. et al. (2014). Ethylene negatively regulates aluminium-induced malate efflux from wheat roots and tobacco cells transformed with *TaALMT1*. *Journal of Experimental Botany* 65 (9): 2415–2426.

Tie, W., Zhou, F., Wang, L. et al. (2012). Reasons for lower transformation efficiency in indica rice using *Agrobacterium tumefaciens*-mediated transformation: lessons from transformation assays and genome-wide expression profiling. *Plant Molecular Biology* 78: 1–18.

Trobacher, C.P., Zarei, A., Liu, J. et al. (2013). Calmodulin-dependent and calmodulin-independent glutamate decarboxylases in apple fruit. *BMC Plant Biology* 13 (1): 1–10.

Virdi, A.S., Singh, S., and Singh, P. (2015). Abiotic stress responses in plants: roles of calmodulin-regulated proteins. *Frontiers in Plant Science* 6: 809.

Wang, C., Fan, L., Gao, H. et al. (2014a). Polyamine biosynthesis and degradation are modulated by exogenous gamma-aminobutyric acid in root-zone hypoxia-stressed melon roots. *Plant Physiology and Biochemistry* 82: 17–26.

Wang, Y., Luo, Z., Huang, X. et al. (2014b). Effect of exogenous γ-aminobutyric acid (GABA) treatment on chilling injury and antioxidant capacity in banana peel. *Scientia Horticulturae* 168: 132–137.

Wilhelmi, L.K. and Preuss, D. (1996). Self-sterility in *Arabidopsis* due to defective pollen tube guidance. *Science* 274 (5292): 1535–1537.

Xie, T., Ji, J., Chen, W. et al. (2020). GABA negatively regulates adventitious root development in poplar. *Journal of Experimental Botany* 71 (4): 1459–1474.

Xie, C., Wang, P., Sun, M. et al. (2021). Nitric oxide mediates γ-aminobutyric acid signaling to regulate phenolic compounds biosynthesis in soybean sprouts under NaCl stress. *Food Bioscience* 44: 101356.

Xing, S.G., Jun, Y.B., Hau, Z.W., and Liang, L.Y. (2007). Higher accumulation of γ-aminobutyric acid induced by salt stress through stimulating the activity of diamine oxidases in *Glycine max* (L.) Merr. roots. *Plant Physiology and Biochemistry* 45 (8): 560–566.

Xu, B., Sai, N., and Gilliham, M. (2021a). The emerging role of GABA as a transport regulator and physiological signal. *Plant Physiology* 187 (4): 2005–2016.

Xu, B., Long, Y., Feng, X. et al. (2021b). GABA signalling modulates stomatal opening to enhance plant water use efficiency and drought resilience. *Nature Communications* 12 (1): 1952.

Yang, A., Cao, S., Yang, Z. et al. (2011a). γ-Aminobutyric acid treatment reduces chilling injury and activates the defence response of peach fruit. *Food Chemistry* 129 (4): 1619–1622.

Yang, R., Chen, H., and Gu, Z. (2011b). Factors influencing diamine oxidase activity and γ-aminobutyric acid content of fava bean (*Vicia faba* L.) during germination. *Journal of Agricultural and Food Chemistry* 59 (21): 11616–11620.

Ye, J., Wang, X., Hu, T. et al. (2017). An InDel in the promoter of *Al-ACTIVATED MALATE TRANSPORTER9* selected during tomato domestication determines fruit malate contents and aluminum tolerance. *The Plant Cell* 29 (9): 2249–2268.

Yin, Y., Yang, R., and Gu, Z. (2014). Calcium regulating growth and GABA metabolism pathways in germinating soybean (*Glycine max* L.) under NaCl stress. *European Food Research and Technology* 239: 149–156.

You, J. and Chan, Z. (2015). ROS regulation during abiotic stress responses in crop plants. *Frontiers in Plant Science* 6: 1092.

Yu, G., Liang, J., He, Z., and Sun, M. (2006). Quantum dot-mediated detection of γ-aminobutyric acid binding sites on the surface of living pollen protoplasts in tobacco. *Chemistry & Biology* 13 (7): 723–731.

Yu, G.H., Zou, J., Feng, J. et al. (2014). Exogenous γ-aminobutyric acid (GABA) affects pollen tube growth via modulating putative Ca^{2+}-permeable membrane channels and is coupled to negative regulation on glutamate decarboxylase. *Journal of Experimental Botany* 65 (12): 3235–3248.

Yuan, D., Wu, X., Gong, B. et al. (2023). GABA metabolism, transport and their roles and mechanisms in the regulation of abiotic stress (hypoxia, salt, drought) resistance in plants. *Metabolites* 13 (3): 347.

Yue, X., Gao, X.Q., Wang, F. et al. (2014). Transcriptional evidence for inferred pattern of pollen tube-stigma metabolic coupling during pollination. *PLoS One* 9 (9): e107046.

Zarei, A., Trobacher, C.P., and Shelp, B.J. (2016). Arabidopsis aldehyde dehydrogenase 10 family members confer salt tolerance through putrescine-derived 4-aminobutyrate (GABA) production. *Scientific Reports* 6 (1): 35115.

Zhang, Q., Xiang, J., Zhang, L. et al. (2014). Optimizing soaking and germination conditions to improve gamma-aminobutyric acid content in Japonica and Indica germinated brown rice. *Journal of Functional Foods* 10: 283–291.

3

GABA and Its Crosstalk with Other Metabolites in Relation to Abiotic Stress Responses in Plants

Akhilesh Kumar Pandey[1] *and Nishtha Srivastava*[2]

[1] *Department of Biotechnology, Faculty of Bioscience and Biotechnology, Invertis University, Bareilly, Uttar Pradesh, 243123, India*
[2] *Department of Applied Sciences & Humanities, Faculty of Applied Sciences, Invertis University, Bareilly, Uttar Pradesh, 243123, India*

Introduction

The 4-carbon, naturally occurring chemical known as gamma-aminobutyric acid (GABA) is found in both producers and consumers (Rodrigues-Corrêa and Fett-Neto 2019). It is ubiquitous and nonproteinaceous in nature. GABA was initially discovered in potato tubers during the examination of several biological constituents (Rashmi et al. 2018). GABA has been thoroughly investigated as a significant animal neurotransmitter in the past few years. The primary method of production is GABA bypass. The evolutionary conservation of GABA shunt enzymes suggests that GABA originated in bacteria ages ago (Seifikalhor et al. 2019). This theory is further supported by findings of GABA in plants and animals. Research revealed that increased reactive oxygen species (ROS) buildup in cells is one way that GABA reacts to a variety of abiotic stressors (Sachdev et al. 2021). Numerous studies demonstrate that in unfavorable circumstances like extreme temperatures, sodium chloride, droughts, poor light, chilly, a lack of oxygen, and UV-B radiation rays, GABA quickly concentrates in crops (Fatima et al. 2023). Researchers have documented aggregation of GABA in response to exposure to various heavy metals in various crops, including zinc (Zn), cadmium (Cd), aluminum (Al), arsenic (As), and calcium (Ca^{2+}) (Mukherjee et al. 2023). Much study is required to identify its significance under stress due to its significant elevation throughout various types of stresses. Through its interactions with distinct signaling chemicals such as hydrogen peroxide (H_2O_2), nitric oxide (NO), calcium ions, mitogen-activated protein kinase (MAPK), and cyclic guanosine-3′,5′-monophosphate (cGMP), the signaling function of GABA has been further investigated in the regulating of how plants grow and develop (Jain et al. 2018). In addition, GABA controls the formation of roots and pollen tubes in plants by means of aluminum-activated malate transporters (ALMTs) (Xu et al. 2021b). Moreover, a particular level of GABA has been shown to affect the genes coding for antioxidants and the plant's protective system; as a result, it seems that

GABA in Plants: Biosynthesis, Plant Development, and Food Security, First Edition. Edited by Samiksha Singh, Durgesh Kumar Tripathi, and Vijay Pratap Singh.

GABA may protect plants from oxidative damages (Gahlowt et al. 2024). GABA appears to function as a harmless osmolyte in sodium stress conditions and can also quench ROS. It has been demonstrated that via regulating physiological reactions, GABA improves resilience to excess salt concentrations in *Nicotiana sylvestris* and barley (Dabravolski and Isayenkov 2024; Kumari et al. 2024). It has also been observed that externally administered GABA improves resilience to stress in crops by controlling osmotic equilibrium. In addition, the research demonstrated that externally supplied GABA effectively inhibits H_2O_2 generation during salt overload and controls the gene transcription of H_2O_2-generating enzymes in *Caragana intermedia* (Zarbakhsh et al. 2024). Furthermore, it was recently demonstrated that seed primed using GABA significantly boosts the degree of sprouting and lessens the effects of salinity in maize and wheat (Cheng et al. 2018; Aljuaid and Ashour 2022). Numerous findings indicate that externally administered GABA shields the plant's tissues against the damaging effects of thermal stress.

Several additional molecules, including proline, polyamines, salicylic acid, jasmonates, etc., in addition to GABA, communicate with stress indicators that control different metabolic events and affect the subsequent response to GABA (Podlešáková et al. 2019). The production of GABA during environmental stress and how it interacts with ROS, or nitrogen oxides, calcium ions, growth regulators, and polyamines in regulating stress and adaptation have been evaluated in the current chapter.

Enzymes in GABA Metabolic Pathways

The biosynthesis of GABA within plant cells is interconnected with several metabolisms like nitrogen, proline, polyamine, and tricarboxylic acid (Singhal et al. 2021). It is usually generated via the GABA shunt pathway in both animals and plants because of the existence of the GAD gene (Li et al. 2021). This property is not exhibited by maximum cyanobacterial species due to the absence of the GAD gene. It can also be formed during oxidative stress circumstances by many routes, including the breakdown of polyamines, as documented in plants, animals, and microorganisms (Raza et al. 2022). The key component of the GABA shunt is glutamate, as demonstrated experimentally by a study, which showed that administered glutamate enhanced the buildup of GABA in *Aphanothece halophytica* during acidic stress (Suhel et al. 2023). The natural production of glutamate via α-ketoglutarate is carried out by an enzyme named glutamate dehydrogenase. GABA shunt in plants and cyanobacteria is mediated by three main enzymes: succinic semialdehyde dehydrogenase (SSADH), glutamate decarboxylase (GAD), and GABA transaminase (GABA-T) (Liao et al. 2022). L-Glutamate is converted by the GADs into GABA; an irreversible phenomenon that takes place in the cytoplasm. Different crops have different types of GADs; like 4 and 9 types of GADs are reported in Prunus and Arabidopsis, respectively (Salvatierra et al. 2016). In plants, GABA transaminase uses pyruvate as an acceptor to change GABA into succinic semialdehyde. Animals use GABA-TK to carry out this process. It has been observed that GABA-TP plays an important function in GABA breakdown in Arabidopsis mutants, exhibiting a hundred times more GABA as a result of GABA-TP deficiency and its impact on crops (Shelp et al. 2021). Yet, substantial levels of GABA were seen in plants when GABA transaminase was inhibited. Succinate is produced from succinic semialdehyde by the SSADH. In addition, research shows that *Synechocystis* sp. produces GABA by means of the metabolism of nitrogen and accumulates GABA when nitrogenous chemicals are available (Jantaro and Kanwal 2017). They observed increased GABA buildup when different forms of nitrogen, like nitrate (NO^{3-}), nitrite (NO^{2-}), ammonium (NH^{4+}), and urea, were available in proper amounts (Royes et al. 2016). Research findings indicate a connection between the carbon pool and the biosynthesis of GABA since the formation of GABA is augmented in response to glucose availability (Sulieman et al. 2024).

There are various methods that plants can use to increase GABA accumulation in tissues under stress, including GAD stimulation by the breakdown of proline, higher levels of protons, polyamines, and calcium-calmodulin (Signorelli et al. 2023). Both prokaryotic and eukaryotic organisms contain polyamines, which are nitrogen-rich compounds, made up of amino groups having lower molecular mass (Krysenko and Wohlleben 2023). The three main polyamines found in crops are spermidine, spermine, and putrescine (Mattoo et al. 2010). Polyamines not only control other vital functions but also play a role in the reactions of plants against several stresses. Plants experiencing stress from abiotic factors have been shown to produce more polyamines via elevating biosynthesis (Minocha et al. 2014). Studies on Arabidopsis have shown that GABA is generated in plant peroxisomes by the breakdown of spermine and putrescine (Shelp et al. 2012). There have been reports of GABA production during salt stress via polyamine breakdown (Xu et al. 2019). Since *S*-adenosyl-methionine is a common component of the pathways involved in the formation of ethylene and polyamines, these two molecules are synthesized via similar pathways (Sauter et al. 2013). During salty conditions, GABA has been shown to increase the enzyme activities of ornithine decarboxylase, *S*-adenosyl-methionine decarboxylase (SAMDC), and arginine decarboxylase in muskmelon young plants, hence affecting the production of polyamines compounds (Hasanuzzaman et al. 2019). Through increasing the rate of action of SAMDC, the primary factor in polyamine biosynthesis, GABA reduces the production of polyamines. The arginase route is another way that GABA controls the synthesis of polyamines (Majumdar et al. 2016).

Role of GABA Under Stressful Conditions in Plants

Crops under stress create GABA through the breakdown of polyamines (Yong et al. 2017). Putrescine was found to be changed into gamma-amino-butyraldehyde through the action of diamine oxidase (DAO), a limiting factor for the reaction (Sita and Kumar 2020). It is subsequently transformed into GABA by amino aldehyde dehydrogenase (AMADH). A study saysthat while DAO action is inhibited in fava beans using the blocker aminoguanidine, the amount of GABA decreases by almost 30% (Yin et al. 2018). Augmented GAD levels in maize and improved DAO function in glycine max are assumed to be accountable for elevated levels of GABA amid stress (Singh et al. 2018) (Figure 3.1). Infected plants likewise experience a buildup of GABA upon viral exposure (Kinnersley and Turano 2000). Stress evokes faster formation of GABA in a diverse range of life forms like bacteria, crops, and animals (Ramesh et al. 2017).

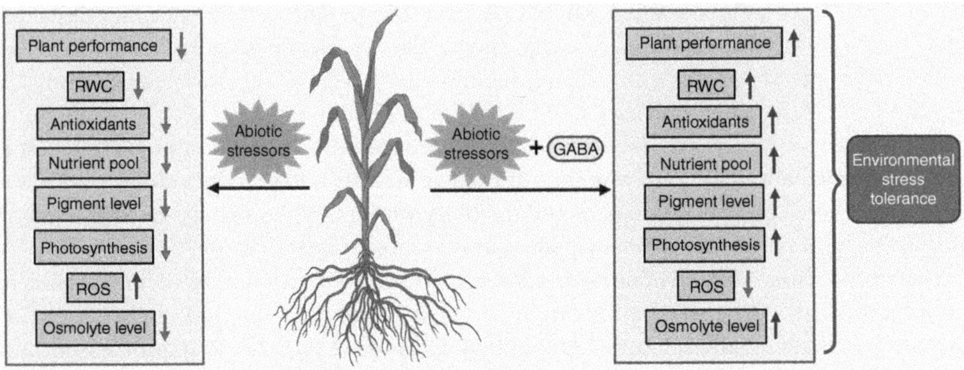

Figure 3.1 Schematic representation of plants under various stressors and in combination with GABA. GABA positively impacts the growth and development of plants through modulation in levels of antioxidants, osmolyte, ROS, pigment, and physiological process.

GABA and Salt Stress

According to earlier research, a large geographical region of the earth's total area (approx. 7%) is affected by salinity, which has led to it becoming a significant element that modifies the morphological, metabolical, and physiological traits of a variety of plants (Okon 2019). According to a number of tests, amid excess salt situations, GABA can be used as a major compound that regulates osmotic adjustment and as a ROS quencher (Ansari et al. 2021). GABA was applied externally, and it reduced the detrimental effects of salts in *Caragana* intermedia by controlling the regulation of those genes that produce peroxidase and hydrogen peroxide (Hayat et al. 2023). GABA likewise improved seed germination potential and decreased salt damage in both crop maize and wheat (Wang et al. 2017). Results revealed that externally administered GABA increases the internal level of GABA via elevating biosynthesis, which confers resilience towardsalt exposure (Rashmi et al. 2018) (Table 3.1). A significant portion of absorbed carbon is redirected toward respirationduring salty conditions (France and Reid 1983). In this scenario, a high level of ATP is needed for the TCA cycle to function, ion rejection, detoxification of ROS, and osmotically active compounds synthesis to preserve osmotic equilibrium; yet, GABA shunt provides the ATP needed for the cell to survive (Bhatla et al. 2018).

GABA and Drought Stress

The most common strain in the present situation is droughts (Seleiman et al. 2021). Plant development is restricted by drought, whereas bentgrass treated with GABA expresses genes implicated in stress management (Tang et al. 2020). It was recently observed that plants like soybean, wheat, sesame, turnip, and Arabidopsis acquire GABA under drought (Suhel et al. 2023). These research findings indicate that in drought circumstances, there is an impressive rise in levels of glucose, palmitic acid, threonine, stearic acid, GABA, and an unknown disaccharide (Killiny 2017). Due to adverse physiological alterations, a mutated

Table 3.1 Crops showing stress response alone and in application with GABA under stress conditions.

Crops	Stress	Stress + GABA	Impact of stress	Stress + GABA induced impacts	References
Tomato	Heavy metals	GABA	Increased ROS, damages membrane integrity, short stature	Diminished ROS, elevated antioxidants pool	Suhel et al. (2023)
Tomato	Salt	GABA	Increased lipid peroxidation andelectrolyte leakage, diminished growth	Increased leaf and root growth, water content, amino acids, andcatalase activity	Bao et al. (2015)
Arabidopsis	Drought	GABA	Increased proline, ROS, and oxidative stress	Increased root/shoot height and biomass, chlorophyll	Xu et al. (2021a)
Arabidopsis	Flood	GABA	Increased protein carbonylation and decreased antioxidant activity	Elevate level of proline, glutamine synthetase, stability of O_2-evolving complex	Wu et al. (2021)
Arabidopsis	Salt	GABA	Decrease fresh and dry mass and Na^+ uptake in shoot	Increase glucose level and antioxidant enzymes activity	Renault et al. (2010)
Carrot	Drought	GABA	Diminished vigor, high level of ROS, low antioxidants	Increased vigor, relative water content, amino acids,catalase activity	Bashir et al. (2021)
Sunflower	Drought	GABA	High level of hydrogen peroxide, malondialdehyde, electrolyte leakage	Shoot height and fresh mass, root fresh mass, chlorophyll	Abdel Razik et al. (2021)
Tobacco	Heavy metal stress	GABA	Oxidative stress, low chlorophyll level, and conductance, diminished PS II	Increased proline, PSIIactivity, and stomatal conductance	Yang et al. (2021)
Mustard	Heavy metal stress	GABA	High proline, reactive oxygen species, and malondialdehyde	Increased chlorophyll and antioxidative machinery	Mahmud et al. (2017)
Barley	Heavy metal stress	GABA	Elevated protein carbonylation and reactive oxygen species	Increased activities of superoxide dismutase andcatalase	Zhao et al. (2020)
Perennial ryegrass	Drought	GABA	Wilt rating and electrolyte leakage	Relative water content, canopy temperature, turf quality,peroxidase activity	Vijayakumari et al. (2016)
Lettuce	Salt	GABA	Decreased germination time, electron transport	Increased photosynthesis electron flow,activities of CAT, POD, and SOD	Kalhor et al. (2018)

Arabidopsis lacking the transcription of the GAD gene demonstrated vulnerability to drought-related stresses (Salvi et al. 2021). Research indicates that GABA plays a part in mitigating oxidative damages by controlling the antioxidant machinery during dry conditions (Ullah et al. 2023).

GABA and Chilling Stress

On Earth, the temperature changes with respect to time, length of day and night, season, and location (Peng et al. 2013). In vegetation, the cooling temperature spans between 0 and 15°C. Low temperature is a key determining factorthat governs the level of plant growth and production (Hasanuzzaman et al. 2012). The application of external GABA was found to reduce cold stress. In addition, it has been shown that GABA may benefit seedlings of tomatoes by raising their proline concentration and helping wheat plants withstand the negative effects of freezing (Gramazio et al. 2020). A further study resulted in an upsurge in GABA concentrations after cold stress. GABA administered externally also minimizes the chilling damage that cold storage does to peach and banana peels (Rastegar et al. 2024). Fruits that have been harvested earlier have a longer after-harvesting time and are more resilient to the effects of cold (Medda et al. 2022). Studies have shown that in both sensitive and tolerant zucchini species, exposure to cold stimulates the GABA shunt (Palma et al. 2014). There has been evidence of increased GABA buildup in barley and wheat seedlings in resistant types via GABA shunt and in susceptible kinds via GAD (Al-Quraan et al. 2019). GABA buildup is also induced by a cold treatment to fungal cells being grown (Ramos-Ruiz et al. 2019). Plants utilize the GABA shunt to protect themselves against insects by participating in the signaling cascade, maintenance of pH, oxidative loss, osmolytes, nitrogen metabolism, and TCA cycle (Seifikalhor et al. 2019). Crops store these nonproteinaceous molecules as a means of adaptation in the face of oxidative stress (Guo et al. 2023) (Table 3.1).

Crosstalk of GABA with Other Metabolites and Chemicals

Elevated GABA buildup is associated with its significant function in bacteria, fungi, and plants under stressful situations (Gahlowtet al. 2024). The goal of recent research has been to better comprehend the signaling pathways that living things use when facing stress. Numerous environmental stressors, including radiation, intense light, saline, drought, and plant–microbes interaction, cause a surplus of ROS to be produced in the cell, leading to damage through oxidation (Zandi and Schnug 2022). Oxidative injury causes peroxidation of lipids moiety in plasmamembranes, DNA strand breaks, denatured proteins, combustion of sugar molecules, deterioration of pigments, and decreases in enzyme function. According to research, ROS are created under environmental stress to carry out the plant's intracellular reaction to the stressor and function as a signaling player in overcoming stress (Mansoor et al. 2022). Inside plants, peroxisomes, Reactive Nitrogen Species (RNS), and ROS undergo proper metabolism that results in the breakdown of polyamines, which produce metabolites like GABA, NO, and H_2O_2 (Corpas et al. 2019). Stress circumstances aggravate this nitrosative metabolism (Corpas and Barroso 2013). The main organelles responsible for producing ROS include peroxisomes, mitochondria, and chloroplasts. Peroxisomes are thought to pick up a signal under stress and increase the production of RNS and ROS (Sandalio and Romero-Puretas 2015). ROS and NO are produced by peroxisomes and function as signaling chemicals in plant cells. This NO is generated through nitration, which controls the expression of peroxisomal enzymes. Enhanced concentrations of GABA after different types of stress suggest that GABA serves a signaling function instead of being used by the TCA cycle. A large rise in ROS is seen in mutants that lack the genes for GABA shunt, suggesting that GABA plays a significant role in the pathways involved in the removal of ROS (Xu et al. 2024). Plants are thought to use the GABA shunt as a biosynthetic and signaling mechanism for adapting to environmental stress (Suhel et al. 2023). However,it is still up for debate how much GABA acts as an intermediate or an alarm chemical among plant cells. It is thought that GABA-binding molecules, which are widely distributed in microorganisms like Pseudomonas, serve as communication signals among prokaryotic and eukaryotic organisms (Sarasa et al. 2020).

GABA with H_2O_2

GABA is generated together with messenger molecules like NO and H_2O_2, but its function across the signaling cascade is unclear. On the one hand, H_2O_2 has detrimental effects on crops. For example, it promotes lentil seedlings to undergo apoptotic (Maccarrone et al. 2000). Research has shown that when crops are under strain, superoxide and H_2O_2 radicals act as both signaling and harmful chemicals (Xie et al. 2019). It has been shown that GABA preadministration reduces the generation of ROS in beta cells of the pancreas, which are induced through the administration of H_2O_2 as a result of a compromised protective mechanism (Jugran et al. 2021). The goal of their research was to find an indigenous antioxidant chemical that can shield it by reducing the effects of oxidative stress. It has been revealed that GABA controls the state of redox equilibrium in beta cells of the pancreas (Tamarit-Rodriguez 2023). In the signaling of abscisic acid (ABA), ROS functions as secondary signals that set a rate restriction (Postiglione and Muday 2020). When a *C. intermedium*is under pressure, GABA is shown to boost the level of expression of three molecules that contribute to the generation of H_2O_2 : amine oxidase (AO), peroxidase, and NADPH oxidase (Asija et al. 2023). This is demonstrated by SSH as well as by examining the similarity among the regions of the transcribed gene. Peroxidases oxidize auxin to create H_2O_2 (Pasternak et al. 2023). External administration of GABA increases the level of expression of the genes of the enzyme NADPH oxidase. After 12 hours of GABA administration, NADPH oxidase activity becomes more noticeable (Hayat et al. 2023). The increased expression of AO and peroxidase in *C. intermedia* under salt stress was previously investigated (Jiménez-Bremont et al. 2023). The increased expression of genes of certain enzymes, specifically those located in peroxisomes, results in an upsurge in H_2O_2 synthesis. External GABA administration significantly decreased the amount of H_2O_2 that accumulated under these circumstances by lowering the level of NADPH oxidase (Fan and Li 2024). Comparable research also found that plants cannot be inhibited from accumulating H_2O_2 by greater levels of GABA, as much as 20 mM (Zhang et al. 2023). It has been observed that GABA regulates proteins associated with transferring signals, polyamine production, hormone transfer, and ROS formation (Yuan et al. 2024). When GABA was treated exogenously amid a sodium chloride stress, the enzyme NADPH oxidase was expressed, resulting in the creation of H_2O_2 that subsequently controlled the generation of H_2O_2 in *C. intermedia* tissues (Khan et al. 2021). Subsequent research will undoubtedly be essential to provide more experimental evidence for it.

GABA with Nitric Oxide (NO)

When crops are under stress, NO plays a significant role in the transmission of signals (Khan et al. 2023). According to a study, polyamines increased NO signaling in *Nicotiana tobaccum* to improve its defense against cold, as well as *Trifolium repens* to increase its ability to withstand drought stress via increased activity of enzymes (Nandy et al. 2023). When soybean seedlings are exposed to UV-B exposure, an external NO donor (sodium nitroprusside) causes elevated GABA production (Seth et al. 2023). However, in water-deficit *T. repens*, external GABA administration has increased NO level in the crops by 90% compared to the untreated group (Hayat et al. 2023). One may argue that GABA contributes to NO buildup, whereas NO is additionally induced by GABA. When NO is produced in soybean seedlings, it triggers cGMP/PKG, a secondary messenger that provokes the transcription of MAPK genes (Xia et al. 2023). It has been observed in tobacco tissue that NO stimulates MAPK by temporary intracellular calcium injection (Zhao et al. 2007). AMADH and DAO are two distinct enzymes whose expression of genes and proteins are activated by MAPK (Tyagi et al. 2023). Putrescine is metabolized by DAO to γ-amino-butyraldehyde; afterward, it converts to GABA via means of AMADH in sprouting soybeans (Yuet al. 2023). After two days of experiencing NO, soybean seedlings showed increased levels of AMADH-expressed proteins (Yin et al. 2023).

GABA with Calcium

GABA synthesis is elevated at low levels of pH amid flooding due to GAD stimulation brought on by a drop in cytoplasmic pH (Komatsu et al. 2023). By activating GAD and raising GABA concentrations, other stressors like heat and cold also work through the Ca^{2+}-calmodulin cascade (Ansari et al. 2021). There are two manners through which a substantial level of proton or an elevated level of calcium may stimulate the GAD (Kinnersley and Turano 2000). GAD function is increased in *Escherichiacoli* and yeast to control oxidative damage under lower pH environments (Boura et al. 2020). Elevated ROS concentrations under different conditions raise cytoplasmic Ca^{2+} level in two ways: first, by activating the calcium valves; second, via producing NO, which raises calcium concentrations by controlling the function of calcium-related proteins that are involved in numerous aspects of stress-adaption, elevated calcium levels reduce stress (Feissner et al. 2009). The majority of plant-based GADs have a calmodulin-binding domain (Lee et al. 2023). There is more proof of GAD-calmodulin connection provided by nuclear magnetic resonance (Gallego et al. 1995). In addition, two chains of proteins at the GAD's calmodulin-binding motif are linked up with calmodulin. The question of whether or not most GADs identified in crops have a calmodulin-binding motif continues to be a mystery. Subsequently, clarifying the role of calmodulin association in GAD function is a fascinating issue. Five GAD genes have been identified in the *Arabidopsis*, and distinct GADs respond differently across various tissues, indicating their tissues-specific localization (Bouché et al. 2004). GABA signaling contributes to the cytosolic Ca^{2+} rise in response to abiotic stressors, including level of light and nutritional insufficiency (Seifikalhor et al. 2019). This Ca^{2+} then interacts with calmodulin to trigger the enzyme GAD (Snedden et al. 1995). GABA metabolism-related enzyme maintains the transcription of genes and hence rises in response to the aforementioned stress, which eventually raises GABA levels. GABA penetrates mitochondria and engages in the TCA cycle (Hertz and Rothman 2016). In addition, GABA might have a role in controlling the functioning of a variety of genes linked to metabolical and signaling ways (Fait et al. 2008). Synthesis of GABA is correlated with a rise in cytoplasmic calcium density; hence, the application of calcium quencher inhibited DAO and AMADH protein and gene expression in the soybean seedlings, thereby suppressing GABA formation (Podlešáková et al. 2019).

Interplay of GABA with Plant Hormones

Hormones produced by plants additionally play a role in stress reactions, as evidenced by changes in how they express gene levels throughout stressful situations (Fahad et al. 2023). When separately auxin and ABA are applied to rice and wheat roots, respectively, GABA synthesis rises (Podlešáková et al. 2019).

GABA with Auxin

Transcription of gene ALMToccurs because of (indole-3-acetic acid) IAA/ABA, which is thought to have a function in regulating Al toxicity (Reyes-Díaz et al. 2023). With the exception of *Triticum aestivum* (TaALMTs), the ALMTs aren't often triggered by Al^{3+} (Abdel-Hameed et al. 2024). These ALMTs are regulated by GABA (Ramesh et al. 2018). During conditions of aluminum toxicity, GABA adversely impacts malate efflux in plants via ALMTs (Sharma et al. 2016). Resistance of roots occurs due to different levels of pH, and Al^{3+} ions are regulated by GABA via ALMTs (Ramesh et al. 2018). Growth hormones cause GABA, NO, and ROS to be produced during the metabolism of polyamines, which raises the level of calcium ions (Podlešáková et al. 2019). Such molecules are responsible for fighting against stressreactions.

GABA with Abscisic Acid

Enzyme copper AO, which catalyzesthe oxidation of putrescin in peroxisomes and generates H_2O_2 as a result, is regulated by ABA in the guard tissues of *Vicia faba* (Qu et al. 2014). H_2O_2 ultimately causes the closing of stomatal pores and raises cytosolic Ca^{2+} concentrations

(Rodrigues and Shan 2022). Research indicates that the inability to seal stomata was seen in Arabidopsis mutants that were deficient in GABA production. Arabidopsis plants are more vulnerable to salt conditions because of an aberration in the AMADH gene that prevents GABA buildup (Biancucci et al. 2015). GABA is produced by the enzyme AMADH via precursor γ-amino-butyraldehyde. According to a different study, ABA causes NO to cause the closure of stomatal cells (Hsu et al. 2021). Thus, research shows that ABA produces GABA and that this GABA plays a part in the closing of stomatal cells (Xu et al. 2021b).

GABA with Ethylene

There are arguments to support the involvement of GABA during ethylene signaling. Ethylene is expressed more when GABA enhances the transcription of the gene for enzymes 1-aminocyclopropane-1-carboxylic acid (ACC) oxidase and 1-aminocyclopropane-1-carboxylic acid synthase (Shi et al. 2010). The simultaneous rise in ethylene, GABA, and Ca^{2+} levels during stressful times is found to be correlated (Kinnersley and Turano 2000). GABA controls the mRNA transcript level of ACC synthase, which, in turn, controls the process of ethylene biosynthesis (Takayama et al. 2017). Since ethylene promotes GABA's effect throughout sodium chloride stress, using aminoethoxy vinyl glycine and silver thiosulfate, inhibitors of ethylene biosynthesis and activity, respectively, substantially lowered the degree of stem extension that GABA had previously blocked (Kinnersley and Turano 2000). The results indicated that ET, H_2O_2, and ABA are related to GABA-induced relief of stress caused by salt (Sheteiwy et al. 2019). Since amino acids were used to promote additional metabolic processes rather than being broken down to produce sugar in crops, it is likely that GABA plays a signaling function in crops (Li et al. 2021).

Mechanisms of Action of GABA in Plants Under Stress

In conditions of cold, GABA modifies the transcription of certain genes; nonetheless, its biochemical and genomic functions remain unclear. Different crop stressors have been shown to enhance GABA synthesis (Podlešáková et al. 2019). The observation that crops with GAD mutations became more responsive supports the significance of GABA under adversity (Chen et al. 2023). During sodium chloride stress, *C. intermedia* has already been shown to express gene GAD at a higher level (Ji et al. 2015). GABA stimulates the expression of proteins involved in polyamine biosynthesis in the roots of *Cattleya* intermedia under sodium chloride stress, and it also lowers the intracellular content of H_2O_2 (Suhel et al. 2023).

GABA is thought to be involved in a number of physiological functions in addition to signaling and overcoming stress. A variety of signaling moleculesare associated with GABA and control distinct physiological reactions, such as polyamines, proline, plant hormones, and calcium ions (Akula and Mukherjee 2020). Researchers have found that genes of GABA-T and GAD are expressed in tomato fruit growth (Takayama et al. 2015). For a proper GABA pool, the aforementioned genes must function. The short stature of the tomato crop was caused by the suppression of the production of GABA-T genes (Gramazio et al. 2020). However, the loss of an inhibitory region in different GAD genes results in an elevated GABA pool, which boosts fruit harvests and plant development (Aghdam et al. 2022). In another study, tomatoes' showed a significant amount of GABA, which lowers the generation of ethylene, making tomatoes the lower ethylene-sensitive fruits (Baranov and Timerbaev 2024). Research has shown that under different challenging conditions, the expression of genes for GABA shunt varies (Ansari et al. 2021). GABA levels are raised by a unique temperature-sensitive transcription factor (MYB55), which binds to the specific regions of the gene GAD (Usman et al. 2023). However, it was shown that in the atstop1 mutant, GAD1 and GABA-T transcription are low, leading to a decreased accumulation of GABA, whereas GAD4 transcription improves (Zarei et al. 2017). This is because WRKY40 and STOP1 suppress the activity associated with the specific GAD genes (Suhel et al. 2023). It has been noticed that atstop1 mutants have reduced the levels of GABA-T and GAD1, leading to a minimal build-up of GABA, but GAD4 synthesis is elevated. ALMT1, an essential gene responsible for aluminum tolerance, is correlated to the level of expression of STOP1 (Daspute et al. 2017). The results suggested that GABA functions via attaching GABA to the regions of the ALMT1 transporter (Long et al. 2020).

Polyamine oxidasesand copper-containing amine oxidase terminally oxidize polyamines to form 4-aminobutanal and 3-aminopropanal, which are the byproducts of this oxidation process, which additionally produces GABA in crops (Alcázar et al. 2010). GABA is produced by another oxidation of aminobutanal by AMADH (Zarei et al. 2015). It has already been documented that active-form AMDAH genes create beta-alanine and GABA when aminobutanal, aminopropanal, and NAD^+ are present (Suhel et al. 2023).

Numerous kinds of plants exhibited tolerance for heat primarily via controlling the expression of genes related to heat shock proteins (HSPs) and ascorbate peroxidases (APXs) (Liu et al. 2019). APXs are connected to stress resistance in crops and play a vital role in antioxidant defense mechanisms (Rajput et al. 2021). The stabilization of proteins involved in heat resistance is mostly mediated by HSPs (Jacob et al. 2017). According to the results, Arabidopsis with overexpressed miR160 exhibits increased heat resistance through the upregulation of various HSPs, and miR398 functions as a critical regulator for fluctuating stress conditions (Saini et al. 2022). Research revealed that miR398 suppresses heat shock transcription factors (HSFs) and HSPs expression in plants. GABA supplementation decreased the expression of miR398s in bentgrass but increased the expression of HSPs, APXs, and HSFs (Feng et al. 2023). Studies have shown that applying GABA topically to the leaves increases the expression of two genes related to heat resistance, i.e. APX and HSP90. These findings demonstrate the role of GABA in regulating temperature via HSF circuits (Mohanta et al. 2018).

ABA, a phyto-haromone, plays a crucial role as a means of communicating molecules both under stressful and nonstressful circumstances (Vishwakarma et al. 2017; Chen et al. 2020). By controlling the cytosolic Ca^{2+} amount, ABA controls stomatal motility (Li et al. 2022). Stress situations cause either an immediate or indirect spike in cytoplasmic Ca^{2+}. GAD functioning is induced by an upsurge in cytoplasmic levels of calcium through Ca^{2+}-CaM signaling, which turns glutamate into GABA (Virdi et al. 2015). GABA buildup on the rise controls the synthesis of malate transport proteins stimulated by aluminum, which, in turn, controls stomatal movement, emergence of seeds, and grain production in stressful environments (Li et al. 2021). Nevertheless, it has been noted that elevated GABA buildup is associated with elevated ALMT production (Xu et al. 2021b) When Arabidopsis is under conditions of drought, gene expression of enzymes spermine synthase, spermidine synthase, arginine decarboxylase, and level of polyamines is significantly elevated; in contrast, these features are diminished in ABA-deficient and ABA-insensitive variants (Valdés-Santiago and Ruiz-Herrera 2014). Spermine and spermidine cause *V. faba* to produce H_2O_2, which raises the level of cytosolic calcium with the closure of stomatal cells (Liu et al. 2023). Research has shown that polyamine breakdown produces GABA, and a variant of *Arabidopsis* lacking in GABA synthesis displays stomata with a deformed morphology (Masson et al. 2017). Furthermore, owing to a decrease in GABA concentration, plants lacking the AMADH were more susceptible to the effects of saltiness than naturally occurring crops (Ullah et al. 2023). The above findings imply a relationship between GABA, ABA, and polyamines and control of stomatal closing (Sequera-Mutiozabal et al. 2016).

Conclusions and Future Perspectives

Research reveals details regarding GABA synthesis along with its function within plants and microorganisms under stress adaptation. GABA shunt in crops and microorganisms is primarily mediated by key enzymes like GABA-T, GAD, and SSADH. The proper pool of nitrogen, carbon, and proline are also connected to GABA metabolic pathways. The GABA concentration is raised by polyamine breakdown, while GABA additionally governs the production of polyamines. Both indigenous and administered GABA contribute significantly to the ability to tolerate various stresses, such as cold, metal, droughts, and floods. Through its interactions with various other messenger molecules, such as Ca^{2+}, NO, H_2O_2, and phytohormones, it significantly lowers the risk of oxidative stress. However, GABA also raises H_2O_2 levels by boosting NADPH oxidase activity; therefore, further research is needed to determine the exact mechanism by which GABA regulates H_2O_2 homeostasis. Research on the functional aspects of GABA and NO cross-talks plays in controlling the development and growth of plants, and stress tolerance is still needed because externally supplied GABA raises the quantity of NO, and externally administered NO raises the basal level of GABA. Changing the transcript level of particular genes, such as those responsible for polyamine production, additionally lessens stressful circumstances. During different stress situations, the functioning of genes associated with GABA metabolic pathways is under or overexpressed. HSPs are also induced to be expressed by GABA. Consequently, this chapter provides a thorough understanding of how GABA balance is preserved in the biochemical, and physiological, which, in turn, allows crops to withstand environmental stresses. However, there is still much to learn about the precise method by which GABA collaborates with different signaling molecules and plant hormones in plants under fluctuating environments.

References

Abdel-Hameed, A.A., Liao, W., Prasad, K.V., and Reddy, A.S. (2024). CAMTAs, a family of calmodulin-binding transcription factors, are versatile regulators of biotic and abiotic stress responses in plants. *Critical Reviews in Plant Sciences* 43: 1–40.

Abdel Razik, E.S., Alharbi, B.M., Pirzadah, T.B. et al. (2021). γ-Aminobutyric acid (GABA) mitigates drought and heat stress in sunflower (*Helianthus annuus* L.) by regulating its physiological, biochemical and molecular pathways. *Physiologia Plantarum* 172 (2): 505–527.

Aghdam, M.S., Flaherty, E.J., and Shelp, B.J. (2022). γ-Aminobutyrate improves the postharvest marketability of horticultural commodities: advances and prospects. *Frontiers in Plant Science* 13: 884572.

Akula, R. and Mukherjee, S. (2020). New insights on neurotransmitters signaling mechanisms in plants. *Plant Signaling & Behavior* 15 (6): 1737450.

Alcázar, R., Altabella, T., Marco, F. et al. (2010). Polyamines: molecules with regulatory functions in plant abiotic stress tolerance. *Planta* 231: 1237–1249.

Aljuaid, B.S. and Ashour, H. (2022). Exogenous γ-aminobutyric acid (GABA) application mitigates salinity stress in maize plants. *Life* 12 (11): 1860.

Al-Quraan, N.A., Al-Ajlouni, Z.I., and Obedat, D.I. (2019). The GABA shunt pathway in germinating seeds of wheat (*Triticum aestivum* L.) and barley (*Hordeum vulgare* L.) under salt stress. *Seed Science Research* 29 (4): 250–260.

Ansari, M.I., Jalil, S.U., Ansari, S.A., and Hasanuzzaman, M. (2021). GABA shunt: a key-player in mitigation of ROS during stress. *Plant Growth Regulation* 94: 131–149.

Asija, S., Seth, T., Umar, S., and Gupta, R. (2023). Polyamines and their crosstalk with phytohormones in the regulation of plant defense responses. *Journal of Plant Growth Regulation* 42 (8): 5224–5246.

Baranov, D. and Timerbaev, V. (2024). Recent advances in studying the regulation of fruit ripening in tomato using genetic engineering approaches. *International Journal of Molecular Sciences* 25 (2): 760.

Bao, H., Chen, X., Lv, S. et al. (2015). Virus-induced gene silencing reveals control of reactive oxygen species accumulation and salt tolerance in tomato by γ-aminobutyric acid metabolic pathway. *Plant, Cell and Environment* 38 (3): 600–613.

Bashir, R., Riaz, H.N., Anwar, S. et al. (2021). Morpho-physiological changes in carrots by foliar γ-aminobutyric acid under drought stress. *Brazilian Journal of Botany* 44: 57–68.

Bhatla, S.C.A., Lal, M., Kathpalia, R., and Bhatla, S.C. (2018). Plant mineral nutrition. In: *Plant Physiology, Development and Metabolism* (ed. S.C. Bhatla and M.A. Lal), 37–81. Singapore: Springer.

Biancucci, M., Mattioli, R., Forlani, G. et al. (2015). Role of proline and GABA in sexual reproduction of angiosperms. *Frontiers in Plant Science* 6: 680.

Bouché, N., Fait, A., Zik, M., and Fromm, H. (2004). The root-specific glutamate decarboxylase (GAD1) is essential for sustaining GABA levels in *Arabidopsis*. *Plant Molecular Biology* 55: 315–325.

Boura, M., Brensone, D., and Karatzas, K.A. (2020). A novel role for the glutamate decarboxylase system in listeria monocytogenes; protection against oxidative stress. *Food Microbiology* 85: 103284.

Chen, K., Li, G.J., Bressan, R.A. et al. (2020). Abscisic acid dynamics, signaling, and functions in plants. *Journal of Integrative Plant Biology* 62 (1): 25–54.

Chen, P., Liu, Q., Sun, B. et al. (2023). Creation and gene expression analysis of a giant embryo rice mutant with high GABA content. *Molecular Breeding* 43 (1): 3.

Cheng, B., Li, Z., Liang, L. et al. (2018). The γ-aminobutyric acid (GABA) alleviates salt stress damage during seeds germination of white clover associated with Na^+/K^+ transportation, dehydrins accumulation, and stress-related genes expression in white clover. *International Journal of Molecular Sciences* 19 (9): 2520.

Corpas, F.J. and Barroso, J.B. (2013). Nitro-oxidative stress vs oxidative or nitrosative stress in higher plants. *New Phytologist* 199 (3): 633–635.

Corpas, F.J., Del Río, L.A., and Palma, J.M. (2019). Plant peroxisomes at the crossroad of NO and H_2O_2 metabolism. *Journal of Integrative Plant Biology* 61 (7): 803–816.

Dabravolski, S.A. and Isayenkov, S.V. (2024). The physiological and molecular mechanisms of silicon action in salt stress amelioration. *Plants* 13 (4): 525.

Daspute, A.A., Sadhukhan, A., Tokizawa, M. et al. (2017). Transcriptional regulation of aluminum-tolerance genes in higher plants: clarifying the underlying molecular mechanisms. *Frontiers in Plant Science* 8: 1358.

Fahad, S., Nie, L., Chen, Y. et al. (2023). Crop plant hormones and environmental stress. *Sustainable Agriculture Reviews* 15: 371–400.

Fait, A., Fromm, H., Walter, D. et al. (2008). Highway or byway: the metabolic role of the GABA shunt in plants. *Trends in Plant Science* 13 (1): 14–19.

Fan, M. and Li, Q. (2024). The metabolism, detrimental effects, and signal transduction mechanism of reactive oxygen species in plants under abiotic stress. *Journal of Biobased Materials and Bioenergy* 18 (3): 359–376.

Fatima, Z., Qayyum, Z., Anjum, B. et al. (2023). Alterations in metabolic profiling of crop plants under abiotic stress. In: *Phytohormones and Stress Responsive Secondary Metabolites* (ed. M. Ozturk, R.A. Bhat, A. Muhammad, and G.H. Dar), 197–233. Academic Press.

Feissner, R.F., Skalska, J., Gaum, W.E., and Sheu, S.S. (2009). Crosstalk signaling between mitochondrial Ca^{2+} and ROS. *Frontiers in Bioscience: A Journal and Virtual Library* 14: 1197.

Feng, D., Jia, X., Yan, Z. et al. (2023). Underlying mechanisms of exogenous substances involved in alleviating plant heat stress. *Plant Stress* 10: 100288.

France, R.C. and Reid, C.P. (1983). Interactions of nitrogen and carbon in the physiology of ectomycorrhizae. *Canadian Journal of Botany* 61 (3): 964–984.

Gahlowt, P., Tripathi, D.K., Singh, S.P. et al. (2024). GABA in plants: developmental and stress resilience perspective. *Physiologia Plantarum* 176 (1): e14116.

Gallego, P.P., Whotton, L., Picton, S. et al. (1995). A role for glutamate decarboxylase during tomato ripening: the characterisation of a cDNA encoding a putative glutamate decarboxylase with a calmodulin-binding site. *Plant Molecular Biology* 27: 1143–1151.

Gramazio, P., Takayama, M., and Ezura, H. (2020). Challenges and prospects of new plant breeding techniques for GABA improvement in crops: tomato as an example. *Frontiers in Plant Science* 11: 577980.

Guo, Z., Gong, J., Luo, S. et al. (2023). Role of gamma-aminobutyric acid in plant defense response. *Metabolites* 13 (6): 741.

Hasanuzzaman, M., Hossain, M.A., da Silva, J.A., and Fujita, M. (2012). Plant response and tolerance to abiotic oxidative stress: antioxidant defense is a key factor. In: *Crop Stress and Its Management: Perspectives and Strategies* (ed. V. Bandi, A.K. Shanker, C. Shanker, and M. Mandapaka), 261–315. New York: Springer.

Hasanuzzaman, M., Alhaithloul, H.A., Parvin, K. et al. (2019). Polyamine action under metal/metalloid stress: regulation of biosynthesis, metabolism, and molecular interactions. *International Journal of Molecular Science* 20 (13): 3215.

Hayat, F., Khan, U., Li, J. et al. (2023). γ aminobutyric acid (GABA): akey player in alleviating abiotic stress resistance in horticultural crops: current insights and future directions. *Horticulturae* 9 (6): 647.

Hertz, L. and Rothman, D.L. (2016). Glucose, lactate, β-hydroxybutyrate, acetate, GABA, and succinate as substrates for synthesis of glutamate and GABA in the glutamine–glutamate/GABA cycle. In: *The Glutamate/GABA-Glutamine Cycle: Amino Acid Neurotransmitter Homeostasis* (ed. A. Schousboe and U. Sonnewald), 9–42. Cham, Switzerland: Springer.

Hsu, P.K., Dubeaux, G., Takahashi, Y., and Schroeder, J.I. (2021). Signaling mechanisms in abscisic acid-mediated stomatal closure. *The Plant Journal* 105 (2): 307–321.

Jacob, P., Hirt, H., and Bendahmane, A. (2017). The heat-shock protein/chaperone network and multiple stress resistance. *Plant Biotechnology Journal* 15 (4): 405–414.

Jain, M., Nagar, P., Goel, P. et al. (2018). Second messengers: central regulators in plant abiotic stress response. In: *Abiotic Stress-Mediated Sensing and Signaling in Plants: An Omics Perspective* (ed. S. Zargar and M. Zargar), 47–94. Singapore: Springer.

Jantaro, S. and Kanwal, S. (2017). Low-molecular-weight nitrogenous compounds (GABA and polyamines). In: *Blue–Green Algae in Algal Green Chemistry* (ed. R.P. Rastogi, D. Madamwar, and A. Pandey), 149–169. Amsterdam, The Netherlands: Elsevier.

Ji, J., Zheng, L., Yue, J. et al. (2015). Identification of two CiGADs from *Caragana* intermedia and their transcriptional responses to abiotic stresses and exogenous abscisic acid. *PeerJ* 5: e3439.

Jiménez-Bremont, J.F., Rodríguez-Kessler, M., Ortega-Amaro, M.A. et al. (2023). Polyamines and metal stress tolerance in plants. In: *Biostimulants in Alleviation of Metal Toxicity in Plants* (ed. S.S. Gill, N. Tuteja, A. Khan Nafees, and R. Gill), 267–294. London: Academic Press.

Jugran, A.K., Rawat, S., Devkota, H.P. et al. (2021). Diabetes and plant-derived natural products: from ethnopharmacological approaches to their potential for modern drug discovery and development. *Phytotherapy Research* 35 (1): 223–245.

Kalhor, M.S., Aliniaeifard, S., Seif, M. et al. (2018). Enhanced salt tolerance and photosynthetic performance: Implication of ɤ-amino butyric acid application in salt-exposed lettuce (*Lactuca sativa* L.) plants. *Plant Physiology and Biochemistry* 130: 157–172.

Khan, M.I., Jalil, S.U., Chopra, P. et al. (2021). Role of GABA in plant growth, development and senescence. *Plant Gene* 26: 100283.

Khan, M., Ali, S., Al Azzawi, T.N., and Yun, B.W. (2023). Nitric oxide acts as a key Signaling molecule in plant development under stressful conditions. *International Journal of Molecular Sciences* 24 (5): 4782.

Killiny, N. (2017). Metabolite signature of the phloem sap of fourteen citrus varieties with different degrees of tolerance to *Candidatus Liberibacter asiaticus. Physiological and Molecular Plant Pathology* 97: 20–29.

Kinnersley, A.M. and Turano, F.J. (2000). Gamma aminobutyric acid (GABA) and plant responses to stress. *Critical Reviews in Plant Sciences* 19 (6): 479–509.

Komatsu, S., Kimura, T., Rehman, S.U. et al. (2023). Proteomic analysis reveals salt-tolerant mechanism in soybean applied with plant-derived smoke solution. *International Journal of Molecular Sciences* 24 (18): 13734.

Krysenko, S. and Wohlleben, W. (2023). Polyamine and ethanolamine metabolism in bacteria as an important component of nitrogen assimilation for survival and pathogenicity. *Medical Sciences* 10 (3): 40.

Kumari, S., Nazir, F., Singh, A. et al. (2024). γ-Aminobutyric acid (GABA) strengthened nutrient accumulation, defense metabolism, growth and yield traits against salt and endoplasmic reticulum stress conditions in wheat plants. *Plant and Soil* 498 (1): 409–429.

Lee, X.Y., Tan, J.S., and Cheng, L.H. (2023). Gamma aminobutyric acid (GABA) enrichment in plant-based food–a mini review. *Food Reviews International* 39 (8): 5864–5885.

Li, L., Dou, N., Zhang, H., and Wu, C. (2021). The versatile GABA in plants. *Plant Signaling &Behavior* 16 (3): 1862565.

Li, Y., Zhang, X., Zhang, Y., and Ren, H. (2022). Controlling the gate: the functions of the cytoskeleton in stomatal movement. *Frontiers in Plant Science* 13: 849729.

Liao, H.S., Chung, Y.H., and Hsieh, M.H. (2022). Glutamate: a multifunctional amino acid in plants. *Plant Science* 318: 111238.

Liu, T., Liu, Z., Li, Z. et al. (2019). Regulation of heat shock factor pathways by γ-aminobutyric acid (GABA) associated with thermotolerance of creeping bentgrass. *International Journal of Molecular Sciences* 20 (19): 4713.

Liu, X.D., Zeng, Y.Y., Zhang, X.Y. et al. (2023). Polyamines inhibit abscisic acid-induced stomatal closure by scavenging hydrogen peroxide. *Physiologia Plantarum* 175: e13903.

Long, Y., Tyerman, S.D., and Gilliham, M. (2020). Cytosolic GABA inhibits anion transport by wheat ALMT1. *New Phytologist* 225 (2): 671–678.

Maccarrone, M., Van Zadelhoff, G., Veldink, G.A. et al. (2000). Early activation of lipoxygenase in lentil (*Lens culinaris*) root protoplasts by oxidative stress induces programmed cell death. *European Journal of Biochemistry* 267 (16): 5078–5084.

Mahmud, J.A., Hasanuzzaman, M., Nahar, K. et al. (2017). γ-aminobutyric acid (GABA) confers chromium stress tolerance in *Brassica juncea* L. by modulating the antioxidant defense and glyoxalase systems. *Ecotoxicology* 26: 675–690.

Majumdar, R., Barchi, B., Turlapati, S.A. et al. (2016). Glutamate, ornithine, arginine, proline, and polyamine metabolic interactions: the pathway is regulated at the post-transcriptional level. *Frontiers in Plant Science* 7: 78.

Mansoor, S., Ali Wani, O., Lone, J.K. et al. (2022). Reactive oxygen species in plants: from source to sink. *Antioxidants* 11 (2): 225.

Masson, P.H., Takahashi, T., and Angelini, R. (2017). Molecular mechanisms underlying polyamine functions in plants. *Frontiers in Plant Science* 8: 14.

Mattoo, A.K., Minocha, S.C., Minocha, R., and Handa, A.K. (2010). Polyamines and cellular metabolism in plants: transgenic approaches reveal different responses to diamine putrescine versus higher polyamines spermidine and spermine. *Amino Acids* 38: 405–413.

Medda, S., Fadda, A., and Mulas, M. (2022). Influence of climate change on metabolism and biological characteristics in perennial woody fruit crops in the Mediterranean environment. *Horticulturae* 8 (4): 273.

Minocha, R., Majumdar, R., and Minocha, S.C. (2014). Polyamines and abiotic stress in plants: a complex relationship. *Frontiers in Plant Science* 5: 175.

Mohanta, T.K., Bashir, T., Hashem, A. et al. (2018). Early events in plant abiotic stress signaling: interplay between calcium, reactive oxygen species and phytohormones. *Journal of Plant Growth Regulation* 37: 1033–1049.

Mukherjee, S., Chatterjee, N., Sircar, A. et al. (2023). A comparative analysis of heavy metal effects on medicinal plants. *Applied Biochemistry and Biotechnology* 195 (4): 2483–2518.

Nandy, S., Mandal, S., Gupta, S.K. et al. (2023). Role of polyamines in molecular regulation and cross-talks against drought tolerance in plants. *Journal of Plant Growth Regulation* 42 (8): 4901–4917.

Okon, O.G. (2019). Effect of salinity on physiological processes in plants. In: *Microorganisms in Saline Environments: Strategies and Functions* (ed. B. Giri and A. Varma), 237–262. Cham, Switzerland: Springer.

Palma, F., Carvajal, F., Jamilena, M., and Garrido, D. (2014). Contribution of polyamines and other related metabolites to the maintenance of zucchini fruit quality during cold storage. *Plant Physiology and Biochemistry* 82: 161–171.

Pasternak, T., Palme, K., and Pérez-Pérez, J.M. (2023). Role of reactive oxygen species in the modulation of auxin flux and root development in *Arabidopsis thaliana*. *The Plant Journal* 114 (1): 83–95.

Peng, S., Piao, S., Ciais, P. et al. (2013). Asymmetric effects of daytime and night-time warming on northern hemisphere vegetation. *Nature* 501 (7465): 88–92.

Podlešáková, K., Ugena, L., Spíchal, L. et al. (2019). Phytohormones and polyamines regulate plant stress responses by altering GABA pathway. *New Biotechnology* 48: 53–65.

Postiglione, A.E. and Muday, G.K. (2020). The role of ROS homeostasis in ABA-induced guard cell signaling. *Frontiers in Plant Science* 11: 968.

Qu, Y., An, Z., Zhuang, B. et al. (2014). Copper amine oxidase and phospholipase D act independently in abscisic acid (ABA)-induced stomatal closure in *Vicia faba* and *Arabidopsis*. *Journal of Plant Research* 127: 533–544.

Rajput, V.D., Harish, S.R.K., Verma, K.K. et al. (2021). Recent developments in enzymatic antioxidant defence mechanism in plants with special reference to abiotic stress. *Biology* 10 (4): 267.

Ramesh, S.A., Tyerman, S.D., Gilliham, M., and Xu, B. (2017). γ-Aminobutyric acid (GABA) signalling in plants. *Cellular and Molecular Life Sciences* 74: 1577–1603.

Ramesh, S.A., Kamran, M., Sullivan, W. et al. (2018). Aluminum-activated malate transporters can facilitate GABA transport. *The Plant Cell* 30 (5): 1147–1164.

Ramos-Ruiz, R., Martinez, F., and Knauf-Beiter, G. (2019). The effects of GABA in plants. *Cogent Food & Agriculture* 5 (1): 1670553.

Rashmi, D., Zanan, R., John, S. et al. (2018). γ-Aminobutyric acid (GABA): biosynthesis, role, commercial production, and applications. *Studies in Natural Products Chemistry* 57: 413–452.

Rastegar, S., Aghaei Dargiri, S., and Mohammadi, M. (2024). Mitigating postharvest chilling injury in Orlando tangelo fruit: potential of melatonin and GABA in enhancing the antioxidant system. *Acta Physiologiae Plantarum* 46 (3): 30.

Raza, A., Salehi, H., Rahman, M.A. et al. (2022). Plant hormones and neurotransmitter interactions mediate antioxidant defenses under induced oxidative stress in plants. *Frontiers in Plant Science* 13: 1–36.

Renault, H., Roussel, V., El Amrani, A. et al. (2010). The Arabidopsis pop2-1 mutant reveals the involvement of GABA transaminase in salt stress tolerance. *BMC Plant Biology* 10: 1–6.

Reyes-Díaz, M., González-Villagra, J., Ulloa-Inostroza, E.M. et al. (2023). Phytohormone involvement in plant responses to soil acidity. In: *Plant Hormones and Climate Change* (ed. G.J. Ahammed and J. Yu), 301–323. Singapore: Springer Nature Singapore.

Rodrigues, O. and Shan, L. (2022). Stomata in a state of emergency: H_2O_2 is the target locked. *Trends in Plant Science* 27 (3): 274–286.

Rodrigues-Corrêa, K.C.D.S. and Fett-Neto, A.G. (2019). Abiotic stresses and non-protein amino acids in plants. *Critical Reviews in Plant Sciences* 38: 411–430.

Royes, L.F., Gabbi, P., Ribeiro, L.R. et al. (2016). A neuronal disruption in redox homeostasis elicited by ammonia alters the glycine/glutamate (GABA) cycle and contributes to MMA-induced excitability. *Amino Acids* 48: 1373–1389.

Sachdev, S., Ansari, S.A., Ansari, M.I. et al. (2021). Abiotic stress and reactive oxygen species: generation, signaling, and defense mechanisms. *Antioxidants* 10 (2): 277.

Saini, N., Nikalje, G.C., Zargar, S.M., and Suprasanna, P. (2022). Molecular insights into sensing, regulation and improving of heat tolerance in plants. *Plant Cell Reports* 41 (3): 799–813.

Salvatierra, A., Pimentel, P., Almada, R., and Hinrichsen, P. (2016). Exogenous GABA application transiently improves the tolerance to root hypoxia on a sensitive genotype of Prunus rootstock. *Environmental and Experimental Botany* 125: 52–66.

Salvi, P., Manna, M., Kaur, H. et al. (2021). Phytohormone signaling and crosstalk in regulating drought stress response in plants. *Plant Cell Reports* 40: 1305–1329.

Sandalio, L.M. and Romero-Puertas, M.C. (2015). Peroxisomes sense and respond to environmental cues by regulating ROS and RNS signalling networks. *Annals of Botany* 116 (4): 475–485.

Sarasa, S.B., Mahendran, R., Muthusamy, G. et al. (2020). A brief review on the non-protein amino acid, gamma-amino butyric acid (GABA): its production and role in microbes. *Current Microbiology* 77: 534–544.

Sauter, M., Moffatt, B., Saechao, M.C. et al. (2013). Methionine salvage and *S*-adenosylmethionine: essential links between sulfur, ethylene and polyamine biosynthesis. *Biochemical Journal* 451 (2): 145–154.

Seifikalhor, M., Aliniaeifard, S., Hassani, B. et al. (2019). Diverse role of γ-aminobutyric acid in dynamic plant cell responses. *Plant Cell Reports* 38: 847–867.

Seleiman, M.F., Al-Suhaibani, N., Ali, N. et al. (2021). Drought stress impacts on plants and different approaches to alleviate its adverse effects. *Plants* 10 (2): 259.

Sequera-Mutiozabal, M., Tiburcio, A.F., and Alcázar, R. (2016). Drought stress tolerance in relation to polyamine metabolism in plants. In: *Drought Stress Tolerance in Plants Physiology and Biochemistry* (ed. M. Hossain, S. Wani, S. Bhattacharjee, et al.), 267–286. Cham, Switzerland: Springer.

Seth, T., Asija, S., Khan, M.I., and Iqbal, N. (2023). Emerging role of nitric oxide in combating abiotic stresses in crops. In: *Nitric Oxide in Developing Plant Stress Resilience* (ed. M.I.R. Khan, N. Iqbal, P. Poór, et al.), 31–68. Amsterdam, The Netherlands: Academic Press.

Sharma, T., Dreyer, I., Kochian, L., and Piñeros, M.A. (2016). The ALMT family of organic acid transporters in plants and their involvement in detoxification and nutrient security. *Frontiers in Plant Science* 7: 1488.

Shelp, B.J., Aghdam, M.S., and Flaherty, E.J. (2021). γ-Aminobutyrate (GABA) regulated plant defense: mechanisms and opportunities. *Plants* 10 (9): 1939.

Shelp, B.J., Bozzo, G.G., Trobacher, C.P. et al. (2012). Hypothesis/review: contribution of putrescine to 4-aminobutyrate (GABA) production in response to abiotic stress. *Plant Science* 193: 130–135.

Sheteiwy, M.S., Shao, H., Qi, W. et al. (2019). GABA-alleviated oxidative injury induced by salinity, osmotic stress and their combination by regulating cellular and molecular signals in rice. *International Journal of Molecular Sciences* 20 (22): 5709.

Shi, S.Q., Shi, Z., Jiang, Z.P. et al. (2010). Effects of exogenous GABA on gene expression of *Caragana* intermedia roots under NaCl stress: regulatory roles for H_2O_2 and ethylene production. *Plant, Cell &Environment* 33 (2): 149–162.

Signorelli, S., Tarkowski, Ł.P., O'Leary, B. et al. (2023). GABA and proline metabolism in response to stress. In: *Hormones and Plant Response* (ed. D.K. Gupta and F.J. Corpas), 291–314. Cham, Switzerland: Springer.

Singh, P., Basu, S., and Kumar, G. (2018). Polyamines metabolism: a way ahead for abiotic stress tolerance in crop plant.In Biochemical, Physiological and Molecular Avenues for Combating Abiotic Stress Tolerance in Plants.WaniSH (ed) London: Academic Press 39–55.

Singhal, R.K., Saha, D., Skalicky, M. et al. (2021). Crucial cell signaling compounds crosstalk and integrative multi-omics techniques for salinity stress tolerance in plants. *Frontiers in Plant Science* 12: 670369.

Sita, K. and Kumar, V. (2020). Role of gamma amino butyric acid (GABA) against abiotic stress tolerance in legumes: a review. *Plant Physiology Reports* 25: 654–663.

Snedden, W.A., Arazi, T., Fromm, H., and Shelp, B.J. (1995). Calcium/calmodulin activation of soybean glutamate decarboxylase. *Plant Physiology* 108 (2): 543–549.

Suhel, M., Husain, T., Pandey, A. et al. (2023). An appraisal of ancient molecule GABA in abiotic stress tolerance in plants, and its crosstalk with other signaling molecules. *Journal of Plant Growth Regulation* 42 (2): 614–629.

Sulieman, S., Sheteiwy, M.S., Abdelrahman, M., and Tran, L.S. (2024). γ-Aminobutyric acid (GABA) in N_2-fixing-legume symbiosis: metabolic flux and carbon/nitrogen homeostasis in responses to abiotic constraints. *Plant Physiology and Biochemistry* 11: 108362.

Takayama, M., Koike, S., Kusano, M. et al. (2015). Tomato glutamate decarboxylase genes SlGAD2 and SlGAD3 play key roles in regulating γ-aminobutyric acid levels in tomato (*Solanum lycopersicum*). *Plant and Cell Physiology* 56 (8): 1533–1545.

Takayama, M., Matsukura, C., Ariizumi, T., and Ezura, H. (2017). Activating glutamate decarboxylase activity by removing the autoinhibitory domain leads to hyper γ-aminobutyric acid (GABA) accumulation in tomato fruit. *Plant Cell Reports* 36: 103–116.

Tamarit-Rodriguez, J. (2023). Metabolic role of GABA in the secretory function of pancreatic β-cells: its hypothetical implication in β-cell degradation in type 2 diabetes. *Metabolites* 13 (6): 697.

Tang, M., Li, Z., Luo, L. et al. (2020). Nitric oxide signal, nitrogen metabolism, and water balance affected by γ-aminobutyric acid (GABA) in relation to enhanced tolerance to water stress in creeping bentgrass. *International Journal of Molecular Sciences* 21 (20): 7460.

Tyagi, A., Ali, S., Ramakrishna, G. et al. (2023). Revisiting the role of polyamines in plant growth and abiotic stress resilience: mechanisms, crosstalk, and future perspectives. *Journal of Plant Growth Regulation* 42 (8): 5074–5098.

Ullah, A., Ali, I., Noor, J. et al. (2023). Exogenous γ-aminobutyric acid (GABA) mitigated salinity-induced impairments in mungbean plants by regulating their nitrogen metabolism and antioxidant potential. *Frontiers in Plant Science* 13: 1081188.

Usman, B., Derakhshani, B., and Jung, K.H. (2023). Recent molecular aspects and integrated omics strategies for understanding the abiotic stress tolerance of Rice. *Plants* 12 (10): 2019.

Valdés-Santiago, L. and Ruiz-Herrera, J. (2014). Stress and polyamine metabolism in fungi. *Frontiers in Chemistry* 1: 42.

Vijayakumari, K., Jisha, K.C., and Puthur, J.T. (2016). GABA/BABA priming: a means for enhancing abiotic stress tolerance potential of plants with less energy investments on defence cache. *Acta Physiologiae Plantarum* 38: 1–4.

Virdi, A.S., Singh, S., and Singh, P. (2015). Abiotic stress responses in plants: roles of calmodulin-regulated proteins. *Frontiers in Plant Science* 6: 809.

Vishwakarma, K., Upadhyay, N., Kumar, N. et al. (2017). Abscisic acid signaling and abiotic stress tolerance in plants: a review on current knowledge and future prospects. *Frontiers in Plant Science* 8: 161.

Wang, Y., Gu, W., Meng, Y. et al. (2017). γ-Aminobutyric acid imparts partial protection from salt stress injury to maize seedlings by improving photosynthesis and upregulating osmoprotectants and antioxidants. *Scientific Reports* 7 (1): 43609.

Wu, Q., Su, N., Huang, X. et al. (2021). Hypoxia-induced increase in GABA content is essential for restoration of membrane potential and preventing ROS-induced disturbance to ion homeostasis. *Plant Communications* 2 (3): 100188.

Xia, Q., Zheng, Y., Wang, L., and Chen, X. (2023). Proposing Signaling molecules as key optimization targets for intensifying the phytochemical biosynthesis induced by emerging nonthermal stress Pretreatments of plant-based foods: afocus on γ-aminobutyric acid. *Journal of Agricultural and Food Chemistry* 71 (34): 12622–12644.

Xie, X., He, Z., Chen, N. et al. (2019). The roles of environmental factors in regulation of oxidative stress in plant. *BioMed Research International* 2019: 9732325.

Xu, J., Liu, T., Yang, S. et al. (2019). Polyamines are involved in GABA-regulated salinity-alkalinity stress tolerance in muskmelon. *Environmental and Experimental Botany* 164: 181–189.

Xu, B., Long, Y., Feng, X. et al. (2021a). GABA signalling modulates stomatal opening to enhance plant water use efficiency and drought resilience. *Nature Communications* 12 (1): 1952.

Xu, B., Sai, N., and Gilliham, M. (2021b). The emerging role of GABA as a transport regulator and physiological signal. *Plant Physiology* 187 (4): 2005–2016.

Xu, B., Feng, X., Piechatzek, A. et al. (2024). The GABA shunt contributes to ROS homeostasis in guard cells of *Arabidopsis*. *New Phytologist* 241 (1): 73–81.

Yang, F., Zhang, H., Wang, Y. et al. (2021). The role of antioxidant mechanism in photosynthesis under heavy metals Cd or Zn exposure in tobacco leaves. *Journal of Plant Interactions* 16 (1): 354–366.

Yin, Y., Cheng, C., and Fang, W. (2018). Effects of the inhibitor of glutamate decarboxylase on the development and GABA accumulation in germinating fava beans under hypoxia-NaCl stress. *RSC Advances* 8 (36): 20456–20461.

Yin, Y., Hu, J., Tian, X. et al. (2023). Nitric oxide mediates melatonin-induced isoflavone accumulation and growth improvement in germinating soybeans under NaCl stress. *Journal of Plant Physiology* 279: 153855.

Yong, B., Xie, H., Li, Z. et al. (2017). Exogenous application of GABA improves PEG-induced drought tolerance positively associated with GABA-shunt, polyamines, and proline metabolism in white clover. *Frontiers in Physiology* 8: 1107.

Yu, Y., Li, M., Li, C. et al. (2023). Accelerated accumulation of γ-aminobutyric acid and modifications on its metabolic pathways in black rice grains by germination under cold stress. *Foods* 12 (6): 1290.

Yuan, D., Wu, X., Jiang, X. et al. (2024). Types of membrane transporters and the mechanisms of interaction between them and reactive oxygen species in plants. *Antioxidants* 13 (2): 221.

Zandi, P. and Schnug, E. (2022). Reactive oxygen species, antioxidant responses and implications from a microbial modulation perspective. *Biology* 11 (2): 155.

Zarbakhsh, S., Shahsavar, A.R., Afaghi, A., and Hasanuzzaman, M. (2024). Predicting and optimizing reactive oxygen species metabolism in *Punica granatum* L. through machine learning: role of exogenous GABA on antioxidant enzyme activity under drought and salinity stress. *BMC Plant Biology* 24 (1): 65.

Zarei, A., Trobacher, C.P., and Shelp, B.J. (2015). NAD^+-aminoaldehyde dehydrogenase candidates for 4-aminobutyrate (GABA) and β-alanine production during terminal oxidation of polyamines in apple fruit. *FEBS Letters* 589 (19): 2695–2700.

Zarei, A., Chiu, G.Z., Yu, G. et al. (2017). Salinity-regulated expression of genes involved in GABA metabolism and signaling. *Botany* 95 (6): 621–627.

Zhang, S., Khan, A., Zhao, L., Feng, N., Zheng, D., and Shen, X. (2023). Effect of GABA on seed germination and seedling growth of rapeseed under salt stress. https://doi.org/10.21203/rs.3.rs-3132215/v1.

Zhao, J., Harper, R., Barchowsky, A., and Di, Y.P. (2007). Identification of multiple MAPK-mediated transcription factors regulated by tobacco smoke in airway epithelial cells. *American Journal of Physiology-Lung Cellular and Molecular Physiology* 293 (2): L480–L490.

Zhao, Y., Song, X., Zhong, D.B. et al. (2020). γ-Aminobutyric acid (GABA) regulates lipid production and cadmium uptake by *Monoraphidium* sp. QLY-1 under cadmium stress. *Bioresource Technology* 297: 122500.

4

GABA as a Signaling Molecule in Plants

Navya Sreepathi[1,2*], *Deepthi Puttegowda*[1*], *Nagma Firdose*[2], *Bhavya Somaplara Gangadharappa*[3], *V B Chandana Kumari*[1], *and Ramith Ramu*[1]

[1] *Department of Biotechnology and Bioinformatics, JSS Academy of Higher Education and Research, Mysuru, Karnataka, 570015, India*
[2] *Department of Pharmacology, JSS Medical College, JSS Academy of Higher Education & Research, Mysuru, Karnataka, 570015, India*
[3] *Department of Biotechnology, M. S. Ramaiah Institute of Technology, Bangalore, Karnataka, 560054, India*

Abbreviations

ABA	Abscisic acid
ABI2	Abscisic acid-insensitive 2
AR	Adventitious root
CBL1	Calcineurin B-like protein 1
CIPK5	Calcineurin B-like-interacting protein kinase 5
Cl^-	Chloride
CO_2	Carbon dioxide
DAF	Days after flowering
DHAR	Dehydroascorbate reductase
GABA	Gamma-aminobutyric acid
GABA-T	Gamma-aminobutyric acid transaminase
GAD	Glutamic acid decarboxylase
GORK	Guard cell outward-rectifying K^+ channel
GR	Glutathione reductase
H_2O_2	Hydrogen peroxide
IBA	Indole-3-butyric acid
JA	Jasmonic acid
K^+	Potassium
NAD^+	Nicotinamide adenine dinucleotide
NADPH	Nicotinamide adenine dinucleotide phosphate
NFT	Nutrient film technique

[*] Equal contribution

GABA in Plants: Biosynthesis, Plant Development, and Food Security, First Edition. Edited by Samiksha Singh, Durgesh Kumar Tripathi, and Vijay Pratap Singh.

RNAi	RNA interference
ROS	Reactive oxygen species
SSA	Succinic semialdehyde
SSR	Succinic semialdehyde reductase
TCAC	Tricarboxylic acid cycle

Introduction

Climate change's potential impact on plant development is a significant barrier to food security and agricultural productivity. Unlike mobile animals, plants are immobile organisms that cannot relocate to more favorable environmental conditions. Consequently, they must adapt to their specific growth circumstances and react to environmental changes at their current location to ensure their survival and growth (Raza et al. 2019). Although known primarily as a neurotransmitter in animals, gamma-aminobutyric acid (GABA) also serves as a signaling molecule in plants, playing crucial roles in various physiological processes (Zhao et al. 2022). In the plant kingdom, GABA has been identified as a multifaceted molecule involved in significant processes such as plant growth, responses, and development to stress (Dabravolski and Isayenkov 2023). Its presence in plants was initially discovered almost 70 years ago in potato tubers (*Solanum tuberosum*). Since then, extensive research has underscored its physiological significance as both a metabolite and a signaling molecule in plants (Bouché et al. 2003). Therefore, agricultural productivity and food security are immediately threatened by the possible effects of climate change on plant growth (Takayama and Ezura 2015; Wang et al. 2023).

GABA, a nonproteinogenic amino acid, is an extensively distributed compound found in both prokaryotic and eukaryotic organisms. Cell-to-cell communication, also known as cell signaling, refers to a molecular process occurring either within the cell (intracellular) or between cells (intercellular), regulating crucial functions such as proliferation, differentiation, growth, survival, and behavioral responses in both prokaryotic and eukaryotic organisms (Hasan et al. 2021). Initially identified in plants, its discovery in animal brains as a significant neurotransmitter redirected the scientific focus, unraveling its diverse roles in plant growth, development, and response to environmental stressors (Li et al. 2021a). With its pivotal act as a signaling molecule and its contribution to enhancing plant tolerance to various stressors, GABA holds promising implications for improving crop quality, prolonging shelf life, and bolstering antioxidant defense systems (Bouché et al. 2003). The spectrum of molecules involved in this signaling cascade encompasses primary metabolites (e.g. glutamate and serine), phospholipids, sugars, and derivatives of primary metabolites, such as nonprotein amino acids such as GABA (Fromm 2020). GABA, a ubiquitous four-carbon compound, is prevalent across various life domains, including animals, plants, bacteria, and fungi (Fromm 2020). It has been generally studied in animal central nervous system as a signaling molecule in neurotransmission, operating via multiple GABA receptors, and similarly in bacteria, its role as the signal for quorum sensing has been reported (Yang and Chiu 2017). These findings raised speculation about GABA as a signaling molecule in plants. GABA in plants plays an essential role in response to environmental stresses by accumulating during adverse conditions and regulating stress-responsive gene expression (Hasanuzzaman et al. 2013; Zhu et al. 2019). Furthermore, it functions as a signaling molecule that affects a number of plant growth and development processes, including blooming, shoot and root development, fruit ripening, and seed germination (Sheng et al. 2018; Liu et al. 2021; Zhao et al. 2023). This chapter explores GABA's complex function as a key signaling molecule in plants, elucidating its intricate involvement in diverse aspects such as stress response, growth, and development, the modulation of stomatal movement, intricate interaction with phytohormones, and precise modulation of ion transport and cellular homeostasis.

GABA in Plants as a Stress Response

In recent times, the observable rise in temperatures, largely attributed to the effects of global warming and various contributing factors, has undeniably become a tangible reality. This current state of affairs represents a pressing challenge for the contemporary scientific community, urging them to explore innovative solutions to address the ensuing complexities (Rao et al. 2016). Agriculture, a sector intricately entwined with climatic conditions, remains particularly susceptible to the implications of these changes. Without appropriate adaptive measures and mitigation strategies, the potential consequences on agricultural systems could be profound and far-reaching (Rihan et al. 2017).

Many abiotic stresses, including drought, salt, high temperatures, chemical pollutants, nutritional deficiencies, and oxidative stress, continue to be significant elements influencing current agricultural practices within the context of environmental limits (Haglund and Dikic 2005; Rao et al. 2016) shown in Figure 4.1. These stresses not only prevent the natural development and growth of crops but also exert a substantial influence on various aspects of plant life, ranging from morphology and physiology to biochemistry and molecular processes (Raza et al. 2019; Hayat et al. 2023).

In the realm of plant biology, the role of GABA has emerged as a critical component in the intricate network governing stress responses (Hayat et al. 2023). GABA, a nonproteinogenic amino acid, has a variety of functions that help plants fight against and adjust to a range of environmental stresses. The essential signaling molecule of GABA coordinates plant reactions to various stresses, including heavy metal toxicity, drought, salt, and severe temperatures, as shown in Figure 4.1 (Li et al. 2021a). By examining its intricate mechanisms and effects, this exploration aims to shed light on the pivotal function of GABA in facilitating resilience and adaptation in the complex world of plant physiology (Thompson et al. 1966). Climate variations may have an effect on a plant's growth, agricultural yield, and food security throughout the course of its existence. Because they are sessile, plants must create an environment that is conducive to their own development at the same location to survive and develop. They also need to keep up with environmental changes (Raza et al. 2019; Chen et al. 2021). GABA is a plant chemical that may adapt to different surroundings. Numerous investigations have documented the accumulation of GABA in plants as a reaction to various environmental stressors, such as insect–pathogen assault, salt, drought, extremes in temperature, reactive oxygen species (ROS), and flooding (Planamente et al. 2013; Scholz et al. 2017; Hasan et al. 2021; Wang et al. 2023). In addition, elevated levels of the amino acid restore homeostasis in plants by revamping

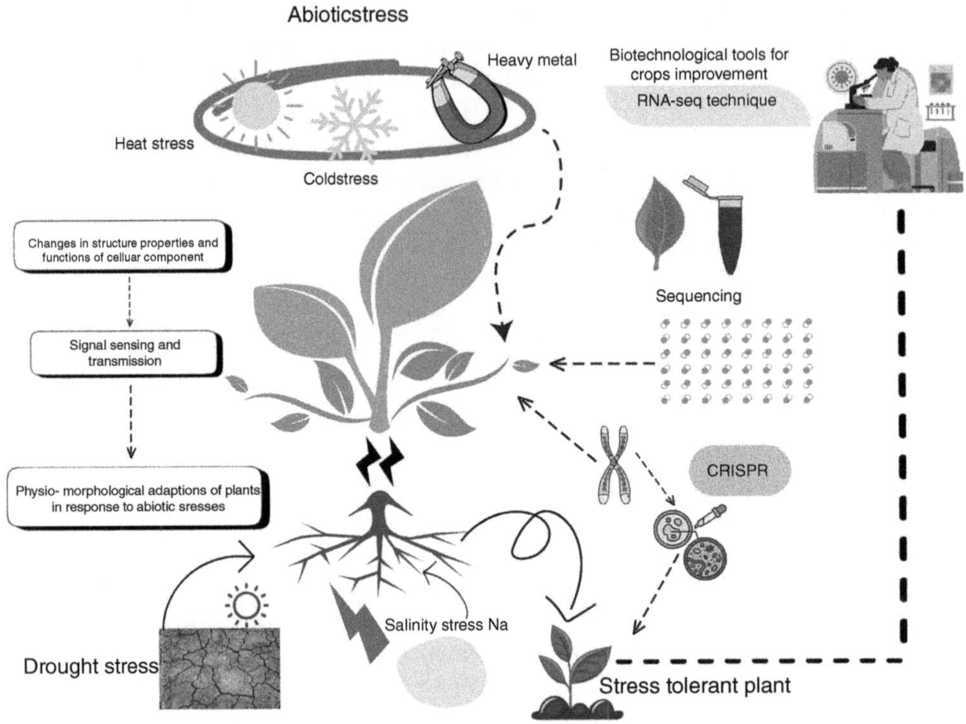

Figure 4.1 Biotechnological methods and fundamental abiotic stress to protect plants from hazardous environmental circumstances.

photosynthesis, impeding ROS generation, activating antioxidant enzymes, and multiple other mechanisms will be induced by GABA to increase the plant stress-tolerance ability (Wu et al. 2021).

GABA as a Drought-Induced Stress Response in Plants

Recent studies have highlighted the crucial role of GABA in plants, particularly during their response to environmental stress, such as drought (Fàbregas and Fernie 2019) (Figure 4.2). When plants encounter drought conditions, they activate a spectrum of mechanisms within their cellular and biochemical machinery (Hasan et al. 2021). These mechanisms are intricately designed to protect essential physiological functions such as cellular metabolism, photosynthesis, and nutrient absorption, even under water-limited circumstances. GABA has emerged as a pivotal component in this intricate interplay (Hasan et al. 2021; Zhao et al. 2023).

During periods of drought stress, plants undergo significant biochemical and physiological adjustments to adapt and endure. Employing various strategies, plants regulate the stomatal aperture to minimize water loss through transpiration and synthesize osmoprotectants to maintain cellular turgor, preventing dehydration (Serraj et al. 1998; Bor et al. 2009; Xu et al. 2021). In addition, they activate stress-responsive genes to reconfigure

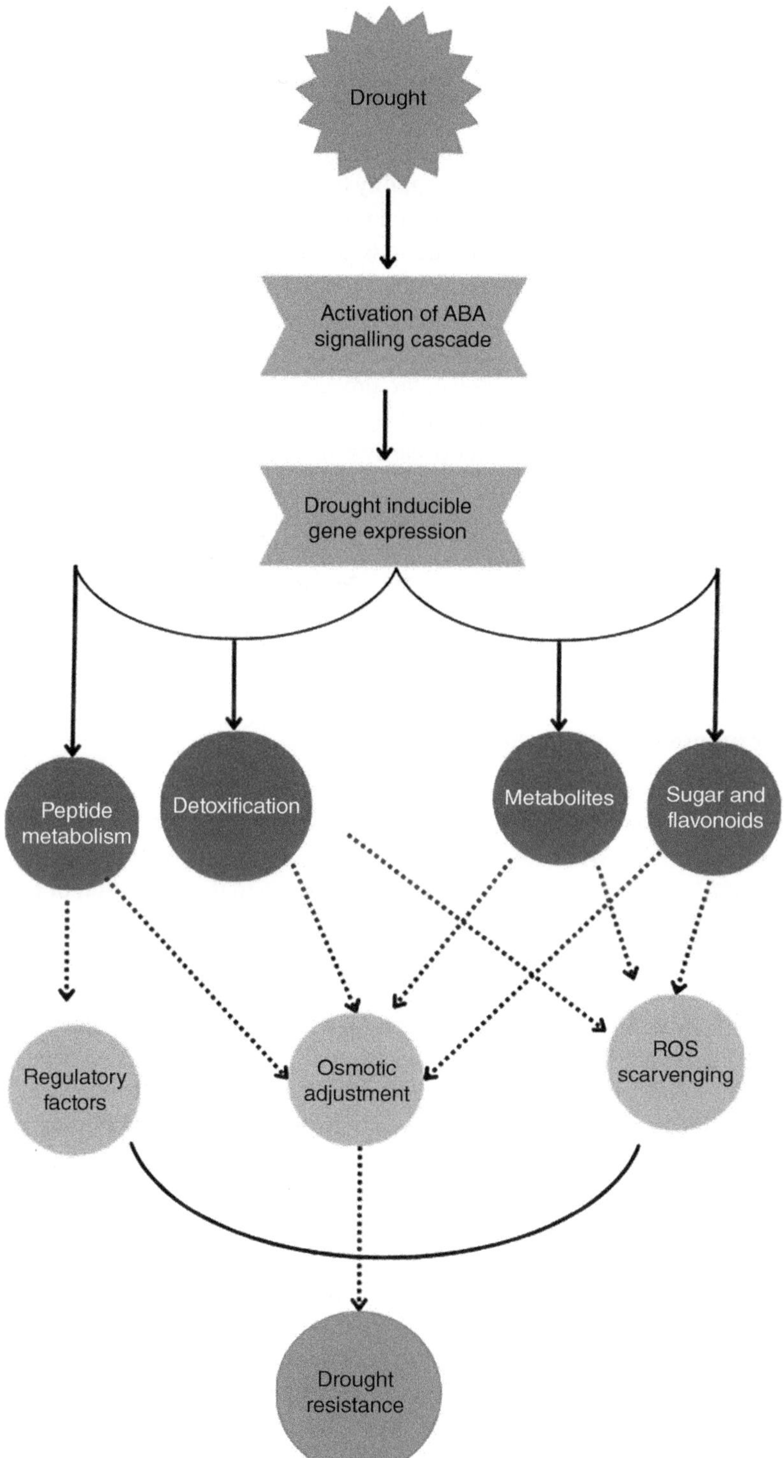

Figure 4.2 Activation of GABA during drought-induced stress in plants.

biochemical pathways and trigger the production of specific proteins and enzymes that scavenge ROS, thus mitigating potential cellular damage (Fàbregas and Fernie 2019). GABA operates as a signaling molecule that modulates multiple stress-responsive processes, encompassing energy metabolism and ROS scavenging, both critical for cell stability, as shown in Figure 4.2 (Hasan et al. 2021; Xu et al. 2021). By coordinating various stress-responsive pathways, GABA strengthens the plant's ability to withstand adverse conditions posed by drought. Moreover, GABA influences stomatal movement, optimizing the plant's water-use efficiency by conserving water during periods of limited water availability (Li et al. 2019; Xu et al. 2021). This mechanism helps plants maintain their turgor pressure and overall water balance, ensuring the continuity of vital physiological processes even during drought stress (Xu et al. 2021).

A comprehensive understanding of the molecular processes behind GABA's role in drought tolerance is crucial for crop production and agriculture. Using this knowledge, breeding programs or genetic engineering can more easily produce drought-resistant plants (Hasan et al. 2021). GABA's role in regulating osmotic balance, influencing stress-related gene expression, and modulating ion transport contributes significantly to the plant's resilience under water-limited conditions (Thompson et al. 1966). By harnessing the potential of GABA-mediated stress responses, the scientific and agricultural communities can make substantial strides in ensuring global food security amidst the escalating challenges of climate variability and water scarcity.

GABA as a Stress Response in Plants Induced by Salinity

GABA is essential for helping plants withstand the stresses brought on by salt. High salinity levels cause plants to react in a number of ways to lessen the negative effects of salt on their physiological processes, and GABA is known to be a key component of this reaction (Munns and Tester 2008). Under salinity stress, GABA contributes to the regulation of ion homeostasis, specifically by modulating the uptake and distribution of ions within plant cells (Gupta and Huang 2014). This function aims to maintain the equilibrium of essential ions and prevent excessive accumulation of toxic ions such as sodium (Na^+) and chloride (Cl^-) which have the potential to disrupt various cellular functions and impede plant growth (Shams and Khadivi 2023). Salinity stress negatively affects the growth, biomass, chlorophyll concentration, and nitrogen metabolism of mung bean plants while increasing oxidative stress and osmolyte accumulation (Priya et al. 2019). By enhancing photosynthetic pigments, the K^+ and K^+/Na^+ ratio, nitrogen metabolism, osmolyte synthesis, and enzyme antioxidant capabilities, exogenous GABA assists in reducing the negative effects of salinity stress (Ullah et al. 2023; Wang et al. 2023).

Furthermore, GABA aids in the synthesis of compatible solutes, including proline, which actively maintains the cellular osmotic balance, ensuring the retention of water within plant cells (Hayat et al. 2023). Proline, as a compatible solute, plays a critical role in osmotic adjustment by accumulating in plant cells and serving as an osmolyte (Sheteiwy et al. 2019). Its accumulation assists in preserving cellular water potential, preventing water loss from the cell into the surrounding environment and thereby enabling the plant to uphold its turgor pressure and overall cellular integrity (dos Santos et al. 2022).

In addition, proline plays a role in stabilizing macromolecules such as proteins and membranes, even under stressful conditions. It protects cellular structures from denaturation and helps maintain the stability of enzymes and other crucial proteins, even in the presence of elevated salt concentrations (Sheteiwy et al. 2019; Hayat et al. 2023). This protective function of proline aids in preserving the structural and functional integrity of cellular components, thereby enhancing the tolerance to salinity stress. Moreover, GABA contributes to the activation of antioxidant defense mechanisms, facilitating the scavenging of ROS generated during salinity stress (Huang et al. 2019) The negative effects of salt stress on plant growth and development are reduced by GABA, which increases the activity of antioxidant enzymes and encourages the manufacture of certain antioxidant molecules (Wang et al. 2023). Proline also has the potential to scavenge ROS, which helps to prevent oxidative damage and maintain plant viability and health in environments where salt is present. By controlling the amounts of ROS, stimulating the production of antioxidant enzymes, and stimulating secondary metabolism, GABA priming improves rice's resistance to osmotic stressors and salt (Sheteiwy et al. 2019; Dabravolski and Isayenkov 2023).

Salinity in the soil hurts a number of plant functions, including seed germination, growth, development, blooming, and fruiting. Salt stress hinders the absorption of nutrients and water and disrupts mitochondrial respiration and the tricarboxylic acid (TCA) cycle by inhibiting the activity of pyruvate dehydrogenase and 2-oxoglutarate dehydrogenase complex, which are essential for ATP supply and ROS detoxification during salt stress (Zhao et al. 2022). In the context of GABA's role in plants under salt stress, it accumulates as an alternate carbon source for the TCA cycle, bypassing salt-inhibited enzymes (Yuan et al. 2023). This occurs via the GABA shunt, which escapes the TCA cycle's 2-oxoglutarate dehydrogenase-catalyzed stage (Carillo 2018). Glutamic acid decarboxylase (GAD) converts glutamate from 2-oxoglutarate into GABA in the GABA shunt route, as shown in Figure 4.3 (Carillo 2018; Yuan et al. 2023).

Following its entry into mitochondria, GABA undergoes catabolization by γ-aminobutyrate transaminase, resulting in succinic semialdehyde (SSA). This compound is then transformed into succinate by succinic semialdehyde dehydrogenase (SSADH). Wheat leaves under salt stress have been researched for this putative mechanism of the response to salt stress, the GABA shunt, and the TCA cycle, as shown in Figure 4.3. The results show that the GABA shunt becomes active during salt stress, and the TCA cycle shifts to the TCA cycle due to GABA activating the salt-tolerant respiratory pathway. Thus, GABA's role in cellular communication and activation of protective pathways facilitates mitochondrial respiration and adaptive processes in plants under salt stress (Carillo 2018; Sheteiwy et al. 2019; Yuan et al. 2023).

Studies focussing on the application of these techniques to decipher salt stress adaptation mechanisms in plants have highlighted several key findings (Dabravolski and Isayenkov 2023; Yuan et al. 2023). One such study revealed that *Arabidopsis* root cells exhibited elevated levels of sugars, amino acids, and specific metabolites, including SSA and GABA, indicating their crucial role in adapting to salt stress adaptation and memory retention (Dabravolski and Isayenkov 2023). Comparative analysis of two closely related halophyte species, *Limonium albuferae* L. and *Limonium doufourii* L., under salt stress conditions showed variations in their metabolic responses, highlighting the activation of

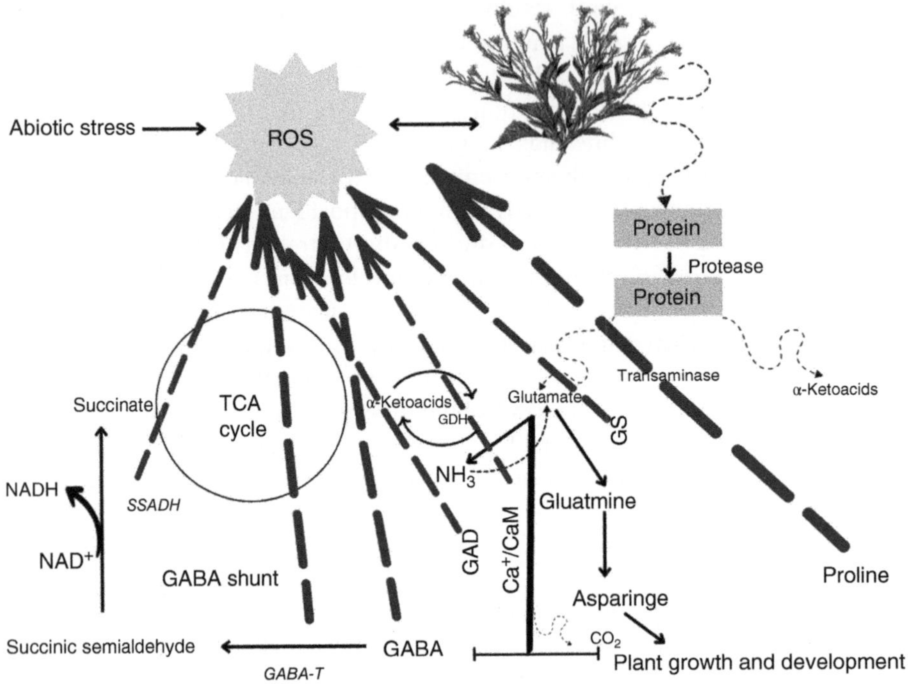

Figure 4.3 GABA's role in response to salinity-induced stress in plants.

the glutamate pathway and subsequent accumulation of GABA and proline (González-Orenga et al. 2019). Furthermore, investigations of wild salt-tolerant barley *Hordeum maritimum* With. and salt-sensitive cultivar *Hordeum vulgare* L. cv. Lamsi highlighted differential amino acid dynamics, particularly an increase in GABA levels in salt-stressed *H. maritimum* compared to the control (Widodo et al. 2009). Analysis of three salt-sensitive invasive freshwater species, American waterweed, Parrot feather watermilfoil, and Water primrose, demonstrated varying responses to salt stress, with notable changes in the levels of carbohydrates and amino acids (Thouvenot et al. 2015). Furthermore, the comparison between salt-sensitive and salt-tolerant sorghum and rice genotypes revealed differences in the accumulation of polyamines, soluble sugars, and GABA, highlighting distinctive metabolic adaptations to salinity stress (Chen et al. 2021). Similarly, investigations in sugar beet roots (*Beta vulgaris* L.) exposed to high salt concentrations indicated significant alterations in amino acid and organic acid levels in response to salt stress (Aycan et al. 2023). Furthermore, research on poplar seedlings (*Populus alba* × *Populus glandulosa*) and tomato plants highlighted the activation of GABA-related enzymes and antioxidants in response to salt stress, suggesting the crucial role of GABA in mediating adaptive responses in various plant species under saline conditions (Chen et al. 2020). The complex nature of these responses is shown by the variety of metabolic adaptations seen in various plant species under salt stress. These adaptations also highlight GABA's role as a major orchestrator in the plant's all-encompassing salinity-coping strategy, providing important information for future tactics to improve salt tolerance in agriculture.

GABA as a Temperature-Induced Stress Response in Plants

Plants are frequently exposed to a myriad of environmental stresses, including fluctuations in temperature, which can significantly impact their growth and development (Hasanuzzaman et al. 2013). In recent years, there has been a growing interest in understanding the molecular mechanisms underlying how plants respond and adapt to temperature stress (Devireddy et al. 2021). The part GABA plays in modulating the stress responses that plants experience in response to temperature changes is one such mechanism that has drawn interest. A variety of complex processes are triggered in plants in response to temperature variations to manage stress and preserve cellular homeostasis (Sweetman et al. 2014).

High temperature is one of the serious factors that limit plant growth and development. In various investigations by researchers about the relationship between GABA and heat stress, it is confirmed that GABA levels are upregulated. For example, the GABA content in heat-treated soybean was maximum; consequently, heat-dried soybean seed expressed the GABA content fivefold more than the control (Takahashi et al. 2013; Li et al. 2021a). Similar results were observed in ripening grapes in response to heat (Sweetman et al. 2014). The probable mechanism through which GABA accumulates in response to heat is that the heat-stressed plants produce increased concentrations of glutamic acid in the cytoplasm, which combines with calmodulin to activate GAD thus inducing GABA shunt (Locy et al. 2000; Li et al. 2021a). Furthermore, accumulated GABA is expected to activate and enhance the heat shock factor (HSF) pathway, which is very important in the adaptive response of plants to heat stress. The HSF pathway's key players, heat shock proteins (HSPs) (HSP17.8, HSP26.7, HSP70, and HSP90.1-b1), HSFs (HSFA-6a, HSFA-2c, and HSFB-2b), and ascorbate peroxidase 3, are shown to have substantially higher transcript levels in response to exogenous or endogenous GABA. The addition of GABA biosynthesis inhibitors downregulated the aforementioned gene expression and HSPs. Therefore, through GABA-induced cell communication, the activated HSF pathway increases thermotolerance in plants (Liu et al. 2019).

Low temperature (less than 16°C) induces devastating effects on plant growth and metabolism by decreasing enzyme activity, disrupting nutrient uptake from the soil, and altering photosynthesis. It also leads to the rigidification of cell membranes, modification in the stability, and degradation of the protein (Hasanuzzaman et al. 2013; Williams et al. 2015; Rihan et al. 2017). In addition, excessive ROS production and buildup in the cell membrane caused by freezing temperatures results in oxidative stress, oxidative modification, and cell damage that ultimately destroys plants (Devireddy et al. 2021). As ROS play a wide range of roles as secondary messengers, including stem cell maintenance, cell division, differentiation, organogenesis, and biotic and abiotic stress responses, cells are in a reduced state when ROS levels are low (Huang et al. 2019; Wu et al. 2021). Similarly, a high concentration of ROS affects cell homeostasis. Thus, plants require a threshold level of ROS for survival (Wu et al. 2021). *Medicago ruthenica* is a perennial herb that is cold-tolerant and has a high protein content. It is suitable for low-input cropping systems and can be used to improve grassland and livestock conditions. However, low temperatures during sowing in alpine areas can affect the growth and overwintering of *M. ruthenica*. By increasing antioxidant enzyme activity, raising osmotic substance content, decreasing damage to the membrane system, and controlling GABA synthesis and degradation, GABA works as an efficient chemical agent to relieve low-temperature stress in plants. This helps plants become less susceptible to cold stress in alpine meadow forages (Li et al. 2022).

ROS in plant cells usually exist in ionic and/or molecular states. The ionic form of ROS includes OH (hydroxyl radical), O_2^- (superoxide anions) and molecular states include H_2O_2 (hydrogen peroxide), $^1O_2$ (singlet oxygen). Each type of ROS formed in plants is associated with different oxidative capacities, which, in turn, affects various physiological and biochemical reactions that regulate genes. In plants, scavenging of ROS does work synergistically in both enzymatic and nonenzymatic patterns to neutralize the formed free radicals due to stress (Huang et al. 2019).

High concentration of GABA in plants exposed to cold stress is often reported in various studies (Hasan et al. 2021; Li et al. 2021a; Wang et al. 2023). In the study, the hypoxia-treated soya sprouts frozen at −18°C expressed 7.21-fold more GABA content than the control (Yang et al. 2016). Similarly, markedly increased GABA accumulation has been observed in barley and wheat seeding on exposure to freezing temperatures (Mazzucotelli et al. 2006). Scientific reports also support the statement that the application of exogenous GABA also increases endogenous GABA concentration and improves low-temperature tolerance in a few plants (Zhu et al. 2019). The potential mechanism through which GABA minifies the injury caused by low temperature is probably by enhancing the plant defense system, which, in turn, reduces ROS content in plants (Malekzadeh et al. 2014). In general, GABA inhibits the accumulation of ROS in plants by activating antioxidant enzymes which include catalase, superoxide dismutase (SOD), and peroxidase (Rajput et al. 2021). Therefore, the activated ROS scavenging, i.e. SOD, rapidly converts OH to H_2O_2, and generated hydrogen peroxide is converted into water and dihydrogen by catalase and peroxidase. Through this mechanism of GABA accumulation and signaling for superoxide, dismutase activation brings homeostasis in plants (under stress) with exceeded levels of ROS (Huang et al. 2019). GABA plays a crucial role in mediating temperature-induced stress responses in plants, with increased levels observed under both high- and low-temperature conditions. In high temperatures, GABA accumulation activates the HSF pathway, enhancing thermotolerance, while in low temperatures, it reduces oxidative damage by promoting antioxidant enzyme activity, thus contributing to plant cold stress resilience.

GABA's Role in Mediating Oxidative Stress-Induced Responses in Plants

GABA has garnered considerable attention in recent years for its crucial role in mediating oxidative stress-induced responses in plants (Li et al. 2021a). As plants are continuously exposed to various environmental stressors, including oxidative stress, understanding the mechanisms by which GABA operates in this context has become increasingly significant (Hasanuzzaman et al. 2013). Recent studies have shed light on the multifaceted functions of GABA in alleviating oxidative stress and enhancing the overall resilience of plants (Hasanuzzaman et al. 2013; Sheteiwy et al. 2019; Li et al. 2021a; Xu et al. 2021). Oxidative stress, characterized by the excessive production of ROS and subsequent disruption of cellular redox balance, can significantly impair various physiological processes in plants, including photosynthesis, respiration, and growth (Huang et al. 2019; Xu et al. 2021).

GABA, acting as a signaling molecule, modulates a cascade of responses aimed at mitigating the detrimental effects of oxidative stress on plant cells. It achieves this through various mechanisms that involve both direct and indirect pathways (Hasan et al. 2021; Ullah et al. 2023). SOD, catalase, and peroxidase are examples of antioxidant enzymes whose activity is increased by GABA in plants under oxidative stress (Malekzadeh et al. 2014). Studies have demonstrated that GABA can effectively increase the scavenging capacity of these enzymes, thereby reducing the accumulation of ROS and minimizing oxidative damage to cellular components (Sheteiwy et al. 2019). In addition, GABA has been found to promote the synthesis of specific antioxidant compounds, including flavonoids, phenolic compounds, and tocopherols, which further contribute to the overall enhancement of the plant's antioxidative defense system (Zhao et al. 2021).

Furthermore, GABA's involvement in regulating the cellular redox state is closely associated with its impact on the glutathione–ascorbate cycle, a vital component of the plant's antioxidant defense mechanism (Hasanuzzaman et al. 2013). GABA has been shown to influence the activity of enzymes involved in this cycle, such as glutathione reductase and dehydroascorbate reductase, thereby maintaining the equilibrium between reduced and oxidized forms of glutathione and ascorbate. This redox homeostasis is essential for preventing oxidative damage to cellular structures and maintaining overall cellular integrity (Jin et al. 2019; Chen et al. 2020).

Recent research has also highlighted the role of GABA in regulating the expression of stress-responsive genes involved in oxidative stress tolerance (Sita and Kumar 2020). GABA has been found to modulate the transcription of various genes associated with the synthesis of stress-related proteins, heat shock proteins, and pathogenesis-related proteins, all of which play crucial roles in enhancing the plant's resilience to oxidative stress (Yuan et al. 2023). Moreover, GABA's signaling role extends to the regulation of gene expression related to the biosynthesis of secondary metabolites, such as polyphenols and terpenoids, which serve as potent antioxidants and contribute to the plant's defense against oxidative stress (Hasanuzzaman et al. 2013).

Different types of stress in plants can cause the buildup of harmful ROS. However, when plants activate the GABA shunt in response to these stresses, it helps to reduce the amount of ROS in the plant tissues (Tripathy and Oelmüller 2012). An enzyme called SSADH plays a key role in this process. Studies on plants with SSADH mutations have shown that these plants have stunted growth, damaged leaves, reduced chlorophyll, and other negative changes. These mutated plants also have high levels of ROS, leading to cell damage and death when they face heat or UV stress (Busch and Fromm 1999). Interestingly, these mutated plants have more of a substance called γ-hydroxybutyrate (GHB), which comes from SSA. When specific inhibitors of GABA-T/POP2 are used or when the POP2 gene is mutated, it helps prevent the buildup of ROS in these mutated plants, leading to improved growth and preventing cell death (Ludewig et al. 2008). Similar observations have been made in tomato plants where the SSADH gene is silenced, showing similar stunted growth, leaf curling, and increased ROS levels as seen in the mutated *Arabidopsis* plants (Scholz et al. 2017). The accumulation of SSA and/or GHB due to the inability of SSADH to convert SSA to succinate seems to cause these negative changes, rather than the increased levels of GABA. In double mutants, increased sensitivity to SSA or GHB further supports the idea that these substances, not GABA, are responsible for the observed negative changes in the mutated plants (Busch and Fromm 1999; Ludewig et al. 2008; Tripathy and Oelmüller 2012). By activating the antioxidative defense system, controlling the cellular redox state, and modifying the production of stress-responsive genes and secondary metabolites, GABA plays a crucial mediating role in oxidative stress-induced responses in plants. To improve plant resilience to environmental stressors and, eventually, support sustainable agriculture and ensure global food security in the face of rising climate variability and environmental challenges, it is essential to comprehend the complex interactions between GABA and oxidative stress responses.

GABA's Role in Mediating Pathogen and Herbivore Attack Stress-Induced Responses in Plants

GABA plays a vital role in mediating stress responses in plants triggered by pathogen and herbivore attacks. When plants face these biotic stressors, GABA acts as a signaling molecule, orchestrating various defense mechanisms to counteract the detrimental impact of stress on plant health (Guo et al. 2023). Recent research has shed light on the multifaceted functions of GABA in bolstering the plant's defense system and enhancing its resilience against pathogen and herbivore attacks (Ludewig et al. 2008; Iqbal et al. 2021). GABA's involvement in regulating the expression of defense-related genes, including those associated with the production of antimicrobial peptides and defense proteins, underscores its pivotal role in strengthening the plant's immune response (Shelp et al. 2021). Also, GABA has been found to modulate the synthesis of secondary metabolites, such as phenolics, alkaloids, and terpenoids, which serve as potent deterrents against pathogens and herbivores (Iqbal et al. 2021). Moreover, GABA's impact on the regulation of ion fluxes and membrane potential in plant cells contributes to the reinforcement of physical barriers and the enhancement of overall cellular integrity, thereby fortifying the plant's defense against pathogen invasion and herbivore feeding (Sita and Kumar 2020). By elucidating the intricate interplay between GABA and stress responses induced by pathogen and herbivore attacks, researchers aim to develop effective strategies to bolster plant immunity and ensure sustainable crop protection in agriculture, fostering global food security in the face of evolving biotic challenges.

GABA Signaling in attKLM Operon (Bacteria)

Pathogenic bacteria exert biotic stress after the natural invasion of the host plant, bacterial T-DNA integrates with the host genome, and thus, transformed plant cells produce opines (class of carbohydrates), which positively controls the synthesis of OC8-HSL (*N*-(3-oxooctanoyl) homoserine lactone) (Hwang et al. 2017). OC8-HSL is a quorum-sensing signaling molecule of *Agrobacterium tumefaciens*, involved in controlling the conjugation and amplification of Ti plasmid and the severity of the tumor. The tumors are formed due to encoding of bacterial T-DNA with the enzyme (Tmr), which modifies the synthesis of auxins and cytokines (key regulators of plant growth and development and fairly important in controlling the formation and development of meristems by altering CK biosynthesis pathway) (Sakakibara et al. 2005; Su et al. 2011). Thus, in response to the stress, GABA

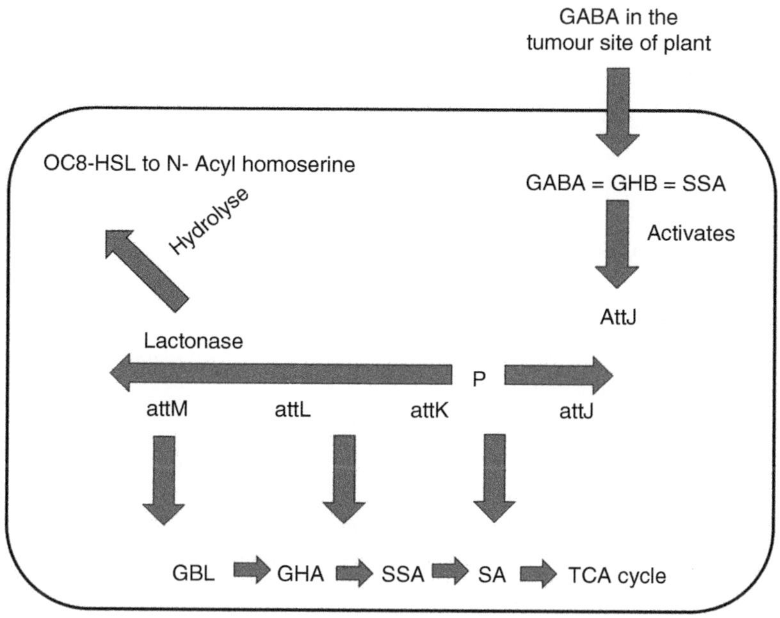

Figure 4.4 Schematic representation of GABA degrades a chemical OC8-HSL.

accumulates in the wounded site due to the stimulation of GAD activity by the calmodulin (H^+ or Ca^{2+}) complex (Scholz et al. 2017). The exogenous GABA accumulated in the wounded site is now imported into the pathogen through *A. tumefaciens* encoding ABC transport system, Bra and Gts (Planamente et al. 2010, 2013).

Generally, plants with rich lines of GABA are found to be least affected by crown gall disease, and researchers have proven it in various studies. For example, in a study, the increased frequency of T-DNA transformation has been observed in tomato plants with low concentrations of GABA (Nonaka et al. 2017). Transgenic tobacco with elevated levels of GABA was found to be less sensitive to *A. tumefaciens* than the wild type, indicating the inhibitory effect of GABA (Chevrot et al. 2006).

A. tumefaciens's attKLM operon, shown in Figure 4.4, participates in an assimilative pathway for succinate semialdehyde (SSA), GHB, and γ-butyrolactone. These substances cause *A. tumefaciens* to accumulate OC8-HSL and activate the attKLM promoter's expression (Chai et al. 2007). The GABA in *A. tumefaciens* with a similar structure of known attKLM operon inducer GHB and SSA initiate the attKLM operon and activate the expression of lactonase AttM, as shown in Figure 4.5. There, lactonase AttM hydrolyzes OC8-HSL to *N*-acyl homoserine, blocks quorum sensing, and suppresses the Ti plasmid conjugation. Thus, GABA acts as a signaling molecule in *A. tumefaciens*–plant interaction.

GABA Signaling Against Viruses, Fungi, and Insect Attack

In the course of growth and development, plants come across a diversity of pathogenic attract, including fungi, viruses, and insects, causing stress in plants, ultimately leading to plant death. In plants under the virally infected scenario, rapid accumulation of cytosolic GABA has been

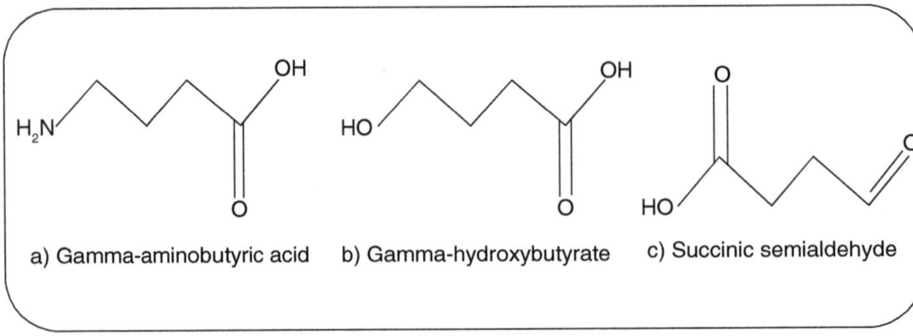

Figure 4.5 Chemical structures of (a) GABA (gamma-aminobutyric acid), (b) GHB (gamma-hydroxybutyric acid), and (c) SSA (succinic semialdehyde).

observed in rice cells infected with *Magnaporthe grisea*, *Jatropha curcas* L. infected with the Jatropha mosaic virus, tomato leaves infected by *Botrytis cinereal*, potato plants infected with potato virus Y, and *Phaseolus vulgaris* infected with *P. syringae* pv. Phaseolicola (Li et al. 2021a). Similarly, recognition of fungi on plants, i.e. *Fusarium graminearum* on *Arabidopsis thaliana, Pyricularia oryzae* on *Oryza sativa* leaf and cell suspension, and *Rhizoctonia solani* on *Glycine max* leaf was proven to trig GABA accumulation and boost host endurance against the pathogen (Shelp et al. 2021). However, the invasion of viruses and fungi into plants is found to cause changes in sugar/carbohydrate metabolism (Morkunas and Ratajczak 2014; Ertunç, 2020). Now the accumulated GABA enters the mitochondria and oxidizes to succinate, which directly fuels the TCA cycle that maintains cell viability (Li et al. 2021a).

Furthermore, in plant defense against herbivore insects, the involvement of GABA has been reported in various studies (Rao et al. 2016; Sita and Kumar 2020; Rajput et al. 2021; Hayat et al. 2023). The herbivorous insect causes wounds on plant tissue and disrupts cells, leading to rapid and systemic jasmonate-independent synthesis and accumulation of GABA. Now GABA acts as an inhibitory neurotransmitter in the neuromuscular junction, which provokes paralysis in insects (Rajput et al. 2021). Thus, in various plant–microbe, – insect interaction, the signaling molecule GABA induces an alternative pathway to protect the plant from biotic stresses.

GABA Signaling in Plant Growth and Development

The multifaceted role of GABA in plant systems encompasses diverse mechanisms, ranging from modulating seed germination and root growth to influencing flowering and fruiting patterns (Takayama and Ezura 2015). GABA signaling is pivotal in the regulation of seed germination, where it aids in breaking seed dormancy and promoting seedling growth (Malekzadeh et al. 2014). It facilitates the transition from seed dormancy to active growth by modulating the balance of hormones and controlling the expression of genes associated with germination and early seedling development (Sheng et al. 2018). Moreover, GABA has been found to regulate root growth and development by influencing root architecture, lateral root formation, and root hair elongation (Li et al. 2021b). It participates in the intricate interplay of hormones and environmental cues to shape root growth patterns and enhance nutrient uptake efficiency, thus contributing to overall plant health and productivity (Devireddy et al. 2021). The biosynthesis of this molecule through the GABA shunt pathway, i.e. irreversible decarboxylation of glutamate catalyzed by GAD in the cytosol, is most accepted (Li et al. 2021a). Certainly, the synthesized GABA mediates interaction in plants. In various reports on GABA, it is estimated to appear as a signaling molecule in pollen tube elongation, fertilization, growth promotion, stomatal regulation, seed germination, and fruit ripening (Yu et al. 2006).

Furthermore, GABA is implicated in the modulation of shoot development, including stem elongation, leaf expansion, and branching (Guo et al. 2021). It affects shoot growth by influencing the balance of auxin and other growth-regulating hormones, thereby regulating cell division and elongation processes (Su et al. 2011). GABA signaling also contributes to the control of leaf size, shape, and senescence, impacting photosynthetic efficiency and carbon assimilation (Zhao et al. 2022). Moreover, it regulates the timing and progression of flowering by modulating the expression of floral development genes and coordinating the transition from vegetative to reproductive growth stages (Chevrot et al. 2006).

The involvement of GABA signaling in fruit development and ripening underscores its importance in determining fruit quality and yield. GABA influences fruit growth, size, and maturation by regulating cell division, expansion, and metabolism (Takayama and Ezura 2015). It also impacts the accumulation of various secondary metabolites and antioxidants, thus affecting fruit flavor, color, and nutritional value (Sweetman et al. 2014). In addition, GABA signaling has been implicated in the regulation of various stress responses, enabling plants to withstand environmental challenges and maintain growth and development under adverse conditions. Overall, the intricate signaling mechanisms of GABA in plant growth and development highlight its indispensable role as a key modulator of various physiological and developmental processes.

GABA Signaling in Seed Germination

It has been shown that GABA affects the balance of hormones such as gibberellins (GAs) and abscisic acid (ABA) to alter seed germination (Zhang et al. 2021). Interacting with these hormones can control germination and dormancy timing. The basic process of seed germination is when a single seed becomes the parent plant species (Sheng et al. 2018).

It begins from imbibition (the process of absorption of water by seeds before germination), and the activated enzymes through imbibition start the rapid breakdown of starch reserve in the endosperm, where starch serves as the main source of carbon during germination and seedling (Wang et al. 2023). The hydrolytic enzyme α-amylase has a central role in starch degradation and mobilization, which will be utilized for further seed growth, as shown in Figure 4.6.

The close relationship between GABA and seed germination, i.e. the concentration of GABA gradually increases during germination, suggests that GABA is involved in germination and is speculated to have an effect on α-amylase activity (Sheng et al. 2018). Following this, the crosstalk between GABA and α-amylase activity has been investigated by treating barley aleurone with GABA, and as a result, enzyme activity was observed after 24 hours of treatment and reached a peak at 48 h, thus activated enzyme by GABA signal transduction brings the starch breakdown to precede seed germination (Sheng et al. 2018).

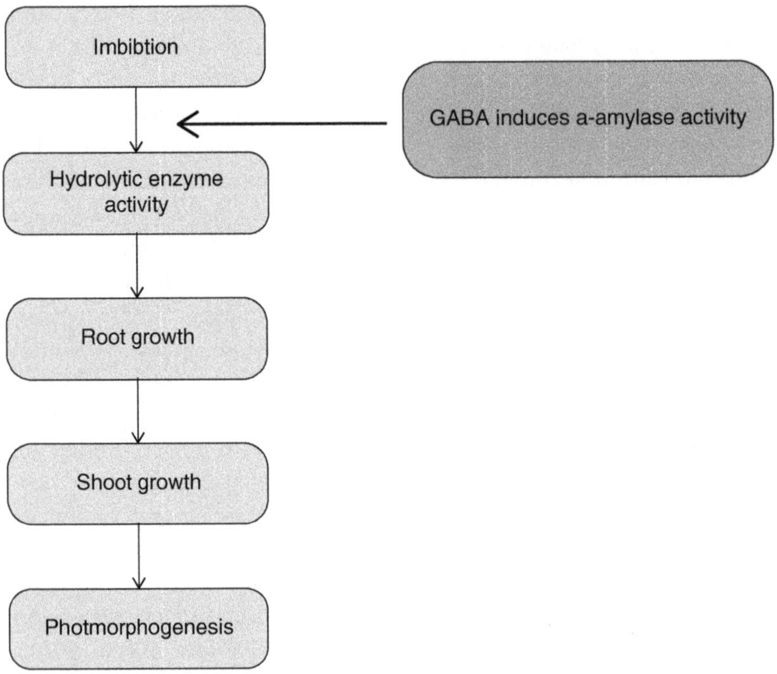

Figure 4.6 Schematic representation of the hydrolytic enzyme α-amylase degrades starch for seed growth.

GABA in ABA (Abscisic Acid) Signaling Pathway

Drought is another abiotic stress in plants, the restrictive factor affects their growth and development. Similar to other stresses described above, drought also promotes the accumulation of GABA, this accumulation has been examined and confirmed in various studies on subjecting excised leaves of soybean (Serraj et al. 1998), turnip (Thompson et al. 1966), sesame (Bor et al. 2009), tomato (Pal et al. 2016), creeping bentgrass (Liu et al. 2019), and *Phyllanthus* species (Filho et al. 2018) to drought. All the abovementioned studies also indicated that GABA accumulation in response to drought stress regulates stomatal opening to prevent excess loss of water (Li et al. 2021a). The GABA accumulated in apple seedlings in response to drought could accumulate a large amount of ABA by regulating the expression of ABA-related genes *PYL4, ABI1, ABI2, HAB1, ABF3,* and *OST1*, which, in turn, activates ABA signaling pathway (Li et al. 2021b).

Plants under normal circumstances regulate their stomal closer through a complex ABA signaling pathway, which takes place in response to the ABA-R receptor present on the membrane of guard cells to ABA (plant hormone/chemical signal) (Munemasa et al. 2015). Thus, GABA signals to the guard cells for stomatal closure that the stress-specific response of GABA prevents water loss and increases the water usage efficiency in plants.

Furthermore, upon GABA-inducing signal for stomatal closure, the hypothetic signaling of GABA on decreasing the stomal opening by altering the cytosolic malate-dependent aluminum-activated malate transporter 9 (ALMT9) (tonoplast localized channel, involved in chloride ion uptake into vacuoles of guard cell during stomatal opening and no documented role in the closure) has been demonstrated by using ALMT9 mutant plants (De Angeli et al. 2013; Xu et al. 2021). In the results, the GABA expressed no antagonism against the stomatal opening, whereas dark-induced stomatal closure in mutant plant retained GABA sensitivity, which means GABA negatively regulates ALMT9-mediated Cl^- ion uptake into vacuoles, lessens the turgidity of the guard cell and reduces stomal opening and interesting altered ALMT9 activity in the guard cell also induced ABA signaling pathway in this experiment (Xu et al. 2021). GABA accumulation in response to drought stress serves a crucial role in regulating stomatal closure, preventing water loss, and enhancing water usage efficiency through modulation of the ABA signaling pathway and alteration of ALMT9 activity.

GABA Signaling for Auxin Biosynthesis for Plant Growth Under Fe Deficiency

Among various reasons for shunted plant growth, the deficiency of iron is a major constraint (Li and Lan 2017). The primary symptom of this deficiency is interveinal chlorosis (i.e., loss of green colouration and development of yellow leaf with dark green veins) which undoubtedly reduces chlorophyll content and subsequently decreases photosynthetic rate and growth (Bermejo and García-López 2009; Dobrikova and Apostolova 2019). In plants, the deficiency of nutrients such as low potassium, boron, and nitrogen could immediately increase the concentration of cytosolic calcium, which, in turn, activates GABA-synthesizing enzyme GAD, similarly, it is speculated that GABA accumulates in the roots and leaves of plant in response iron deficiency. It is a well-known fact that GABA modulates phytohormones (Shelp et al. 2021) and positively regulates auxin synthesis in unstressed conditions, and interestingly auxin is the primary player in controlling iron deficiency (Hijaz et al. 2018). Thus, it is postulated that the accumulated GABA under Fe deficient condition prominently simulates the accumulation of IAA (indole acetic acid) in the shoot apex and induces auxin biosynthesis, and these synthesized auxins regulate various aspects of plant growth and development (Su et al. 2011). On demonstrating this hypothesis by treating Fe-deficient *Cucumis sativus* L. (cucumber) with GABA, the significant increase in the transcript levels of IAA-related genes including auxin biosynthesis gene YUC4 (encoding for YUC protein), auxin-inducible gene *IAA1*, a maker of IAA signal and transport gene *PIN1* has been observed and this treatment also increased the activity of ferric-chelate reductase and transcription levels of Fe uptake-related genes, i.e., HA1, FRO2, and IRT1 in roots. Apparently, in this experiment, the usage of auxin transport inhibitors abolished the changes induced by GABA under deficient conditions. This concludes that GABA signaling for auxin biosynthesis enhances plant growth (Guo et al. 2020).

GABA in Root and Shoot Development

GABA influences root and shoots development by modulating cell division, elongation, and differentiation. It has been found to regulate root architecture, lateral root formation, and primary root growth, as well as shoot branching and leaf expansion, thereby affecting overall plant architecture and morphology (Li et al. 2021b). In root development, GABA signaling has been implicated in regulating root growth, architecture, and overall root system establishment. Studies have shown that GABA can influence primary root elongation, lateral root formation, and root hair development, thereby shaping the overall structure of the root system (Verma et al. 2022). GABA's impact on root development is often associated with its role in modulating ion transport, nutrient uptake, and water balance, ensuring the proper functioning of root cells under varying environmental conditions (Chevrot et al. 2006). GABA signaling contributes to the regulation of root-to-shoot communication, facilitating the transmission of developmental and stress-related signals between the root and shoot systems (Sita and Kumar 2020). This communication is crucial for coordinating the adaptive responses of plants to environmental cues, such as nutrient availability, water availability, and biotic and abiotic stresses (Yuan et al. 2023). GABA's involvement in root-to-shoot signaling helps synchronize the development and growth of roots and shoots, enabling plants to optimize resource allocation and enhance their overall fitness and resilience.

In shoot development, GABA participates in the control of shoot growth, leaf expansion, and the regulation of shoot architecture. It influences shoot elongation, leaf initiation, and the development of specialized structures, such as flowers and fruits (Dengler 2006). GABA's role in shoot development is closely intertwined with its impact on phytohormone signaling pathways, particularly auxins, cytokinins, and GAs, which are key regulators of shoot growth and organogenesis (Castro-Camba et al. 2022). By modulating the cross-talk between GABA and these phytohormones, plants can fine-tune their developmental programs and optimize shoot morphogenesis in response to internal and external signals (Gramazio et al. 2020). Moreover, GABA signaling contributes to the modulation of shoot branching patterns, lateral bud outgrowth, and axillary meristem activation, thereby influencing shoot architecture and overall plant form (Chevrot et al. 2006). Its role in shoot development is also associated with the regulation of photosynthetic activity, carbon assimilation, and energy allocation, which are critical for sustaining shoot growth and ensuring optimal plant productivity.

An experimental system using the embryogenic callus of hybrid *Liriodendron* was employed to investigate the influence of GABA on somatic embryogenesis induction and maturation. The results indicated a positive correlation between GABA application and the enhancement of SE occurrence and development (Chen et al. 2022). Furthermore, an augmentation in the root length of plantlets that germinated from the somatic embryos of hybrid *Liriodendron* was observed, similar to the impact of a lower GABA concentration on the stimulation of adventitious root (AR) growth (Chen et al. 2022). In addition, the application of exogenous GABA was found to contribute to the improvement of somatic embryo quality and a reduction in the proportion of abnormal embryos. GA, a crucial class of diterpenoid phytohormones, is known for its regulatory roles in inducing flowering, as well as breaking dormancy in plants (Castro-Camba et al. 2022). In addition, studies have revealed its inhibitory influence on AR growth in poplar. However, it has been observed that the inhibitory impact of GA on ARs does not involve interference with the auxin signaling pathway, but rather operates through the perturbation of polar auxin transport (Su et al. 2011). Interestingly, investigations into the effects of GABA on the rooting of woody plants have indicated that GABA promotes GA accumulation in poplars. Furthermore, it has been demonstrated that GABA also inhibits polar auxin transport in the case of *M. xiaojinensis*, suggesting a potential interaction between GABA, GA, and auxin, along with other phytohormones, during the process of AR formation. While exogenous GABA was found to have an inhibitory effect on AR formation, characterized by delayed rooting time, reduced root number, and decreased total root length in *in vitro* shoots of both *M. xiaojinensis* and tobacco (Guan et al. 2020). Notably, the application of indole-3-butyric acid (IBA) led to an increase in the expression of MxPIN members, which was subsequently hindered by the presence of GABA (Mao et al. 2018). These findings suggest that the accumulation of GABA is contingent on the combined influence of the loss of juvenility and IBA (Wu et al. 2020). Furthermore, GABA appears to exert its influence on AR formation as a co-actor by impeding polar auxin transport. Consequently, these insights contribute to an improved understanding of the intricate regulatory network involving GABA, juvenility, and auxin signaling in the context of AR formation (Łatka et al. 2020).

GABA Influence in Pollen Tube Elongation

Sexual reproduction is important in flowering plants to complete its life cycle. The event of polar growth of pollen tubes initiates after the landing of pollen grains on the surface of the stigma, and thus, formed tubes deliver sperm cells into the ovule (Ling et al. 2013). The concentration of GABA in the flower can function as a signal for pollen tube elongation and orientation by forming a gradient along the tube's path through the pistil and reaching its highest concentration near the micropyle, through which the pollen tube penetrates the ovule. Proper guidance of pollen tube elongation is crucial for plant mating (Yu et al. 2006). Although the involvement of GABA signaling in pollen tube elongation is known, the molecular mechanisms through which it communicates remain complicated, through scientific reports it is confirmed GABA binds to $GABA_B$ receptors present on the surface of the protoplast membrane, which regulates Ca^{2+} and K^+ ions as it is G couple protein receptor which can regulate Ca^{2+} and K^+ permeable channels (Li et al. 2021a; Wu et al. 2021). It is well well-known fact that calcium ions mediate multiple interactions that occur between pollen tubes and pistil cells to deliver the sperm. Thus, the activated calcium ion dynamics by the interaction of GABA-$GABA_B$ give guidance for the pollen tube to deliver sperm cells to the ovule in the embryo sac (Yu et al. 2006). The scientific reports also mention the dose-dependent effect of exogenous GABA on pollen tube elongation, where low concentrations promote tube growth and, contrarily, dose inhibits the process (Ling et al. 2013).

GABA in Flowering, Fruit Development, and Ripening

GABA has been implicated in the regulation of flowering time and flower development. It can interact with various flowering regulatory genes and pathways, influencing the transition from vegetative to reproductive growth and controlling the timing of flowering in response to environmental cues (Takayama and Ezura 2015). Also, has a role in fruit development and ripening processes. It affects fruit size, shape, and quality by regulating cell division and expansion during fruit development. GABA also influences the accumulation of sugars, organic acids, and aroma compounds, contributing to the flavor and nutritional content of ripe fruits (Sweetman et al. 2014). Its interplay with hormonal pathways, including ethylene, auxin, and GAs, underscores its significance in the modulation of these developmental processes (Castro-Camba et al. 2022). Elucidating the intricate molecular mechanisms underlying GABA-mediated regulation of flowering, fruit development, and ripening is crucial for a comprehensive understanding of plant growth and reproductive processes.

The enzyme gamma-aminobutyric acid transaminase (GABA-T) plays a crucial role in the metabolic conversion of GABA to succinate during the ripening of fruits. A recent study employed a sophisticated technique known as RNA interference (RNAi) to selectively suppress the expression of different GABA-T isoforms. Intriguingly, the investigation highlighted the pivotal role of the GABA-T1 isoform in orchestrating GABA metabolism in tomato plants (Malekzadeh et al. 2014). The suppression of GABA-T1 resulted in a cascade of significant physiological changes, including the development of dwarfism and infertility in the plants. Moreover, the suppression of GABA-T1 led to a substantial accumulation of GABA levels in the fruits. These findings shed light on the critical physiological functions of GABA-T1 in tomatoes, deepening our comprehension of the intricate biochemical processes involved in GABA metabolism during fruit development and ripening. This study contributes valuable insights into the specific functions of GABA-T and its isoforms in the context of fruit ripening, enhancing our understanding of the regulatory mechanisms that underlie the complex metabolic pathways in plants (Malekzadeh et al. 2014).

GABA accumulation in plants, as documented in numerous studies, including those focused on tomato plants. These reports underscore the essential role of calcium ions (Ca^{2+}) and calmodulin in modulating the activity of GAD, an enzyme involved in GABA biosynthesis (Malekzadeh et al. 2014; Gramazio et al. 2020). An elevation in intracellular Ca^{2+} levels, likely triggered by salinity stress, is believed to stimulate GAD activity, thereby leading to increased GABA levels in tomato fruits (Malekzadeh et al. 2014). In contrast, the intricate regulatory mechanisms governing GABA degradation in plants have not been fully elucidated. To gain a comprehensive understanding of the mechanisms underlying GABA accumulation in fruits, further research is needed, including the identification and characterization of all genes involved in the GABA shunt pathway, along with in-depth investigations into posttranslational modifications and metabolomic studies. Such comprehensive endeavors are crucial for unraveling the intricate processes governing GABA metabolism in plants, particularly in response to environmental stressors like salinity.

The application of GABA in lemon trees through foliar spray has demonstrated promising effects on fruit growth and yield, alongside improvements in various quality attributes during harvest (Badiche et al. 2023). Through the application of GABA at varying concentrations via foliar spray, the study observed significant enhancements in fruit growth, particularly an increase in fruit diameter (Sweetman et al. 2014). Moreover, the application of GABA positively influenced crop yield, leading to a notable rise in the number of fruits per tree. Notably, GABA application also impacted several key quality attributes of the fruits, including total soluble solids, total acidity, firmness, and color (Takayama and Ezura 2015). These improvements collectively suggest that GABA application can serve as a valuable strategy for augmenting the growth, yield, and overall quality of lemon trees (Badiche et al. 2023). The findings of this study offer promising insights into the potential of GABA as a beneficial agent in promoting citrus fruit production and enhancing fruit quality, thus providing valuable implications for agricultural practices and fruit cultivation techniques. GABA's multifaceted roles in flowering, fruit development, and ripening, along with its involvement in crucial metabolic pathways and potential applications in enhancing fruit quality, underscore its significance in plant growth and agricultural practices.

GABA-Mediated Regulation of Stomatal Aperture in Plants

GABA also plays a role in the modulation of stomatal movement in plants. Stomata are small openings primarily found on the surfaces of plant leaves and stems that regulate gas exchange, allowing for the uptake of carbon dioxide (CO_2) for photosynthesis while facilitating the release of oxygen and water vapor (Munemasa et al. 2015). Understanding the precise mechanisms underlying GABA-mediated regulation of stomatal movement is critical for unraveling the intricate signaling networks that govern plant responses to changing environmental conditions (Xu et al. 2021). Insights into the molecular and physiological processes involved in GABA-mediated stomatal regulation could offer valuable implications for enhancing plant water-use efficiency, drought tolerance, and overall stress responses (De Angeli et al. 2013). Harnessing the potential of GABA-mediated signaling pathways in the context of stomatal movement holds promise for the development of innovative strategies aimed at improving crop productivity, resilience, and sustainability in the face of increasingly challenging environmental conditions and climate change.

GABA Regulation in Stomatal Closure

Beyond its role in various physiological processes, GABA has emerged as a key player in the modulation of stomatal movement, exerting its influence on the complex cellular and molecular mechanisms underlying this critical plant function (Munemasa et al. 2015). Stomata, composed of specialized guard cells, act as dynamic gatekeepers, regulating the exchange of gases and water vapor between the plant and its external environment (Xu et al. 2021). Stomata, the microscopic pores located on the surfaces of plant leaves and stems, play a vital role in regulating gas exchange, facilitating the uptake of CO_2 essential for photosynthesis, and enabling the release of oxygen and water vapor (De Angeli et al. 2013). The controlled opening and closing of stomata are crucial for maintaining an optimal balance of gas exchange and water loss, thus influencing overall growth, development, and response of plants to external stimuli. Emerging research has uncovered the involvement of GABA, a multifunctional signaling molecule, in the intricate regulation of stomatal movement in plants (Munemasa et al. 2015).

GABA-mediated regulation of the stomatal aperture represents an essential aspect of plant responses to fluctuating environmental conditions, including light intensity, humidity levels, and water availability (Hildebrandt 2018). Research has revealed the presence of GABA receptors in guard cells, implying a direct mode of action for GABA in influencing stomatal movement (Badiche et al. 2023). Activation of these GABA receptors triggers a series of downstream events, leading to alterations in ion fluxes and subsequent changes in guard cell turgor pressure, ultimately resulting in the controlled opening or closure of stomata (Pantin et al. 2013). The delicate balance between GABA and other endogenous signaling molecules, such as ABA, serves as a key determinant in maintaining stomatal homeostasis, thereby enabling plants to adapt to varying environmental stimuli and stressors (Pantin et al. 2013). GABA-induced changes in stomatal conductance and aperture have been linked to alterations in plant water-use efficiency, emphasizing the significance of GABA in optimizing water conservation strategies in plants. By influencing the sensitivity of stomata to environmental cues and stress signals, GABA helps regulate transpiration rates, thereby minimizing water loss and maximizing water-use efficiency under challenging conditions (Hossain et al. 2018).

The research demonstrates that GABA does not directly instigate alterations in the size of stomatal pores, but rather counteracts changes in pore dimensions. Application of GABA to *Arabidopsis thaliana* leaves resulted in the reduction of stomatal opening and closure, thereby decreasing transpirational water loss and enhancing water-use efficiency (Lee et al. 2009). Moreover, the study indicated that GABA impedes the influence of other signaling molecules implicated in the regulation of stomatal function, including ABA and H_2O_2 (Zhang et al. 2021). These observations emphasize the role of GABA as a key modulator of stomatal aperture, thereby contributing to the regulation of plant water loss.

GABA Signaling for GORK Channels Under Hypoxia

Waterlogging causes severe effects on the yield of the plant, it is estimated that nearly 12% of cultivation land has been affected by this, resulting in a 20% loss in crop productivity. This stress reduces the photosynthetic rate and antioxidant enzyme activity, and ultimately reduces growth by inducing hypoxic conditions in plants (Adem et al. 2020). Under hypoxic conditions, the activity of diamine oxidase and GAD increases rapidly, which, in turn, elevates GABA content in the root nodules (Li et al. 2021a). However, a marked increase in the GABA concentration by several orders of magnitude in individuals under waterlogged has been identified by researchers, but the mechanism by which GABA protects plants is largely unanswered. Under water-logged hypoxic conditions, the reduced oxygen availability impacts H^+ ATPase pumping, leading to membrane depolarization and potassium ion loss via outward rectifying guard cell outward-rectifying potassium (GORK) channels (Wu et al. 2021). These channels function as ligand-gated channels because they have binding motifs and domains. It has been demonstrated that signaling molecules such as GABA and G-proteins function as ligands to modulate K^+ efflux channels in the mammalian system (Adem et al. 2020). Similarly, it is suspected that GABA can act as an operating ligand for the GORK channel controlling potassium efflux. In this context, recent bioinformatics analysis suggests that GORK channels possess F, W, E, and L conserved consensus sequences, which can operate as GABA binding motifs in ALMT plant channels. It directly points to the possibility of GABA operating GORK channels and regulating potassium ion homeostasis in the cytosol of plants cell. Thus, guard cells are kept open for the elimination of excess accumulated water, and this GABA signaling (as an operating ligand) gradually enables plants to adapt to hypoxic conditions under water-logged conditions (Yang et al. 2016; Wu et al. 2021).

GABA in Ion-Exchange Regulation

The intricate mechanisms by which GABA influences stomatal movement involve a complex interplay between GABA receptors, ion channels, and second messengers. GABA signaling pathways, including both GABA signaling receptors and ion channels, regulate the influx and efflux of ions such as potassium (K^+) and chloride (Cl^-) within guard cells, the specialized cells surrounding the stomatal pore (Munemasa et al. 2015). These intricate ion fluxes trigger changes in turgor pressure, resulting in the controlled opening or closure of stomata in response to various internal and external cues, including light, humidity, and drought stress (Fàbregas and Fernie 2019).

The role of GABA in ion-exchange regulation within guard cells is critical for maintaining stomatal homeostasis and facilitating efficient gas exchange while minimizing water loss (Xu et al. 2021). GABA functions as a key regulator of ion channel activities, affecting the movement of ions across the plasma membrane and thereby modulating the osmotic pressure within guard cells (Sheteiwy et al. 2019). The intricate molecular machinery underlying GABA-mediated ion-exchange regulation involves the integration of GABA receptors, ion transporters, and regulatory proteins, orchestrating a highly coordinated response to environmental stimuli.

The GABA receptors in guard cells, belonging to the GABA-gated anion channel family, play a pivotal role in mediating the influx of anions, primarily chloride ions, in response to GABA binding. The activation of these receptors leads to the depolarization of the plasma membrane, triggering subsequent changes in ion fluxes and turgor pressure. Notably, GABA receptors exhibit specificity in their response to GABA concentrations, demonstrating distinct sensitivities to varying GABA levels, thereby fine-tuning stomatal responses based on GABA availability (Hübner and Holthoff 2013). Furthermore, GABA signaling pathways modulate the activities of various ion transporters, including K^+ inward-rectifying channels and anion channels, such as slow anion channel-associated 1 and ALMT12 (de Avellar et al. 2004). These ion channels facilitate the influx and efflux of ions, crucial for maintaining cellular osmotic balance and controlling the swelling or shrinking of guard cells. GABA-induced activation of K^+ channels promotes K^+ efflux, leading to a decrease in guard cell turgor and subsequent stomatal closure. On the other hand, the activation of anion channels results in the efflux of anions, causing a depolarization of the plasma membrane and promoting stomatal opening (Waadt et al. 2022).

In addition to its direct effects on ion channels, GABA signaling cascades also interact with second messenger systems, including Ca^{2+} and ROS, which further amplify the downstream signaling events involved in the stomatal movement (Wu et al. 2021). GABA-induced changes in intracellular Ca^{2+} concentrations modulate the activities of Ca^{2+}-dependent protein kinases and phosphatases, influencing the phosphorylation status of ion channels and regulatory proteins (Hossain et al. 2015). This intricate interplay between GABA signaling, Ca^{2+} fluxes, and protein phosphorylation events orchestrates the dynamic regulation of ion transport processes and guard cell responses. GABA can influence the concentration of cytosolic calcium ions in plant cells (Chevrot et al. 2006).

GABA can modulate calcium channels and transporters, influencing the influx and efflux of calcium ions across the plasma membrane and the endomembrane system. This modulation contributes to the activation of calcium-dependent signaling pathways and the coordination of various physiological processes, including signal transduction and stress responses (Waadt et al. 2022). In addition, GABA has been demonstrated to influence potassium ion transport and homeostasis in plant cells, regulating potassium channels and transporters to impact potassium uptake, distribution, and compartmentalization within the cell (Ragel et al. 2019). Such regulation of potassium transport is vital for maintaining cellular osmotic balance, enzyme activation, and the regulation of cell turgor pressure, critical for various cellular functions and stress responses (Dobrikova and Apostolova 2019). Furthermore, GABA can affect chloride ion transport and distribution in plant cells by modulating chloride channels and transporters, which, in turn, influence the influx and efflux of chloride ions across the plasma membrane and tonoplast (Waadt et al. 2022). This regulatory role of GABA in chloride ion transport is crucial for upholding cellular osmolarity, pH balance, and stomatal movement, as well as facilitating the detoxification of excess chloride ions under saline conditions (Ullah et al. 2023). Moreover, GABA-mediated regulation of ROS levels within guard cells impacts the redox status and signaling cascades associated with stomatal movement (Hossain et al. 2015). ROS, acting as important signaling molecules, integrate GABA-induced signals with other environmental cues, contributing to the coordination of stomatal responses during stress conditions (Chevrot et al. 2006). The fine-tuned balance between ROS production and scavenging mechanisms dictates the sensitivity of guard cells to GABA-mediated signaling, ultimately determining the extent of stomatal closure or opening in response to various stressors (Xu et al. 2021). Overall, the intricate interplay between GABA signaling, ion channels, and second messengers within guard cells highlights the multifaceted regulatory mechanisms underlying stomatal movement (De Angeli et al. 2013). The dynamic nature of GABA-mediated ion-exchange regulation serves as a central component of plant adaptive responses, allowing for the integration of diverse environmental cues and the fine-tuning of stomatal behavior in various growth and stress conditions (Munemasa et al. 2015). Understanding the nuances of GABA-mediated ion-exchange regulation is crucial for unraveling the complexities of plant physiology and facilitating the development of novel strategies for enhancing plant resilience, stress tolerance, and water-use efficiency in the face of fluctuating environmental challenges.

GABA's Interplay with Diverse Signaling Pathways in Plants

Furthermore, GABA has been found to interact with other signaling molecules and phytohormones, such as ABA and jasmonic acid (JA), in modulating stomatal movement. The cross-talk between GABA and these signaling pathways highlights the interconnected nature of plant responses to diverse environmental stimuli, underscoring the complexity of the regulatory networks governing stomatal aperture (Lim et al. 2015). The interplay between GABA signaling and phytohormonal pathways, including ABA and JA, underscores the complexity of the mechanisms governing stomatal movement in response to biotic and abiotic stimuli (Xu et al. 2021). GABA's multifaceted roles in the intricate signaling networks associated with stomatal regulation highlight its potential as a target for enhancing plant resilience, stress tolerance, and water-use efficiency in the face of rapidly changing environmental conditions (Liu et al. 2023). Leveraging the intricate knowledge of GABA-mediated stomatal modulation could pave the way for the development of novel strategies for sustainable agriculture and improved crop productivity in the context of climate change and resource limitations (Muhammad Aslam et al. 2022).

Under conditions of drought stress, the biosynthesis and signaling of the plant hormone ABA are triggered, allowing the plant to initiate a series of adaptive responses to mitigate the effects of water scarcity (Xu et al. 2021). In contrast, instances of pathogen attack or herbivore-induced wounding prompt an increase in the levels of jasmonate, leading to the activation of jasmonate signaling pathways (Lee et al. 2009). These distinct hormonal responses to various environmental stimuli underscore the intricate and dynamic nature of plant defense mechanisms, enabling plants to mount tailored and context-specific defense strategies to combat a diverse array of stressors (Guan and Liu 2020). The studies focus on the intricate interplay between the hormone JA and the process of stomatal closure (Munemasa et al. 2015). Notably, the study reveals that the occurrence of wounding triggers the activation of the JA signaling cascade, ultimately leading to the phosphorylation and subsequent activation of the potassium efflux channel GORK within guard cells. This activation of GORK is facilitated by the Ca^{2+} sensor-kinase complex calcineurin B-like protein 1-calcineurin B-like-interacting protein kinase 5 (Adem et al. 2020). Furthermore, the study highlights the inhibitory role of the ABA-coreceptor abscisic acid-insensitive 2 (ABI2) in the activation of GORK, underscoring the existence of cross-talk between the JA and ABA signaling pathways (Liu et al. 2021; Zhang et al. 2021). These findings significantly contribute to our understanding of the underlying molecular mechanisms governing JA-induced stomatal closure as a response to wounding. The combined efflux of anions and potassium leads to a reduction in guard cell turgor, ultimately causing stomatal closure (Munemasa et al. 2015). Alongside ABA, the plant hormone jasmonate is recognized as a significant initiator of stomatal closure (Lim et al. 2015). Recent discussions have highlighted the involvement of jasmonate in the context of stomatal function as a component of plant innate immunity and its implications under elevated CO_2 levels (Hossain et al. 2018).

Conclusion and Future Prospectives

GABA has emerged as a pivotal signaling molecule in plants, orchestrating diverse physiological and developmental processes, and mediating intricate responses to various environmental cues and stresses. Its multifaceted roles encompass a wide array of functions, including stress tolerance, growth regulation, and developmental processes, highlighting its significance in plant biology. As a signaling molecule, GABA operates through complex molecular networks, involving receptors, ion channels, and interconnected pathways that collectively modulate fundamental cellular processes. The extensive involvement of GABA in the intricate interplay of hormonal signaling pathways, including ABA, JA, and other phytohormones, its crucial role in integrating diverse environmental signals and fine-tuning plant responses for optimal adaptation and survival. GABA's regulatory influence extends to various ion transport systems, including calcium, potassium, and chloride channels, modulating their activities to maintain cellular homeostasis, osmotic balance, and turgor pressure, which is critical for sustaining normal cellular functions and coping with environmental fluctuations. Its ability to modulate ion fluxes underscores its significance in regulating cellular processes, including signal transduction, stress responses, and metabolic activities, reflecting its importance in maintaining cellular stability and functional integrity.

Besides, GABA's versatile functions in developmental processes, such as flowering, fruit development, and senescence, further emphasize its critical role in orchestrating the intricate molecular events underlying these complex phenomena. Its involvement in flowering time regulation, floral organ development, and fruit quality enhancement highlights its potential as a key regulator of plant growth and reproductive development. In light of its multifaceted roles, elucidating the intricate mechanisms and signaling pathways underlying GABA-mediated responses in plants holds substantial promise for advancing our understanding of plant biology and improving agricultural practices. Leveraging the comprehensive knowledge of GABA signaling pathways could pave the way for the development of innovative strategies for enhancing stress tolerance, crop productivity, and sustainability in the face of escalating environmental challenges and climate uncertainties. Furthermore, harnessing the potential of GABA signaling modulation may offer novel avenues for the development of crop varieties with improved resilience, nutritional quality, and yield, thus contributing to global efforts for sustainable agriculture and food security.

References

Adem, G.D., Chen, G., Shabala, L. et al. (2020). GORK channel: a master switch of plant metabolism? *Trends in Plant Science* 25 (5): 434–445.

Aycan, M., Erkilic, E.G., Ozgen, Y. et al. (2023). The response of sugar beet (*Beta vulgaris* L.) genotypes at different ploidy levels to salt (NaCl) stress. *International Journal of Plant Biology* 14 (1): 199–217.

Badiche, F., Valverde, J.M., Martínez-Romero, D. et al. (2023). Preharvest use of γ-aminobutyric acid (GABA) as an innovative treatment to enhance yield and quality in lemon fruit. *Horticulturae* 9 (1): 93.

Bermejo, F. and García-López, S. (2009). A guide to diagnosis of iron deficiency and iron deficiency anemia in digestive diseases. *World Journal of Gastroenterology* 15 (37): 4638.

Bor, M., Seckin, B., Ozgur, R. et al. (2009). Comparative effects of drought, salt, heavy metal and heat stresses on gamma-aminobutryric acid levels of sesame (*Sesamum indicum* L.). *Acta Physiologiae Plantarum* 31: 655–659.

Bouché, N., Lacombe, B., and Fromm, H. (2003). GABA signaling: a conserved and ubiquitous mechanism. *Trends in Cell Biology* 13 (12): 607–610.

Busch, K.B. and Fromm, H. (1999). Plant succinic semialdehyde dehydrogenase. Cloning, purification, localization in mitochondria, and regulation by adenine nucleotides. *Plant Physiology* 121 (2): 589–598.

Carillo, P. (2018). GABA shunt in durum wheat. *Frontiers in Plant Science* 9: 100.

Castro-Camba, R., Sánchez, C., Vidal, N., and Vielba, J.M. (2022). Interactions of gibberellins with phytohormones and their role in stress responses. *Horticulturae* 8 (3): 241.

Chai, Y., Tsai, C.S., Cho, H., and Winans, S.C. (2007). Reconstitution of the biochemical activities of the AttJ repressor and the AttK, AttL, and AttM catabolic enzymes of *Agrobacterium tumefaciens*. *Journal of Bacteriology* 189 (9): 3674–3679.

Chen, W., Meng, C., Ji, J. et al. (2020). Exogenous GABA promotes adaptation and growth by altering the carbon and nitrogen metabolic flux in poplar seedlings under low nitrogen conditions. *Tree Physiology* 40 (12): 1744–1761.

Chen, T., Shabala, S., Niu, Y. et al. (2021). Molecular mechanisms of salinity tolerance in rice. *The Crop Journal* 9 (3): 506–520.

Chen, T., Yang, D., Fan, R. et al. (2022). γ-Aminobutyric acid a novel candidate for rapid induction in somatic embryogenesis of *Liriodendron* hybrid. *Plant Growth Regulation* 96 (2): 293–302.

Chevrot, R., Rosen, R., Haudecoeur, E. et al. (2006). GABA controls the level of quorum-sensing signal in *Agrobacterium tumefaciens*. *Proceedings of the National Academy of Sciences of the United States of America* 103 (19): 7460–7464.

Dabravolski, S.A. and Isayenkov, S.V. (2023). The role of the γ-aminobutyric acid (GABA) in plant salt stress tolerance. *Horticulturae* 9 (2): 230.

De Angeli, A., Zhang, J., Meyer, S., and Martinoia, E. (2013). AtALMT9 is a malate-activated vacuolar chloride channel required for stomatal opening in *Arabidopsis*. *Nature Communications* 4 (1): 1804.

de Avellar, I.G., Magalhães, M.M., Silva, A.B. et al. (2004). Reevaluating the role of 1, 10-phenanthroline in oxidative reactions involving ferrous ions and DNA damage. *Biochimica et Biophysica Acta (BBA) – General Subjects* 1675 (1–3): 46–53.

Dengler, N.G. (2006). The shoot apical meristem and development of vascular architecture. *Botany* 84 (11): 1660–1671.

Devireddy, A.R., Tschaplinski, T.J., Tuskan, G.A. et al. (2021). Role of reactive oxygen species and hormones in plant responses to temperature changes. *International Journal of Molecular Sciences* 22 (16): 8843.

Dobrikova, A.G. and Apostolova, E.L. (2019). Damage and protection of the photosynthetic apparatus under cadmium stress. In: *Cadmium Toxicity and Tolerance in Plants* (ed. M. Hasanuzzaman, M.N.V. Prasad, and M. Fujita), 275–298. Cambridge, MA: Academic Press.

dos Santos, T.B., Ribas, A.F., de Souza, S.G. et al. (2022). Physiological responses to drought, salinity, and heat stress in plants: a review. *Stresses* 2 (1): 113–135.

Ertunç, F. (2020). Physiology of virus-infected plants. In: *Applied Plant Virology* (ed. L.P. Awasthi), 199–205. London: Academic Press.

Fàbregas, N. and Fernie, A.R. (2019). The metabolic response to drought. *Journal of Experimental Botany* 70 (4): 1077–1085.

Filho, E.G., Braga, L.N., Silva, L.M. et al. (2018). Physiological changes for drought resistance in different species of *Phyllanthus*. *Scientific Reports* 8 (1): 15141.

Fromm, H. (2020). GABA signaling in plants: targeting the missing pieces of the puzzle. *Journal of Experimental Botany* 71 (20): 6238–6245.

González-Orenga, S., Ferrer-Gallego, P.P., Laguna, E. et al. (2019). Insights on salt tolerance of two endemic *Limonium* species from Spain. *Metabolites* 9 (12): 294.

Gramazio, P., Takayama, M., and Ezura, H. (2020). Challenges and prospects of new plant breeding techniques for GABA improvement in crops: tomato as an example. *Frontiers in Plant Science* 11: 577980.

Guan, N. and Liu, L. (2020). Microbial response to acid stress: mechanisms and applications. *Applied Microbiology and Biotechnology* 104 (1): 51–65.

Guan, L., Li, Y., Huang, K., and Cheng, Z.M. (2020). Auxin regulation and MdPIN expression during adventitious root initiation in apple cuttings. *Horticulture Research* 7: 143.

Guo, Z., Du, N., Li, Y. et al. (2020). Gamma-aminobutyric acid enhances tolerance to iron deficiency by stimulating auxin signaling in cucumber (*Cucumis sativus* L.). *Ecotoxicology and Environmental Safety* 192: 110285.

Guo, Y., Ren, G., Zhang, K. et al. (2021). Leaf senescence: progression, regulation, and application. *Molecular Horticulture* 1 (1): 1–25.

Guo, Z., Gong, J., Luo, S. et al. (2023). Role of gamma-aminobutyric acid in plant defense response. *Metabolites* 13 (6): 741.

Gupta, B. and Huang, B. (2014). Mechanism of salinity tolerance in plants: physiological, biochemical, and molecular characterization. *International Journal of Genomics* 2014.

Haglund, K. and Dikic, I. (2005). Ubiquitylation and cell signaling. *The EMBO Journal* 24 (19): 3353–3359.

Hasan, M.M., Alabdallah, N.M., Alharbi, B.M. et al. (2021). GABA: a key player in drought stress resistance in plants. *International Journal of Molecular Sciences* 22 (18): 10136.

Hasanuzzaman, M., Nahar, K., and Fujita, M. (2013). Extreme temperature responses, oxidative stress and antioxidant defense in plants. In: *Abiotic Stress-Plant Responses and Applications in Agriculture*, vol. 13 (ed. K. Vahdati and C. Leslie), 169–205. Rijeka: InTech.

Hayat, F., Khan, U., Li, J. et al. (2023). γ-Aminobutyric acid (GABA): a key player in alleviating abiotic stress resistance in horticultural crops: current insights and future directions. *Horticulturae* 9 (6): 647.

Hijaz, F., Nehela, Y., and Killiny, N. (2018). Application of gamma-aminobutyric acid increased the level of phytohormones in *Citrus sinensis*. *Planta* 248: 909–918.

Hildebrandt, T.M. (2018). Synthesis versus degradation: directions of amino acid metabolism during *Arabidopsis* abiotic stress response. *Plant Molecular Biology* 98: 121–135.

Hossain, M.A., Bhattacharjee, S., Armin, S.M. et al. (2015). Hydrogen peroxide priming modulates abiotic oxidative stress tolerance: insights from ROS detoxification and scavenging. *Frontiers in Plant Science* 6: 420.

Hossain, M.A., Li, Z.G., Hoque, T.S. et al. (2018). Heat or cold priming-induced cross-tolerance to abiotic stresses in plants: key regulators and possible mechanisms. *Protoplasma* 255: 399–412.

Huang, H., Ullah, F., Zhou, D.X. et al. (2019). Mechanisms of ROS regulation of plant development and stress responses. *Frontiers in Plant Science* 10: 800.

Hübner, C.A. and Holthoff, K. (2013). Anion transport and GABA signaling. *Frontiers in Cellular Neuroscience* 7: 177.

Hwang, H.H., Yu, M., and Lai, E.M. (2017). *Agrobacterium*-mediated plant transformation: biology and applications. *The Arabidopsis Book* 15: e0186.

Iqbal, Z., Iqbal, M.S., Hashem, A., and Abd_Allah EF, Ansari MI. (2021). Plant defense responses to biotic stress and its interplay with fluctuating dark/light conditions. *Frontiers in Plant Science* 12: 631810.

Jin, X., Liu, T., Xu, J. et al. (2019). Exogenous GABA enhances muskmelon tolerance to salinity-alkalinity stress by regulating redox balance and chlorophyll biosynthesis. *BMC Plant Biology* 19 (1): 1–5.

Łątka, K., Jończyk, J., and Bajda, M. (2020). γ-Aminobutyric acid transporters as relevant biological target: their function, structure, inhibitors and role in the therapy of different diseases. *International Journal of Biological Macromolecules* 158: 750–772.

Lee, C.W., Efetova, M., Engelmann, J.C. et al. (2009). *Agrobacterium tumefaciens* promotes tumor induction by modulating pathogen defense in *Arabidopsis thaliana*. *The Plant Cell* 21 (9): 2948–2962.

Li, W. and Lan, P. (2017). The understanding of the plant iron deficiency responses in strategy I plants and the role of ethylene in this process by omic approaches. *Frontiers in Plant Science* 8: 40.

Li, Z., Huang, T., Tang, M. et al. (2019). iTRAQ-based proteomics reveals key role of γ-aminobutyric acid (GABA) in regulating drought tolerance in perennial creeping bentgrass (*Agrostis stolonifera*). *Plant Physiology and Biochemistry* 145: 216–226.

Li, L., Dou, N., Zhang, H., and Wu, C. (2021a). The versatile GABA in plants. *Plant Signaling & Behavior* 16 (3): 1862565.

Li, X., Xu, X., Shen, F. et al. (2021b). γ-Aminobutyric acid participates in the adult-phase adventitious rooting recalcitrance. *Journal of Plant Growth Regulation* 40: 1981–1991.

Li, Y., Yu, X., and Ma, K. (2022). Physiological effects of γ-aminobutyric acid application on cold tolerance in *Medicago ruthenica*. *Frontiers in Plant Science* 13: 958029.

Lim, C.W., Baek, W., Jung, J. et al. (2015). Function of ABA in stomatal defense against biotic and drought stresses. *International Journal of Molecular Sciences* 16 (7): 15251–15270.

Ling, Y., Chen, T., Jing, Y. et al. (2013). γ-Aminobutyric acid (GABA) homeostasis regulates pollen germination and polarized growth in *Picea wilsonii*. *Planta* 238: 831–843.

Liu, T., Liu, Z., Li, Z. et al. (2019). Regulation of heat shock factor pathways by γ-aminobutyric acid (GABA) associated with thermotolerance of creeping bentgrass. *International Journal of Molecular Sciences* 20 (19): 4713.

Liu, C., Wang, H., Zhang, X. et al. (2021). Activation of the ABA signal pathway mediated by GABA improves the drought resistance of apple seedlings. *International Journal of Molecular Sciences* 22 (23): 12676.

Liu, S., Zenda, T., Tian, Z., and Huang, Z. (2023). Metabolic pathways engineering for drought or/and heat tolerance in cereals. *Frontiers in Plant Science* 14: 1111875.

Locy, R.D., Wu, S.J., Bisnette, J. et al. (2000). The regulation of GABA accumulation by heat stress in *Arabidopsis*. In: *Plant Tolerance to Abiotic Stresses in Agriculture: Role of Genetic Engineering* (ed. J.H. Cherry, R.D. Locy, and A. Rychter), 39–52. Dordrecht, The Netherlands: Springer.

Ludewig, F., Hüser, A., Fromm, H. et al. (2008). Mutants of GABA transaminase (POP2) suppress the severe phenotype of succinic semialdehyde dehydrogenase (ssadh) mutants in *Arabidopsis*. *PLoS One* 3 (10): e3383.

Malekzadeh, P., Khara, J., and Heydari, R. (2014). Alleviating effects of exogenous gamma-aminobutiric acid on tomato seedling under chilling stress. *Physiology and Molecular Biology of Plants* 20: 133–137.

Mao, J.P., Zhang, D., Zhang, X. et al. (2018). Effect of exogenous indole-3-butanoic acid (IBA) application on the morphology, hormone status, and gene expression of developing lateral roots in *Malus hupehensis*. *Scientia Horticulturae* 232: 112–120.

Mazzucotelli, E., Tartari, A., Cattivelli, L., and Forlani, G. (2006). Metabolism of γ-aminobutyric acid during cold acclimation and freezing and its relationship to frost tolerance in barley and wheat. *Journal of Experimental Botany* 57 (14): 3755–3766.

Morkunas, I. and Ratajczak, L. (2014). The role of sugar signaling in plant defense responses against fungal pathogens. *Acta Physiologiae Plantarum* 36: 1607–1619.

Muhammad Aslam, M., Waseem, M., Jakada, B.H. et al. (2022). Mechanisms of abscisic acid-mediated drought stress responses in plants. *International Journal of Molecular Sciences* 23 (3): 1084.

Munemasa, S., Hauser, F., Park, J. et al. (2015). Mechanisms of abscisic acid-mediated control of stomatal aperture. *Current Opinion in Plant Biology* 28: 154–162.

Munns, R. and Tester, M. (2008). Mechanisms of salinity tolerance. *Annual Review of Plant Biology* 59: 651–681.

Nonaka, S., Someya, T., Zhou, S. et al. (2017). An *Agrobacterium tumefaciens* strain with gamma-aminobutyric acid transaminase activity shows an enhanced genetic transformation ability in plants. *Scientific Reports* 7 (1): 42649.

Pal, S., Zhao, J., Khan, A. et al. (2016). Paclobutrazol induces tolerance in tomato to deficit irrigation through diversified effects on plant morphology, physiology and metabolism. *Scientific Reports* 6 (1): 39321.

Pantin, F., Renaud, J., Barbier, F. et al. (2013). Developmental priming of stomatal sensitivity to abscisic acid by leaf microclimate. *Current Biology* 23 (18): 1805–1811.

Planamente, S., Vigouroux, A., Mondy, S. et al. (2010). A conserved mechanism of GABA binding and antagonism is revealed by structure-function analysis of the periplasmic binding protein Atu2422 in *Agrobacterium tumefaciens*. *The Journal of Biological Chemistry* 285 (39): 30294–30303.

Planamente, S., Moréra, S., and Faure, D. (2013). In planta fitness-cost of the Atu4232-regulon encoding for a selective GABA-binding sensor in *Agrobacterium*. *Communicative & Integrative Biology* 6 (3): e23692.

Priya, M., Sharma, L., Kaur, R. et al. (2019). GABA (γ-aminobutyric acid), as a thermo-protectant, to improve the reproductive function of heat-stressed mungbean plants. *Scientific Reports* 9 (1): 7788.

Ragel, P., Raddatz, N., Leidi, E.O. et al. (2019). Regulation of K^+ nutrition in plants. *Frontiers in Plant Science* 10: 281.

Rajput, V.D., Harish, S.R.K., Verma, K.K. et al. (2021). Recent developments in enzymatic antioxidant defence mechanism in plants with special reference to abiotic stress. *Biology* 10 (4): 267.

Rao, N.S., Shivashankara, K.S., and Laxman, R.H. (ed.) (2016). *Abiotic Stress Physiology of Horticultural Crops*. New Delhi, India: Springer.

Raza, A., Razzaq, A., Mehmood, S.S. et al. (2019). Impact of climate change on crops adaptation and strategies to tackle its outcome: a review. *Plants* 8 (2): 34.

Rihan, H.Z., Al-Issawi, M., and Fuller, M.P. (2017). Advances in physiological and molecular aspects of plant cold tolerance. *Journal of Plant Interactions* 12 (1): 143–157.

Sakakibara, H., Kasahara, H., Ueda, N. et al. (2005). *Agrobacterium tumefaciens* increases cytokinin production in plastids by modifying the biosynthetic pathway in the host plant. *Proceedings of the National Academy of Sciences of the United States of America* 102 (28): 9972–9977.

Scholz, S.S., Malabarba, J., Reichelt, M. et al. (2017). Evidence for GABA-induced systemic GABA accumulation in *Arabidopsis* upon wounding. *Frontiers in Plant Science* 8: 388.

Serraj, R., Shelp, B.J., and Sinclair, T.R. (1998). Accumulation of γ-aminobutyric acid in nodulated soybean in response to drought stress. *Physiologia Plantarum* 102 (1): 79–86.

Shams, M. and Khadivi, A. (2023). Mechanisms of salinity tolerance and their possible application in the breeding of vegetables. *BMC Plant Biology* 23 (1): 139.

Shelp, B.J., Aghdam, M.S., and Flaherty, E.J. (2021). γ-Aminobutyrate (GABA) regulated plant defense: mechanisms and opportunities. *Plants* 10 (9): 1939.

Sheng, Y., Xiao, H., Guo, C. et al. (2018). Effects of exogenous gamma-aminobutyric acid on α-amylase activity in the aleurone of barley seeds. *Plant Physiology and Biochemistry* 127: 39–46.

Sheteiwy, M.S., Shao, H., Qi, W. et al. (2019). GABA-alleviated oxidative injury induced by salinity, osmotic stress and their combination by regulating cellular and molecular signals in rice. *International Journal of Molecular Sciences* 20 (22): 5709.

Sita, K. and Kumar, V. (2020). Role of gamma amino butyric acid (GABA) against abiotic stress tolerance in legumes: a review. *Plant Physiology Reports* 25: 654–663.

Su, Y.H., Liu, Y.B., and Zhang, X.S. (2011). Auxin–cytokinin interaction regulates meristem development. *Molecular Plant* 4: 616–625.

Sweetman, C., Sadras, V.O., Hancock, R.D. et al. (2014). Metabolic effects of elevated temperature on organic acid degradation in ripening *Vitis vinifera* fruit. *Journal of Experimental Botany* 65 (20): 5975–5988.

Takahashi, Y., Sasanuma, T., and Abe, T. (2013). Accumulation of gamma-aminobutyrate (GABA) caused by heat-drying and expression of related genes in immature vegetable soybean (edamame). *Breeding Science* 63 (2): 205–210.

Takayama, M. and Ezura, H. (2015). How and why does tomato accumulate a large amount of GABA in the fruit? *Frontiers in Plant Science* 6: 612.

Thompson, J.F., Stewart, C.R., and Morris, C.J. (1966). Changes in amino acid content of excised leaves during incubation I. The effect of water content of leaves and atmospheric oxygen level. *Plant Physiology* 41 (10): 1578–1584.

Thouvenot, L., Deleu, C., Berardocco, S. et al. (2015). Characterization of the salt stress vulnerability of three invasive freshwater plant species using a metabolic profiling approach. *Journal of Plant Physiology* 175: 113–121.

Tripathy, B.C. and Oelmüller, R. (2012). Reactive oxygen species generation and signaling in plants. *Plant Signaling & Behavior* 7 (12): 1621–1633.

Ullah, A., Ali, I., Noor, J. et al. (2023). Exogenous γ-aminobutyric acid (GABA) mitigated salinity-induced impairments in mungbean plants by regulating their nitrogen metabolism and antioxidant potential. *Frontiers in Plant Science* 13: 1081188.

Verma, P.K., Verma, S., and Pandey, N. (2022). Root system architecture in rice: impacts of genes, phytohormones and root microbiota. *3 Biotech* 12 (9): 239.

Waadt R, Seller CA, Hsu PK, Takahashi Y, Munemasa S, Schroeder JI. Publisher correction: plant hormone regulation of abiotic stress responses *Nature Reviews Molecular Cell Biology*, (2022), 23(7):516. Doi: https://doi.org/10.1038/s41580-022-00479-6.

Wang, Y., Cao, H., Wang, S. et al. (2023). Exogenous γ-aminobutyric acid (GABA) improves salt-inhibited nitrogen metabolism and the anaplerotic reaction of the tricarboxylic acid cycle by regulating GABA-shunt metabolism in maize seedlings. *Ecotoxicology and Environmental Safety* 254: 114756.

Widodo, P.J.H., Newbigin, E.D., Tester, M. et al. (2009). Metabolic responses to salt stress of barley (*Hordeum vulgare* L.) cultivars, Sahara and Clipper, which differ in salinity tolerance. *Journal of Experimental Botany* 60 (14): 4089–4103.

Williams, C.M., Henry, H.A., and Sinclair, B.J. (2015). Cold truths: how winter drives responses of terrestrial organisms to climate change. *Biological Reviews* 90 (1): 214–235.

Wu, X., Jia, Q., Ji, S. et al. (2020). Gamma-aminobutyric acid (GABA) alleviates salt damage in tomato by modulating Na^+ uptake, the GAD gene, amino acid synthesis and reactive oxygen species metabolism. *BMC Plant Biology* 20: 1–21.

Wu, Q., Su, N., Huang, X. et al. (2021). Hypoxia-induced increase in GABA content is essential for restoration of membrane potential and preventing ROS-induced disturbance to ion homeostat. *Plant Communications* 2: 1–12.

Xu, B., Long, Y., Feng, X. et al. (2021). GABA signalling modulates stomatal opening to enhance plant water use efficiency and drought resilience. *Nature Communications* 12 (1): 1952.

Yang, R., Feng, L., Wang, S. et al. (2016). Accumulation of γ-aminobutyric acid in soybean by hypoxia germination and freeze–thawing incubation. *Journal of the Science of Food and Agriculture* 96 (6): 2090–2096.

Yang, N.J. and Chiu, I.M. (2017). Bacterial signalling to the nervous system through toxins and metabolites. *Journal of Molecular Biology* 429 (5): 587–605.

Yu, G., Liang, J., He, Z., and Sun, M. (2006). Quantum dot-mediated detection of γ-aminobutyric acid binding sites on the surface of living pollen protoplasts in tobacco. *Chemistry and Biology* 13 (7): 723–731.

Yuan, D., Wu, X., Gong, B. et al. (2023). GABA metabolism, transport and their roles and mechanisms in the regulation of abiotic stress (hypoxia, salt, drought) resistance in plants. *Metabolites* 13 (3): 347.

Zhang, Y., Kilambi, H.V., Liu, J. et al. (2021). ABA homeostasis and long-distance translocation are redundantly regulated by ABCG ABA importers. *Science Advances* 7 (43): eabf6069.

Zhao, S., Zhang, Q., Liu, M. et al. (2021). Regulation of plant responses to salt stress. *International Journal of Molecular Sciences* 22 (9): 4609.

Zhao, W., Zhao, H., Wang, H., and He, Y. (2022). Research progress on the relationship between leaf senescence and quality, yield and stress resistance in horticultural plants. *Frontiers in Plant Science* 13: 1044500.

Zhao, Q., Ma, Y., Huang, X. et al. (2023). GABA application enhances drought stress tolerance in wheat seedlings (*Triticum aestivum* L.). *Plants.* 12 (13): 2495.

Zhu, X., Liao, J., Xia, X. et al. (2019). Physiological and iTRAQ-based proteomic analyses reveal the function of exogenous γ-aminobutyric acid (GABA) in improving tea plant (*Camellia sinensis* L.) tolerance at cold temperature. *BMC Plant Biology* 19 (1): 1–20.

5

GABA and Drought Stress

Petronia Carillo[1] *and Andrea Carra*[2]

[1] *Department of Environmental, Biological and Pharmaceutical Sciences and Technologies, University of Campania "Luigi Vanvitelli", Caserta, 81100, Italy*
[2] *Council for Agricultural Research and Economics, Research Centre for Forestry and Wood, Casale Monferrato, 15033, Italy*

Introduction

The first report on γ-aminobutyric acid (GABA) was that of Steward (1949), who isolated it from potato tubers. However, it was Eugene Roberts who brought attention to this nonprotein amino acid as an important compound in the brain in the 1950s (Spiering 2018). GABA is, in fact, a primary inhibitory neurotransmitter for the central nervous system in the mature brain, able to decrease neuronal excitability through nerve transmission inhibition; moreover, it plays a role in the differentiation of the nervous system, maturation of synapses and cell death (Owens and Kriegstein 2002). Avoli and Krnjević (2015) reported that the GABA receptors, which are ligand-gated ion channels, may function in animals to induce neuronal diversity during brain development and adulthood. More recently, the dysfunction of GABAergic pathways has been strongly associated with neurodevelopmental disorders and may be related to Angelman syndrome and autism spectrum disorder (Babij et al. 2023). However, GABA is not only important for animals, as it also significantly enters the pool of free amino acids of Monera, Protists, Fungi, and Plantae, conserving the enzymes involved in its metabolism in the different organisms of the five kingdoms (Michaeli and Fromm 2015; Shelp et al. 1999; Nugraha et al. 2022).

The first clear evidence that GABA plays a key signaling role also in plants came thanks to the characterization of a mutant for a gene encoding GABA transaminase (4-aminobutyrate:2-oxoglutarate aminotransferase, GABA-T) in Arabidopsis (*pop2*) which, not being able to degrade GABA, was unable to contribute to the formation of its gradient up to the micropyle, thus determining an excess of GABA enough to inhibit the growth of the pollen tube to the ovary (Palanivelu et al. 2003). In addition, GABA is also involved

GABA in Plants: Biosynthesis, Plant Development, and Food Security, First Edition. Edited by Samiksha Singh, Durgesh Kumar Tripathi, and Vijay Pratap Singh.

in gene expression regulation, emission of ethylene, and elongation of the stem (Kathiresan et al. 1998; Palanivelu et al. 2003; Lancien and Roberts 2006). However, it has also been found that the increase of GABA content occurs together with a decrease in glutamine and protein synthesis and degradation of proteins, as evidenced by Bouché and Fromm (2004) and references therein. GABA may be accumulated in plants in concomitance with several stresses exceeding the concentration of proteinogenic amino acids (Kinnersley and Turano 2000), even if the reasons for its accumulation are still not fully understood (Fait et al. 2008; Roberts 2007; Renault et al. 2013; Michaeli and Fromm 2015). However, under stress, the activity of GABA-T decreases causing an accumulation of GABA that can be used for providing NADH and succinate for the Krebs cycle when stress is reduced or ceases, thus allowing growth to resume (Bouché et al. 2003; Studart-Guimarães et al. 2007). In fact, *her1* and *gaba-t/pop2-1* deficient mutants of GABA-T, unable to provide anaplerotic succinate to root tissues after salt stress relief, show defects in the composition of the cell wall and inhibition of root elongation and hypocotyl development (Mirabella et al. 2008; Renault et al. 2013). However, GABA not only provides anaplerotic precursors to the tricarboxylic acid cycle (TCA), but it plays vital functions in cell signaling for plant development, growth, and plant defense responses to (a)biotic stresses (Bouché et al. 2004; Fait et al. 2008), and its changes are related to nitrogen sensing (Mazzucotelli et al. 2006) and nitrogen starvation (Recht et al. 2014). In particular, GABA may regulate the expression of 14-3-3-genes, by a mechanism involving calcium, ethylene, and abscisic acid (ABA) signaling pathways, thus controlling carbon and nitrogen metabolism enzymes (e.g., glyceraldehyde-3-P DH, starch synthase III, nitrate reductase and glutamine synthetase) targeted by 14-3-3 proteins (Lancien and Roberts 2006). GABA also plays a role as a long-distance signal for enhancing nitrate uptake in *Brassica napus* (Beuve et al. 2004). It may also be used in Arabidopsis as a source of nutrition for the foliar pathogen *Pseudomonas syringae,* thus inhibiting its growth and the expression of pathogenesis genes (Vest 2013). Moreover, being involved in the biotic stress transduction pathway of E-2-hexenal, enzymatically generated from oxylipins (Mirabella et al. 2008), and functioning as an inhibitory neurotransmitter, its synthesis, and accumulation could also be a toxic mechanism enacted by plants to cope with herbivore insects' attacks (Scholz et al. 2017). Interestingly, in moths and butterflies, it was observed a convergent substitution of amino acids in duplicated copies of the gene coding for the GABA receptor (*Resistance to dieldrin, Rdl*), thus generating GABA receptors resistant to terpenoids, essential for plant–insect coevolution (Guo et al. 2023).

Ramesh et al. (2015) found that aluminum-activated malate transporters (ALMTs) were inhibited by GABA at low micromolar concentrations but activated by malate, thus causing membrane hyperpolarization and excitability decrease (Hedrich et al. 2016). The same authors demonstrated that GABA was also able to regulate the ALMTs present on the plasma membrane and tonoplast in several plants regulating anion fluxes (e.g., Arabidopsis, barley, grapevine, rice, and wheat) (Ramesh et al. 2015). Being ALMTs ubiquitous in plant cells and involved in mechanisms like signaling pathways for local and systemic biotic and abiotic stress response, as well as the control of physiological functions such as pollen tube growth and gas exchanges, they emerged as good candidate receptors for GABA signaling in plants (Ramesh et al. 2015; Gilliham and Tyerman 2016; Ramesh et al. 2017).

GABA Shunt in Plants

The GABA metabolic pathways or GABA shunt consists of three enzymatic reactions bypassing two steps of the Krebs or TCA cycle (Bouché and Fromm 2004). The cytosolic Ca^{2+}-calmodulin-activated and pyridoxal-5′-phosphate-dependent enzyme glutamate decarboxylase (GAD) decarboxylates glutamate, synthesizing GABA. This latter is then transported into the mitochondrial matrix where GABA-T catabolizes it to succinic semialdehyde (SSA), which is subsequently oxidized by succinic semialdehyde dehydrogenase (SSADH) to succinate or, in the cytosol, reduced by γ-hydroxybutyrate dehydrogenase (GHBDH) to γ-hydroxybutyrate (GHB) (Bouché et al. 2003; Fait et al. 2005; Renault et al. 2010). The catalytic activity of GAD is positively and rapidly regulated by calcium/calmodulin (Bouché et al. 2004), or acidification of the cytosol (Allan et al. 2009) under stress. GAD deficiency causes an increased production of reactive oxygen species (ROS), with consequent oxidative stress and tissue necrosis in Arabidopsis even under standard light conditions (Bouché et al. 2003). However, under salinity, also GABA-T may regulate the levels of GABA by synthesizing it through a backward reaction from SSA (Akçay et al. 2012) or by not degrading it as in Arabidopsis GABA-T deficient mutants *pop2-1* (Renault et al. 2010).

The GABA permease GABP transports the GABA synthesized and/or accumulated in the cytosol to the mitochondria (Michaeli et al. 2011), where it, after being converted to SSA and then succinate, bypasses three Krebs cycle sites of production of NADH and enters the TCA cycle, thus altering the redox status of the cell (Bouché and Fromm 2004; Fait et al. 2008). To avoid the accumulation of SSA, which can inhibit GABA-T, the activity of this latter enzyme is coordinated with that of SSADH, and the expression of the two enzymes coding genes *POP2* and *SSADH* appears upregulated by high levels of GABA under abiotic stresses (Renault et al. 2010). *GABA-T*-deficient mutants are particularly susceptible to salt stress (Renault et al. 2010), while *SSADH* knockout mutants undergo excess ROS production causing apoptosis, in addition to oversensitivity to abiotic stresses and dwarfism (Fait et al. 2008). Therefore, GABA synthesis and its shunt are essential for controlling ROS production and limiting necrotic cell death.

GABA Accumulates in Plants Under Drought Stress

GABA accumulates in plants in response to abiotic and biotic stresses, including, cold, heat, hypoxia, salinity, mechanical damage, and insect attacks (Bown and Shelp 1997; Shelp et al. 1999; Renault et al. 2010; Woodrow et al. 2017; Suhel et al. 2023). GABA has also been shown to accumulate in several plant species under water stress (Thompson et al. 1966; Raggi 1994; Kinnersley and Turano 2000; Urano et al. 2009; Michaeli and Fromm 2015; Pires et al. 2016; Fàbregas and Fernie 2019).

GABA accumulation under stress conditions can be explained by several mechanisms. In plants, GAD isoforms are encoded by a small gene family and most of them are activated by Ca^{2+}/Calmodulin (Ca^{2+}/CaM) binding to a C-terminal activation domain. Ca^{2+} is an early messenger quickly mobilized in the cytoplasm upon perception of different abiotic stresses, including drought. Thus, GABA biosynthesis can be induced by GAD activation by drought-elicited increases in Ca^{2+} cytoplasmic concentrations (Kinnersley and Turano 2000). In addition, some contribution may be given by upregulation of specific members of the *GAD* gene family under water stress (Yong et al. 2017; Fracasso et al. 2021; Cheng et al. 2023). However, the extent of this contribution is difficult to evaluate because GADs have distinct localization, different basal levels of transcription and vary for their responsiveness to water stress. For instance, in Arabidopsis, GAD2 is the major expressed isoform in leaves (Mekonnen et al. 2016; Xu et al. 2021a) but is poorly responsive to drought stress. By contrast, GAD3 and GAD4 have a very low basal level of expression in leaves but are strongly upregulated under stress conditions (Schmid et al. 2005; Shelp et al. 2012). The regulatory pathways controlling *GAD* gene family expression are also poorly understood. Urano et al. (2009) reported that Arabidopsis *GAD4* expression was upregulated in response to dehydration in the *NCED3* knockout mutant *nc3-2*, which is unable to accumulate ABA under drought stress, but not in the wild type. This suggests that GAD upregulation in stress conditions may be under the control of an ABA-independent signaling pathway and that it may be repressed by ABA. Interestingly, GAD4 upregulation and GABA accumulation in the drought-treated *nc3-2* mutant was associated with increased accumulation of raffinose family oligosaccharides, including raffinose and galactinol, which are well-known osmoprotectants and membrane stabilizers and have been proposed to play a ROS scavenging action (Nishizawa et al. 2008; Urano et al. 2009). Galactinol accumulated more in creeping bentgrass plants treated with GABA than in those treated with ABA or salicylic acid, reinforcing the idea that the raffinose oligosaccharide pathway may be upregulated by GABA (Li et al. 2017). Accordingly, *GOLS1*, encoding a galactinol synthase, was upregulated in GABA-treated poplar roots exposed to salt stress (Ji et al. 2018).

As GAD isoforms have optimum activity at acidic pHs (around 5.8), while the physiological cytoplasmic pH is approximately neutral, GABA accumulation under stress conditions may also be due to GAD activation by transient cytosol acidification (Kinnersley and Turano 2000). However, changes in cytoplasmic pH during drought stress are not well known and are difficult to measure. In guard cells, for instance, cytosol alkalinization rather than acidification has been detected prior to ABA-mediated stomatal closure in drought stress conditions, which would result in reduced GABA accumulation. However, it is possible that in milder conditions or in other cell types, pH fluctuations favor GAD activity.

Other mechanisms of accumulation may involve the regulation of additional enzymatic processes. Particularly interesting is the regulation of the TCA cycle, which is obviously closely intertwined with the GABA shunt. In wheat plants exposed to salt stress, the reduction in abundance and activity of mitochondrial oxoglutarate dehydrogenase (mtOGDC) favors the conversion of oxoglutarate to glutamic acid, which can enter the GABA shunt (Che-Othman et al. 2020). However, if this also happens under drought conditions remains to be confirmed.

GABA Accumulation Increases Drought Tolerance

Numerous studies have shown that treatments with GABA confer increased drought tolerance to plants. For instance, creeping bentgrass (*Agrostis stolonifera* L.) plants sprayed with GABA prior to a drought stress treatment displayed increased leaf relative water content and chlorophyll concentration, and reduced electrolyte leakage (an estimate of membrane integrity) compared to controls (Li et al. 2017). These physiological traits were associated with increased concentrations of disaccharides (maltose and cellobiose), amino acids (including proline, 5-oxoproline, and threonine), organic acids (malonic acid and malic acid), and beta-sitosterol, a phytosterol with antioxidant properties known to increase in concentration and to play a role in stabilizing plant cell membranes in stress conditions (Li et al. 2017). An increase in relative water content was also reported in leaves of white clover plants treated with GABA by irrigation and subjected to PEG-induced drought stress. GABA accumulation was associated with increased proline accumulation and increased activity of the rate-limiting enzyme of proline biosynthesis Delta 1-Pyrroline-5-Carboxylate Synthetase (P5CS) (Yong et al. 2017). Previous and subsequent studies also found that exogenous GABA induced the accumulation of proline in *B. napus* (Lei et al. 2016) and tobacco leaves (He et al. 2021), enhancing the resistance to abiotic stresses of plants. Furthermore, apple seedlings pretreated with GABA by irrigation displayed increased tolerance to a subsequent drought stress treatment, as indicated by higher relative water content and reduced malondialdehyde (MDA) content, an indicator of membrane damage caused by lipid peroxidation. This response was associated with reduced ROS accumulation, increased activity of the ROS-scavenging enzymes peroxidase (POD), superoxide dismutase (SOD), and catalase (CAT), and with increased expression of genes of the abscisic acid (ABA) biosynthesis and signal transduction pathways, which resulted in increased ABA accumulation (Liu et al. 2021). In addition, some genes encoding known drought-responsive proteins were upregulated, including *RESPONSIVE TO DEHYDRATION 22* (*RD22*) and a *late embryogenesis abundant* (*LEA*) one. Upregulation of genes of ABA biosynthesis and signaling, as well as of genes encoding drought protective proteins of the RD and LEA groups, were also observed in seedlings of the apple rootstock M.9-T337 pre-treated with GABA before being exposed to drought (Cheng et al. 2023). In this experiment, a dose effect was observed. Among GABA concentrations of 0.5, 1, 5, and 10 mM, 1 mM elicited the maximum advantage in terms of net photosynthetic rate, relative leaf water content, water use efficiency (WUE), and delay in relative electrolyte leakage, while 10 mM had an inhibitory effect. Interestingly, an ethylene response factor (ERF), *MdERF1*, was also strongly upregulated by 1 mM GABA. Increased drought tolerance indicated by higher yield, leaf area, and relative water content, associated with increased proline accumulation and higher activity of ROS-scavenging enzymes was confirmed by a two-year experiment in field conditions in a semiarid environment in which snap bean (*Phaseolus vulgaris* L.) plants were exposed to mild drought (Abd El-Gawad et al. 2021).

Overall, these studies indicate that treatments with GABA represent an interesting agronomical practice aimed at minimizing the effects of drought on crop yield and that increasing GABA accumulation by breeding or biotechnological approaches may represent a promising strategy to obtain drought-resilient genotypes.

The mechanisms by which GABA mitigates the effects of drought stress are complex and only partially known; among these, the alleviation of oxidative stress is considered of primary importance. ROS accumulate in plants under drought stress conditions, damaging cellular membranes and proteins with detrimental effects on physiological processes and, ultimately, crop productivity. GABA was reported to display per se ROS-scavenging activity (Smirnoff and Cumbes 1989; Shelp et al. 1999). GABA ability to stabilize and protect membranes and macromolecules by detoxifying hydrogen peroxide, singlet oxygen, and superoxide anion radicals was reported to be higher than that of proline at equimolar concentrations (Liu et al. 2011; Molina-Rueda et al. 2015). The glutamate derivatives GABA and proline are both accumulated under water and salt stresses (Li et al. 2016). However, the more efficient scavenging activity of GABA may explain its preferential synthesis over proline in tobacco leaves under water stress (Liu et al. 2011) and durum wheat under salinity (Woodrow et al. 2017). Carillo (2018) also suggested that, in durum wheat under stress-induced stomatal closure, the CO_2 released during GABA synthesis by GAD decarboxylation of glutamate is able to support/restart the carboxylation activity of the Calvin cycle thus reducing the pressure on the electron transport chain while reducing ROS production (e.g., $^1O_2$) and photodamage. In addition, the last reaction of the GABA shunt, the oxidation of SSA to succinate catalyzed by SSADH, is coupled with the production of NADH, which plays an important role in maintaining the redox balance of the cell. Arabidopsis knockout mutants for SSADH display increased ROS accumulation and the development of necrotic lesions when exposed to high light or heat stress (Bouché et al. 2003). In this view, it may be a functional and active GABA shunt more than GABA accumulation per se that contributes to antioxidant defense. NADH production from the GABA shunt can be particularly important in stress conditions when some NADH-producing reactions of the TCA cycle can be inhibited as, for instance, the conversion of 2-oxoglutarate into succinate by oxoglutarate dehydrogenase (mtOGDC) under salt stress (Che-Othman et al. 2020).

In drought conditions, GABA also upregulates the expression of drought-responsive genes, which results in the accumulation of drought-protective metabolites and in increased expression and activity of ROS-scavenging enzymes (Salah et al. 2019; Abd El-Gawad et al. 2021; Liu et al. 2021). The activation of the ABA pathway by GABA may explain, in part, some favorable physiological responses to drought as stomatal closure and reduced water loss by transpiration. However, GABA has also been shown to act per se as a signal molecule able to regulate stomatal movement.

GABA Signaling and the Regulation of Stomatal Opening

GABA was suggested to play a role as a signal molecule in plants based on similarity with animal systems, in which it acts as an inhibitory neurotransmitter in the central nervous system by binding GABA-A and GABA-B receptors. As mentioned above, the first evidence of a signaling role in plants came from the observation that pollen tube development was impaired in the Arabidopsis *pop2* mutant, which has reduced expression of GABA-T (Palanivelu et al. 2003). The modulatory effect of GABA treatments on plant gene expression patterns also hints at the role of GABA in signaling. For instance, in Arabidopsis seedlings grown on media supplemented with GABA, the expression of several members of the 14-3-3 gene family was downregulated (Lancien and Roberts 2006). In view of the manifold roles played by 14-3-3 proteins in the responses to different stress conditions, including drought, this suggested that GABA may regulate the expression of stress-related genes through 14-3-3 proteins acting as intermediates. The involvement of GABA in the regulation of 14-3-3 proteins was confirmed by the observation that in drought-stressed plants of the Arabidopsis *gad1/2* mutant, which is impaired for

GABA biosynthesis, several members of the 14-3-3 gene family were upregulated compared to wild-type (Mekonnen et al. 2016).

Plants do not have homologs of the animal GABA receptors, and the search for proteins that could recognize and transduce GABA as a signal in plants went on for a long time. Early experiments with agonists and antagonists led to the hypothesis that these might be glutamate receptor-like proteins (GLRs), which are transmembrane channel proteins that, in their N-terminal region, show some homology with animal $GABA_B$ receptors (Kinnersley and Lin 2000; Kinnersley and Turano 2000). Only 15 years later, it was shown that GABA accumulation could downregulate the activity of ALMT1, a member of a different class of membrane proteins, the aluminum-activated malate transporters, in roots of bread wheat (*Triticum aestivum* L.) exposed to acidic conditions (Ramesh et al. 2015). In *Xenopus laevis* oocytes, the transport activity of ALMT1 was inhibited by GABA and by the GABA analog muscimol, while the inhibition was attenuated by bicuculline, a GABA competitive antagonist. ALMT1, as well as other ALMTs from different plant species, including Arabidopsis, rice, barley, and grapevine, possess a small region sharing homology with a region of animal $GABA_A$ receptors involved in ligand binding, and site-directed mutagenesis of this region results in loss of GABA regulation (Ramesh et al. 2015). These results strongly suggested that ALMT1 was a GABA receptor in plants. Subsequently, it was shown that TaALMT1 from bread wheat was able to facilitate GABA transport through the plasma membrane (Ramesh et al. 2018), and that GABA-mediated inhibition of TaALMT1 activity occurs from the cytosolic side but not from the extracellular side (Long et al. 2020).

The discovery of the regulation of TaALMT1 by GABA opened exciting perspectives because some ALMTs from Arabidopsis were previously shown to participate in the control of stomala movement, which is one of the most important responses to drought stress in land plants (Meyer et al. 2010; De Angeli et al. 2013; Eisenach et al. 2017). Stomatal movement in plants is regulated mainly by ABA, but other signals are also involved. Evidence for GABA-mediated control of stomatal movement was provided by experiments with Arabidopsis *gad* knockout mutants in which GABA accumulation is severely reduced. In drought conditions, *gad1/2* mutants displayed increased stomatal pore aperture and increased stomatal conductance associated with a wilted phenotype (Mekonnen et al. 2016). This drought-sensitive phenotype was suppressed in the *gad1/2 x pop2* mutant, obtained by crossing with the *pop2* mutant, which accumulates high levels of GABA. Stomatal opening in *gad1/2* mutant was speculated to be due to reduced activity of an unidentified ALMT and/or to increased activation of H^+-ATPase by 14-3-3 proteins. Indeed, in *gad1/2* plants, some members of the 14-3-3 gene family were more upregulated by drought than in WT plants, and the proton extrusion by roots was increased (Mekonnen et al. 2016). Even under nonlimiting water conditions, *gad1/2* mutants displayed higher transpiration than WT, which was associated with increased stomatal density and aperture. However, since the Arabidopsis *gad2* mutant displays increased transpiration without significant modifications in stomatal density, the effect of GABA on leaf transpiration rate was ascribed primarily to the regulation of stomatal movement (Xu et al. 2021a).

Experiments with Arabidopsis leaf epidermal peels and detached leaves fed with artificial xylem sap showed that GABA inhibits light-induced stomatal pore opening (Xu et al. 2021a). GABA feeding through petioles also slowed the increase in the transpiration rate of detached tomato leaves following the transition from dark to light (Fracasso et al. 2021). Interestingly, in excised epidermal peels, but not in intact leaves, GABA also antagonized stomatal closure following the transition from light to dark and in response to low (e.g., 2.5 μM) ABA concentrations, suggesting that the GABA signal might be perceived differently in different cell types (Xu et al. 2021a). Higher (e.g., 25 μM) ABA concentrations, however, induced complete stomatal closure also in the presence of GABA. ALMT9, a tonoplast-localized anion channel expressed in guard cells, previously characterized for playing a role in stomatal opening in response to dark-to-light transition (De Angeli et al. 2013), was found to be required for GABA-mediated regulation of stomatal aperture. Differently from *gad2* mutants, *gad2/almt9* double mutants did not show a constitutively open stomata phenotype despite both being impaired in GABA accumulation. Furthermore, in leaf epidermal peels of *almt9* mutants, GABA did not antagonize stomatal opening following dark-to-light transition. The effect of GABA was resumed when the *almt9* mutant was complemented with a native form of ALMT9 but not with ALMT9F243C/Y245C, a mutagenized variant that carries substitutions in the amino acid motif displaying homology with a GABA-binding region of mammalian $GABA_A$ receptors. Detached leaf incubation experiments with Arabidopsis WT and *almt9* complemented with native ALMT9 showed that GABA supplementation also resulted in increased WUE (Xu et al. 2021a). These results support the role of GABA as a signal molecule, indicating that its accumulation in guard cells is perceived by ALMT9 to adjust stomatal opening as a plant response to water-limiting conditions, which potentially results in increased drought tolerance.

However, the agronomic potential of the increase in WUE allowed by GABA would deserve further study at the whole plant level and in field conditions. In fact, while GABA appears to act as a "brake" on stomatal movement, other studies have highlighted a direct correlation between the rapidity of stomatal movement, which would allow quicker adaptation to changing environmental conditions, and WUE (Lawson and Blatt 2014).

Several molecular and physiological aspects of GABA signaling mediated by ALMTs remain to be elucidated. Differently from the case of ALMT1 from wheat, in vitro, electrophysiological experiments with heterologous expression of ALMT9 failed to detect modifications of transport activity by GABA (Xu et al. 2021a; Jáslan and De Angeli 2022). Although this observation is not conclusive, due to potential limitations inherent to the heterologous experimental systems adopted (e.g., incorrect protein folding, lack of interacting factors), it is possible that the regulatory action of GABA is played through signaling intermediates, which may be other ALMTs or proteins belonging to other classes (Gilliham and Xu 2022). Most ALMTs from Arabidopsis, as well as a number of ALMTs from other plant species and some proteins with unknown functions, display the 12 aa motif with homology to animal GABAA receptors (Ramesh et al. 2017), thus, they could potentially be, to some extent, regulated by GABA. Arabidopsis ALMT12 (also known as QUAC1, quick-activating anion channel), a plasma membrane-localized transporter involved in stomatal closure in response to light-to-dark transition and ABA, is an interesting candidate for further study. In fact, this protein has been shown to be required for GABA-mediated inhibition of stomatal closure in response to dark in epidermal peels and for muscimol-mediated inhibition of pollen tube growth and is regulated by GABA in pollen grain protoplasts (Domingos et al. 2019; Xu et al. 2021a).

Conclusion

In addition to the regulation of stomatal movement, GABA recognition and signal transduction by ALMTs may be involved in other stress-related processes. ALMTs are widely expressed in different plant tissues and organs, including ALMT9, which is highly expressed in the mesophyll and in vascular tissues of shoots and roots (Kovermann et al. 2007; De Angeli et al. 2013; Baetz et al. 2016). Thus, the GABA/ALMT9 regulatory circuit may control specific processes in these tissues. In this view, it may be noted that the activity of ion channels has considerable regulatory potential because the consequent changes in membrane voltage dynamics affect signaling pathways by altering the concentration of second messengers (ROS, Ca^{2+}, nitric oxide, H^{+}), which can result in transcriptional reprogramming (Klejchova et al. 2021). In the Arabidopsis *almt9-1* mutant, for instance, more than 300 genes were differentially expressed in shoots, and more than 50 were differentially expressed in roots (Baetz et al. 2016), compared to WT. Among these, interestingly, there were several genes encoding proteins involved in ABA signaling. Thus, another theme for future investigations is the interplay between ABA and GABA signaling, which appears to be complex and may be important for the control of plant responses to drought. GABA was shown to act as a modulator of ABA sensitivity in guard cells (Xu et al. 2021a, 2021b) and to induce the expression of genes involved in ABA biosynthesis and transduction pathways (Liu et al. 2021; Cheng et al. 2023). Furthermore, it was suggested that the GABA shunt may be repressed by ABA through the downregulation of GAD genes (Urano et al. 2009). Thus, the potential involvement of ALMT channels in the modulation of gene expression operated by GABA would deserve further investigation. In addition, both GABA and ALMTs have been involved in the determination of stomatal density, which is a key trait related to drought tolerance in plants, but a mechanistic relationship between these molecular players in this process is still to be established (Mekonnen et al. 2016; Ye et al. 2021). In this context, for instance, it would be interesting to test whether ALMT15 from tomato, which has been shown to control stomatal density (Ye et al. 2021), is regulated by GABA.

References

Abd El-Gawad, H.G., Mukherjee, S., Farag, R. et al. (2021). Exogenous γ-aminobutyric acid (GABA)-induced signaling events and field performance associated with mitigation of drought stress in *Phaseolus vulgaris* L. *Plant Signaling & Behavior* 16: https://doi.org/10.1080/15592324.2020.1853384.

Akçay, N., Bor, M., Karabudak, T. et al. (2012). Contribution of gamma amino butyric acid (GABA) to salt stress responses of *Nicotiana sylvestris* CMSII mutant and wild type plants. *Journal of Plant Physiology* 169: https://doi.org/10.1016/j.jplph.2011.11.006.

Allan, W.L., Clark, S.M., Hoover, G.J., and Shelp, B.J. (2009). Role of plant glyoxylate reductases during stress: a hypothesis. *Biochemical Journal* 423.

Avoli, M. and Krnjević, K. (2015). The long and winding road to gamma-amino-butyric acid as neurotransmitter. *Canadian Journal of Neurological Sciences* 43: 219.

Babij, R., Ferrer, C., Donatelle, A. et al. (2023). Gabrb3 is required for the functional integration of pyramidal neuron subtypes in the somatosensory cortex. *Neuron* 111: https://doi.org/10.1016/j.neuron.2022.10.037.

Baetz, U., Eisenach, C., Tohge, T. et al. (2016). Vacuolar chloride fluxes impact ion content and distribution during early salinity stress. *Plant Physiology* 172: https://doi.org/10.1104/pp.16.00183.

Beuve, N., Rispail, N., Lainé, P. et al. (2004). Putative role of γ-aminobutyric acid (GABA) as a long-distance signal in up-regulation of nitrate uptake in *Brassica napus* L. *Plant, Cell and Environment* 27: https://doi.org/10.1111/j.1365-3040.2004.01208.x.

Bouché, N. and Fromm, H. (2004). GABA in plants: just a metabolite? *Trends in Plant Science* 9: 110–115. https://doi.org/10.1016/j.tplants.2004.01.006.

Bouché, N., Fait, A., Bouchez, D. et al. (2003). Mitochondrial succinic-semialdehyde dehydrogenase of the γ-aminobutyrate shunt is required to restrict levels of reactive oxygen intermediates in plants. *Proceedings of the National Academy of Sciences of the United States of America* 100: https://doi.org/10.1073/pnas.1037532100.

Bouché, N., Fait, A., Zik, M., and Fromm, H. (2004). The root-specific glutamate decarboxylase (GAD1) is essential for sustaining GABA levels in Arabidopsis. *Plant Molecular Biology* 55: https://doi.org/10.1007/s11103-004-0650-z.

Bown, A.W. and Shelp, B.J. (1997). The metabolism and functions of γ-aminobutyric acid. *Plant Physiology* 115: https://doi.org/10.1104/pp.115.1.1.

Carillo, P. (2018). GABA shunt in durum wheat. *Frontier in Plant Science* 9: 100.

Cheng, P., Yue, Q., Zhang, Y. et al. (2023). Application of γ-aminobutyric acid (GABA) improves fruit quality and rootstock drought tolerance in apple. *Journal of Plant Physiology* 280: https://doi.org/10.1016/j.jplph.2022.153890.

Che-Othman, M.H., Jacoby, R.P., Millar, A.H., and Taylor, N.L. (2020). Wheat mitochondrial respiration shifts from the tricarboxylic acid cycle to the GABA shunt under salt stress. *New Phytologist* 225: https://doi.org/10.1111/nph.15713.

De Angeli, A., Zhang, J., Meyer, S., and Martinoia, E. (2013). AtALMT9 is a malate-activated vacuolar chloride channel required for stomatal opening in Arabidopsis. *Nature Communications* 4: https://doi.org/10.1038/ncomms2815.

Domingos, P., Dias, P.N., Tavares, B. et al. (2019). Molecular and electrophysiological characterization of anion transport in Arabidopsis *thaliana* pollen reveals regulatory roles for pH, Ca^{2+} and GABA. *New Phytologist* 223: https://doi.org/10.1111/nph.15863.

Eisenach, C., Baetz, U., Huck, N.V. et al. (2017). ABA-induced stomatal closure involves ALMT4, a phosphorylation-dependent vacuolar anion channel of Arabidopsis. *Plant Cell* 29: https://doi.org/10.1105/tpc.17.00452.

Fàbregas, N. and Fernie, A.R. (2019). The metabolic response to drought. *Journal of Experimental Botany* 70.

Fait, A., Yellin, A., and Fromm, H. (2005). GABA shunt deficiencies and accumulation of reactive oxygen intermediates: insight from Arabidopsis mutants. *FEBS Letters* 579: https://doi.org/10.1016/j.febslet.2004.12.004.

Fait, A., Fromm, H., Walter, D. et al. (2008). Highway or byway: the metabolic role of the GABA shunt in plants. *Trends in Plant Science* 13: https://doi.org/10.1016/j.tplants.2007.10.005.

Fracasso, A., Vallino, M., Staropoli, A. et al. (2021). Increased water use efficiency in miR396-downregulated tomato plants. *Plant Science* 303: https://doi.org/10.1016/j.plantsci.2020.110729.

Gilliham, M. and Tyerman, S.D. (2016). Linking metabolism to membrane signaling: the GABA–malate connection. *Trends in Plant Science* 21: 295.

Gilliham, M. and Xu, B. (2022). γ-Aminobutyric acid may directly or indirectly regulate Arabidopsis ALMT9. *Plant Physiology* 190.

Guo, L., Qiao, X., Haji, D. et al. (2023). Convergent resistance to GABA receptor neurotoxins through plant–insect coevolution. *Nature Ecology & Evolution* 7: https://doi.org/10.1038/s41559-023-02127-4.

He, G., Zhang, H., Liu, S. et al. (2021). Exogenous γ-glutamic acid (GABA) induces proline and glutathione synthesis in alleviating CD-induced photosynthetic inhibition and oxidative damage in tobacco leaves. *Journal of Plant Interactions* 16: https://doi.org/10.1080/17429145.2021.1944676.

Hedrich, R., Salvador-Recatalà, V., and Dreyer, I. (2016). Electrical wiring and Long-distance plant communication. *Trends in Plant Science* 21: 376.

Jáslan, J. and De Angeli, A. (2022). Heterologous expression reveals that GABA does not directly inhibit the vacuolar anion channel AtALMT9. *Plant Physiology* 189: 469.

Ji, J., Yue, J., Xie, T. et al. (2018). Roles of γ-aminobutyric acid on salinity-responsive genes at transcriptomic level in poplar: involving in abscisic acid and ethylene-signalling pathways. *Planta* 248: https://doi.org/10.1007/s00425-018-2915-9.

Kathiresan, A., Miranda, J., Chinnappa, C.C., and Reid, D.M. (1998). γ-Aminobutyric acid promotes stem elongation in *Stellaria longipes*: the role of ethylene. *Plant Growth Regulation* 26: https://doi.org/10.1023/A:1006107815064.

Kinnersley, A.M. and Lin, F. (2000). Receptor modifiers indicate that 4-aminobutyric acid (GABA) is a potential modulator of ion transport in plants. *Plant Growth Regulation* 32: https://doi.org/10.1023/A:1006305120202.

Kinnersley, A.M. and Turano, F.J. (2000). Gamma aminobutyric acid (GABA) and plant responses to stress. *CRC Critical Reviews in Plant Sciences* 19: https://doi.org/10.1080/07352680091139277.

Klejchova, M., Silva-Alvim, F.A.L., Blatt, M.R., and Alvim, J.C. (2021). Membrane voltage as a dynamic platform for spatiotemporal signaling, physiological, and developmental regulation. *Plant Physiology* 185: 1523.

Kovermann, P., Meyer, S., Hörtensteiner, S. et al. (2007). The Arabidopsis vacuolar malate channel is a member of the ALMT family. *Plant Journal* 52: https://doi.org/10.1111/j.1365-313X.2007.03367.x.

Lancien, M. and Roberts, M.R. (2006). Regulation of *Arabidopsis thaliana* 14-3-3 gene expression by γ-aminobutyric acid. *Plant, Cell & Environment* 29: https://doi.org/10.1111/j.1365-3040.2006.01526.x.

Lawson, T. and Blatt, M.R. (2014). Stomatal size, speed, and responsiveness impact on photosynthesis and water use efficiency. *Plant Physiology* 164: 1556–1570. https://doi.org/10.1104/pp.114.237107.

Lei, P., Xu, Z., Liang, J. et al. (2016). Poly(γ-glutamic acid) enhanced tolerance to salt stress by promoting proline accumulation in *Brassica napus* L. *Plant Growth Regulation* 78: https://doi.org/10.1007/s10725-015-0088-0.

Li, Z., Yu, J., Peng, Y., and Huang, B. (2016). Metabolic pathways regulated by γ-aminobutyric acid (GABA) contributing to heat tolerance in creeping bentgrass (*Agrostis stolonifera*). *Scientific Reports* 6: https://doi.org/10.1038/srep30338.

Li, Z., Yu, J., Peng, Y., and Huang, B. (2017). Metabolic pathways regulated by abscisic acid, salicylic acid and γ-aminobutyric acid in association with improved drought tolerance in creeping bentgrass (*Agrostis stolonifera*). *Physiologia Plantarum* 159: https://doi.org/10.1111/ppl.12483.

Liu, C., Zhao, L., and Yu, G. (2011). The dominant glutamic acid metabolic flux to produce gamma-amino butyric acid over proline in *Nicotiana tabacum* leaves under water stress relates to its significant role in antioxidant activity. *Journal of Integrative Plant Biology* 53: 608–618. https://doi.org/10.1111/j.1744-7909.2011.01049.x.

Liu, C., Wang, H., Zhang, X. et al. (2021). Activation of the aba signal pathway mediated by GABA improves the drought resistance of apple seedlings. *International Journal of Molecular Sciences* 22: https://doi.org/10.3390/ijms222312676.

Long, Y., Tyerman, S.D., and Gilliham, M. (2020). Cytosolic GABA inhibits anion transport by wheat ALMT1. *New Phytologist* 225: https://doi.org/10.1111/nph.16238.

Mazzucotelli, E., Tartari, A., Cattivelli, L., and Forlani, G. (2006). Metabolism of γ-aminobutyric acid during cold acclimation and freezing and its relationship to frost tolerance in barley and wheat. *Journal of Experimental Botany* 57: https://doi.org/10.1093/jxb/erl141.

Mekonnen, D.W., Flugge, U.I., and Ludewig, F. (2016). Gamma-aminobutyric acid depletion affects stomata closure and drought tolerance of *Arabidopsis thaliana*. *Plant Science* 245: 25–34. https://doi.org/10.1016/j.plantsci.2016.01.005.

Meyer, S., Mumm, P., Imes, D. et al. (2010). AtALMT12 represents an R-type anion channel required for stomatal movement in Arabidopsis guard cells. *Plant Journal* 63: https://doi.org/10.1111/j.1365-313X.2010.04302.x.

Michaeli, S. and Fromm, H. (2015). Closing the loop on the GABA shunt in plants: are GABA metabolism and signaling entwined? *Frontier in Plant Science* 6: 419.

Michaeli, S., Fait, A., Lagor, K. et al. (2011). A mitochondrial GABA permease connects the GABA shunt and the TCA cycle, and is essential for normal carbon metabolism. *Plant Journal* 67: https://doi.org/10.1111/j.1365-313X.2011.04612.x.

Mirabella, R., Rauwerda, H., Struys, E.A. et al. (2008). The Arabidopsis her1 mutant implicates GABA in E-2-hexenal responsiveness. *Plant Journal* 53: https://doi.org/10.1111/j.1365-313X.2007.03323.x.

Molina-Rueda, J.J., Garrido-Aranda, A., and Gallardo, F. (2015). Glutamate decarboxylase. In: *Amino Acids in Higher Plants* (ed. G.O. Osuji and W.C. Madu), 129–141. Oxford: CAB International.

Nishizawa, A., Yabuta, Y., and Shigeoka, S. (2008). Galactinol and raffinose constitute a novel function to protect plants from oxidative damage. *Plant Physiology* 147: https://doi.org/10.1104/pp.108.122465.

Nugraha, R.Y.B., Jeelani, G., and Nozaki, T. (2022). Physiological roles and metabolism of γ-aminobutyric acid (GABA) in parasitic protozoa. *Trends in Parasitology* 38: 462–477.

Owens, D.F. and Kriegstein, A.R. (2002). Is there more to GABA than synaptic inhibition? *Nature Reviews. Neuroscience* 3: 715.

Palanivelu, R., Brass, L., Edlund, A.F., and Preuss, D. (2003). Pollen tube growth and guidance is regulated by POP2, an Arabidopsis gene that controls GABA levels. *Cell* 114: https://doi.org/10.1016/S0092-8674(03)00479-3.

Pires, M.V., Pereira Júnior, A.A., Medeiros, D.B. et al. (2016). The influence of alternative pathways of respiration that utilize branched-chain amino acids following water shortage in Arabidopsis. *Plant, Cell & Environment* 39: https://doi.org/10.1111/pce.12682.

Raggi, V. (1994). Changes in free amino acids and osmotic adjustment in leaves of water-stressed bean. *Physiologia Plantarum* 91: https://doi.org/10.1111/j.1399-3054.1994.tb02970.x.

Ramesh, S.A., Tyerman, S.D., Xu, B. et al. (2015). GABA signalling modulates plant growth by directly regulating the activity of plant-specific anion transporters. *Nature Communications* 6: 7879. https://doi.org/10.1038/ncomms8879.

Ramesh, S.A., Tyerman, S.D., Gilliham, M., and Xu, B. (2017). γ-Aminobutyric acid (GABA) signalling in plants. *Cellular and Molecular Life Sciences* 74: 1577–1603.

Ramesh, S.A., Kamran, M., Sullivan, W. et al. (2018). Aluminum-activated malate transporters can facilitate GABA transport. *Plant Cell* 30: 1147–1164. https://doi.org/10.1105/tpc.17.00864.

Recht, L., Töpfer, N., Batushansky, A. et al. (2014). Metabolite profiling and integrative modeling reveal metabolic constraints for carbon partitioning under nitrogen starvation in the green algae *Haematococcus pluvialis*. *Journal of Biological Chemistry* 289: https://doi.org/10.1074/jbc.M114.555144.

Renault, H., Roussel, V., El Amrani, A. et al. (2010). The Arabidopsis pop2-1 mutant reveals the involvement of GABA transaminase in salt stress tolerance. *BMC Plant Biology* 10: https://doi.org/10.1186/1471-2229-10-20.

Renault, H., El Amrani, A., Berger, A. et al. (2013). γ-Aminobutyric acid transaminase deficiency impairs central carbon metabolism and leads to cell wall defects during salt stress in Arabidopsis roots. *Plant, Cell & Environment* 36: https://doi.org/10.1111/pce.12033.

Roberts, M.R. (2007). Does GABA act as a signal in plants? Hints from molecular studies. *Plant Signaling & Behavior* 2: https://doi.org/10.4161/psb.2.5.4335.

Salah, A., Zhan, M., Cao, C. et al. (2019). γ-Aminobutyric acid promotes chloroplast ultrastructure, antioxidant capacity, and growth of waterlogged maize seedlings. *Scientific Reports* 9: https://doi.org/10.1038/s41598-018-36334-y.

Schmid, M., Davison, T.S., Henz, S.R. et al. (2005). A gene expression map of *Arabidopsis thaliana* development. *Nature Genetics* 37: https://doi.org/10.1038/ng1543.

Scholz, S.S., Malabarba, J., Reichelt, M. et al. (2017). Evidence for GABA-induced systemic GABA accumulation in Arabidopsis upon wounding. *Frontiers in Plant Science* 8: https://doi.org/10.3389/fpls.2017.00388.

Shelp, B.J., Bown, A.W., and McLean, M.D. (1999). Metabolism and functions of gamma-aminobutyric acid. *Trends in Plant Science* 4: 446.

Shelp, B.J., Bozzo, G.G., Zarei, A. et al. (2012). Strategies and tools for studying the metabolism and function of γ-aminobutyrate in plants. II. Integrated analysis. *Botany* 90: 781.

Smirnoff, N. and Cumbes, Q.J. (1989). Hydroxyl radical scavenging activity of compatible solutes. *Phytochemistry* 28: https://doi.org/10.1016/0031-9422(89)80182-7.

Spiering, M.J. (2018). The discovery of GABA in the brain. *Journal of Biological Chemistry* 293: 19159.

Steward, F. (1949). γ-Aminobutyric acid: a constituent of potato tubers? *Science* 1979 (110): 439–440.

Studart-Guimarães, C., Fait, A., Nunes-Nesi, A. et al. (2007). Reduced expression of succinyl-coenzyme a ligase can be compensated for by up-regulation of the γ-aminobutyrate shunt in illuminated tomato leaves. *Plant Physiology* 145: https://doi.org/10.1104/pp.107.103101.

Suhel, M., Husain, T., Pandey, A. et al. (2023). An appraisal of ancient molecule GABA in abiotic stress tolerance in plants, and its crosstalk with other signaling molecules. *Journal of Plant Growth Regulation* 42: 614.

Thompson, J.F., Stewart, C.R., and Morris, C.J. (1966). Changes in amino acid content of excised leaves during incubation I. The effect of water content of leaves and atmospheric oxygen level. *Plant Physiology* 41: https://doi.org/10.1104/pp.41.10.1578.

Urano, K., Maruyama, K., Ogata, Y. et al. (2009). Characterization of the ABA-regulated global responses to dehydration in Arabidopsis by metabolomics. *Plant Journal* 57: https://doi.org/10.1111/j.1365-313X.2008.03748.x.

Vest, A. (2013). *The Role of JAZs and GABA in Plant Immune Response*. Exeter: University of Exeter.

Woodrow, P., Ciarmiello, L.F., Annunziata, M.G. et al. (2017). Durum wheat seedling responses to simultaneous high light and salinity involve a fine reconfiguration of amino acids and carbohydrate metabolism. *Physiologia Plantarum* 159: https://doi.org/10.1111/ppl.12513.

Xu, B., Long, Y., Feng, X. et al. (2021a). GABA signalling modulates stomatal opening to enhance plant water use efficiency and drought resilience. *Nature Communications* 12: https://doi.org/10.1038/s41467-021-21694-3.

Xu, B., Sai, N., and Gilliham, M. (2021b). The emerging role of GABA as a transport regulator and physiological signal. *Plant Physiology* 187: https://doi.org/10.1093/plphys/kiab347.

Ye, J., Wang, X., Wang, W. et al. (2021). Genome-wide association study reveals the genetic architecture of 27 agronomic traits in tomato. *Plant Physiology* 186: https://doi.org/10.1093/plphys/kiab230.

Yong, B., Xie, H., Li, Z. et al. (2017). Exogenous application of GABA improves PEG-induced drought tolerance positively associated with GABA-shunt, polyamines, and proline metabolism in white clover. *Frontiers in Physiology* 8: https://doi.org/10.3389/fphys.2017.01107.

6

The Role of GABA on Programmed Cell Death and Senescence in Plants

Fazilet Özlem Albayrak[1], *Filiz Vardar*[2], *and Nihal Gören-Sağlam*[3]

[1] *Department of Biology, Faculty of Science, Mersin University, Mersin, 33100, Turkey*
[2] *Department of Biology, Faculty of Science, Marmara University, Istanbul, 34722, Turkey*
[3] *Department of Biology, Faculty of Science, Istanbul University, Istanbul, 34452, Turkey*

Introduction

Gamma-aminobutyric acid (GABA) is a nonproteinaceous amino acid known in various organisms such as bacteria, fungi, animals, and plants in many aspects of the plant life cycle (Michaeli and Fromm 2015; Suhel et al. 2023). The function of GABA as an inhibitory neurotransmitter in mammals is well known. However, it was first reported from potato tubers in 1949 (Li et al. 2021a). In plants, GABA has functions in regulating plant growth and development, facilitating photosynthesis, maintaining C/N and pH balance, cell elongation, fruit ripening, seed germination, sexual reproduction, patterning, and cell identity (Häusler et al. 2014; Li et al. 2021a; Guo et al. 2023). GABA is mentioned as an exciting new player in the mating game, though it has an important mission in regulating pollen tube growth (Yang 2003). GABA is also mentioned as a significant sensor for the energetic state of cells (Michaeli and Fromm 2015). GABA could also transmit calcium signals through the receptors to trigger the downstream signaling cascades (Guo et al. 2023). In animals, the signal role of GABA has been well-known for many years, however, GABA has been mentioned as a signal molecule in plants in recent years (Ramesh et al. 2017). In plants, focus on GABA is increased especially after its intracellular and/or extracellular rapid accumulation under stress conditions (Roberts 2007; Suhel et al. 2023). GABA could regulate the transport of proteins and channels which have important roles under stress conditions. Moreover, GABA could regulate the malate efflux, which is mediated by ALMTs (Aluminum-activated Malate Transporters), and modulate the stomatal opening under drought stress (Kaspal et al. 2021). GABA has also a close relation with senescence and programmed cell death (PCD) in plants.

Death, both at the cellular and the organism level, is a certain fate being in balance with cell division in all prokaryotic and eukaryotic organisms. During growth and development, nonfunctional, overproduced, retired, abnormal, or damaged cells are eliminated by PCD. It has been described as a genetically regulated progression in the course of development,

GABA in Plants: Biosynthesis, Plant Development, and Food Security, First Edition. Edited by Samiksha Singh, Durgesh Kumar Tripathi, and Vijay Pratap Singh.

defense, and homeostasis (Van Hautegem et al. 2015; Huysmans et al. 2017). General hallmarks of PCD are shrinkage, condensation, vacuolization, and fragmentation of cytoplasm and chromatin with a cascade of specific proteases and nucleases (Nath and Lu 2015; Petrov et al. 2015). While most of the detailed research is on animal PCD (or more common name, apoptosis), it has also been described in other kingdoms, principally in plants. Numerous comparative studies declared that plants and animals have both common and diverse characteristics of PCD (Nath and Lu 2015; Petrov et al. 2015). One of the underlying reasons is the existence of cell walls in plants blocking phagocytosis. Considering the cell wall generation, plants have evolved their own pathways in PCD (Danon et al. 2004). According to recent studies, plant genomes do not code certain homologs of animal cell death regulators. Plant PCD plays crucial roles in plant life and occurs in two situations: during growth, development and during stress conditions. PCD has a vital role in the constitution of various types of cells and tissues, vegetative and generative organs such as tracheary element differentiation, aerenchyma formation, leaf senescence, sex determination, pollen–pistil interaction, functional megaspore formation, and anther wall development (Fukuda 2000; Gunawardena et al. 2001; Kacprzyk et al. 2011; Van Hautegem et al. 2015). PCD also takes place in response to biotic–abiotic stresses, such as pathogen attack (more common name hypersensitive response), heavy metal stress, osmotic stress, nanoparticle accumulation, low–high temperature, or irradiation (Wu et al. 2014; Petrov et al. 2015; Yanık and Vardar 2015, 2019).

Senescence is a programmed developmental stage that causes the death of a cell, tissue, organ, or whole plant (Lim and Nam 2007). It is a process in which not only a part of the plant dies but also various macromolecules are recycled into developing parts of the plant (Sarwat et al. 2013; Schippers et al. 2015). Senescence is vital in the development of a plant in terms of transporting and recycling important nutrients such as phosphorus, nitrogen, potassium, and sulfur. These nutrients are transported from senescent leaves to actively growing tissues, thus promoting plant growth and reproduction (Balazadeh et al. 2008; Zhang and Zhou 2013). Some researchers have shown that senescence occurs not only as a result of unfavorable environmental conditions (drought, temperature, deficiency of nitrogen, light/dark conditions, and disease and pathogen attacks) but also genetically in healthy plants grown under optimal growing conditions (He and Gan 2002; Woo et al. 2019). Leaf senescence is an extremely well-organized, complex process that includes more than leaf death and constitutes the last stage of leaf development (Andrade Galan et al. 2023), progressing with age, different internal and environmental adverse conditions (Feller and Fischer 1994). Senescence causes the destruction of macromolecules and the transport of the formed products to developing organs of the plant in a coordinated manner (Andrade Galan et al. 2023). These events result in leaf yellowing and death. Leaf senescence is one of the important stages in plant leaf development and includes the expression of senescence-related genes (SAG = senescence-associated genes). During senescence, leaf cells undergo fairly regular changes in gene expression, metabolism, and structure (Lim et al. 2007).

GABA Pathways

GABA pathways have important roles in maintaining the primary plant metabolism, such as redistributing C and N metabolism and maintaining intracellular pH homeostasis (Guo et al. 2023). GABA is synthesized by evolutionarily conserved enzymes via two pathways

(Michaeli and Fromm 2015; Yuan et al. 2023). GABA is metabolized in plants through the GABA shunt, which is a bypass of the TCA cycle. The other synthesis pathway of GABA is through the polyamine pathway (Li et al. 2021a).

GABA shunt is connected with three enzymatic pathways located in mitochondria and cytosol (Renault et al. 2010). The amination of alfa-ketoglutarate to glutamate is catalyzed by glutamate dehydrogenase (GDH, EC 1.4.1.4), then glutamate decarboxylase (GAD, EC 4.1.1.15), which is affected by a decrease in pH and Ca^{2+}, catalyzes the decarboxylation of glutamate to GABA (Bouche and Fromm 2004; Akçay et al. 2012). GABA is then transported into the mitochondrial matrix and converted to succinic semialdehyde, and after the oxidation to succinate, it enters the TCA cycle (Häusler et al. 2014).

Glutamate acts as a precursor in GABA synthesis; therefore, the synthesis is strongly related to the TCA cycle, and GABA could be assumed as a bridge in the regulating C–N balance (Guo et al. 2023). Glutamate directly links C and N metabolism and also can act as a signaling molecule (Liao et al., 2022). In an alternative way, succinic semialdehyde can be converted to γ-hydroxybutyrate (Shelp et al. 2012). The amino group of GABA is then transferred to pyruvate by the GABA transaminase activity in the mitochondria, providing succinic semialdehyde and alanine. Then, by the activity of succinic semialdehyde dehydrogenase, succinate is maintained. Succinate is further metabolized in the TCA cycle (Häusler et al. 2014). GABA influx can be controlled by transporters ALMTs, AAP3, GAT1, or ProT2, and GABA efflux controlled by ALMTs (Li et al. 2021b). Under various environmental conditions, an important amount of cytosolic GABA could enter mitochondria through GABA permease (Michaeli et al. 2011).

GABA can also be synthesized via polyamine (PA) and linked to polyamine catabolism. The synthesis can be via spermidine and putrescine degradation (Michaeli and Fromm 2015) (Figure 6.1). As it is known that polyamines have essential roles in plant immunity by serving as important modulators of cell redox status (Gerlin et al. 2021), GABA has multiple effects on plant metabolism. In the alternative GABA pathway, plastids and peroxisomes are mentioned as the main organelles (Dabravolski and Isayenkov 2023). Since the nonenzymatic reaction of proline has been discovered in recent years, other methods are indicated for GABA synthesis (Yuan et al. 2023). Moreover, the significance of proline in GABA formation in plants has not been directly proven. GABA accumulation by activation of the GABA shunt is mentioned as crucial for stress responses (Dabravolski and Isayenkov 2023).

The Roles of GABA under Stress Conditions

GABA has important effects on multiple physiological processes; among them, research related to stress responses has been promising. Researchers have reported the protective role of GABA in stress tolerance to abiotic and biotic stress factors. GABA is accumulated rapidly in large quantities under the stress conditions (Guo et al. 2023; Hayat et al. 2023; Yuan et al. 2023). In recent years, omics and reverse-genetic studies have shown a correlation between GABA content and stress tolerance in plants (Dabravolski and Isayenkov 2023).

The participation of GABA in C and N metabolism could have an effect on the resistance to various stresses. GABA metabolic pathways also have an effect on reducing ROS accumulation and improving tolerance to oxidative stress. GABA accumulation could reduce ROS content in the cells and could improve tolerance to oxidative stress in plants. GABA

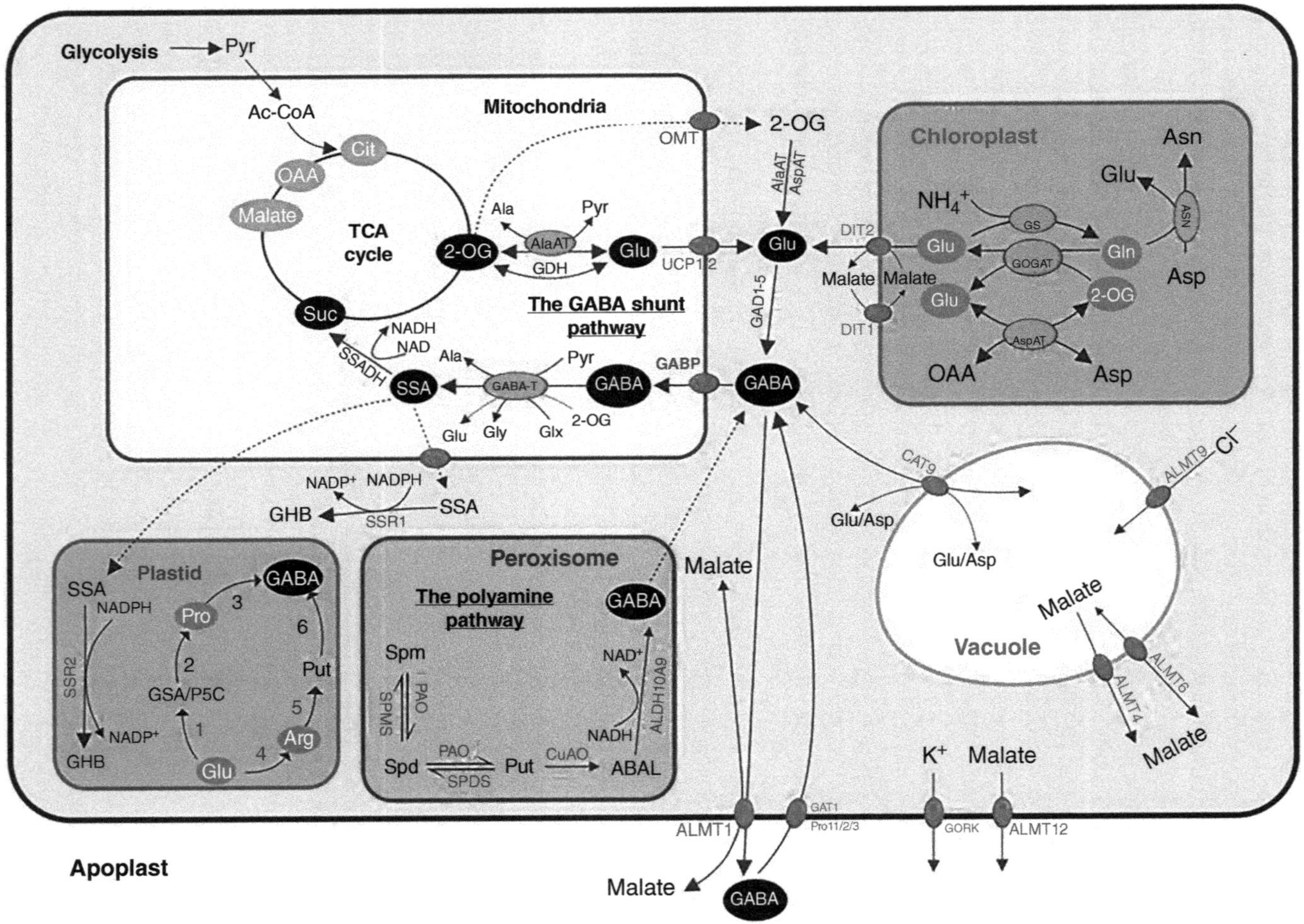

Figure 6.1 GABA pathway model in *Arabidopsis* (Meng 2023).

could act as a buffer and activate H^+-ATPase. Moreover, calcium signals could participate in GABA accumulation under various stress conditions (Guo et al. 2023). Also, in stress tolerance, the probability of other anion channels with important roles that could be gated by GABA has been reported (Kaspal et al. 2021).

Several studies have shown the relationship between plant hormones, polyamines, and GABA (Podlešáková et al. 2019). Under several abiotic stress conditions, change in the phytohormone levels could result in changes in the GABA levels, leading to an interaction between pathways and multiple players, which could mediate stress tolerance (Kaspal et al. 2021). Moreover, GABA could modulate the phytohormone cross-talk (Dabravolski and Isayenkov 2023).

Under salt stress, many interconnected metabolic processes are involved in this mechanism. GABA could help to decrease salt stress effects by affecting phytohormone biosynthesis and signaling. GABA could increase tolerance to salt stress by functioning in harmony with the phytohormones. According to previous studies, GABA could have a distinctive and considerable function in regulating stress signaling under salt stress (Dabravolski and Isayenkov 2023).

It is mentioned that exogenous GABA application could enhance tolerance under multiple stresses by reducing ROS content and enhancing membrane stability (Dabravolski and Isayenkov 2023). Moreover, exogenous application of GABA could lead to enhanced tolerance to various stress factors such as drought, salt, and heat stresses by lowering the oxidative damage, and improving the antioxidant capacity (Hayat et al. 2023). Exogenous GABA could be suggested for improving the tolerance under field conditions and could combat changeable environmental situations (Guo et al. 2023). It was also mentioned that GABA treatment could exhibit higher SOS1 expression, which is important in salt tolerance mechanisms (Kaspal et al. 2021). Foliar GABA application is suggested as a promising strategy for mitigating salt stress by regulating plant growth, chlorophyll content, antioxidant enzyme activities, and also related gene expressions (Aljuaid and Ashour 2022).

Therefore, the properties of GABA could lead to a new era in crop improvement for sustainable agriculture. The molecular studies should be done comprehensively to understand the protective role of GABA. Moreover, GABA could be suggested as a valuable tool for enhancing crop quality and yield (Hayat et al. 2023). The journey in the GABA signaling mechanism could lead to new perspectives in mediating plant stress response.

GABA as a Signal Molecule

It is known that amino acids function as protein constituents and precursors for anabolism. Moreover, in plants, they could act as signaling molecules (Häusler et al. 2014). The signaling role of GABA is well-known in animals for a long time. However, GABA in plants is argued to have a function as a signaling molecule or not. Previously, it was hypothesized that GABA and calcium could interact to signal various environmental stresses. Moreover, it is thought that it could act as a signaling molecule for N status in the germinating seedlings (Roberts 2007). After the discovery of the putative GABA binding site and the GABA regulation of the anion channels, it is mentioned as a signaling molecule (Kaspal et al. 2021). The contribution of signaling networks to plant growth and survival has gained attention (Bor and Turkan 2019).

GABA is known to have a role in the participation of the tolerance mechanisms against various abiotic stress conditions as a signaling molecule (Yuan et al. 2023). Between GABA, ROS and reactive nitrogen species, an indirect interaction has been mentioned. As known in plants, reactive nitrogen species and ROS have important roles in the signaling process, and in this process, GABA could have a remarkable effect (Bor and Turkan 2019). GABA also has an effect on the downstream signaling cascades, which could affect the transport of Ca^{2+} via ALMTs (Guo et al. 2023). Moreover, GABA could have important interactions with other signaling molecules (Kaspal et al. 2021).

With the help of GABA-deficient plants, the questions in distinguishing between signaling and metabolic functions of GABA can be clearly identified (Häusler et al. 2014). GABA signaling mechanism can also lead to a new era. Therefore, further studies should be done to identify GABA-mediated improvements in plant metabolism.

GABA-Mediated Avoidance from PCD

Free radicals are oxygen-containing molecules with unpaired electrons. They are produced during photosynthesis in chloroplast, respiration in mitochondria, and photorespiration in peroxisomes continuously. Moreover, they are also generated in the course of other metabolic processes in the cell (Caverzan et al. 2016). Under constant circumstances, free radicals are removed by the antioxidant defense system. Oxidative stress appears during the imbalance between the formation and elimination of free radicals, which can potentially damage the cells.

Triplet oxygen is a free radical containing two impaired electrons having high reaction potential with molecules in the binary state to form ROS (Gill and Tuteja, 2010). The common free radicals in plant cells are superoxide ($O_2^{-\cdot}$), hydroxyl ($OH^{\cdot}$), perhydroxyl ($HO_2^{\cdot}$), also nonradical molecules such as hydrogen peroxide (H_2O_2) and singlet oxygen ($^1O_2$) (Sharma et al. 2012). Although ROS are highly reactive and toxic, they can play a role as endogenous signaling factors at lower concentrations adjusting adaptation of stress status or triggering PCD (Nath and Lu 2015). Overproduction of ROS and, as a result, oxidative stress has adverse effects on several cellular systems such as oxidized proteins, lipid peroxidation, damaged nucleic acids, and inhibited enzyme activity leading to PCD (Mishra et al. 2011; Sharma et al. 2012; Reape et al. 2015; Yanık et al. 2020).

In common, the activation of the PCD pathway is initiated by the damage of the outer membrane of mitochondria by ROS under stress conditions. As a result, cytochrome c (cyt c), which is discharged from the inner membrane of mitochondria to the cytoplasm is a key event in PCD. The efflux of cyt c to cytoplasm induces interruption of electron transport in mitochondria provoking the formation of toxic ROS levels. Besides, cyt c can also give a reaction with free radicals forming extremely reactive and mutagenic radicals accelerating ROS accumulation. Overproduction of ROS results in opening the permeability transition pore (PTP) in mitochondria. By the PTPs, the mitochondrial signals are moved to the nucleus causing changes in gene expression (Reape et al. 2015). It can be concluded that ROS acting as signaling molecules, causing PTP opening, which leads to the formation of more ROS. They also give feedback that amplifies PCD-activating signals (Jabs 1999; Reape et al. 2015). During PCD, ROS induces Ca^{2+} ion efflux from the endoplasmic reticulum.

Several researchers agreed that enhanced ROS generation and cyt c release trigger caspase-like activities (such as metacaspases, vacuolar processing enzymes [VPE], phytaspases, caspases, and 26S proteasome β subunit 1) in plants (Cai and Gallois 2015) which are important proteolytic enzymes in PCD.

In normal metabolic conditions, enzymatic and nonenzymatic antioxidants balance the ROS levels. In common, the antioxidant defense system contains ascorbate peroxidase (APX), superoxide dismutase (SOD), glutathione reductase (GR), glutathione *S*-transferase (GST) catalase (CAT), monodehydroascorbate reductase (MDHAR), glutathione peroxidase (GPX), dehydroascorbate reductase (DHAR), and guaiacol peroxidase (GPX). Besides, glutathione (GSH), ascorbic acid (ASH), oxidized glutathione (GSSG), alkaloids, carotenoids, phenolic compounds, and α-tocopherols which are nonenzymatic antioxidants also scavenge the adverse effects of ROS (Gill and Tuteja 2010). It is obvious that antioxidant strategies are crucial to preserve the redox homeostasis in plant cells.

GABA is one of the endogenous signal molecules that can rapidly accumulate to maintain ion homeostasis in terms of stress tolerance (Su et al. 2019). It can be stated that GABA has a crucial role in regulating plant growth, development, and plant senescence (Jalil et al. 2017, 2019a; Sita and Kumar 2020). GABA considerably enhanced plant morphological and physiological characteristics under environmental stress conditions with regard to photosynthetic capacity, osmoprotectants' accumulation such as proline (Li et al., 2016b; Ma et al. 2019), polyamine metabolism (Seifikalhor et al. 2020), and regulating the expression of genes involved in signal transduction and stress-sensitive processes (Podlešáková et al. 2019). It fundamentally overcomes stress-induced ROS accumulation by activating antioxidant defense mechanism (Chen et al. 2018; Kumar et al. 2019). For this reason, plant responses mediated by GABA may help formulate strategies that will enhance growth and development in stressful environments and finally reduce plant productivity. Several researches support GABA-mediated alleviation during abiotic stress conditions in plants.

Song et al. (2010) treated barley seedlings with combined stress of H^+ and Al^{3+} for 24 hours with and without GABA. According to their results, Al toxicity increased the production and accumulation of ROS (Kochian et al. 2004), which triggers oxidative damage in cells leading to PCD (Sharma and Dubey 2007; Gupta et al. 2013). The content of carbonylated proteins in the roots of GABA-treated seedlings was significantly decreased compared to under stresses without GABA treatment. Moreover, the higher levels of antioxidant enzymes (SOD, CAT, GPX) were monitored in GABA-treated barley roots compared to roots under stress without GABA. The lower contents of MDA, H_2O_2, and O_2^- were observed in GABA-treated barley seedlings. These results suggest that GABA is able to mitigate oxidative stress induced by H^+ and Al^{3+} in barley. Several studies revealed that Al-induced PCD is related to elevated ROS levels leading to DNA fragmentation, caspase-like activities, and cyt c release, which are the main characteristics of PCD (Vardar et al. 2015, 2016; Aytürk and Vardar 2015; Gölge and Vardar 2020).

Similarly, Zhao et al. (2020) reported that under Cd stress conditions, GABA application increased protein and GSH content, promoted lipid synthesis, and upregulated the α-amylase activity, however, it decreased contents of starch, ROS, and Cd^{2+} accumulation in microalgae (*Monoraphidium* sp. QLY-1). It has been also known that dose-dependent Cd^{2+} in different organisms may cause PCD depending on ROS accumulation (López et al. 2003; Templeton and Liu 2010).

Kalhor et al. (2018) studied the GABA effects on germination, photosynthetic performance, and oxidative damage in salt-treated lettuce. Based on their results adverse effects of salinity on germination, plant growth, electrolyte leakage, and photosynthetic capacity were ameliorated by GABA application. Although proline levels increased by NaCl applications, combined treatment of salt with GABA caused a significant decrease. CAT, APX, and SOD activities were also increased by GABA exposure resulting in regulated H_2O_2 levels. The researchers concluded that GABA has a multifaceted role in minimizing the detrimental effects of salinity on lettuce through the improvement of photosynthetic efficiency and regulation of oxidative stress. Besides, Kaur and Zhawar (2021) reported the ameliorative effects of GABA on salinity–alkalinity stress (SAS) in two rice cultivars CSR 43 (tolerant) and Pusa 44 (susceptible). According to their results, increased soluble GPX, lignin, flavonoids in shoots were observed in both susceptible and tolerant rice under GABA. The overall study indicated GABA improves secondary antioxidants and cellular reserves like starch and sucrose. As in other abiotic stress conditions, Fedoreyeva et al. (2022) reported the salinity and PCD relation in wheat varieties. According to their results, Na accumulation was accompanied by an increase in ROS triggering autophagy production and PCD in the absence of GABA application. PCD markers revealed DNA fragmentation, cells with a surface location of phosphatidylserine, and the release of cyt c into the cytoplasm.

In a similar way, Hasan et al. (2021) indicated drought stress, which is one of the most serious problems that plants face around the world and the effects of GABA application in the alleviation of drought. Most of the researchers report that overproduction of ROS causes damage to plant structure in drought stress (Hasan et al. 2020). The treatment of exogenous GABA enhanced the activities of MDHAR, CAT, GPX, APX, and GR enzymes and decreased H_2O_2 production, resulting in a more effective tolerance to drought in plants (Razik et al. 2020). These data indicate that GABA increases tolerance to oxidative stress caused by abiotic stress factors such as drought by activating glyoxalase and antioxidant pathways. Hameed et al. (2013) also analyzed two wheat genotypes with varying degrees of drought tolerance in relation to PCD and related biochemical changes under drought stress. According to their results, drought-induced PCD resulted in evident by DNA fragmentation in sensitive genotypes. However, drought-tolerant genotypes showed higher GPX, SOD, APX, and CAT activities under drought stress to prevent the plant from going to PCD.

The Role of GABA on Leaf Senescence

The breakdown of spermine (Spm) and spermidine (Spd) strongly feeds the GABA pool. Leaf senescence is a tightly regulated process of removing and recycling cellular components that allows plants to optimize resource use and adapt to changing environmental conditions. GABA shunt is seen as an important metabolic pathway in the senescence process (Ansari et al. 2014). Therefore, there is a strong relationship between GABA and senescence. Research on the GABA senescence relationship has increased in recent years. Studies have shown that the application of Spm and Spd on wheat leaves under light increases the GABA content (Gören-Sağlam and Çekiç 2019). Recent research has shown that GABA, an abundant and conserved molecule in plants, is a new signaling molecule in leaf senescence. GABA is known to be involved in many physiological processes and stress responses

(Li et al. 2021). The protective effect of GABA against stress conditions in plants also brings to mind that it may have an effect on stress-induced senescence. GABA treatment has been shown to be an efficient approach in improving storage performance and maintaining postharvest quality in many plants due to the increase of antioxidant enzymes (Ramos-Ruiz et al. 2019). Under drought stress conditions, GABA accumulation prevents water loss by regulating stomatal opening and closing (Mekonnen et al. 2016).

GABA transaminase (GABA-T) has a significant role in the regulation of plant development. GABA-T catalyzes the reduction of GABA to SSA in the GABA shunt. Many senescence-associated genes (SAGs) have been identified in various plants. Among SAGs, GABA-T has been identified as a leaf senescence gene (Ansari et al. 2005). GABA-T expression has a critical function in nitrogen metabolism during leaf senescence (Jalil et al. 2019b). Loss of GABA-T in *Arabidopsis thaliana* causes premature leaf senescence under stress conditions and GABA-T has an important role in leaf development and senescence of *A. thaliana* (Jalil et al. 2017, 2019b) (Figure 6.2).

With the occurrence of stress conditions, GABA accumulates rapidly in plants, and these conditions can lead to leaf senescence (Bouche and Fromm, 2004; Ansari et al., 2005). Considering this situation, it is mentioned that the GABA pathway potentially regulates the aging metabolism (Al-Quraan and Al-Share 2016, Jalil et al. 2017, 2019b).

In a study conducted with creeping bent grass, it has been shown that the exogenous application of GABA suppresses the heat-induced leaf senescence and increases the tolerance to heat. Results showed that heat-induced chlorophyll degradation may be due to

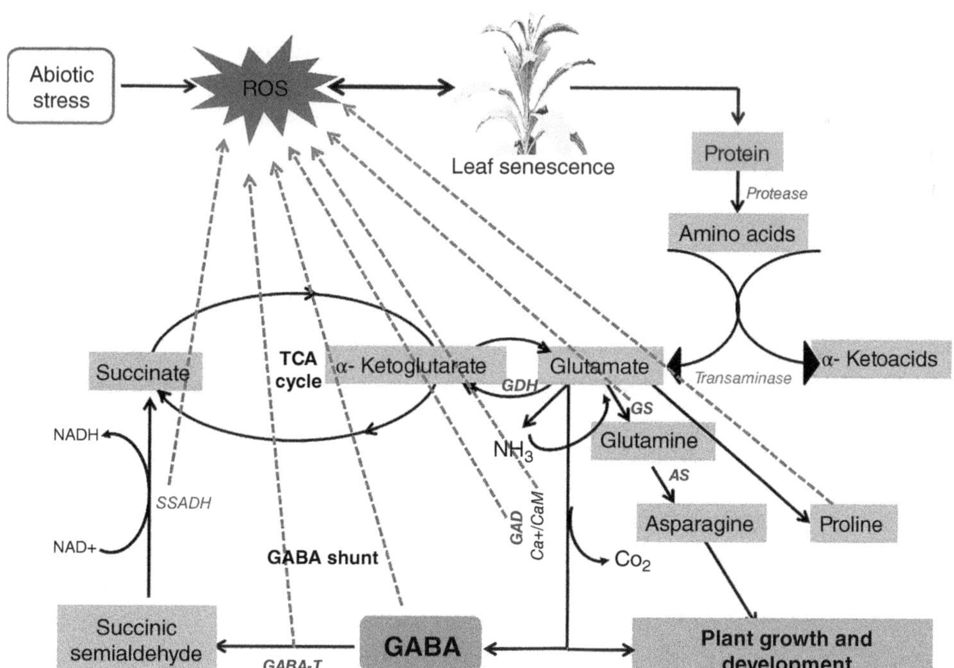

Figure 6.2 Relationship between GABA shunt and ROS mechanism during leaf senescence. *Source:* Khan et al. (2021)/with permission of Elsevier.

chlorophyllase regulation, controlling heat-induced leaf senescence (Rossi et al. 2020). Plants treated with GABA under heat stress conditions showed that GABA can increase heat tolerance by regulating many plant processes (Li et al. 2016a).

Pavlíková et al. (2014) showed that SAG tobacco plants had an improved Zn tolerance compared with WT plants. The amino acid analysis concluded that enhanced Zn tolerance in SAG plants was related to GABA accumulation, proline, and methionine (Pavlíková et al. 2014). It has been reported that SAG tobacco plants showed increased cytokinin content and accumulation of GABA, methionine, and proline when exposed to excess zinc in the soil (Pavlíková et al. 2014; Kaspal et al. 2021).

GABA application increased anti-oxidative enzymes in the anthurium-cut flowers and peaches to protect them from postharvest chilling damage (Yang et al. 2011; Aghdam et al. 2016). GABA has also been shown to increase antioxidant enzyme activity and thus enhance disease resistance in citrus fruits (Song et al. 2010; Sheng et al. 2017). GABA treatment maintains redox balance and supports the strengthening of the cell wall to delay senescence by increasing the activity of enzymes in ROS metabolism and the phenylpropanoid pathway, antioxidants, and antifungal compounds in blueberry fruits (Ge et al. 2018). It has been shown that GABA accumulation increased significantly in control fruit during storage (Li et al. 2019).

Applications of exogenous GABA and Spd have been shown to significantly extend vase life and prevent protein degradation and chilling injury, and enhancing the detoxification and scavenging capacity of H_2O_2 and ROS leads to alleviation of the negative results of senescence and chilling injury. In general, GABA and Spd play important roles in preventing chilling injury and delaying senescence due to their antiaging properties and ability to increase proline content, increase cellular antioxidant capacity, and scavenge H_2O_2 and ROS (Mohammadi et al. 2021).

It has also been reported that GAD activity increases in senescent plants when protein synthesis decreases, glutamine synthesis is inhibited, and protein degradation increases (Ansari and Chen 2009). Which GAD gene will be expressed may vary depending on plant development processes and growth conditions. It has been reported that ubiquitous GAD2 expression increases in Arabidopsis during senescence (Podlešáková et al. 2019). MeJA could significantly enhance the expression levels of *FaGAD, FaGABA-T, and FaSSADH* genes, which are involved in the GABA shunt pathway. It has also been mentioned that enhanced GABA shunt pathway activity and GABA levels (Vaezi et al. 2022).

Results of a study aiming to investigate the effect of foliar application of salicylic acid (SA) and GABA on postharvest life and some qualitative characteristics of cut tuberose flowers, the vase life of GABA and SA, water uptake, fresh weight, ion leakage, total dissolved solids, chlorophyll showed that it positively affected the protein and antioxidant enzyme activity. Then, senescence could be delayed (Babarabie et al. 2019).

It is known that nitrogen is the most important nutrient to be recycled during the senescence process. During leaf senescence GABA shunt plays an integral role in carbon/nitrogen metabolism via the TCA cycle, and it is also important for plants (Feller and Fischer, 1994). N, found in the structure of macromolecules, is converted into glutamine and asparagine amino acids. These amino acids are also converted into glutamate and α-keto acids with the contribution of transaminases (Ansari et al. 2014). Metabolite recycling pathways (GABA-T and TCA cycle) significantly affect the balance of C–N in the seed (Fait et al. 2011), resulting in changes in storage reserve accumulation and fatty acid metabolism. GABA shunt is a critical signaling and metabolic pathway in the senescence process. It is also of critical importance in agricultural production. Therefore, it is crucial to investigate mutants lacking GABA shunt enzymes to know the exact mechanism of nutrient remobilization in leaf senescence (Ansari et al. 2014). Proper functioning of the GABA shunt appears to be necessary to restrict ROS accumulation (Ansari et al. 2021).

The GABA shunt is a metabolic pathway involving nitrogen assimilation of amino acids during the process of plant growth and leaf senescence (Ansari et al. 2005, 2014; Shi et al. 2010). Degradation of GABA appears as another alternative pathway that supports the preservation of mitochondrial respiration during senescence. Under stress conditions and in senescent leaves, GABA accumulates through glutamate metabolism (Kamranfar et al. 2018).

GABA may increase with senescence. The possible reason for this increase is that it is released due to protein degradation by proteases. Therefore, it can be confirmed that it is one of the best senescence markers in many monocot and dicot model plant studies (Masclaux et al. 2000; Diaz et al. 2005; Watanabe et al. 2013; Avila-Ospina et al. 2017). On the other hand, researchers have suggested that GABA has a strong connection to senescence in plants. To unravel the connections between GABA and senescence, it appears that comprehensive studies on GABA kinetics, dose responses, and GABA-induced changes in SAG expressions in plants are needed. Using strategies that alter GABA-regulated dynamics in plant senescence could be very useful in altering plant growth and productivity to achieve agricultural sustainability (Khan et al. 2021). Moreover, the signaling role of GABA should be associated with the senescence process, and it could lead to novel perspectives for engineering plants under various stress conditions.

Conclusions

In conclusion, GABA plays a critical role in the control of environmental stress tolerance in plants. GABA can boost the antioxidative enzyme activities and the defense mechanism. In this way, ROS-mediated damage of the outer mitochondrial membrane is prevented, which triggers PCD. Also, obtained data suggest that genetic traits such as better capacity to scavenge ROS and higher endogenous GABA content may play positive roles in mitigating heat-induced senescence and oxidative damage. Considering GABA's roles in regulating cytosolic pH, abiotic stress conditions, and primary C and N metabolism, which are the main processes associated with senescence, it can be concluded that GABA has a strong link to plant senescence. The signaling role of GABA could lead to a new era in regulating the senescence process. Using strategies that alter GABA-regulated dynamics in plant senescence and PCD could be very useful in altering plant growth and productivity to ensure agricultural sustainability.

References

Aghdam, M.S., Naderi, R., Jannatizadeh, A. et al. (2016). Enhancement of postharvest chilling tolerance of anthurium cut flowers by γ-aminobutyric acid (GABA) treatments. *Scientia Horticulturae* 198: 52–60.

Akçay, N., Bor, M., Karabudak, T. et al. (2012). Contribution of gamma aminobutyric acid (GABA) to salt stress responses of *Nicotiana sylvestris* CMSII mutant and wild type plants. *Journal of Plant Physiology* 169 (5): 452–458.

Aljuaid, B.S. and Ashour, H. (2022). Exogenous γ-aminobutyric acid (GABA) application mitigates salinity stress in maize plants. *Life* 12 (11): 1860.

Al-Quraan, N.A. and Al-Share, A.T. (2016). Characterization of the γ-aminobutyric acid shunt pathway and oxidative damage in Arabidopsis thaliana pop 2 mutants under various abiotic stresses. *Biologia Plantarum* 60: 132–138.

Andrade Galan, A.G., Doll, J., Saile, S.C. et al. (2023). The Non-JAZ TIFY protein TIFY8 of Arabidopsis thaliana interacts with the HD-ZIP III transcription factor REVOLUTA and regulates leaf senescence. *International Journal of Molecular Sciences* 24 (4): 3079.

Ansari, M.I. and Chen, S.C.G. (2009). Biochemical characterization of gamma aminobutyric acid (GABA): pyruvate transaminase during rice leaf senescence. *International Journal of Integrative Biology* 6: 1–27.

Ansari, M.I., Lee, R.H., and Chen, S.C.G. (2005). A novel senescence-associated gene encoding γ-aminobutyric acid (GABA):pyruvate transaminase is upregulated during rice leaf senescence. *Physiologia Plantarum* 123 (1): 1–8.

Ansari, M.I., Hasan, S., and Jalil, S.U. (2014). Leaf senescence and GABA shunt. *Bioinformation* 10 (12): 730–732.

Ansari, M.I., Jalil, S.U., Ansari, S.A., and Hasanuzzaman, M. (2021). GABA shunt: a key-player in mitigation of ROS during stress. *Plant Growth Regulation* 94: 131–149.

Avila-Ospina, L., Clément, G., and Masclaux-Daubresse, C. (2017). Metabolite profiling for leaf senescence in barley reveals decreases in amino acids and glycolysis intermediates. *Agronomy* 7: 15.

Aytürk, Ö. and Vardar, F. (2015). Aluminum-induced caspase-like activities in some Gramineae species. *Advances in Food Sciences* 37: 71–75.

Babarabie, M., Zarei, H., and Eskandari, A. (2019). The impact of pre-harvest treatment with gamma-aminobutyric acid (GABA) and salicylic acid on vase life and post-harvest traits of tuberose cut flowers. *Acta Scientiarum Polonorum Hortorum Cultus* 18 (4): 83.

Balazadeh, S., Parlitz, S., Mueller-Roeber, B., and Meyer, R.C. (2008). Natural developmental variations in leaf and plant senescence in Arabidopsis thaliana. *Plant Biology* 10: 136–147.

Bor, M. and Turkan, I. (2019). Is there a room for GABA in ROS and RNS signalling? *Environmental and Experimental Botany* 161: 67–73.

Bouche, N. and Fromm, H. (2004). GABA in plants; just a metabolite? *Trends in Plant Science* 9: 110–115.

Cai, Y.M. and Gallois, P. (2015). Programmed cell death regulation by plant proteases with caspase-like activity. In: *Plant Programmed Cell Death* (ed. A.N. Gunawardena and P.F. McCabe), 191–202. Cham, Switzerland: Springer.

Caverzan, A., Casassola, A., and Brammer, S.P. (2016). Reactive oxygen species and antioxidant enzymes involved in plant tolerance to stress. In: *Recent Advances and Future Perspectives* (ed. A. Shanker), 463–480. Rijeka, Croatia: InTech.

Chen, H., Liu, T., Xiang, L. et al. (2018). GABA enhances muskmelon chloroplast antioxidants to defense salinity- alkalinity stress. *Russian Journal of Plant Physiology* 654: 674–679.

Dabravolski, S.A. and Isayenkov, S.V. (2023). The role of the γ-aminobutyric acid (GABA) in plant salt stress tolerance. *Horticulturae* 9: 230.

Danon, A., Rotari, V.I., Gordon, A. et al. (2004). Ultraviolet-C exposure induces programmed cell death in *Arabidopsis*, which is mediated by caspase-like activities and which can be suppressed by caspase inhibitors, p35 and defender against apoptotic death. *The Journal of Biological Chemistry* 279: 779–787.

Diaz, C., Purdy, S., Christ, A. et al. (2005). Characterization of markers to determine the extent and variability of leaf senescence in *Arabidopsis*. A metabolic profiling approach. *Plant Physiology* 138: 898–908.

Fait, A., Nesi, A.N., Angelovici, R. et al. (2011). Targeted enhancement of glutamate-to-γ-aminobutyrate conversion in Arabidopsis seeds affects carbon-nitrogen balance and storage reserves in a development-dependent manner. *Plant Physiology* 157 (3): 1026–1042.

Fedoreyeva, L.I., Lazareva, E.M., Shelepova, O.V. et al. (2022). Salt-induced autophagy and programmed cell death in wheat. *Agronomy* 12: 1909.

Feller, U. and Fischer, A. (1994). Nitrogen metabolism in senescing leaves. *Critical Reviews in Plant Sciences* 13 (3): 241–273.

Fukuda, H. (2000). Programmed cell death of tracheary elements as a paradigm in plants. *Plant Molecular Biology* 44: 245–253.

Ge, Y., Duan, B., Li, C. et al. (2018). γ-Aminobutyric acid delays senescence of blueberry fruit by regulation of reactive oxygen species metabolism and phenylpropanoid pathway. *Scientia Horticulturae* 240: 303–309.

Gerlin, L., Baroukh, C., and Genin, S. (2021). Polyamines: double agents in disease and plant immunity. *Trends in Plant Science* 26 (10): 1061–1071.

Gill, S.S. and Tuteja, N. (2010). Reactive oxygen species and antioxidant machinery in abiotic stress tolerance in crop plants. *Plant Physiology and Biochemistry* 48: 909–930.

Gölge, B.H. and Vardar, F. (2020). Temporal analysis of Al-induced programmed cell death in barley (*Hordeum vulgare* L.) roots. *Caryologia* 73 (1): 45–55.

Gören-Sağlam N., Çekiç F.O. Roles of some plant growth regulators on natural and dark-induced senescence in wheat leaf segments. *Notulae Botanicae Horti Agrobotanici Cluj-Napoca*, 2019, 47(4):1230-1237. DOI:https://doi.org/10.15835/nbha47411614

Gunawardena, A.H.L.A.N., Pearce, D.M.E., Jackson, M.B. et al. (2001). Rapid changes in cell wall pectic polysaccharides are closely associated with early stages of aerenchyma formation, a spatially localized form of programmed cell death in roots of maize (Zea mays L.) promoted by ethylene. *Plant, Cell & Environment* 24 (12): 1369–1375.

Guo, Z., Gong, J., Luo, S. et al. (2023). Role of gamma-aminobutyric acid in plant defense response. *Metabolites* 13 (6): 741.

Gupta, N., Gaurav, S.S., and Kumar, A. (2013). Molecular basis of aluminium toxicity in plants: a review. *American Journal of Plant Sciences* 4: 21–37.

Hameed, A., Goher, M., and Iqbal, N. (2013). Drought induced programmed cell death and associated changes in antioxidants, proteases, and lipid peroxidation in wheat leaves. *Biologia Plantarum* 57: 370–374.

Hasan, M.M., Alharby, H.F., Uddin, M.N. et al. (2020). Magnetized water confers drought stress tolerance in *Moringa* biotype via modulation of growth, gas exchange, lipid peroxidation and antioxidant activity. *Polish Journal of Environmental Studies* 29: 1625–1636.

Hasan, M.M., Alabdallah, N.M., Alharbi, B.M. et al. (2021). GABA: a key player in drought stress resistance in plants. *International Journal of Molecular Sciences* 22: 10136.

Häusler, R.E., Ludewig, F., and Krueger, S. (2014). Amino acids – a life between metabolism and signaling. *Plant Science* 229: 225–237.

Hayat, F., Khan, U., Li, J. et al. (2023). γ-Aminobutyric acid (GABA): a key player in alleviating abiotic stress resistance in horticultural crops: current insights and future directions. *Horticulturae* 9: 647.

He, Y. and Gan, S. (2002). A gene encoding an acyl hydrolase is involved in leaf senescence in Arabidopsis. *The Plant Cell* 14 (4): 805–815.

Huysmans, M., Lema, S.A., Coll, N.S., and Nowack, M.K. (2017). Dying two deaths – programmed cell death regulation in development and disease. *Current Opinion in Plant Biology* 35: 37–44.

Jabs, T. (1999). Reactive oxygen intermediates as mediators of programmed cell death in plants and animals. *Biochemical Pharmacology* 57: 231–245.

Jalil, S.U., Ahmad, I., and Ansari, M.I. (2017). Functional loss of GABA transaminase (GABA-T) expressed early leaf senescence under various stress conditions in *Arabidopsis thaliana*. *Current Plant Biology* 9–10: 11–22.

Jalil, S.U., Khan, M.A.A., and Ansari, M.I. (2019a). Tremendous role of GABA shunt metabolic pathway in plant system. *Journal of Biological and Chemical Research* 36: 188–196.

Jalil, S.U., Khan, M.I.R., and Ansari, M.I. (2019b). Role of GABA transaminase in the regulation of development and senescence in *Arabidopsis thaliana*. *Current Plant Biology* 19: 100119.

Kacprzyk, J., Daly, C.T., and McCabe, P.F. (2011). The botanical dance of death: programmed cell death in plants. In: *Advances in Botanical Research* (ed. J.C. Kader and M. Delseny), 169–261. Burlington, NJ: Elsevier.

Kalhor, M.S., Aliniaeifard, S., Seif, M. et al. (2018). Enhanced salt tolerance and photosynthetic performance: implication of γ-aminobutyric acid application in salt-exposed lettuce (*Lactuca sativa* L.) plants. *Plant Physiology and Biochemistry* 130: 157–172.

Kamranfar, I., Xue, G.P., Tohge, T. et al. (2018). Transcription factor RD 26 is a key regulator of metabolic reprogramming during dark-induced senescence. *New Phytologist* 218 (4): 1543–1557.

Kaspal, M., Kanapaddalagamage, M.H., and Ramesh, S.A. (2021). Emerging roles of γ aminobutyric acid (GABA) gated channels in plant stress tolerance. *Plants* 10 (10): 2178.

Kaur, R. and Zhawar, V.K. (2021). Regulation of secondary antioxidants and carbohydrates by gamma-aminobutyric acid under salinity–alkalinity stress in rice (*Oryza sativa* L.). *Biologia Futura* 72 (2): 229–239.

Khan, M.I.R., Jalil, S.U., Chopra, P. et al. (2021). Role of GABA in plant growth, development and senescence. *Plant Gene* 26: 100283.

Kochian, L.V., Hoekenga, O.A., and Piñeros, M.A. (2004). How do crop plants tolerate acid soils? Mechanisms of aluminum tolerance and phosphorous efficiency. *Annual Review of Plant Biology* 55: 459–493.

Kumar, N., Gautam, A., Dubey, A.K. et al. (2019). GABA mediated reduction of arsenite toxicity in rice seedling through modulation of fatty acids, stress responsive amino acids and polyamines biosynthesis. *Ecotoxicology and Environmental Safety* 173: 15–27.

Li, Z., Yu, J., Peng, Y., and Huang, B. (2016a). Metabolic pathways regulated by γ-aminobutyric acid (GABA) contributing to heat tolerance in creeping bentgrass (*Agrostissto lonifera*). *Scientific Reports* 6 (1): 30338.

Li, M., Guo, S., Yang, X. et al. (2016b). Exogenous gamma aminobutyric acid increases salt tolerance of wheat by improving photosynthesis and enhancing activities of antioxidant enzymes. *Biologia Plantarum* 60: 123–131.

Li, T., Shi, D., Wu, Q. et al. (2019). Sodium *para*-aminosalicylate delays pericarp browning of litchi fruit by inhibiting ROS-mediated senescence during postharvest storage. *Food Chemistry* 278: 552–559.

Li, L., Dou, N., Zhang, H., and Wu, C. (2021a). The versatile GABA in plants. *Plant Signaling & Behavior* 16 (3): e1862565.

Li, Z., Tang, M., Hassan, M.J. et al. (2021b). Adaptability to high temperature and stay-green genotypes associated with variations in antioxidant, chlorophyll metabolism, and γ-aminobutyric acid accumulation in creeping bentgrass species. *Frontiers in Plant Science* 12: 750728.

Liao, H.S., Chung, Y.H., and Hsieh, M.H. (2022). Glutamate: a multifunctional amino acid in plants. *Plant Science* 318: 111238.

Lim, P.O. and Nam, H.G. (2007). Aging and senescence of the leaf organ. *Journal of Plant Biology* 50: 291–300.

Lim, P.O., Kim, H.J., and Gil Nam, H. (2007). Leaf senescence. *Annual Review of Plant Biology* 58 (1): 115–136.

López, E., Figueroa, S., Oset-Gasque, M.J., and González, M.P. (2003). Apoptosis and necrosis: two distinct events induced by cadmium in cortical neurons in culture. *British Journal of Pharmacology* 138: 901–911.

Ma, Y., Wang, P., Gu, Z. et al. (2019). Ca^{2+} involved in GABA signal transduction for phenolics accumulation in germinated hulless barley under NaCl stress. *Food Chemistry: X* 2: 100023.

Masclaux, C., Valadier, M.H., Brugière, N. et al. (2000). Characterization of the sink/source transition in tobacco (*Nicotiana tabacum* L.) shoots in relation to nitrogen management and leaf senescence. *Planta* 211: 510–518.

Mekonnen, D.W., Flügge, U.I., and Ludewig, F. (2016). Gamma-aminobutyric acid depletion affects stomata closure and drought tolerance of Arabidopsis thaliana. *Plant Science* 245: 25–34.

Meng, Y. (2023). The role of GABA in plant salinity and hypoxia responses. Doctoral dissertation.

Michaeli, S. and Fromm, H. (2015). Closing the loop on the GABA shunt in plants: are GABA metabolism and signaling entwined? *Frontiers in Plant Science* 6: 419.

Michaeli, S., Fait, A., Lagor, K. et al. (2011). A mitochondrial GABA permease connects the GABA shunt and the TCA cycle, and is essential for normal carbon metabolism. *The Plant Journal* 67: 485–498.

Mishra, S., Jha, A.B., and Dubey, R.S. (2011). Arsenite treatment induces oxidative stress, upregulates antioxidant system, and causes phytochelatin synthesis in rice seedlings. *Protoplasma* 248: 565–577.

Mohammadi, M., Aelaei, M., and Saidi, M. (2021). Pre-harvest spray of GABA and spermine delays postharvest senescence and alleviates chilling injury of gerbera cut flowers during cold storage. *Scientific Reports* 11 (1): 14166.

Nath, K. and Lu, Y.A. (2015). Paradigm of reactive oxygen species and programmed cell death in plants. *Journal of Cell Science & Therapy* 6: 2.

Pavlíková, D., Zemanová, V., Procházková, D. et al. (2014). The long-term effect of zinc soil contamination on selected free amino acids playing an important role in plant adaptation to stress and senescence. *Ecotoxicology and Environmental Safety* 100: 166–170.

Petrov, V., Hille, J., Mueller-Roeber, B., and Gechev, T.S. (2015). ROS-mediated abiotic stress-induced programmed cell death in plants. *Frontiers in Plant Science* 6: 69.

Podlešáková, K., Ugena, L., Spíchal, L. et al. (2019). Phytohormones and polyamines regulate plant stress responses by altering GABA pathway. *New Biotechnology* 48 (1): 53–65.

Ramesh, S.A., Tyerman, S.D., Gilliham, M., and Xu, B. (2017). γ-Aminobutyric acid (GABA) signalling in plants. *Cellular and Molecular Life Sciences* 74 (9): 1577–1603.

Ramos-Ruiz, R., Martinez, F., and Knauf-Beiter, G. (2019). The effects of GABA in plants. *Cogent Food & Agriculture* 5: 1.

Razik, E.S.A., Alharbi, B.M., Pirzadah, T.B. et al. (2020). Aminobutyric acid (GABA) mitigates drought and heat stress in sunflower (*Helianthus annuus* L.) by regulating its physiological, biochemical and molecular pathways. *Physiologia Plantarum* 172: 505–527.

Reape, T.J., Brogan, N.P., and McCabe, P.F. (2015). Mitochondrion and chloroplast regulation of plant programmed cell death. In: *Plant Programmed Cell Death* (ed. A.N. Gunawardena and M.F. McCabe), 33–54. New York: Springer.

Renault, H., Roussel, V., El Amrani, A. et al. (2010). The *Arabidopsis* pop2-1 mutant reveals the involvement of GABA transaminase in salt stress tolerance. *BMC Plant Biology* 10: 1–16.

Roberts, M.R. (2007). Does GABA act as a signal in plants? Hints from molecular studies. *Plant Signaling & Behavior* 2 (5): 408–409.

Rossi, S., Chapman, C., and Huang, B. (2020). Suppression of heat-induced leaf senescence by γ-aminobutyric acid, proline, and ammonium nitrate through regulation of chlorophyll degradation in creeping bentgrass. *Environmental and Experimental Botany* 177: 104116.

Sarwat, M., Naqvi, A.R., Ahmad, P. et al. (2013). Phytohormones and microRNAs as sensors and regulators of leaf senescence: assigning macro roles to small molecules. *Biotechnology Advances* 31 (8): 1153–1171.

Schippers, J.H., Schmidt, R., Wagstaff, C., and Jing, H.C. (2015). Living to die and dying to live: the survival strategy behind leaf senescence. *Plant Physiology* 169 (2): 914–930.

Seifikalhor, M., Aliniaeifard, S., Bernard, F. et al. (2020). γ-Aminobutyric acid confers cadmium tolerance in maize plants by concerted regulation of polyamine metabolism and antioxidant defense systems. *Scientific Reports* 10: 3356.

Sharma, P. and Dubey, R.S. (2007). Involvement of oxidative stress and role of antioxidative defense system in growing rice seedlings exposed to toxic concentrations of aluminum. *Plant Cell Reports* 26: 2027–2038.

Sharma, P., Jha, A.B., Dubey, R.S., and Pessarakli, M. (2012). Reactive oxygen species, oxidative damage, and antioxidative defense mechanism in plants under stressful conditions. *Journal of Botany* 2012: 217037.

Shelp, B., Bozzo, G., Trobacher, C. et al. (2012). Strategies and tools for studying the metabolism and function of γ-aminobutyrate in plants. I. Pathway structure. *Botany* 90: 651–668.

Sheng, L., Shen, D., Luo, Y. et al. (2017). Exogenous γ-aminobutyric acid treatment affects citrate and amino acid accumulation to improve fruit quality and storage performance of postharvest citrus fruit. *Food Chemistry* 216: 138–145.

Shi, S.Q., Shi, Z., Jiang, Z.P. et al. (2010). Effects of exogenous GABA on gene expression of Caragana intermedia roots under NaCl stress: regulatory roles for H2O2 and ethylene production. *Plant, Cell & Environment* 33 (2): 149–162.

Sita, K. and Kumar, V. (2020). Role of gamma amino butyric acid (GABA) against abiotic stress tolerance in legumes: a review. *Plant Physiology Reports* 25: 654–663.

Song, H., Xu, X., Wang, H. et al. (2010). Exogenous γ-aminobutyric acid alleviates oxidative damage caused by aluminium and proton stresses on barley seedlings. *Journal of the Science of Food and Agriculture* 90: 1410–1416.

Su, N., Wu, Q., Chen, J. et al. (2019). GABA operates upstream of H^+-ATPase and improves salinity tolerance in *Arabidopsis* by enabling cytosolic K^+ retention and Na^+ exclusion. *Journal of Experimental Botany* 70: 6349–6361.

Suhel, M., Husain, T., Pandey, A. et al. (2023). An appraisal of ancient molecule GABA in abiotic stress tolerance in plants, and its crosstalk with other signaling molecules. *Journal of Plant Growth Regulation* 42 (2): 614–629.

Templeton, D.M. and Liu, Y. (2010). Multiple roles of cadmium in cell death and survival. *Chemico-Biological Interactions* 188: 267–275.

Vaezi, S., Asghari, M., Farokhzad, A. et al. (2022). Exogenous methyl jasmonate enhances phytochemicals and delays senescence in harvested strawberries by modulating GABA shunt pathway. *Food Chemistry* 393: 133418.

Van Hautegem, T., Waters, A.J., Goodrich, J., and Nowack, M.K. (2015). Only in dying, life: programmed cell death during plant development. *Trends in Plant Science* 20: 102–113.

Vardar, F., Akgül, N., Aytürk, Ö., and Aydın, Y. (2015). Assessment of aluminum induced genotoxicity with comet assay in wheat, rye and triticale roots. *Fresenius Environmental Bulletin* 24: 3352–3358.

Vardar, F., Çabuk, E., Aytürk, Ö., and Aydın, Y. (2016). Determination of aluminum induced programmed cell death characterized by DNA fragmentation in *Gramineae* species. *Caryologia* 69: 111–115.

Watanabe, M., Balazadeh, S., Tohge, T. et al. (2013). Comprehensive dissection of spatiotemporal metabolic shifts in primary, secondary, and lipid metabolism during developmental senescence in *Arabidopsis*. *Plant Physiology* 162: 1290–1310.

Woo, H.R., Kim, H.J., Lim, P.O., and Nam, H.G. (2019). Leaf senescence: systems and dynamics aspects. *Annual review of plant biology* 70 (1): 347–376.

Wu, L., Chen, H., Curtis, C., and Fu, Z.Q. (2014). Go in for the kill. *Virulence* 5: 710–721.

Yang, Z. (2003). GABA, a new player in the plant mating game. *Developmental Cell* 5 (2): 185–186.

Yang, A., Cao, S., Yang, Z. et al. (2011). γ-Aminobutyric acid treatment reduces chilling injury and activates the defence response of peach fruit. *Food Chemistry* 129 (4): 1619–1622. https://doi.org/10.1016/J.FOODCHEM.2011.06.018.

Yanık, F. and Vardar, F. (2015). Toxic effects of aluminum oxide (Al_2O_3) nanoparticles on root growth and development in *Triticum aestivum*. *Water, Air, and Soil Pollution* 226: 296.

Yanık, F. and Vardar, F. (2019). Assessment of silver nanoparticle-induced morphological, biochemical and physiological alterations in wheat roots. *Annali di Botanica* 9: 83–94.

Yanık, F., Çetinbaş-Genç, A., and Vardar, F. (2020). Abiotic stress induced programmed cell death in plants. In: *Plant Life Under Changing Environment* (ed. D.K. Tripathi, V.P. Singh, D.V. Chauhan, et al.). London: Academic Press.

Yuan, D., Wu, X., Gong, B. et al. (2023). GABA metabolism, transport and their roles and mechanisms in the regulation of abiotic stress (hypoxia, salt, drought) resistance in plants. *Metabolites* 13 (3): 347.

Zhang, H. and Zhou, C. (2013). Signal transduction in leaf senescence. *Plant Molecular Biology* 82 (6): 539–545.

Zhao, Y., Song, X., Zhong, D. et al. (2020). γ-Aminobutyric acid (GABA) regulates lipid production and cadmium uptake by *Monoraphidium* sp. QLY-1 under cadmium stress. *Bioresource Technology* 297: 122500.

7

GABA and Nodulation in Plants

Shubhra Khare[1], *Ajey Singh*[2], *Km Niharika*[2], *Nimisha Amist*[3], *Zeba Azim*[4], *Rangoli Krishna*[1], *Nishtha Srivastava*[1], *and Narsingh Bahadur Singh*[4]

[1] *Department of Applied Sciences and Humanities, Invertis University, Bareilly, Uttar Pradesh, 243123, India*
[2] *Department of Botany, University of Lucknow, Lucknow, Uttar Pradesh, 226007, India*
[3] *Department of Botany, University of Allahabad, Uttar Pradesh, 211002, India*
[4] *Department of Botany, Ewing Christian College, University of Allahabad, Uttar Pradesh, 211003, India*

Introduction

All organisms inherent a very significant feature that regulates the rates of metabolic processes in response to changes in the internal and/or external environment. This adaptability is necessary for maintaining homeostasis of the intercellular environment, which is essential for sustaining a proficient functional state in the organism. For proper survival and growth of the plant, several micro- and macronutrients are required in which nitrogen (N) element is a major limiting nutrient for the growth of most plant species in different ecosystems. Available atmospheric N is not directly utilized by plants because the troposphere N is present in organic form, and plants only uptake the inorganic form of the N (Barker and Bryson 2016). After photosynthesis, N acquisition and assimilation are the second most essential biochemical phenomena required for the proper growth and development of plants (Mengesha 2021).The majority of the plants obtain N from the rhizospheric soil in the form of ammonium ions or nitrate. Fixation of nitrogen is essential to transform the organic form of N into inorganic forms by several physical and biochemical methods. One of the biological fixations of N includes the symbiotic association of plants with microbes. Moreover, some other organisms, including cyanobacteria and bacteria, transform the gaseous pool of N into inorganic form due to the presence of nitrogenase, which is a crucial N-fixing enzyme (Mishra et al. 2019). Fixation of N by symbiotic association between plants and prokaryotes is most prominent among all the endosymbiotic relations (Singh et al. 2021;Ladha et al. 2022). Under some hostile conditions, this biochemical process is regulated by the N-feedback mechanism of the complete plant and helps in fulfilling the N demand of the host plant. This phenomenon is chiefly regulated by the metabolites derived from shoots and delivered by phloem. N fixation by the symbiotic

GABA in Plants: Biosynthesis, Plant Development, and Food Security, First Edition. Edited by Samiksha Singh, Durgesh Kumar Tripathi, and Vijay Pratap Singh.

association in leguminous plants is an important process established from the complex interaction between the host plant and the microorganism. Through this biological process, about 50 million tons of N is sustainably and cleanly available in various agricultural fields (Roy et al. 2020). There is a consensus that the cultivation of legumes can significantly reduce the intensive reliance on chemical N fertilizers and their associative environmental problems (Ladha et al. 2022). This symbiotic mechanism occurs in a distinctive organ called a "nodule" that forms on the roots of leguminous plants. This nodule is formed due to infection and harmonized differentiation by compatible diazotrophic rhizobia belonging to the Rhizobiaceae family (Schulte et al. 2021).

Rhizobium is a gram-negative proteobacterium that lives in the nodules formed on the root of legume plants, and these unique organs have facilitated atmospheric N fixation in exchange for carbon. Several atmospheric constraints, such as heat, salinity, and water scarcity, prompt variability amongthe metabolite profiles of nodules that can reflect the essential capability of legumes to augment the symbiosis of nodules under hostile environments. Numerous plant metabolites have been found to function in regulating downstream stress responses of symbiosis through multiple sensitive tuning mechanisms. These metabolites can be asparagine, proline, malate, trehalose, and aminobutyric acid (GABA) (Sulieman et al. 2024). Chemically, 4-aminobutyrate is known as GABA which shows a prominent role in the symbiotic association of leguminous plant and rhizobia (Sulieman and Schulze 2010a;Roswanjaya2022). In addition, it has been stated that the amendment of the pool size of GABA is the chief metabolic mechanism utilized by a wide range of plants exposed to any ecological stress (Li et al. 2016a; Tarkowski et al. 2020). This metabolite-mediated carbon/nitrogen balance is intensely allied with interconnected mechanisms such as osmoregulation, nitrogen accumulation, oxygen protection, pH regulation, and cellular redox control. Moreover, previous findings have been focused on the metabolic and mitigating mechanism of GABA in the acclimatization of the plant exposed to stress, but fewer reports have been reported on the potent role of this metabolite in nodule symbiosis (Podlešáková et al. 2019; Kaspal et al. 2021; Shelp et al.2021). The present chapter emphasizes on nodulation in legume crops and the role of GABA in nodule functioning, along with the developmental attributes of the plants.

Nodulation in Leguminous Plants

Legumes and certain other plant species have the potential to establish a symbiotic association with rhizobia that fixes N_2 and allows them to thrive in environments poor in N. According to Roy et al. (2020), this endosymbiotic relationship has the potential to sustainably produce approximately 50 million tons of N for use in a variety of agricultural fields. Through this symbiotic relationship, plants can supply photosynthetic carbon to rhizobia, and this microbe fixes atmospheric N for plant growth and development (Figure 7.1). Previous findings stated that growing legumes can drastically cut down on the heavy dependence on the chemical N fertilizers and the associated environmental issues (Ladha et al. 2022).

The symbiotic process takes place in a specific unique organ called a "nodule," which is established on the roots of legume plants belonging to the family Leguminosae. It is caused by

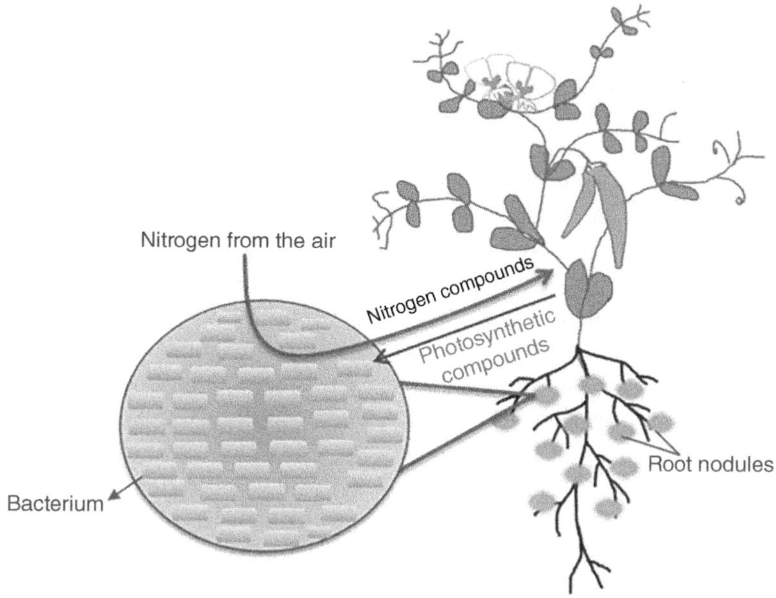

Figure7.1 Diagrammatic representation showing a symbiotic association between bacterium of root nodules and leguminous plant.

infestation and harmonized differentiation by a symbiotic bacterium that belongs to the family Rhizobiaceae and is also known as rhizobia (Schulte et al. 2021). It is well-known that other inhabitants reside in the nodules of leguminous roots besides rhizobial bacteria (Martínez-Hidalgo and Hirsch 2017). The majority of microbes that fix atmospheric N in leguminous plant root nodules belong to class α-proteobacteria and the genera *Rhizobium, Phyllobacterium, Microvirga, Azorhizobium, Bradyrhizobium, Devosia, AllorhizobiumMesorhizobium, Ensifer* (Mayhood and Mirza 2021). Legumes are also modulated by β-rhizobia, which belong to members of the class β-proteobacteria, family Burkholderiaceae (Mukhtar et al.2023).

Beijerinck (1888) provided the first account of the isolation of yellow and brown *Bacillus* strains from root nodules. Subsequent research using culture-dependent methods has revealed a significantly wider range of bacteria than just rhizobia linked to root nodules (Mayhood and Mirza 2021;Deng et al. 2022). *Mycobacterium, Pantoea, Pseudomonas, Enterobacter, Acinetobacter, Paenibacillus, Bacillus, Agrobacterium, Micromonospora*, and other *Bacillus* species are the predominant non-rhizobial bacteria found in legume nodules (Martínez-Hidalgo and Hirsch 2017;Swarnalakshmi et al. 2019;Muresu et al. 2022). The development of the symbiotic organ, the nodule, is undoubtedly the crucial event for the formation of the symbiosis, even though root nodule symbiosis is achieved by a series of regulatory mechanisms that are regulated by plants and rhizobia (Suzaki et al. 2015). Initiating the process with a mutual exchange of chemicals, especially flavonoids, isoflavones, etc., between the rhizobia and leguminous host plant (Singh et al. 2021). This root nodulation is a very complex mechanism that results in the formation of a specialized organelle–analogous structure known as a "symbiosome" (Schulte et al. 2021). This mechanism starts from the root hair apex, where symbiotic rhizobia infiltrates, dividing cortical cells via an intracellular tube-like structure that is specific to plants and is known as the infection thread. Rhizobia invasion must be successful for nodule formation to proceed (Liang 2020). During rhizobia, infection modifies differentiated root cortical cells' destiny during nodulation, and then these modified cortical cells start differentiating to form nodule primordial (Singh et al. 2021). During the first nodulation stage, host plants identify their symbiotic partners with precision and permit them to infiltrate root tissue. Within the infected plant cells of the nodule, this transient plant vesicle houses the N_2-fixing bacteroids (Roy et al. 2020). This N_2-fixing organelle is fully enclosed by a special membrane that is produced from the plasmalemma (PBM, peribacteroid membrane), allowing for tightly coordinated regulation by the host (Booth et al. 2021). Furthermore, by controlling the trafficking and homeostasis of mutual metabolic exchange between the micro- and macrosymbionts with the assistance of certain particular candidate transporters, the PBM can contribute to the provision of a suitable biochemical environment (Brear et al. 2020).

Subsequently, the catalytic mechanism of the nitrogenase enzyme transforms dinitrogen gas into ammonium ion, which is easily utilized by plants and assembled by the symbiotic form of resident endosymbiotic bacteria, known as "bacteroids" (Valentine et al. 2017). In exchange for a reduced carbon source such as sucrose produced through photosynthetic processes and other necessary mineral nutrients such as phosphate, sulfate, and molybdate, the later reducedN product is usually released by the metabolically active rhizobialbacteroids (Brear et al. 2020). Owing to the distinctive qualities of the endosymbiotic association of legumes and rhizobia, this platform has drawn a lot of attention from researchers, particularly in recent climate change problems that have a detrimental impact on N_2 fixation. As a result, the majority of current research is accomplished to analyze the regulatory mechanisms behind N_2-fixing functional symbiosis in abiotic conditions (Lambert et al. 2020). The metabolic alterations brought on by environmental and developmental cues are under such conditions orchestrated by an interconnected network of biochemical, physiological, molecular, and cellular processes (Abdelrahman et al. 2018).

The main goals of the current research are to sustain the carbon and energy metabolism and/or lessen the detrimental effects of stresses (such as oxidative burst) by the coordination of most of these alterations (Schulte et al. 2021). These investigations have the potential to significantly advance our conceptual and mechanistic understandings of nodule symbiosis, which will aid in the development of cultivars that are more effective at fixing nitrogen under stressed circumstances (Sulieman et al. 2022). With the significant development of high-throughput omic/molecular technologies and related databases, this strategy appears to be feasible (Farouk et al. 2023).

Functioning of γ-Aminobutyric Acid in Plants

GABA is a four-carbon, ubiquitous nonproteinogenic amino acid that contains an amino group on the γ-C (Shelp et al.2021; Sulieman et al.2024). It is found in every living entity from prokaryotes to eukaryotes. In animal cells, it is reported as a potent neurotransmitter inhibitor, whereas in plants, it is a chief metabolite that regulates several mechanisms responsible for growth and development (Kumar and Khanum 2012; Hussain et al. 2018).

In plants, it represents a predominant metabolite that substantially regulates various processes related to development and stress tolerance. It is synthesized metabolically by the GABA shunt pathway in the plants by bypassing the Krebs cycle, which occurs in the mitochondrion of the cell (Ansari et al. 2021). Moreover, the polyamine metabolic pathway can be an alternate pathway utilized by the plant to synthesize GABA (Majumdar et al. 2016).

GABA has a diverse role in the development of plants, especially at the time of the entrance of the pollen tube in the ovule where it helps in the elongation of the pollen tube. It is also found to play a significant role in the germination of seedlings and fruit ripening as well it acts as a signaling molecule that helps plant gene transformation mediated by *Agrobacterium tumefaciens* (Lang et al. 2016). In previous findings, it is reported to accumulate in plants under unfavorable biotic and abiotic conditions such as drought, low temperature salinity, insect, and pathogen infection, etc (Yang et al. 2011; Yong et al. 2017). In stressful conditions, a higher concentration of GABA ameliorates the adverse effects of stresses by up surging photosynthetic reactions and activating antioxidant enzymes which inhibit reactive oxygen species generation (Ramos-Ruiz et al. 2019;Srivastava et al. 2021). It also regulates the mechanism of the stomatal aperture under water deficit conditions. To compensate N demand of the plant, N_2 fixation by a symbiotic association between leguminous plants and rhizobia is a very interesting and prominent endosymbiotic microbiological process (Ladha et al. 2022). Ample amounts of GABA are stated in nodules of common legume beans such as *Glycine max, Vigna*sps, etc. (Serraj et al. 1998;Kaul et al. 2020). Recent findings reported that the accumulation of GABA under environmental constraints was due to several biochemical and transcriptional processes (Shelp et al. 2009). In previous studies, under water-deficient conditions, GABA was found to accumulate in every plant tissue but its highest amount was reported in the nodule of the soybean plant (Teixeira et al. 2020). However, the catalytic activity of glutamate decarboxylase (GAD) is more pronounced in roots and leaves as compared to nodules, but due to oxygen-deficient conditions inside the nodule activate the activity of GAD more efficiently. Therefore, when GABA is treated exogenously, it might increase the endogenous concentration of GABA that provides resistance and improves the tolerance capacity of the treated plant to grow under abiotic stress (Liao et al. 2022).

In previous findings on legume plants such as mungbean, soybean, etc., salinity tends to affect seedling germination, growth, biomass, nutrient balance, photosynthetic reactions, respiration, nodulation, and ultimately overall growth and developmental attributes of these plants (Manchanda and Garg 2008; Khare et al. 2020). However, its concentration decreases to almost one-fourth under high salt concentration which in turn adversely affects the quality and yield of the crops (Sehrawat et al.2019). There are plenty of findings about the significant functioning of GABA in mitigating diverse environmental cues, for example, a stimulatory effect of this nonproteinous metabolite on nodulation on root of the legume plants such as common bean and soybean in harsh environmental conditions (Antoniw and Sprent 1978a, 1978b; Miyashita and Good 2008; Li et al. 2016b; Yong et al. 2018; Sita and Kumar 2020).

The GABA shunt pathway seems to be a largely conserved metabolic bypass among all the biochemical processes in response to various environmental inputs. Thus, in addition to the additional carbon costs used in the metabolic stress acclimations include a stimulated production of secondary metabolites and the removal of anions from the cell (Bandehagh and Taylor 2020). Therefore, it may be reported that GABA, via its chief synthetic method, is concurrently generated to evade the decarboxylation reaction of the TCA cycle under stress and to buttress endosymbiont association with energy and carbon skeletons for augmenting catalytic activity of nitrogenase enzyme and assimilation of N (Souza et al. 2016; Sulieman et al.2024). Strong links exist between the carbon–nitrogen balance due to GABA and several interconnected functions, such as osmoregulation, cellular redox management, pH regulation, oxygen (O_2) protection, and N storage (Signorelli et al. 2021). Moreover, GABA may have roles in signaling and transport along with the metabolism of the plant. GABA contributes to the amazing metabolic adaptability of plants, which may help nodules reduce stressful experiences (Ramesh et al. 2017).

Functioning of GABA in Nodulation

In the detection of amino acids by ^{15}N nuclear magnetic resonance, an ample amount of GABA is examined in detached nodules of tested legume plants among all the amino acids (Sulieman and Schulze 2010a, 2010b;Scharff et al. 2003). According to previous studies, high levels of this nonproteinous metabolite has been recorded from diverse bacteroids excised from different bacterial strains of*Rhizobium leguminosarum, Bradyrhizobiumjaponicum, and Sinorhizobiummeliloti*(Streeter 1987; Miller et al. 1991). Recent findings corroborate this finding in different varieties of pea plants (Coda et al. 2010;Dunn 2015;Ullah et al. 2023). After incubating the nodulated roots of plants in $^{15}N_2$ for 30 minutes, they found that *R. leguminosarum's* cytosol and bacteroids fraction had a quick increment in level of GABA (Prell et al. 2009). These results are supported by the bacteroids' activation of the key enzymes involved in GABA metabolism. For a very long period, GABA has been absent from any indication of function, irrespective of its concentration in nodules.

This nonprotein amino acid is produced by all living things via the action of glutamate decarboxylase, and in prokaryotes, it is synthesized by glutamate via the GABA shunt pathway (White et al. 2009). This reaction is a bypass product of the 2-oxoglutarate dehydrogenase complex, in which GABA is formed by decarboxylation of glutamate catalyzed by glutamate decarboxylase (Figure 7.2). Another enzyme, GABA aminotransferase removes the amino residue of GABA and forms the leading succinate semialdehyde, which is then oxidized to succinate by succinate-semialdehyde dehydrogenase. Previous studies proved the presence of a GABA shunt pathway in different strains of rhizobia (Bertrand et al. 2016;Nilsson et al. 2021). However, this mechanism was never proven to function during symbiotic reactions. In a recent report, a nonclassical GABA shunt was found to be functional in rhizobialbacteroid (Sulieman et al. 2024). The chief amendment reported in the transformation of glutamate to GABA is the absence of the major enzyme which is glutamate decarboxylase in rhizobia (Dunn 2015;Ramesh et al. 2017) (Figure 7.2). This study was corroborated by oxygen dependency and less activity in nodules of *G. max* and *Vigna unguiculata* (Carter 2015;Jin et al. 1990).

In the functioning of nodules, many studies reported the signaling and metabolic activities of GABA. In previous studies, variable quantities of this crucial signaling molecule

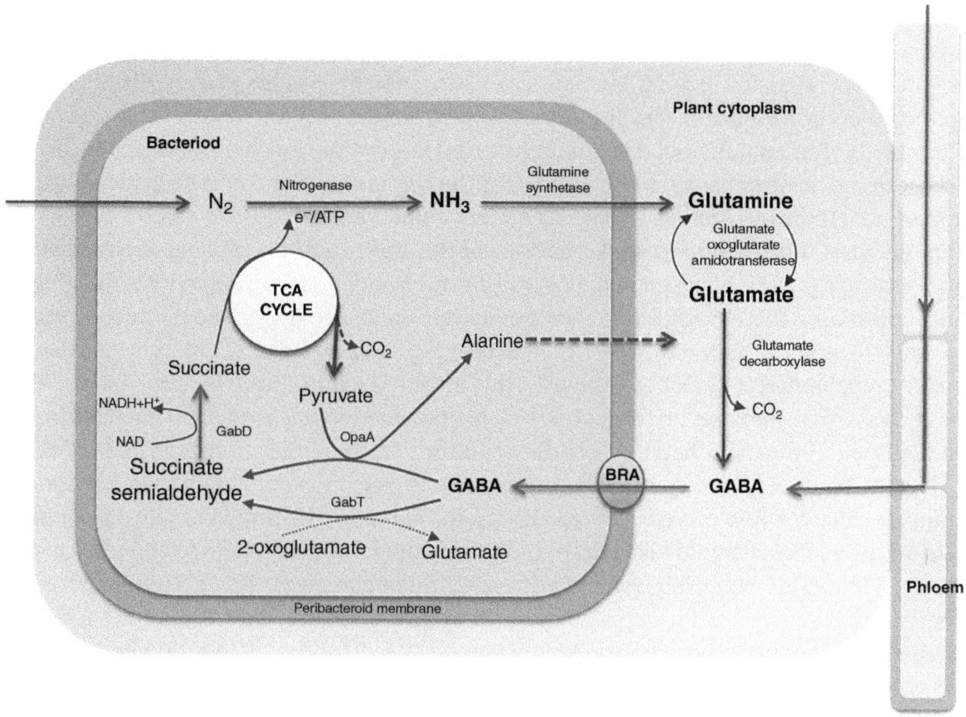

Figure 7.2 Diagrammatic illustration of the proposed metabolic routes of γ-aminobutyric acid (GABA) metabolism and the key enzymes involved in symbiotic nodules. Enzymes involved include: GabD, succinate semialdehyde dehydrogenase; OpaA, omega amino acid pyruvate aminotransferase; GabT, GABA oxoglutarate aminotransferase, and transporter: BRA, Branched-chain amino acid transporter.

have been found in the phloem of several nodule-bearing plants such as*G. max, V. unguiculata, Lupinus, and Medicago sativa* (Sulieman and Schulze 2010b).

Under normal conditions, a higher concentration of GABA was examined with a strong tendency for substantial augmentation in phloem and nodules beneath partial excision of nodule which can be used to enhance nodule activity. Partial detachment of nodules is a way that affects the balance between the source and sink of the symbiotic plant resulting in an N-deficient condition. After the exogenous application of GABA into the phloem sap of the leguminous plants, its substantial role in whole-plant-based stimulatory mechanisms was reported, which ultimately improved the activity of nodules (Huang et al. 2023;Booth et al. 2021). The strong association between N metabolism and GABA has been perceived in nonleguminous plants such as the mustard plant, where GABA derived from the shoot was observed translocating to the root fraction (Beuve et al. 2004;Fatma et al. 2022). The influx of Nitrate during N-deficient conditions is synergistically related to the amount of GABA in the phloem (Beuve et al. 2004). In previous findings, metabolic activities and signaling of GABA provide a correlation with the metabolism of N in *Arabidopsis thaliana* (Barbosa et al. 2010). It is also found that exogenous application of GABA significantly regulates the uptake of nitrate and the metabolism of *Arabidopsis* plantlets. Moreover, the catalytic activity of enzymes involved in carbon and nitrogen metabolism, such as phosphoenolpyruvate carboxylase, nitrate reductase, glutamate synthase, glutamine synthetase, and other enzymes were regulated when GABA was given exogenously (Baslam et al. 2020). Scientists reported that the significant role of GABA derived from the shoot increases the proficiency of symbiotic fixation (Gantait and Mukherjee 2021). A cycling biochemical mechanism of certain amino acids in between bacteroids and the cytoplasm of the nodule could strongly support such findings (Prell and Poole 2006;Mellor and Werner 2018).

For decades, it has been reported that none of the amino acids is capable of passing the PBM, which is present between plant cytoplasm and bacteroids (Mellor and Werner 2018). In recent studies, this evidence has been extensively enfeebled due to the products regarding the separation method and the succeeding measurements (Morère-Le Paven et al. 2024;Sulieman et al. 2024). Primarily, this model was introduced by Kahn et al. 1985, and recently, one study on nodules of pea plants provided strong evidence that additional cycling of an amino acid between plant cytoplasm and bacteroid or symbiosome (White et al. 2009;Booth et al. 2021). Initially, glutamate and one of its metabolites were reported as amino acids for this process, but recent scientific reports identified GABA as one the chief amino acids that could take part in the cycling model between nodule and plant cytoplasm (White et al. 2009;Yemelyanov et al. 2023;Sulieman et al. 2024). This theory was experimentally proved by the application of $^{15}N_2$ and examined the GABA in the bacteroid fraction of the nodules. Additional evidence was given by Hosie et al. (2002), who observed and reported the transporter that is branched-chain amino acid transporter (Bra), which was responsible for the transportation of GABA across the peribacteroidal membrane (Figure 7.2). For the suggested cycling model of amino acid from the cytosol to bacterioid, branched-chain amino acid was found to play a very significant role (De Bandt et al. 2022;Ruocco et al. 2023). In a biological system, GABA is synthesized from the catalytic action of glutamate decarboxylase enzyme within the GABA shunt. In this chapter, the final product of this shunt is succinate, which is transported to the tricarboxylic acid cycle of bacteroids of nodules to supply an ample amount of energy (Figure 7.2). The end product, succinate, was reported as an outstanding substrate that plays a very significant role in providing the maximum amount of N fixation in the bacteroids growing in laboratory conditions (Kahn et al. 1985;Schulte et al. 2022).

Conclusions and Future Prospects

GABA is said to be very mobile in legume crops, particularly inside the phloem sap of the plant, and its maximal concentration was discovered in the root nodules. According to recent research, leguminous plants use this nonamino acid metabolite for biochemical mechanisms as well as a signaling molecule. It is not yet confirmed how GABA functions physiologically in nodules and found to accumulate in response to stress and environments that were created to increase nitrogenase activity. It has possible involvement in a hypothetical amino acid cycle in nodules between bacteroids and the cytoplasm and can be supporting the functioning of the nodule. Future research should make an effort to resolve these problems, find more instances, and understand the processes in which GABA mediates communication between the microsymbiont and the host plant.

References

Abdelrahman, M., El-Sayed, M.A., Hashem, A. et al. (2018). Metabolomics and transcriptomics in legumes under phosphate deficiency in relation to nitrogen fixation by root nodules. *Frontiers in Plant Science* 9: 922.

Ansari, M.I., Jalil, S.U., Ansari, S.A., and Hasanuzzaman, M. (2021). GABA shunt: a key-player in mitigation of ROS during stress. *Plant Growth Regulation* 94: 131–149.

Antoniw, L.D. and Sprent, J.I. (1978a). Growth and nitrogen fixation of *Phaseolus vulgaris* L. at two irradiances II. Nitrogen fixation. *Annals of Botany* 42 (2): 399–410.

Antoniw, L.D. and Sprent, J.I. (1978b). Primary metabolites of *Phaseolus vulgaris* nodules. *Phytochemistry* 17: 675–678.

Bandehagh, A. and Taylor, N.L. (2020). Can alternative metabolic pathways and shunts overcome salinity induced inhibition of central carbon metabolism in crops? *Frontiers in Plant Science* 11: 1072.

Barbosa, J.M., Singh, N.K., Cherry, J.H., and Locy, R.D. (2010). Nitrate uptake and utilization is modulated by exogenous γ-aminobutyric acid in *Arabidopsis thaliana* seedlings. *Plant Physiology and Biochemistry* 48: 443–450.

Barker, A.V. and Bryson, G.M. (2016). Nitrogen. In: *Handbook of Plant Nutrition* (ed. A.V. Barker and D.J. Pilbeam), 37–66. Baco Raton, FL: CRC Press.

Baslam, M., Mitsui, T., Sueyoshi, K., and Ohyama, T. (2020). Recent advances in carbon and nitrogen metabolism in C3 plants. *International Journal of Molecular Sciences* 22 (1): 318.

Beijerinck, M.W. (1888). The root-nodule bacteria. *BotanischeZeitung* 46: 725–804.

Bertrand, A., Bipfubusa, M., Dhont, C. et al. (2016). Rhizobial strains exert a major effect on the amino acid composition of alfalfa nodules under NaCl stress. *Plant Physiology and Biochemistry* 108: 344–352.

Beuve, N., Rispail, N., Laine, P. et al. (2004). Putative role of γ-aminobutyric acid (GABA) as a long-distance signal in upregulation of nitrate uptake in *Brassica napus* L. *Plant, Cell & Environment* 27: 1035–1046.

Booth, N.J., Smith, P.M., Ramesh, S.A., and Day, D.A. (2021). Malate transport and metabolism in nitrogen-fixing legume nodules. *Molecules* 26 (22): 6876.

Brear, E.M., Bedon, F., Gavrin, A. et al. (2020). GmVTL1a is an iron transporter on the symbiosome membrane of soybean with an important role in nitrogen fixation. *New Phytologist* 228 (2): 667–681.

Carter,A.M. (2015).Importance of ureide partitioning processes for growth and productivity of nodulatedsoybean (Glycine max[L.]Merr.).Doctoral dissertation.Washington State University.

Coda, R., Rizzello, C.G., and Gobbetti, M. (2010). Use of sourdough fermentation and pseudo-cereals and leguminous flours for the making of a functional bread enriched of γ-aminobutyric acid (GABA). *International Journal of Food Microbiology* 137 (2–3): 236–245.

De Bandt, J.P., Coumoul, X., and Barouki, R. (2022). Branched-chain amino acids and insulin resistance, from protein supply to diet-induced obesity. *Nutrients* 15 (1): 68.

Deng, Z., Ren, Z., Sun, S., and Wang, Y. (2022). Theoretical design and process control of neonicotinoids insecticides suitable for synergistic degradation with the rubisco enzyme from rhizobia and carbon-fixing bacteria in soil. *Environmental Science and Pollution Research* 29: 12355.

Dunn, M.F. (2015). Key roles of microsymbiont amino acid metabolism in rhizobia-legume interactions. *Critical Reviews in Microbiology* 41 (4): 411–451.

Farouk, S., Osman, A.M., Awadallah, S.M., and Abdelrahman, A.S. (2023). The added value of using artificial intelligence in adult chest X-rays for nodules and masses detection in daily radiology practice. *The Egyptian Journal of Radiology and Nuclear Medicine* 54 (1): 142.

Fatma, M., Bashri, G., Rasheed, F. et al. (2022). Nitric oxide synthesis affects potassium and nitrogen homeostasis in plants for salt tolerance. In: *Role of Potassium in Abiotic Stress* (ed. N. Iqbal and S. Umar), 263–281. Singapore: Springer.

Gantait, S. and Mukherjee, E. (2021). Tissue culture-based genetic improvement of fava bean (*Vicia faba* L.): analysis on previous achievements and future perspectives. *Applied Microbiology and Biotechnology* 105 (18): 6531–6546.

Hosie, A.H.F., Allaway, D., Galloway, C.S. et al. (2002). *Rhizobium leguminosarum* has a second general amino acid permease with unusually broad substrate specificity and high similarity to branched-chain amino acid transporters (bra/LIV) of the ABC family. *Journal of Bacteriology* 184: 4071–4080.

Huang, X.J., Jian, S.F., Wan, S. et al. (2023). Exogenous γ-aminobutyric acid (GABA) alleviates nitrogen deficiency by mediating nitrate uptake and assimilation in *Andrographis paniculata* seedlings. *Plant Physiology and Biochemistry* 198: 107700.

Hussain, G., Rasul, A., Anwar, H. et al. (2018). Role of plant derived alkaloids and their mechanism in neurodegenerative disorders. *International Journal of Biological Sciences* 14 (3): 341.

Jin, H.N., Dilworth, M.J., and Glenn, A.R. (1990). 4-Aminobutyrate is not available to bacteroids of cowpea *Rhizobium*MNF2030 in snake bean nodules. *Archives of Microbiology* 153: 455–462.

Kahn, M.L., Kraus, J., and Somerville, J.E. (1985). A model of nutrient exchange in the *Rhizobium*-legume symbiosis. In: *Nitrogen Fixation Research Progress* (ed. H.J. Evans, P.J. Bottomley, and W.E. Newton), 193–199. Dordrecht, The Netherlands: MartinusNijhoff.

Kaspal, M., Kanapaddalagamage, M.H., and Ramesh, S.A. (2021). Emerging roles of γ aminobutyric acid (GABA) gated channels in plant stress tolerance. *Plants* 10 (10): 2178.

Kaul, T., Sony, S.K., Raman, N.M. et al. (2020). How crisp is CRISPR? CRISPR-Cas-mediated crop improvement with special focus on nutritional traits. *Advancement in Crop Improvement Techniques* 2020: 159–197.

Khare, S., Singh, N.B., Singh, A. et al. (2020). Plant secondary metabolites synthesis and their regulations under biotic and abiotic constraints. *Journal of Plant Biology* 63: 203–216.

Kumar, G.P. and Khanum, F. (2012). Neuroprotective potential of phytochemicals. *Pharmacognosy Reviews* 6 (12): 81.

Ladha, J.K., Peoples, M.B., Reddy, P.M. et al. (2022). Biological nitrogen fixation and prospects for ecological intensification in cereal-based cropping systems. *Field Crops Research* 283: 108541.

Lambert, I., Pervent, M., Le Queré, A. et al. (2020). Responses of mature symbiotic nodules to the whole-plant systemic nitrogen signaling. *Journal of Experimental Botany* 71 (16): 5039–5052.

Lang, J., Gonzalez-Mula, A., Taconnat, L. et al. (2016). The plant GABA signaling downregulates horizontal transfer of the *Agrobacterium tumefaciens* virulence plasmid. *The New Phytologist* 210 (3): 974–983.

Li, M., Guo, S., Yang, X. et al. (2016a). Exogenous gamma-aminobutyric acid increases salt tolerance of wheat by improving photosynthesis and enhancing activities of antioxidant enzymes. *BiologiaPlantarum* 60: 123–131.

Li, Z., Yu, J., Peng, Y., and Huang, B. (2016b). Metabolic pathways regulated by γ-aminobutyric acid (GABA) contributing to heat tolerance in creeping bentgrass (*Agrostis stolonifera*). *Scientific Reports* 6 (1): 30338.

Liang,J. (2020).The *Rhizobium leguminosarum* Norway-Lotus burttii interaction as a model for studying how rhizobia enter plant cells.Doctoral dissertation.Ludwig-Maximilians-Universität.

Liao, H.S., Chung, Y.H., and Hsieh, M.H. (2022). Glutamate: a multifunctional amino acid in plants. *Plant Science* 318: 111238.

Majumdar, R., Barchi, B., Turlapati, S.A. et al. (2016). Glutamate, ornithine, arginine, proline, and polyamine metabolic interactions: the pathway is regulated at the post-transcriptional level. *Frontiers in Plant Science* 7: 78.

Manchanda, G. and Garg, N. (2008). Salinity and its effects on the functional biology of legumes. *ActaPhysiologiaePlantarum* 30: 595–618.

Martínez-Hidalgo, P. and Hirsch, A.M. (2017). The nodule microbiome: N_2-fixing rhizobia do not live alone. *Phytobiomes Journal* 1 (2): 70–82.

Mayhood, P. and Mirza, B.S. (2021). Soybean root nodule and rhizosphere microbiome: distribution of rhizobial and nonrhizobial endophytes. *Applied and Environmental Microbiology* 87 (10): e0288420.

Mellor, R.B. and Werner, D. (2018). Legume nodule biochemistry and function. In: *Molecular Biology of Symbiotic Nitrogen Fixation* (ed. P.M. Gresshof), 111–129. Baco Raton, FL: CRC Press.

Mengesha, M. (2021). Effect and roles of nitrogen supply on photosynthesis. *International Journal of Photochemistry and Photobiology* 5 (2): 19.

Miller, R.W., McRae, D.G., and Joy, K. (1991). Glutamate and γ-aminobutyrate metabolism in isolated *Rhizobium meliloti*bacteroids. *Molecular Plant-Microbe Interactions* 4: 37–45.

Mishra, A.K., Kaushik, M.S., and Tiwari, D.N. (2019). Nitrogenase and hydrogenase: enzymes for nitrogen fixation and hydrogen production in cyanobacteria. In: *Cyanobacteria* (ed. A.K. Mishra, D.N. Tiwari, and A.N. Rai), 173–191. Cambridge, MA: Academic Press.

Miyashita, Y. and Good, A.G. (2008). Contribution of the GABA shunt to hypoxia-induced alanine accumulation in roots of *Arabidopsis thaliana*. *Plant & Cell Physiology* 49: 92–102.

Morère-Le Paven, M.C., Clochard, T., and Limami, A.M. (2024). NPF and NRT2 from *Pisum sativum*potentially involved in nodule functioning: lessons from *Medicagotruncatula* and *Lotus japonicus*. *Plants* 13 (2): 322.

Mukhtar, M., Braydan, A., Drehiba, F.A. et al. (2023). Evaluation of efficiency of two local *Rhizobium leguminosarum*isolates on root nodulation and growth of Faba bean (*Vicia faba* L.). *Scientific Journal for Faculty of Science-Sirte University* 3 (2): 149–153.

Muresu, R., Porceddu, A., Concheri, G. et al. (2022). Legumes of the Sardinia Island: knowledge on symbiotic and endophytic bacteria and interactive software tool for plant species determination. *Plants* 11 (11): 1521.

Nilsson, J.F., Castellani, L.G., Draghi, W.O. et al. (2021). Global transcriptome analysis of *Rhizobium favelukesii* LPU83 in response to acid stress. *FEMS Microbiology Ecology* 97 (1): 235.

Podlešáková, K., Ugena, L., Spíchal, L. et al. (2019). Phytohormones and polyamines regulate plant stress responses by altering GABA pathway. *New Biotechnology* 48: 53–65.

Prell, J. and Poole, P. (2006). Metabolic changes of rhizobia in legume nodules. *Trends in Microbiology* 14 (4): 161–168.

Prell, J., Bourdes, A., Karunakaran, R. et al. (2009). Pathway of γ-aminobutyrate metabolism in *Rhizobium leguminosarum* 3841 and its role in symbiosis. *Journal of Bacteriology* 191 (7): 2177–2186.

Ramesh, S.A., Tyerman, S.D., Gilliham, M., and Xu, B. (2017). γ-Aminobutyric acid (GABA) signalling in plants. *Cellular and Molecular Life Sciences* 74: 1577–1603.

Ramos-Ruiz, R., Martinez, F., and Knauf-Beiter, G. (2019). The effects of GABA in plants. *Cogent Food & Agriculture* 5 (1): 1670553.

Roswanjaya,Y.P. (2022).Conserved mechanisms in nodulation and arbuscular mycorrhization in the non-legume *Parasponiaandersoniiy*. Doctoral dissertation.Wageningen University and Research.

Roy, K., Vrba, J., Kaushik, S.J., and Mraz, J. (2020). Nutrient footprint and ecosystem services of carp production in European fishponds in contrast to EU crop and livestock sectors. *Journal of Cleaner Production* 270: 122268.

Ruocco, C., Malavazos, A., Ragni, M. et al. (2023). Amino acids contribute to adaptive thermogenesis. New insights into the mechanisms of action of recent drugs for metabolic disorders are emerging. *Pharmacological Research* 195: 106892.

Scharff, A.M., Egsgaard, H., Hansen, P.E., and Rosendahl, L. (2003). Exploring symbiotic nitrogen fixation and assimilation in pea root nodules by in vivo 15N nuclear magnetic resonance spectroscopy and liquid chromatographymass spectrometry. *Plant Physiology* 131: 367–378.

Schulte, C.C., Borah, K., Wheatley, R.M. et al. (2021). Metabolic control of nitrogen fixation in *Rhizobium*-legume symbioses. *Science Advances* 7 (31): 2433.

Schulte, C.C., Ramachandran, V.K., Papachristodoulou, A., and Poole, P.S. (2022). Genome-scale metabolic modelling of lifestyle changes in *Rhizobium leguminosarum*. *mSystems* 7 (1): e0097521.

Sehrawat, N., Yadav, M., Sharma, A.K. et al. (2019). Salt stress and mungbean [*Vignaradiata* (L.) Wilczek]: effects, physiological perspective and management practices for alleviating salinity. *Archives of Agronomy and Soil Science* 65 (9): 1287–1301.

Serraj, R., Shelp, B.J., and Sinclair, T.R. (1998). Accumulation of γ-aminobutyric acid in nodulated soybean in response to drought stress. *PhysiologiaPlantarum* 102 (1): 79–86.

Shelp, B.J., Allan, W.L., and Faure, D. (2009). Role of γ-aminobutyrate and γ-hydroxybutyrate in plant communication. In: *Plant-Environment Interactions, Signaling and Communication in Plants* (ed. F. Baluška), 73–84. Berlin; Heidelberg: Springer.

Shelp, B.J., Aghdam, M.S., and Flaherty, E.J. (2021). γ-Aminobutyrate (GABA) regulated plant defense: mechanisms and opportunities. *Plants* 10 (9): 1939.

Signorelli, S., Tarkowski, Ł.P., O'Leary, B. et al. (2021). GABA and proline metabolism in response to stress. In: *Hormones and Plant Response* (ed. D.K. Gupta and F. Corpas), 291–314. Cham, Switzerland: Springer.

Singh, A., Singh, N.B., Yadav, V. et al. (2021). Nod factor signaling in legume-*Rhizobium*symbiosis: specificity and molecular genetics of nod factor signaling. In: *Abiotic Stress and Legumes* (ed. V.P. Singh, S. Singh, D.K. Tripathi, et al.), 33–67. Cambridge, MA: Academic Press.

Sita, K. and Kumar, V. (2020). Role of gamma amino butyric acid (GABA) against abiotic stress tolerance in legumes: a review. *Plant Physiology Reports* 25: 654–663.

Souza, S.C., Mazzafera, P., and Sodek, L. (2016). Flooding of the root system in soybean: biochemical and molecular aspects of N metabolism in the nodule during stress and recovery. *Amino Acids* 48: 1285–1295.

Srivastava, V., Mishra, S., Chowdhary, A.A. et al. (2021). The γ-aminobutyric acid (GABA) towards abiotic stress tolerance. In: *Compatible Solutes Engineering for Crop Plants Facing Climate Change* (ed. S.H. Wani, M.P. Gangola, and B.R. Ramadoss), 171–187. Cham, Switzerland: Springer International Publishing.

Streeter, J.G. (1987). Carbohydrate, organic acid and amino acid composition of bacteroids and cytosol from soybean nodules. *Plant Physiology* 85: 768–773.

Sulieman, S. and Schulze, J. (2010a). Phloem derived γ-aminobutyric acid (GABA) is involved in upregulating nodule N2 fixation efficiency in the model legume *Medicagotruncatula*. *Plant, Cell & Environment* 33: 2162–2172.

Sulieman, S. and Schulze, J. (2010b). The efficiency of nitrogen fixation of the model legume *Medicagotruncatula*(Jemalong A17) is low compared to *Medicago sativa*. *Journal of Plant Physiology* 167: 683–692.

Sulieman, S., Abdelrahman, M., and Tran, L.S.P. (2022). Carbon metabolic adjustment in soybean nodules in response to phosphate limitation: a metabolite perspective. *Environmental and Experimental Botany* 196: 104810.

Sulieman, S., Sheteiwy, M.S., Abdelrahman, M., and Tran, L.S.P. (2024). γ-Aminobutyric acid (GABA) in N2-fixing-legume symbiosis: metabolic flux and carbon/nitrogen homeostasis in responses to abiotic constraints. *Plant Physiology and Biochemistry* 207: 108362.

Suzaki, T., Yoro, E., and Kawaguchi, M. (2015). Leguminous plants: inventors of root nodules to accommodate symbiotic bacteria. *International Review of Cell and Molecular Biology* 316: 111–158.

Swarnalakshmi, K., Rajkhowa, S., Senthilkumar, M., and Dhar, D.W. (2019). Influence of endophytic bacteria on growth promotion and protection against diseases in associated plants. In: *Microbial Interventions in Agriculture and Environment: Volume 3: Soil and Crop Health Management* (ed. D.P. Singh and R. Prabha), 263–287. Berlin: Springer.

Tarkowski, Ł.P., Signorelli, S., and Höfte, M. (2020). γ-Aminobutyric acid and related amino acids in plant immune responses: emerging mechanisms of action. *Plant Cell and Environment* 43 (5): 1103–1116.

Teixeira, W.F., Soares, L.H., Fagan, E.B. et al. (2020). Amino acids as stress reducers in soybean plant growth under different water-deficit conditions. *Journal of Plant Growth Regulation* 39: 905–919.

Ullah, A., Ali, I., Noor, J. et al. (2023). Exogenous γ-aminobutyric acid (GABA) mitigated salinity-induced impairments in mungbean plants by regulating their nitrogen metabolism and antioxidant potential. *Frontiers in Plant Science* 13: 1081188.

Valentine, A.J., Kleinert, A., and Benedito, V.A. (2017). Adaptive strategies for nitrogen metabolism in phosphate deficient legume nodules. *Plant Science* 256: 46–52.

White, J.P., Prell, J., Ramachandran, V.K., and Poole, P.S. (2009). Characterization of a γ-aminobutyric acid transport system of *Rhizobium leguminosarum*bv. Viciae 3841. *Journal of Bacteriology* 191 (5): 1547–1555.

Yang, A., Cao, S., Yang, Z. et al. (2011). γ-Aminobutyric acid treatment reduces chilling injury and activates the defence response of peach fruit. *Food Chemistry* 129: 1619–1622.

Yemelyanov, V.V., Puzanskiy, R.K., and Shishova, M.F. (2023). Plant life with and without oxygen: ametabolomics approach. *International Journal of Molecular Sciences* 24 (22): 16222.

Yong, B., Xie, H., Li, Z. et al. (2017). Exogenous application of GABA improves PEG-induced drought tolerance positively associated with GABA-shunt, polyamines, and proline metabolism in white clover. *Frontiers in Physiology* 8: 1107.

Yong, T.W., Chen, P., Dong, Q. et al. (2018). Optimized nitrogen application methods to improve nitrogen use efficiency and nodule nitrogen fixation in a maize-soybean relay intercropping system. *Journal of Integrative Agriculture* 17 (3): 664–676.

8

GABA and Wounding Stress in Plants

Neeraj Kumar Dubey[1], Vijay Bahadur Yadav[1], Kunwar Deelip Singh[2], Satyendra Kumar Yadav[2], Ran Vijay Singh[2], Amarjeet Singh[3], and Jogeswar Panigrahi[4]

[1] *Department of Botany, Rashtriya Snatkottar Mahavidyalaya, Jaunpur, Uttar Pradesh, 222001, India*
[2] *Department of Zoology, Rashtriya Snatkottar Mahavidyalaya, Jaunpur, Uttar Pradesh, 222001, India*
[3] *Department of Botany, Post Graduate College, Ghazipur, Uttar Pradesh, 233001, India*
[4] *Department of Biotechnology, Berhmpur University, Bhanja, Bihar, 760007, Odisha, India*

Introduction

Being sessile in nature, plants have to face different stresses during their life cycle, from germination to seed setting. Different stresses reduce crop productivity up to several folds and damage human profits. These stresses are derived due to different factors. These factors may be categorized as biotic factors, where viruses, bacteria, fungi, and herbivorous insects attack plants, and abiotic factors like heat, salinity, drought, cold, etc., which influence the normal physiological process (Dubey et al. 2013; Dresselhaus and Hückelhoven 2018). Ultimately, these biotic factors reduce the photosynthetic area by damaging the photosynthetic cell, while abiotic factors reduce the photosynthetic process, and together they reduce net photosynthetic activity. Wound or injury stress is common because it may be created by biotic herbivorous insects or pests and by abiotic mechanical damages (Higley et al. 1993). Different signaling molecules like phytohormone affect the plant life cycle (Aly et al. 2014; Hasanuzzaman et al. 2018). Among them GABA is an important nonproteinogenic, four-carbon amino acid that plays an important role in both plants and animals (Khan et al. 2021). GABA also plays an important role in herbivorous wounds and other abiotic stress management (Kinnersley and Turano 2000). GABA is also very important for vertebrate and invertebrate animals (Lu et al. 2009). The role of GABA in several health benefits like depression, antidiabetic, and hypotensive effects have been reported in the case of humans (Nikmaram et al. 2017). In the case of plants, GABA helps in growth, development, leaf senescence, and abiotic stress management (Khan et al. 2021). The wound is an important stress created by both biotic and abiotic factors (Khan et al. 2021). Biotic factors, such as grazing by insects, animals, or manmade damage, while abiotic factors, like wind speed or soil-mediated wounds, can create a problem for plants

GABA in Plants: Biosynthesis, Plant Development, and Food Security, First Edition. Edited by Samiksha Singh, Durgesh Kumar Tripathi, and Vijay Pratap Singh.

for the successful completion of the life cycle (Savatin et al. 2014). Injury or wound causes nutrient loss and provides a path to the microbe's entry (Savatin et al. 2014; Khan et al. 2021). The role of GABA in the adaptation and management of stresses like salinity, drought, heavy metals, hypoxia, temperature, plant-insect/microbe interaction, and carbon and nitrogen metabolism have been studied (Seifikalhor et al. 2019). These stresses produce reactive oxygen species (ROS), which can lead to death and damage to cells. GABA helps in improving hypoxia and drought-mediated tolerance in plants (Shelp et al. 2021). In addition, GABA accumulated during stress can bind directly to the aluminum-activated malate transporter (ALMT) and the guard cell outward, rectifying K^+ (GORK) channel, thereby improving stress tolerance (Shelp et al. 2021). This book chapter summarizes the role of GABA in the management of biotic and abiotic stress, especially in wound management.

GABA: An Important Molecule for Plant

GABA provokes signals in response to environmental stresses and causes proline accumulation, which maintains cellular structures and reduces environmental stress (Kinnersley and Turano 2000; Sita and Kumar 2020; Mohammadi et al. 2021). In the animal system, GABA functions as an inhibitory neurotransmitter and enables animals to escape from upcoming environmental stresses (Kinnersley and Turano 2000). GABA also helps in biotic stress management, like those of fungus attack (Kinnersley and Turano 2000). Cytosolic acidification, either under oxygen deprivation or through metabolic and/or mechanical disruption, leads to the activation of glutamate decarboxylase and GABA synthesis. GABA is also involved in C/N balance (Fait et al. 2005). GABA also played important role in other biological process like plant growth and development like fruit ripening, root growth, seed germination (Bouchereau et al. 1999; Li et al. 2021), and pollen development (Palanivelu et al. 2003) in differentiation of the Pinus vascular system (Molina-Rueda et al. 2015) and also involved in communication between plants and other organisms such as bacteria, fungi, and certain invertebrates (Shelp et al. 2006). These are also involved in *Agrobacterium tumefaciens*-mediated plant gene transformation. Stress factors like flooding, cold, salt, heat, salt, touch, wind, rain, etc., also induce GABA synthesis (Kinnersley and Turano 2000). GABA accumulates by the GABA-shunt pathway in plant cells, which is regulated by three key enzymes, including GABA transaminase (GABA-T), glutamate decarboxylase (GAD), and succinic semialdehyde dehydrogenase (SSADH) (Kinnersley and Turano 2000). Under stress due to changes in cellular pH changes, the activity of GAD increases, which leads to the formation of more GABA (Shiu et al. 2020). Uses of GABA along with spermine (SPER) showed antiaging factors and delayed the senescence, elevating the chilling tolerance in many species like gerbera (Mohammadi et al. 2021). It increased the fresh weight, solution uptake, and increased capacity for detoxification (Mohammadi et al. 2021). GABA also influences the differential expression of phytohormonal genes (Ji et al. 2018). More than 20 gene expressions are influenced by the exogenous application of GABA (Ji et al. 2018). Furthermore, accumulations of JA (jasmonic acid), SA (salicylic acid), and ABA (abscisic acid) in sweet oranges have been shown during the exogenous application of GABA (Hijaz et al. 2018). The wound is another important factor, and it also induces acidification of the cytosol, which leads to the accumulation of GABA in the treated and local leaves (Cholewa et al. 1997). Systemic accumulations of GABA in systemic leaves were also reported (Scholz et al. 2015). GABA deters feeding on a synthetic diet, as shown for *Choristoneura rosaceana* (Ramputh and Bown 1996). Transgenic *Nicotiana tabacum* plants constitutively expressed GAD enzyme with elevated GABA have shown resistance to both *H. virescens* larvae and *Meloidogyne hapla* (MacGregor et al. 2003; McLean et al. 2003; Bown et al. 2006). Wound-induced plants' GABA deters herbivorous insects (Bown et al. 2006). Further GABA synthesis is also induced during herbivorous insect attacks (Scholz et al. 2015). Increased GABA elevates and improves plant health by antioxidant enzymes, removing ROS, and trying to manage drought stress (Li et al. 2021).

GABA and Abiotic Stress

Low temperature induces GABA production in plants like barley, wheat, *Brachypodium sylvaticum* grass, and tea roots (Mazzucotelli et al. 2006; Yang et al. 2016; Toubiana et al. 2020; Kaspal et al. 2021; Li et al. 2021). Exogenous application of GABA increases the endogenous GABA and enhances the tolerance against cold in tomato, banana, and tea plants (Malekzadeh et al. 2014; Wang et al. 2014, Zhu et al. 2019; Li et al. 2021). GABA also plays an important role in managing heat stress in rice, soybean, etc. (Takahashi et al. 2013; Nayyar et al. 2014). The exogenous application of GABA in rice (*Oryza sativa*) and other plants improved heat stress management (Takahashi et al. 2013; Nayyar et al. 2014; Li et al. 2021). GABA also protects the reproductive part of the mungbean plant during heat stress (Priya et al. 2019). GABA concentration also increases in drought stress like in those of plant turnip, soybean, tomato, and black cumin (Thompson et al. 1966; Serraj et al. 1998; Pal et al. 2016; Rezaei-Chiyaneh et al. 2018). Exogenous GABA treatment showed wax biosynthesis, delayed leaf senescence, and increased tolerant against drought and heat like bent grass (*Agrostis stolonifera*), black cumin (*Nigella sativa* L.) (Rezaei-Chiyaneh et al. 2018; Li et al. 2019). Flood and hypoxia also induce GABA synthesis in maize seedlings, radish, fava bean, and soybean (Streeter and Thompson 1972; Yang et al. 2013, 2015; Salah et al. 2019; Li et al. 2021). Similarly, salt stress is another important abiotic stress for the plant, which reduces the plant's productivity by a maximum amount (Renault et al. 2010). Accumulations of GABA during high salinity in many plant species have been observed (Renault et al. 2010). GABA transaminase (GABA-T) is one of the important enzymes involved in the first step of GABA metabolism, and mutant analysis of GABA-T has shown its importance in salt stress management (Renault et al. 2010). Double mutant of *Arabidopsis*, pop2-5 (pollen–pistil incompatibility) have increased accumulation abilities of GABA, showed salt-tolerant nature in studied plants by changes in H^+-ATPase activity, Na^+ uptake, K^+ efflux, and concentration of ROS (Su et al. 2019; Li et al. 2021). Further exogenous application of GABA enhances salt tolerance in plants like maize, muskmelon, barley, and tomato (Wang et al. 2017; Jin et al. 2019; Ma et al. 2019; Wu et al. 2020; Li et al. 2021). Furthermore, heavy metals are another abiotic stress in plant physiology (Kang et al. 2015). GABA content increases during stress of chromium (Cr) in rice, and zinc (Zn) and copper (Cu) stress in soybean and *N. tabacum* (Dubey et al. 2010; Kang et al. 2015; Daş et al. 2016; Li et al. 2021). GABA application during arsenic (As) stress activates the antioxidant enzyme system, stops the accumulation of As, and provides tolerance against As (III) in rice seedlings (Daş et al. 2016; Li et al. 2021).

Biotic Stress and Wound-Mediated GABA Fluctuation

Biotic stress is another important factor affecting the plant life cycle (Dubey et al. 2013; Shelp et al. 2021). Studies have shown that GABA is involved in the interaction between plants and microbes (Lang et al. 2016). Infections of several pathogens, like those of viral or fungal, induce GABA production in different plants (Shelp et al. 2006). GABA increases during biotic stress like attacking pathogens and insects, and it helps plants by increasing photosynthetic activity, reducing oxidative stress by inducing ROS scavengers and stomata metabolism to upcoming biotic and abiotic stresses (Shelp et al. 2006; Takahashi et al. 2008; Sidhu et al. 2010; Seifi et al. 2013; Li et al. 2021). During plant–microbe interactions due to GAD enzyme activity, GABA concentration increases, and it maintains cell viability by maintaining C : N ratio (Wang et al. 2019). Furthermore, due to pathogen attack, there is a change in ROS level, and GABA induces the activity of antioxidant enzymes like catalase, peroxidase, and superoxide dismutase, and thus, stops the pathogen-induced cell death (Yang et al. 2017; Sidhu et al. 2010). Thus, GABA helps in the development of plant resistance against pathogens (Yang et al. 2017; Li et al. 2021). Furthermore, GABA is also involved in the development of defense against herbivorous insect attacks through the inhibitory neurotransmitter method (Bown et al. 2006; Scholz et al. 2015; Tarkowski et al. 2020; Li et al. 2021). GABA stops *Botrytis* fungus growth by inhibiting the spread of mycelia while it stops the insect attack by the inhibitory effect and induction of the defense system (Seifi et al. 2013; Ramos-Ruiz et al. 2019). Exogenous application of GABA increases resistance in tomatoes against *B. cinerea* fungus (Seifi et al. 2013). Increased GABA levels during wound and attack of insects like *Choristoneura rosaceana* and *Spodoptera littoralis* in the leaves of *Glycine max* and *Arabidopsis thaliana* have been reported (Wallace et al. 1984; Ramputh and Shelp 1996; Scholz et al. 2015). Authors have suggested that increased GABA, either in leaf or artificial diet, stopped the growth of rearing larvae (Ramputh and Shelp 1996). Accumulations of GABA after insect-mediated wounding have been shown by different authors (Wallace et al. 1984; Bown et al. 2002; Scholz et al. 2015; Tarkowski et al. 2020). However, it has been shown that electrophysiological signaling induces the GABA concentration, and wounding or insect feeding is not directly responsible for GABA induction (Scholz et al. 2017; Tarkowski et al. 2020). Not only feeding but crawling of tobacco budworms on a leaf of tobacco also induces GABA production (Bown et al. 2002; MacGregor et al. 2003; Tarkowski et al. 2020). The released insect only prefers wild plants in comparison to higher GABA levels present in mutant plants (Bown et al. 2002; MacGregor et al. 2003; Tarkowski et al. 2020). Wound-induced 10-day-old seedlings have shown increases in aminoaldehyde dehydrogenase activity along with GABA concentration in wound tissue (Petřivalský et al. 2007). And about 300% of GABA increases in wound plant tissue (Petřivalský et al. 2007). It is suggested that insect or wound-induced GABA works as a defensive strategy against phytophagous insects by inhibiting the insect nervous system (Tarkowski et al. 2020). Transgenic plants having GAD OX lines, accumulating higher GABA, show less nematode egg masses in comparison to wild-type plants during infestation with the root-knot nematode (McLean et al. 2003; Tarkowski et al. 2020). Treatment of carbon dioxide and lactic acid bacteria induces GABA in pulses plants (Nikmaram et al. 2017). Treated pulses plants with a high concentration of GABA enhanced the nutritional properties of pulses (Nikmaram et al. 2017). These pulses can be used for medicinal purposes for GABA supplements (Nikmaram et al. 2017).

Transgenic Plants Expressing GABA and Effect on Herbivorous Performance

GABA is an important signaling molecule in biological organisms (Scholz et al. 2017). It acts as a neurotransmitter in invertebrates, and it also plays an important role in the development and functioning of nervous systems in animals, while in plants, it plays different important functions, from seed germination to pollen growth and management of biotic and abiotic stresses (Han et al. 2007; Scholz et al. 2017). Different authors have suggested that the exogenous application of GABA enhances tolerance against multiple biotic and abiotic stresses (Shelp et al. 2021). Transgenic strategy is one of the important strategies to express the foreign gene in the desired crop to get the desired trait in less time (Phillips 2008; Rani and Usha 2013; Zhang et al. 2016; Dubey et al. 2019; Yin et al. 2021). Different transgenic plants have been generated to overcome biotic and abiotic stresses (Parmar et al. 2017; Dubey et al. 2019). Alteration in the expression of GABA-related genes changes the level of GABA in transgenic plants (Bown and Shelp 2016). These plants showed resistance to several biotic stresses, such as insects and pathogens (Bown and Shelp 2016; Deng et al. 2020). Number of authors have reported that a high level of GABA deters the feeding behavior of insects (Ramputh and Bown 1996; Bown et al. 2002; Seifikalhor et al. 2019). GABA works in a systemic way and induces plant defense within few minutes of insect attack or wound condition (Seifikalhor et al. 2019). Transgenic tobacco plants with elevated GABA reduce the *A. tumefaciens* infection and also provide resistance to root-knot nematode in comparison to wild-type plants (Chevrot et al. 2006; McLean et al. 2003). Transgenic plants with more GABA have fewer nematode eggs than wild-type plants (McLean et al. 2003). Koike et al. (2013) have shown that overproduction of GABA by suppression of γ-aminobutyric acid (GABA) transaminases induces dwarfism and infertility in the tomato. High levels of GABA by deregulated forms of AtGAD1 in maize plants retarded growth and severe chlorosis (Rajani et al. 2021). GABA slows the browning and develops resistance against bacterial growth in cut apples (Zhao et al. 2021). Transcriptome analysis reveals that GABA alters the level of browning enzymes and phenolic substances and induces resistant genes (Zhao et al. 2021). These results of GABA expression can be used in crop improvement program against pathogens and pests like bacteria, fungi, insects, and nematodes (Zhao et al. 2021). Furthermore, GABA-related expressions also enhance the quality of fruits and pulses (Nikmaram et al. 2017; Zhao et al. 2021).

References

Aly, R., Dubey, N.K., Yahyaa, M. et al. (2014). Gene silencing of CCD7 and CCD8 in *Phelipanche aegyptiaca* by tobacco rattle virus system retarded the parasite development on the host. *Plant Signaling & Behavior* 9 (8): e29376.

Bouchereau, A., Aziz, A., Larher, F., and Martin-Tanguy, J. (1999). Polyamines and environmental challenges: recent development. *Plant Science* 140 (2): 103–125.

Bown, A.W., Hall, D.E., and MacGregor, K.B. (2002). Insect footsteps on leaves stimulate the accumulation of 4-aminobutyrate and can be visualized through increased chlorophyll fluorescence and superoxide production. *Plant Physiology* 129 (4): 1430–1434.

Bown, A.W., MacGregor, K.B., and Shelp, B.J. (2006). Gamma-aminobutyrate: defense against invertebrate pests? *Trends in Plant Science* 11 (9): 424–427.

Bown, A.W. and Shelp, B.J. (2016). Plant GABA: not just a metabolite. *Trends in Plant Science* 21 (10): 811–813.

Chevrot, R., Rosen, R., Haudecoeur, E. et al. (2006). GABA controls the level of quorum-sensing signal in *Agrobacterium tumefaciens*. *Proceedings of the National Academy of Sciences of the United States of America* 103 (19): 7460–7464.

Cholewa, E., Bown, A.W., Cholewinski, A.J. et al. (1997). Cold-shock-stimulated γ-aminobutyric acid synthesis is mediated by an increase in cytosolic Ca^{2+}, not by an increase in cytosolic H+. *Canadian Journal of Botany* 75 (3): 375–382.

Daş ZA, Dimlioğlu G, Bor M, Özdemir F (2016) Zinc induced activation of GABA-shunt in tobacco (*Nicotiana tabaccum* L.). *Environmental and Experimental Botany* 122:78–84. https://doi.org/10.1016/j.envexpbot.2015.09.006.

Deng, X., Xu, X., Liu, Y. et al. (2020). Induction of γ-aminobutyric acid plays a positive role to *Arabidopsis* resistance against *Pseudomonas syringae*. *Journal of Integrative Plant Biology* 62 (11): 1797–1812.

Dresselhaus, T. and Hückelhoven, R. (2018). Biotic and abiotic stress responses in crop plants. *Agronomy* 8 (11): 267.

Dubey S, Misra P, Dwivedi S, Chatterjee S, Bag SK, Mantri S, Asif MH, Rai A, Kumar S, Shri M, et al. (2010) Transcriptomic and metabolomic shifts in rice roots in response to Cr (VI) stress. *BMC Genomics* 11:648. https://doi.org/10.1186/1471-2164-11-648.

Dubey, N.K., Goel, R., Ranjan, A. et al. (2013). Comparative transcriptome analysis of *Gossypium hirsutum* L. in response to sap sucking insects: aphid and whitefly. *BMC Genomics* 14 (1): 1–20.

Dubey, N.K., Singh, P.K., Yadav, S.K., and Singh, K.D. (2019). Overview of insecticidal genes used in crop improvement program. In: *OMICS-Based Approaches in Plant Biotechnology*, vol. 13 (ed. R. Banerjee, G. Vijay Kumar, and S.P. Jeevan Kumar), 19–53. Hoboken, NJ: John Wiley & Sons, Inc.

Fait, A., Yellin, A., and Fromm, H. (2005). GABA shunt deficiencies and accumulation of reactive oxygen intermediates: insight from *Arabidopsis* mutants. *FEBS Letters* 579 (2): 415–420.

Han, D., Kim, H.Y., Lee, H.J. et al. (2007). Wound healing activity of gamma-aminobutyric acid (GABA) in rats. *Journal of Microbiology and Biotechnology* 17 (10): 1661–1669.

Hasanuzzaman, M., Bhuyan, M.H., Mahmud, J.A. et al. (2018). Interaction of sulfur with phytohormones and signaling molecules in conferring abiotic stress tolerance to plants. *Plant Signaling & Behavior* 13 (5): e1477905.

Higley, L.G., Browde, J.A., and Higley, P.M. (1993). Moving towards new understandings of biotic stress and stress interactions. *International Crop Science* I.1: 749–754.

Hijaz, F., Nehela, Y., and Killiny, N. (2018). Application of gamma-aminobutyric acid increased the level of phytohormones in *Citrus sinensis*. *Planta* 248: 909–918.

Ji J, Yue J, Xie T, Chen W, Du C, Chang E, Chen L, Jiang Z, Shi S (2018) Roles of γ-aminobutyric acid on salinity-responsive genes at transcriptomic level in poplar: involving in abscisic acid and ethylene-signalling pathways. *Planta* 248(3):675–690. https://doi.org/10.1007/s00425-018-2915-9.

Jin X, Liu T, Xu J, Gao Z, Hu X (2019) Exogenous GABA enhances muskmelon tolerance to salinity-alkalinity stress by regulating redox balance and chlorophyll biosynthesis. *BMC Plant Biology* 19:48. https://doi.org/10.1186/s12870-019-1660-y.

Kang, S.M., Radhakrishnan, R., You, Y.H. et al. (2015). *Enterobacter asburiae* KE17 association regulates physiological changes and mitigates the toxic effects of heavy metals in soybean. *Plant Biology* 17: 1013–1022.

Kaspal, M., Kanapaddalagamage, M.H., and Ramesh, S.A. (2021). Emerging roles of γ aminobutyric acid (GABA) gated channels in plant stress tolerance. *Plants* 10 (10): 2178.

Khan, M.I., Jalil, S.U., Chopra, P. et al. (2021). Role of GABA in plant growth, development and senescence. *Plant Gene* 26: 100283.

Kinnersley, A.M. and Turano, F.J. (2000). Gamma aminobutyric acid (GABA) and plant responses to stress. *Critical Reviews in Plant Sciences* 19 (6): 479–509.

Koike, S., Matsukura, C., Takayama, M. et al. (2013). Suppression of γ-aminobutyric acid (GABA) transaminases induces prominent GABA accumulation, dwarfism and infertility in the tomato (*Solanum lycopersicum* L.). *Plant and Cell Physiology* 54 (5): 793–807.

Lang, J., Gonzalez-Mula, A., Taconnat, L. et al. (2016). The plant GABA signaling downregulates horizontal transfer of the *Agrobacterium tumefaciens* virulence plasmid. *The New Phytologist* 210 (3): 974–983.

Li Z, Huang T, Tang M, Cheng B, Peng Y, Zhang X (2019) iTRAQ-based proteomics reveals key role of γ-aminobutyric acid (GABA) in regulating drought tolerance in perennial creeping bentgrass (*Agrostis stolonifera*). *Plant Physiology and Biochemistry* 145:216–226. https://doi.org/10.1016/j.plaphy.2019.10.018.

Li, L., Dou, N., Zhang, H., and Wu, C. (2021). The versatile GABA in plants. *Plant Signaling & Behavior* 16 (3): 1862565.

Lu, W., He, L., Xue, C. et al. (2009). Advances in research of gamma-aminobutyric acid receptors of insects. *Chinese Bulletin of Entomology* 46 (1): 152–158.

Ma, Y., Wang, P., Wang, M. et al. (2019). GABA mediates phenolic compounds accumulation and the antioxidant system enhancement in germinated hulless barley under NaCl stress. *Food Chemistry* 270: 593–601.

MacGregor, K.B., Shelp, B.J., Peiris, S., and Bown, A.W. (2003). Overexpression of glutamate decarboxylase in transgenic tobacco plants deters feeding by phytophagous insect larvae. *Journal of Chemical Ecology* 29: 2177–2182.

Malekzadeh, P., Khara, J., and Heydari, R. (2014). Alleviating effects of exogenous gamma-aminobutiric acid on tomato seedling under chilling stress. *Physiology and Molecular Biology of Plants* 20: 133–137.

Mazzucotelli, E., Tartari, A., Cattivelli, L., and Forlani, G. (2006). Metabolism of γ-aminobutyric acid during cold acclimation and freezing and its relationship to frost tolerance in barley and wheat. *Journal of Experimental Botany* 57 (14): 3755–3766.

McLean, M.D., Yevtushenko, D.P., Deschene, A. et al. (2003). Overexpression of glutamate decarboxylase in transgenic tobacco plants confers resistance to the northern root-knot nematode. *Molecular Breeding* 11: 277–285.

Mohammadi, M., Aelaei, M., and Saidi, M. (2021). Pre-harvest spray of GABA and spermine delays postharvest senescence and alleviates chilling injury of gerbera cut flowers during cold storage. *Scientific Reports* 11 (1): 14166.

Molina-Rueda, J.J., Pascual, M.B., Pissarra, J., and Gallardo, F. (2015). A putative role for γ-aminobutyric acid (GABA) in vascular development in pine seedlings. *Planta* 241: 257–267.

Nayyar, H., Kaur, R., Kaur, S., and Singh, R. (2014). γ-Aminobutyric acid (GABA) imparts partial protection from heat stress injury to rice seedlings by improving leaf turgor and upregulating osmoprotectants and antioxidants. *Journal of Plant Growth Regulation* 33: 408–419.

Nikmaram, N., Dar, B.N., Roohinejad, S. et al. (2017). Recent advances in γ-aminobutyric acid (GABA) properties in pulses: an overview. *Journal of the Science of Food and Agriculture* 97 (9): 2681–2689.

Pal S, Zhao J, Khan A, Yadav NS, Batushansky A, Barak S, Rewald B, Fait A, Lazarovitch N, Rachmilevitch S (2016) Paclobutrazol induces tolerance in tomato to deficit irrigation through diversified effects on plant morphology, physiology and metabolism. *Scientific Reports* 6(1):39321. https://doi.org/10.1038/srep39321.

Palanivelu, R., Brass, L., Edlund, A.F., and Preuss, D. (2003). Pollen tube growth and guidance is regulated by POP2, an *Arabidopsis* gene that controls GABA levels. *Cell* 114 (1): 47–59.

Parmar, N., Singh, K.H., Sharma, D. et al. (2017). Genetic engineering strategies for biotic and abiotic stress tolerance and quality enhancement in horticultural crops: a comprehensive review. *3 Biotech* 7: 1–35.

Petřivalský, M., Brauner, F., Luhová, L. et al. (2007). Aminoaldehyde dehydrogenase activity during wound healing of mechanically injured pea seedlings. *Journal of Plant Physiology* 164 (11): 1410–1418.

Phillips, T. (2008). Genetically modified organisms (GMOs): transgenic crops and recombinant DNA technology. *Nature Education* 1 (1): 213.

Priya, M., Sharma, L., Kaur, R. et al. (2019). GABA (γ-aminobutyric acid), as a thermo-protectant, to improve the reproductive function of heat-stressed mungbean plants. *Scientific Reports* 9: 7788.

Rajani, M.S., Bedair, M.F., Li, H., and Duff, S.M.G. (2021). Phenotypic effects from the expression of a deregulated AtGAD1 transgene and GABA pathway suppression mutants in maize. *PLoS One* 16 (12): e0259365.

Ramos-Ruiz, R., Martinez, F., and Knauf-Beiter, G. (2019). The effects of GABA in plants. *Cogent Food & Agriculture* 5 (1): 1670553.

Ramputh, A.I. and Bown, A.W. (1996). Rapid [gamma]-aminobutyric acid synthesis and the inhibition of the growth and development of oblique-banded leaf-roller larvae. *Plant Physiology* 111 (4): 1349–1352.

Ramputh, A. and Shelp, A.W. (1996). Rapid y-aminobutyric acid synthesis and the inhibition of the growth and development of oblique-banded leaf-roller larvae. *Plant Physiology* 111 (4): 1349–1352.

Rani, S.J. and Usha, R. (2013). Transgenic plants: types, benefits, public concerns and future. *Journal of Pharmacy Research* 6 (8): 879–883.

Renault, H., Roussel, V., El Amrani, A. et al. (2010). The *Arabidopsis* pop2-1mutant reveals the involvement of GABA transaminase in salt stress tolerance. *BMC Plant Biology* 10 (1): 1–6.

Rezaei-Chiyaneh E, Seyyedi SM, Ebrahimian E, Moghaddam SS, Damalas CA (2018) Exogenous application of gamma-aminobutyric acid (GABA) alleviates the effect of water deficit stress in black cumin (*Nigella sativa* L.). *Industrial Crops and Products*;112:741–748. https://doi.org/10.1016/j.indcrop.2017.12.067

Salah A, Zhan M, Cao C, Han Y, Ling L, Liu Z, Li P, Ye M, Jiang Y (2019) γ-Aminobutyric acid promotes chloroplast ultrastructure, antioxidant capacity, and growth of waterlogged maize seedlings. *Scientific Reports* 9(1):484. https://doi.org/10.1038/s41598-018-36334-y.

Savatin, D.V., Gramegna, G., Modesti, V., and Cervone, F. (2014). Wounding in the plant tissue: the defense of a dangerous passage. *Frontiers in Plant Science* 5: 470.

Scholz, S.S., Reichelt, M., Mekonnen, D.W. et al. (2015). Insect herbivory-elicited GABA accumulation in plants is a wound-induced, direct, systemic, and jasmonate-independent defense response. *Frontiers in Plant Science* 6: 1128.

Scholz, S.S., Malabarba, J., Reichelt, M. et al. (2017). Evidence for GABA-induced systemic GABA accumulation in *Arabidopsis* upon wounding. *Frontiers in Plant Science* 8: 388.

Seifi HS, Curvers K, De Vleesschauwer D, Delaere I, Aziz A, Hofte M (2013) Concurrent overactivation of the cytosolic glutamine synthetase and the GABA shunt in the ABA-deficient sitiens mutant of tomato leads to resistance against *Botrytis cinerea*. *The New Phytologist* 199 (2): 490–504. https://doi.org/10.1111/nph.12283.

Seifikalhor, M., Aliniaeifard, S., Hassani, B. et al. (2019). Diverse role of γ-aminobutyric acid in dynamic plant cell responses. *Plant Cell Reports* 38: 847–867.

Serraj R, Shelp BJ, Sinclair TR (1998)Accumulation of γ-aminobutyric acid in nodulated soybean in response to drought stress. *Physiologia Plantarum* 102:79–86. https://doi.org/10.1034/j.1399-3054.1998.1020111.x.

Shelp, B.J., Bown, A.W., and Faure, D. (2006). Extracellular γ-aminobutyrate mediates communication between plants and other organisms. *Plant Physiology* 142 (4): 1350–1352.

Shelp, B.J., Aghdam, M.S., and Flaherty, E.J. (2021). γ-Aminobutyrate (GABA) regulated plant defense: mechanisms and opportunities. *Plants* 10 (9): 1939.

Shiu, M.S., Shyu, Y.T., and Wu, S.J. (2020). Flooding stress and high-pressure treatment enhance the GABA content of the vegetable soybean (Glycine max Merr.). *Agriculture* 10 (5): 175.

Sidhu OP, Annarao S, Pathre U, Snehi SK, Raj SK, Roy R, Tuli R, Khetrapal CL (2010) Metabolic and histopathological alterations of Jatropha mosaic begomovirus-infected *Jatropha curcas* L. by HR-MAS NMR spectroscopy and magnetic resonance imaging. *Planta* 232 (1): 85–93. https://doi.org/10.1007/s00425-010-1159-0.

Sita, K. and Kumar, V. (2020). Role of gamma amino butyric acid (GABA) against abiotic stress tolerance in legumes: a review. *Plant Physiology Reports* 25: 654–663.

Streeter JG, Thompson JF (1972) Anaerobic accumulation of γ-aminobutyric acid and alanine in radish leaves (*Raphanus sativus* L.). *Plant Physiology*;49:572–578. https://doi.org/10.1104/pp.49.4.572.

Su, N., Wu, Q., Chen, J. et al. (2019). GABA operates upstream of H^+-ATPase and improves salinity tolerance in *Arabidopsis* by enabling cytosolic K^+ retention and Na^+ exclusion. *Journal of Experimental Botany* 70 (21): 6349–6361.

Takahashi H, Matsumura H, Kawai-Yamada M, Uchimiya H (2008) The cell death factor, cell wall elicitor of rice blast fungus (*Magnaporthe grisea*) causes metabolic alterations including GABA shunt in rice cultured cells. *Plant Signaling & Behavior* 3 (11): 945–953. https://doi.org/10.4161/psb.6112

Takahashi, Y., Sasanuma, T., and Abe, T. (2013). Accumulation of gamma-aminobutyrate (GABA) caused by heat-drying and expression of related genes in immature vegetable soybean (edamame). *Breeding Science* 63 (2): 205–210.

Tarkowski LP, Signorelli S, Hofte M (2020) γ-Aminobutyric acid and related amino acids in plant immune responses: emerging mechanisms of action. *Plant, Cell and Environment* 43 (5): 1103–1116. https://doi.org/10.1111/pce.13734.

Thompson JF, Stewart CR, Morris CJ (1966) Changes in amino acid content of excised leaves during incubation I. The effect of water content of leaves and atmospheric oxygen level. *Plant Physiology* 41:1578–1584. https://doi.org/10.1104/pp.41.10.1578

Toubiana, D., Sade, N., Liu, L. et al. (2020). Correlation-based network analysis combined with machine learning techniques highlight the role of the GABA shunt in *Brachypodium sylvaticum* freezing tolerance. *Scientific Reports* 10 (1): 4489.

Wallace, W., Secor, J., and Schrader, L.E. (1984). Rapid accumulation of γ-aminobutyric acid and alanine in soybean leaves in response to an abrupt transfer to lower temperature, darkness, or mechanical manipulation. *Plant Physiology* 75 (1): 170–175.

Wang, Y., Luo, Z., Huang, X. et al. (2014). Effect of exogenous γ-aminobutyric acid (GABA) treatment on chilling injury and antioxidant capacity in banana peel. *Scientia Horticulturae* 168: 132–137.

Wang Y, Gu W, Meng Y, Xie T, Li L, Li J, Wei S (2017) γ-Aminobutyric acid imparts partial protection from salt stress injury to maize seedlings by improving photosynthesis and upregulating osmoprotectants and antioxidants. *Scientific Reports* 7:43609. https://doi.org/10.1038/srep43609

Wang G, Kong J, Cui D, Zhao H, Niu Y, Xu M, Jiang G, Zhao Y, Wang W (2019) Resistance against *Ralstonia solanacearum* in tomato depends on the methionine cycle and the γ-aminobutyric acid metabolic pathway. *The Plant Journal* 97:1032–1047. https://doi.org/10.1111/tpj.14175.

Wu, X., Jia, Q., Ji, S. et al. (2020). Gamma-aminobutyric acid (GABA) alleviates salt damage in tomato by modulating Na(+) uptake, the GAD gene, amino acid synthesis and reactive oxygen species metabolism. *BMC Plant Biology* 20: 465.

Yang R, Guo Q, Gu Z (2013) GABA shunt and polyamine degradation pathway on γ-aminobutyric acid accumulation in germinating fava bean (*Vicia faba* L.) under hypoxia. *Food Chemistry*136(1):152–159. https://doi.org/10.1016/j.foodchem.2012.08.008.

Yang R, Guo Y, Wang S, Gu Z (2015) Ca^{2+} and aminoguanidine on γ-aminobutyric acid accumulation in germinating soybean under hypoxia-NaCl stress. *Journal of Food and Drug Analysis* 23(2):287–293. https://doi.org/10.1016/j.jfda.2014.07.004

Yang, R., Feng, L., Wang, S. et al. (2016). Accumulation of γ-aminobutyric acid in soybean by hypoxia germination and freeze–thawing incubation. *Journal of the Science of Food and Agriculture* 96 (6): 2090–2096.

Yang J, Sun C, Zhang Y, Fu D, Zheng X, Yu T (2017) Induced resistance in tomato fruit by γ-aminobutyric acid for the control of alternaria rot caused by *Alternaria alternata*. *Food Chemistry* 221: 1014–1020. https://doi.org/10.1016/j.foodchem.2016.11.061

Yin, Y., Wang, C., Xiao, D. et al. (2021). Advances and perspectives of transgenic technology and biotechnological application in forest trees. *Frontiers in Plant Science* 12: 786328.

Zhang, C., Wohlhueter, R., and Zhang, H. (2016). Genetically modified foods: a critical review of their promise and problems. *Food Science and Human Wellness* 5 (3): 116–123.

Zhao, P., Li, W., Zhen, C. et al. (2021). Transcriptomic analysis of the effects of γ-aminobutyric acid treatment on browning and induced disease resistance in fresh-cut apples. *Postharvest Biology and Technology* 181: 111686.

Zhu, X., Liao, J., Xia, X. et al. (2019). Physiological and iTRAQ-based proteomic analyses reveal the function of exogenous γ-aminobutyric acid (GABA) in improving tea plant (*Camellia sinensis* L.) tolerance at cold temperature. *BMC Plant Biology* 19 (1): 1–20.

9

GABA in Plant Stress Response and Tolerance Mechanisms

Şerife Palabıyık, İrem Çetinkaya, Tülay Öztürk, and Melike Bor

Department of Biology, Faculty of Science, Ege University, Izmir, 35000, Turkey

Introduction

Gamma-aminobutyric acid (GABA) is a nonproteinic amino acid that was first isolated from potato tubers in 1949 before its discovery in animal brain extracts (Steward et al. 1949). GABA is localized in mammalian neuron terminals and interacts with $GABA_A$ (ionotropic) and $GABA_B$ (metabotropic) receptors for the inhibition of neurotransmission. Membrane hyperpolarization prevents the transport of Cl^- into mature neurons and the activation of excitatory neurotransmitters in the brain, which is called the "relaxing effect" (Ramesh et al. 2017). Although GABA receptors have been identified in the nervous system, where they regulate brain function and development, GABAergic receptors have been identified to function beyond neuronal cells in different tissues such as other human organs. Although the role of GABA has been well-defined in animals for about 60 years, it is still an ongoing story in plants. It was mostly identified as a carbon–nitrogen metabolite; however, recent studies indicated its plausible function as a signaling molecule. Its concentration in plant tissues is variable and a rapid increase in quantity was observed following exposure to abiotic and biotic stresses, which was almost 1000-fold. GABA metabolism contributes to plant growth, development, and stress adaptation. In addition, the identification of the regulation (inhibition) of aluminum-activated malate transporters by GABA and its effects on tissue growth and stomatal behavior proved GABA's function as a signaling molecule (Ramesh et al. 2015). Besides the identification of GABA-regulated ion channels, researchers found that these channels are also regulated by drugs known to act on GABA receptors in animals.

GABA biosynthesis and metabolism occur in sequential reactions and are called GABA shunt in all eukaryotes (Jalil et al. 2021; Khan et al. 2021). GABA is synthesized from glutamate by the activity of glutamate decarboxylase (GAD) (Bouché et al. 2003) (Figure 9.1). GAD is a cytosolic enzyme that catalyzes the irreversible decarboxylation of glutamate via the Ca^{2+} calmodulin (CaM) protein complex. Unlike in other organisms, the GAD enzyme in plants contains a CaM-binding domain that is activated by acidic pH

GABA in Plants: Biosynthesis, Plant Development, and Food Security, First Edition. Edited by Samiksha Singh, Durgesh Kumar Tripathi, and Vijay Pratap Singh.

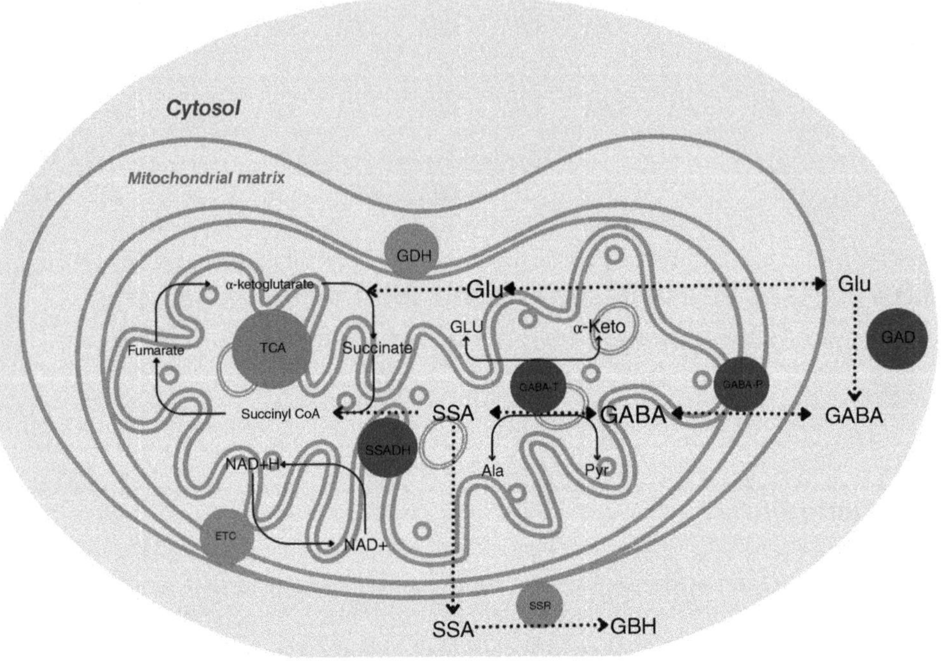

Figure 9.1 A general illustration of GABA shunt taking place in the cytosol and mitochondria. GABA is synthesized by GAD from glutamate in the cytosol and transported to mitochondria, and by a series of reactions, converted to succinate to enter TCA by this shortcut. **GDH**, glutamate dehydrogenase; **Glu**, glutamate; **GABA**, gamma amino butyric acid; **GABA-P**, GABA permease; **GABA-T**, GABA transaminase; **SSA**, succinic semialdehyde; **SSADH**, succinic semialdehyde dehydrogenase; **TCA**, tricarboxyl acetic acid cycle; **ETC**, electron transport chain; **SSR**, succinic semialdehyde reductase; **GBH**; gamma hydroxy butyrate.

(Shelp et al. 2017). GABA accumulates in response to cytosolic acidification. Therefore, it has been suggested that high GABA concentrations at low pH in plant tissues regulate cytosolic pH through Ca^{2+}/CaM-stimulated GAD activity, which consumes protons by converting glutamate to GABA (Ramesh et al. 2015). Apart from the synthesis of GABA within the cell, its translocation across the plasma membrane is carried out by the GABA permease enzyme, which belongs to the amino acid/polyamine/organocation transporter family (Khan et al. 2021). GABA catabolism depends on the GABA transaminase (GABA-T) enzyme, which has two types: α-ketoglutarate or glyoxylate-dependent and pyruvate-dependent (GABA-TP) (Takayama and Ezura 2015) (Figure 9.1). As a result of the activity of GABA-T, succinate semialdehyde (SSA) is synthesized from GABA in mitochondria, and it is either converted to succinate to enter the tricarboxylic acid (TCA) cycle by the activity of succinic semialdehyde dehydrogenase (SSADH) or metabolized in the cytosol. In plant cells, SSA can alternatively be converted to gamma-hydroxybutyric acid in the presence of NADP and H^+. This reaction occurs in the cytosol and/or chloroplast by the enzyme glyoxylate reductase (also known as succinic semialdehyde reductase—SSR), which becomes more remarkable under stress conditions (Simpson et al. 2008; Hildebrandt et al. 2015) (Figure 9.1).

Besides the above-given biosynthetic route, polyamine metabolism and proline are also associated with GABA biosynthesis in plants. GABA can be synthesized as a result of the degradation of polyamines putrescine (Put) and spermidine (Spd) (Li et al. 2021). Put is formed by the activity of ornithine decarboxylase or the enzyme arginine decarboxylase, which decarboxylates arginine. Spd, on the other hand, is formed from Put with *S*-adenosyl-Met (dcSAM) via the Spd synthase enzyme. Diamine oxidases and polyamine oxidases catabolize Put and Spd to form Δ1-pyrroline. GABA is synthesized from Δ1-pyrroline by the activity of pyrroline dehydrogenase (PDH) (Khan et al. 2021). Under oxidative stress, GABA biosynthesis can also occur via a nonenzymatic proline reaction. The proline levels increase rapidly in plant cells under stress conditions, which is correlated with the increased reactive oxygen species (ROS) levels. Proline interacts with the $OH^{\cdot}$ radical, leading to the formation of pyrroline and Δ1-pyrroline, respectively. GABA biosynthesis occurs from Δ1-pyrroline through the activity of the PDH enzyme (Signorelli et al. 2015; Khan et al. 2021).

Studies with *Arabidopsis thaliana* mutants provided information that GABA is among the important molecules for plant growth and development (Bor and Turkan 2019; Pelvan et al. 2021). The growth and orientation of the pollen tube is one of the first defined roles of GABA in plants (Palanivelu et al. 2003; Bor and Turkan 2019). A GABA gradient along the pollen tube is formed, and the highest concentration is reached near the micropyle, where the tube penetrates to the ovule (Palanivelu et al. 2003; Li et al. 2021). Pollen–pistil interaction 2 (POP2) encodes pyruvate-dependent GABA-TP, and the GABA gradient was disrupted in its mutants, which led to the inhibition of pollen tube growth, although GABA accumulation was observed in flowers (Palanivelu et al. 2003). GABA also plays a role in fruit ripening; its accumulation occurs in immature fruits, and as the fruits ripen, GABA levels decrease rapidly. In addition, GABA affects seed formation and germination, starch degradation, primary root, and hypocotyl elongation (Bouché and Fromm 2004; Ramos-Ruiz et al. 2019; Li et al. 2021).

The effective assimilation of C and N is essential for the growth and productivity of the plants. In general, GABA functions in nitrogen metabolism, storage, and/or transport. The balance between carbon and nitrogen is also maintained through the GABA shunt by transferring carbon from glutamate to succinate, which will enter TCA as a shortcut (Bouché and Fromm 2004; Ramos-Ruiz et al. 2019; Kaspal et al. 2021; Li et al. 2021). *Arabidopsis* can grow well in a non-nitrogen and GABA-containing culture medium, which indicates that GABA is an efficient nitrogen source. Barbosa et al. (2010) showed that GABA promotes nitrate uptake and facilitates plant adaptation to low nitrogen conditions. GABA application increases nitrate reductase activity and the phosphorylation of the active site of the enzyme is predicted to be associated with this effect (Barbosa et al. 2010; Bor and Turkan 2019). In addition, malfunction of the GABA-shunt leads to significant changes in sucrose and starch content and affects carbon metabolism of the cell wall (Bouché and Fromm 2004; Ramos-Ruiz et al. 2019; Kaspal et al. 2021; Li et al. 2021). All these and similar studies conducted, especially in *Arabidopsis*, prove that GABA has a wide range of physiological functions in the control of growth and development processes in plants (Bor and Turkan 2019).

The amount of GABA increases very rapidly as a result of exposure to any biotic and/or abiotic stress factor (Akçay et al. 2012; Dimlioglu et al. 2015; Pelvan et al. 2021; Li et al. 2021). Hypotheses suggesting that GABA also serves as a signaling molecule in plants have not been proven for many years due to the failure to detect a receptor or protein that interacts with GABA. However, the discovery of GABA's negative regulatory role on the voltage-dependent anion channels, aluminum-activated malate transporters (ALMTs) has made a very important contribution to our understanding of GABA signaling in wheat (Ramesh et al. 2015). ALMTs are members of the anion transporters family, and they are located in the plasma membrane with diverse expression patterns in different plant tissues (Khan et al. 2021). Compared to GABA receptors in other organisms, the GABA-binding domains of ALMTs share only a 12 amino acid region (FLFPVWAGEDVH) with mouse $GABA_A$ receptors (Ramesh et al. 2015; Fromm 2020). GABA negatively regulates these channels and stops anion flow (Ramesh et al. 2015; Ma et al. 2019; Fromm 2020; Khan et al. 2021; Xu et al. 2021). Another important finding regarding the interactions of ALMTs with GABA is related to the control of stomatal movements (Mekonnen et al. 2016). GABA increases rapidly under drought stress conditions, which binds to ALMT6 and prevents the entry of anions into the vacuole. In addition, it has been predicted that GABA may have an effect on stomatal movements by suppressing the genes encoding regulatory 14-3-3 proteins that play a role in light-induced stomatal opening (Lancien and Roberts 2006; Mekonnen et al. 2016).

In this review, we aimed to gather the most recent information regarding the role of GABA in stress responses and tolerance mechanisms in plants. Accumulation of GABA to high levels has been recognized as a general response to several environmental stress conditions; however, it is still not clear whether this is related to the activation of the GABA-biosynthetic route or inhibition of the GABA-catabolism route. Therefore, we would like to emphasize both the results from the exogenous GABA treatments and the endogenous organization of the GABA shunt under different environmental stress factors for a better understanding of GABA's function in plants. First, we will focus on abiotic stress in relation to GABA and GABA shunt, and then we will discuss these under the effect of different biotic stressors.

Abiotic Stress and GABA

Abiotic stressors such as salt, drought, high temperature, cold, heavy metals, flooding, nutrient deficiency, and toxicity have a negative impact on plant growth and productivity, which become more remarkable in a rapidly changing environment and climate. These stressors all affect agricultural lands irreversibly worldwide, and based on FAO reports, more than 96.5% of agricultural lands are reported to be influenced adversely by abiotic stresses (Cramer et al. 2011; Kopecká et al. 2023). The global climate change that we face as a result of global warming caused by increasing greenhouse gas emissions affects plants all over the world. However, the impact might be more pronounced in certain areas (e.g., high water demand in arid areas and forest fires in Mediterranean regions). Plants that are well-adapted to such harsh conditions have evolved different and efficient stress-coping mechanisms, and for a long time, plant scientists have tried to understand and analyze these mechanisms to introduce them to crop plants. Traits such as abiotic stress tolerance are polygenic in character, and it is difficult to increase tolerance by manipulating them. Hence, increased yield in crop plants always comes with a penalty for agronomic characteristics. Instead of this approach, targeting increasing yield under stress conditions by using common metabolic routes and precursors might be a better option, which depends on understanding how these routes and biosynthetic pathways are controlled along with their similarities and differences in diverse plants. Consequently, GABA and GABA-shunt might be considered by this approach due to their common and conserved nature among plants. Before presenting the details for the forenamed stresses and GABA, first, we would like to emphasize the interaction between Ca^{2+} signaling and GABA, which is a common mode of action in several abiotic stresses (Figure 9.2).

The CaM-binding domain of most GADs provides a link between GABA and Ca^{2+} under different abiotic stress conditions (Reddy et al. 2011; Shelp et al. 2012). As a general response to various environmental stresses, Ca^{2+} rapidly increases in the cytoplasm, leading to a series of signaling and regulatory processes for the transcription of defense genes in plant cells (Fromm 2020). The GAD is very sensitive to Ca^{2+} fluctuations, and the activation of GAD by increased Ca^{2+} levels enhances GABA biosynthesis, which subsequently affects the activity of ALMT channels. Wu et al. (2021) reported that, under hypoxia conditions, Ca^{2+} channels induced by ROS can be controlled by the increased GABA concentrations. They also showed that GABA can act as a ligand and directly control the movements and conductivity of the guard cell channels, which act on the potassium extrusion (guard cell outward rectifying K^+ channel) (Wu et al. 2021). On the other hand, many recent reports indicated that several stress-induced transcriptional changes also influence GAD activity. The expression of CaM-dependent AtGAD4 is induced by drought, oxygen deficiency, cold, and salt stresses, while CaM-independent CsGAD2 is induced by the combination of hypoxia with mechanical damage (Shelp et al. 2012; Gong et al. 2017).

Salt Stress and GABA

Salt stress is among the major abiotic stress factors that affect plant growth and development not only due to its osmotic effect but also its ion toxicity effect. According to Munns and Tester (2008), salt stress tolerance in plants depends on three components: osmotic stress tolerance, ability to export Na^+ and Cl^-, and tissue tolerance to Na^+ and Cl^- accumulation. Although the way salt stress is sensed by plant cells is still unknown, the three processes mentioned have been studied in detail. Sequesterilization of Na^+ and Cl^- in the vacuole and the accumulation of compatible solutes such as proline, glycine betaine, sugar oils, etc., in the cytoplasm to adjust osmotic pressure for the maintenance of normal metabolic processes are necessary to cope with salt stress (Munns and Tester 2008; Akçay et al. 2012; Kopecká et al. 2023). In general, Na^+ exclusion from the cytoplasm via plasma membrane Na^+/H^+ transporter (salt overly sensitive 1—SOS1) takes place by the activation of SOS-pathway components (Sekmen et al. 2014; Ma et al. 2022; Kopecká et al. 2023). Besides these well-defined defense strategies, halophytes exhibit extreme salt tolerance phenotypes with their special anatomic structures (salt gland and bladders) for efficient inter and intracellular partitioning of Na^+ and Cl^- (Munns and Tester 2008).

Various functions have been attributed to GABA regarding salt stress tolerance, and studies have generally reported that exogenous GABA treatment reduces membrane damage, alleviates adverse salinity effects, enhances antioxidative capacity, and increases tissue tolerance (Figure 9.2) (Barbosa et al. 2010; Xiang et al. 2016; Cheng et al. 2018; Wu et al. 2020).

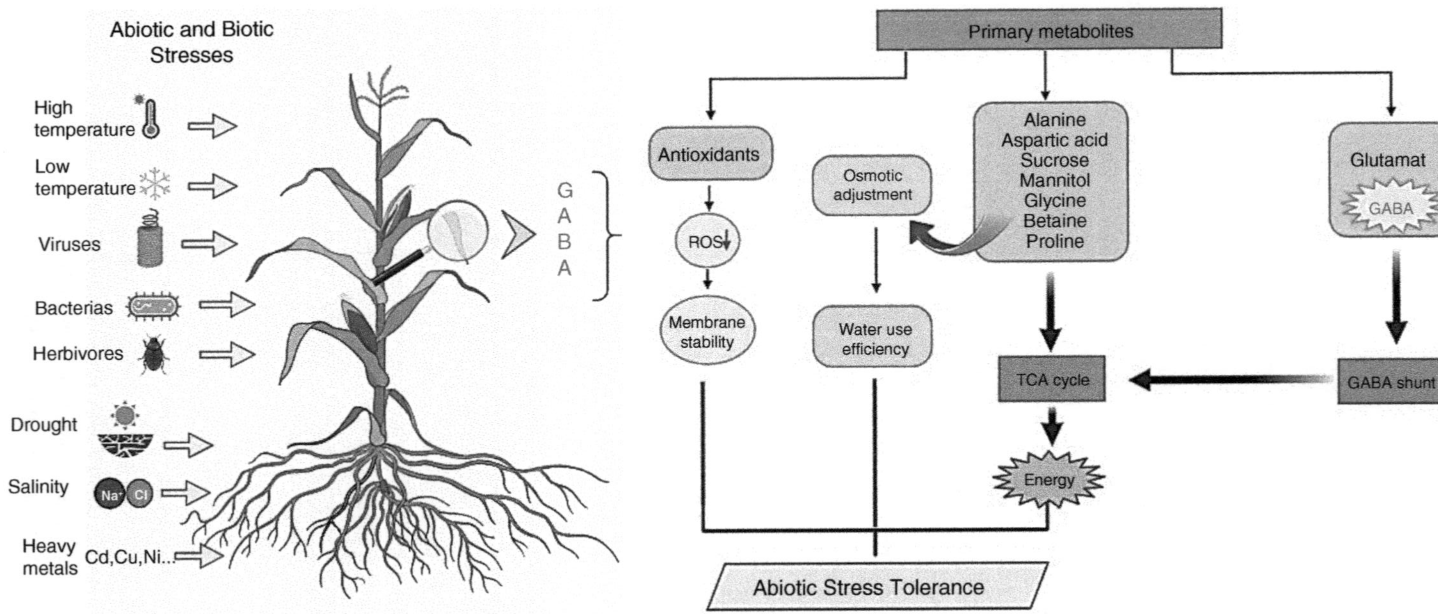

Figure 9.2 General aspects of abiotic and biotic stress tolerance interacting with GABA and GABA shunt.

Barbosa et al. (2010) found that antioxidant enzyme activities and nitrate uptake were enhanced by GABA. Accordingly, exogenous GABA treatment improved salt stress responses in tomatoes via decreased Na^+ transport and increased amino acid content (Wu et al. 2020). Cheng et al. (2018) found germination of clover seeds was possible at high salt levels by the exogenous GABA treatment. They reported that antioxidative defense, starch catabolism, osmotic adjustment, and maintenance of high K^+ levels were all enhanced by GABA. The accumulation of ROS is highly induced under salinity, which damages plasma and organelle membranes and inhibits enzyme activities. Not only the activation of antioxidative defense but also the limitation of ROS formation is important for salt tolerance. With this respect, exogenous GABA was reported to inhibit H_2O_2 accumulation in *Caragana intermedia* roots (Shi et al. 2010).

Salt stress increased GABA accumulation in tomato, tobacco, tea, and *A. thaliana* (Renault et al. 2010; Yin et al. 2010; Zhang et al. 2011). This rapid increment in GABA levels, as a general response to salinity, has long been accepted as evidence for GABA's signaling role. However, an increase in GABA content does not always take place with the activation of its biosynthesis (Bor and Turkan 2019). GABA-T is important for GABA metabolism, and Renault et al. (2010) reported that GABA-T-deficient *Arabidopsis* mutants were oversensitive to salt stress. The GABA-T enzyme, the first component of the TCA connection with the GABA-shunt, is very sensitive to oxidative stress, and increased GABA levels may sometimes be associated with inhibition of GABA-T activity. Moreover, GABA accumulation might result from the inhibition of GABA-T by the effect of high NADH and ADP in mitochondria under salinity (Allan et al. 2008; Akçay et al. 2012). To better define the role of this molecule as a protective or signaling component, the increase in its level must be analyzed along with the biosynthetic and GABA-shunt constituents.

Akçay et al. (2012) found that *Nicotiana sylvestris* CMSII mutant (lacks mitochondrial complex I) performed better under long-term salinity as compared to wild-type plants, which was related to altered glutamate dehydrogenase and GAD activities along with fluctuations in GABA levels. These results provided evidence for the involvement of GABA in the maintenance of C/N balance under salt stress rather than being a signaling molecule. Accordingly, GABA shunt is known to provide an alternative carbon source for TCA, and increased GABA levels facilitated resilience against salt stress in wheat (Shelp et al. 2012). In a different study, salt stress tolerance through GABA was found to be related to enhanced nitric oxide (NO) levels and regulation of $Na^+ : K^+$ balance in muskmelon plants (Xu et al. 2021). Furthermore, GABA-T overexpression in *Morus multicaulis* leads to decreased GABA levels and sensitivity to salt stress (Zhang et al. 2022).

Drought and GABA

Among the environmental stresses maybe drought is the most remarkable one due to its severe effects on crop yield and performance. Climate change further increases the effect of drought on plants by increasing the evaporation and transpiration rates, changing rainfall regimes, and the water retention capacity of the soil (Imran et al. 2021). Drought affects germination rates, root and shoot growth, primary metabolism, the relative water content of the leaves, and nutrient use efficiency, which all cause yield reductions in agricultural plants (Figure 9.2). Control of water loss through the leaves is the most important concept of drought tolerance. Efficient osmotic adjustment, control of stomatal movements, and deep rooting systems are among the well-known mechanisms in tolerant plants (Pelvan et al. 2021). Recently, evidence has been provided for the control of stomatal movements by GABA under drought stress (Xu et al. 2021). Xu et al. (2021) showed that GABA fine-tunes stomatal pore size by interacting with guard cell located ALMT9, which resulted in increased water-use efficiency and reduced transpiration for improved drought tolerance. Accordingly, the sensitivity of *Arabidopsis* gad1/2 mutant to water deficit has been proposed to be related to defective stomatal aperture and closure related to low GABA levels (Mekonnen et al. 2016).

In addition to the direct effects of GABA accumulation under drought stress, exogenous GABA treatments also alleviated the negative effects of drought in several plants, including tomato, apple, white clover, grapevine, strawberry, and lettuce, not only by affecting defense processes but also increasing endogenous GABA levels (Kalhor et al. 2018; Malabarba et al. 2019; Gramazio et al. 2020; Golnari et al. 2021; Liu et al. 2021; Zhou et al. 2021; Hayat et al. 2023). Alleviation of drought stress by exogenous GABA treatments was reported to be related to increased relative water content, antioxidative capacity, accumulation of osmoprotective molecules and reduced lipid peroxidation, electrolyte leakage (Vijayakumari and Puthur 2016; Yong et al. 2017). Yong et al. (2017) found that exogenous GABA leads to enhanced GABA-T and reduced GAD activities, which also affect glutamate and GABA levels. Moreover, the decline in drought stress tolerance was remarkable in *Arabidopsis* gad mutants with low GABA concentrations (Mekonnen et al. 2016).

Accumulation of glutamate is closely related to GABA shunt which might also be the case for proline. Drought-induced proline content is a common phenomenon for several plants, and as indicated before, the increased oxidative stress might lead to nonenzymatic GABA synthesis (Signorelli et al. 2015). Accordingly, Pelvan et al. (2021) proposed that on the first day of drought treatment, decreased proline levels might be related to GABA accumulation, which is a result of the above-mentioned nonenzymatic process, which might link GABA shunt and proline metabolism.

High Temperature and GABA

High temperature is among the major environmental constraints that limit the growth and development of plants since most of the metabolic processes take place at certain optimum temperatures. Especially extremely high temperatures and exposure to them for a long period of time severely affect photosynthesis, fertility, metabolism, defense, plant growth, and development (Figure 9.2) (Hayes et al. 2021). Up to date, several stress coping mechanisms for high-temperature stress have been defined, including biosynthesis of protective molecules (antioxidants, transporters, heat shock proteins, osmoprotectant molecules, etc.) and biochemical alterations in plasma and organelle membrane compositions (adjustment of fatty acid content, etc.) (Hasanuzzaman et al. 2013).

High-temperature stress adversely affects fertility in plants, and GABA's protective role might function in protecting reproductive organs with its well-defined role in the growth and orientation of the pollen tube (Palanivelu et al. 2003; Bor and Turkan 2019). There is a GABA gradient along the pollen tube, and GABA levels reach their highest concentration near the micropyle, where the tube penetrates the ovule (Palanivelu et al. 2003; Li et al. 2021). *POP2* encodes pyruvate-dependent GABA-T, and pollen tube growth was inhibited along with the disrupted GABA gradient in *Arabidopsis POP2* mutants (Palanivelu et al. 2003). Priya et al. (2019) found that heat stress leads to decreased GABA concentrations in mung bean plants, while exogenous GABA (1 mM) rescued pollen germination and viability along with improved heat stress tolerance (higher efficiency in photosynthesis, higher leaf water content enhanced levels of proline and trehalose with less damage to membranes). On the contrary, endogenous GABA levels were high in heat-stressed sesame plants, which were proposed to function in perception rather than protection (Bor et al. 2009).

Thermomorphogenesis, which is defined as the alteration of morphology and development such as early flowering, expansion of leaf, and deep rooting under high temperature, might also be the process where GABA shunt may function. Hypersensitivity to heat stress was reported in *Arabidopsis* SSADH mutants with higher GABA and H_2O_2 accumulation (Fait et al. 2005; Miyashita and Good 2008). Upon exposure to 42 °C rather than other amino acids, GABA levels were highly induced (1.85 and 3.24 μmol/g FM at the 2nd and 24th hours) in *Vigna unguiculata* L. (Shelp et al. 2021). On the other hand, studies with different plants indicated that the improved high-temperature stress tolerance upon exogenous GABA treatments was dependent on enhanced levels of enzymatic and nonenzymatic antioxidants (Nayyar et al. 2014; Li et al. 2018, 2021; Liu et al. 2019).

Cold Stress and GABA

Temperatures below the optimal growth conditions are a stressful environmental factor for plant growth and regulation, affecting yield and performance. The current cold stress definition is categorized into chilling and freezing stresses since the temperatures between 0 and 20 °C are referred to as "chilling" while the ones below 0 °C are known as "freezing" (Guo et al. 2018). As is the case for heat stress, tolerance to cold stress is related to the ability of a plant cell to accommodate rapid transitions in plasma and organelle membrane structures (Karabudak et al. 2014; Zhu 2016; Guo et al. 2018). Besides that, the accumulation of quaternary amines, sugars, polyols, and amino acids to avoid dehydration by facilitating osmotic adjustment and the protection of proteins and membranes is another strategy to cope with cold stress (Mazzucotelli et al. 2006; Maruyama et al. 2009; Kidokoro et al. 2017). In relation to its rapid increase and molecular properties under cold stress, GABA was proposed to be a compatible osmoprotectant molecule (Shelp et al. 1999; Mazzucotelli et al. 2006). However, clear evidence has not been provided to date, and this hypothesis is still under debate.

GABA content was increased in wheat and barley seedlings under cold stress along with the induced GABA-shunt-related transcripts and freezing degrees leading to the accumulation of GABA in *Brachypodium sylvaticum* (Mazzucotelli et al. 2006; Toubiana et al. 2020; Li et al. 2021). Besides a few reports on low levels of GABA under cold stress, a vast number of studies indicated that GABA and GABA shunt are responsive to cold stress. Whether by exogenous GABA treatments or induced endogenous accumulation, there is evidence of GABA's involvement in cold stress tolerance (Figure 9.2). Several studies indicated the accumulation of GABA was positively related to enhanced antioxidative activities, proline accumulation, decreased lipid peroxidation, and ROS levels along with enhanced NO (Li et al. 2021). GABA and GABA-shunt components are probably also related to cold stress sensing in plant cells. Ca^{2+} signaling facilitated through glutamate receptors (GLRs) is important for cold stress sensing, and the glutamate receptor channel-like gene AtGLR3.4 is reported to be stimulated by cold stress. Therefore, cold stress-induced fluctuations in cytosolic calcium levels might affect GAD activity for GABA biosynthesis, leading to the accumulation of glutamate (Meyerhoff et al. 2005; Weiland et al. 2016; Zhu 2016; Guo et al. 2018).

Heavy Metal Stress and GABA

Heavy metals such as Fe, Mn, Cu, Ni, Zn, Cd, Pb, Hg, and Co are accumulated at high levels in agricultural lands due to increased sewage and industrial waste disposals, fertilizer applications, and gas emissions (Ghori et al. 2019). Severe effects of heavy metals arise from their interaction with enzymes and proteins either by substitution with essential metals or by changing their valency by the oxidative status of the cell. The integrity of membranes is influenced by heavy metals, which lead to the differentiation in photosynthesis, respiration, and several metabolic processes. Mitigation of adverse effects of heavy metal stress either by exogenous GABA treatment or by induced endogenous GABA and GABA-shunt components has been reported in different plants (Figure 9.2) (Dubey et al. 2010; Akcan-Das et al. 2016; Yang et al. 2020; Li et al. 2021).

Exogenous GABA treatment increased organic acid (malate, citrate, etc.) contents in apples, which might facilitate the detoxification of metal ions, and Cd levels in apple seedlings were proposed to be controlled by control of the Cd-transport-related transcripts (Li et al. 2020, 2021). Similarly, antioxidative enzyme activities were increased by exogenous GABA treatment in barley seedlings, which mitigated the adverse effects of Al toxicity, especially on membranes (Song et al. 2010). However, several points of such interactions should be analyzed in more detail, and in the case of heavy metal acquisition or tolerance, GABA might have different functions. For instance, GABA's role as a signaling molecule through ALMT transporters was first defined by Ramesh et al. (2015), which provided evidence that the inhibition of malate efflux by exogenous GABA decreased tolerance to Al^{+3} in wheat roots (Ramesh et al. 2015).

Akcan-Das et al. (2016) found a discrepancy in GABA levels and GABA shunt in response to high and medium levels of Zn. Along with glutamate content and GAD activity, GABA levels induced in 10 μmol Zn-treated tobacco plants provided maintenance of growth and photosynthetic efficiency, while 100 μmol Zn-reduced GABA content, implicating adverse effects of Zn toxicity (Akcan-Das et al. 2016). Rice seedlings were severely affected by arsenic, while exogenous GABA led to increased GABA accumulation in the long-term induced tolerance to arsenic (Kumar et al. 2017). Variations were reported about GABA content and its remedial effects in different plants against heavy metal stress. For instance, Cd toxicity in duckweed reduced endogenous GABA content, and this was not reversed by exogenous treatments (Yang et al. 2020). The induced GABA levels do not always refer to increased tolerance against heavy metal stress, probably due to the interaction of these metal ions with GABA-regulated processes at different levels and concentrations.

Biotic Stress and GABA

Plants encounter different biotic interactions with several organisms, including bacteria, viruses, fungi, nematodes, insects, mites, and parasitic plants, which has a major negative impact on their growth, development, and survival. Therefore, it is usually recognized as biotic stress; hence, coping with another organism's defense, invasive properties, and metabolism is a difficult and challenging task. Plant diseases that arise from these biotic stress factors have different symptoms, such as leaf spots, nutrient deficiency-related symptoms, wilting, local lesions, vascular wilts, stunning, malformations, and direct damage due to feeding and sucking (Walters et al. 2010; Schumann and D'Arcy 2010; Iqbal et al. 2021). Plants evolved constitutive and inducible defense processes for protection against biotic stressors. Recognition of the biotic stress factor and the activation of the transcriptional program facilitating the defense processes is well explained in the "zigzag" model of Jones and Dangl (2006) (Bigeard et al. 2015; Vo et al. 2021). Briefly, it explains a series of interacting events through the pathogen-associated molecular pattern-triggered immunity and effector-triggered immunity (ETI) (Bigeard et al. 2015). Upon activation of defense, biosynthesis or release of conjugated forms of various sorts and types of molecules, including hormones (ABA, jasmonic acid-JA, salicylic acid-SA, and ethylene-ET), secondary metabolites (phenolic, phytoalexins), peptides, proteins, ROS, etc. is activated by defense-related processes through gene transcription (Jain et al. 2019).

General studies showed that GABA levels increased in response to biotic stress in relation to enhanced GAD activities (Figure 9.2) (Li et al. 2021). GABA accumulation was reported to be triggered as soon as the recognition of the pathogen in *Phaseolus vulgaris* L., *Solanum lycopersicon* L., and *Glycine max* L. (O'Leary et al. 2016; Copley et al. 2017; Wang et al. 2019). Infection of *A. thaliana* L. leaves with *Pseudomonas syringae* DC 3000-induced GAD expression leading to increased GABA content while increased GABA-T activity was found in rice upon *Pyricularia oryzae* infection (Wu et al. 2006; Ward et al. 2010). On the contrary, increased GABA levels upon exposure to biotic stress are not always accompanied by the GAD biosynthetic route. Forlani et al. (2014) reported that *P. oryzae* lysate-treated rice cell suspensions had high GABA content without the activation of GAD, and high polyamine catabolism was reported to be involved in GABA accumulation against *Aspergillus flavus* infection in maize (Majumdar et al. 2019) and *Botrtytis cinerea* infection in *Vitis vinifera* L. (Hatmi et al. 2015).

Several aspects of GABA and GABA shunt were analyzed, and all indicated a direct or indirect involvement in defense against biotic stress. Transgenic tobacco plants with GAD overexpression were resistant to insect larvae with high GABA levels (MacGregor et al. 2003). Quorum-sensing in *Agrobacterium tumefaciens* was inhibited by GABA accumulation (Haudecoeur et al. 2009). GABA accumulation elicited by herbivory was a jasmonate-independent defense response (Scholz et al. 2015). Yang et al. (2017) found that GABA-activated antioxidative capacity was important for scavenging ROS in tomato plants against *Alternaria alternate* pathogen. However, ROS accumulation is a strategy for limiting the invasion of pathogens, which is known as systemic acquired resistance, and plants use this strategy very efficiently (Iqbal et al. 2021). In light of this, how high GABA content for decreasing ROS by the activation of antioxidant enzymes fits this process is not clearly defined. Accordingly, Dimlioglu et al. (2015) reported that harpin (an ETI inducer protein isolated from *Erwinia amylovora*) activated defense responses through GABA and GABA shunt were not mediated by hydrogen peroxide (H_2O_2) accumulation. On the other hand, exogenous GABA treatment inhibited the infection of necrotrophic fungus *B. cinerea* by the activation of GABA shunt to limit cell death caused by induced H_2O_2 at the infection site in tomato plants (Seifi et al. 2013). Probably, the fine-tuning of ROS in the infected tissues is among the most critical functions of GABA in response to biotic stress factors (Tarkowski et al. 2020).

Interestingly, plant-derived GABA is used by pathogens as a carbon or nitrogen source. *P. syringae* pv. Tomato DC3000 and *Cladosporium fulvum* both use the apoplastic GABA content in tomatoes for a successful infection (Solomon and Oliver 2002; Rico and Preston 2008; Park et al. 2010). Dover and Halpern (1972) reported that a mutant *Escherichia coli* strain with high GABA-T activity was able to metabolize GABA from the plant source. However, this phenomenon might be related to the lifestyle of the pathogen, as indicated by Tarkowski et al. (2020), since nematodes and insects did not prefer a GABA-rich diet. Ramputh and Shelp (1996) reported that *Choristoneura rosaceana* larvae feeding on soybean plants were sensitive to GABA accumulation. Tobacco budworm larvae feed better on wild-type plants as compared to high GABA-accumulating tobacco mutants (MacGregor et al. 2003; Tarkowski et al. 2020).

Conclusion and Future Prospects

GABA and GABA-shunt components are among the interesting stress-responsive molecules in plant science. For decades, the exact mode of action of this nonprotein amino acid has been studied and each recent finding has led to several new questions. Besides the scientific curiosity, this versatile or Swiss army knife molecule (as indicated by different authors) has a high potential for improving crop yield and performance, especially under different environmental constraints. Stress resilience is the most necessary component of sustainable agriculture, and GABA plays an important function in mitigating stress responses by common or unique routes, as indicated in this review for both abiotic and biotic stresses. From this point of view, its potential for triggering both abiotic and biotic stress tolerance in crop plants is unlimited. However, exogenous GABA applications for coping with abiotic and biotic stress impact are limited and unsustainable. Increasing the endogenous GABA content and fine-tuning GABA-shunt in plants by different new approaches (like gene editing, TALEN, CRISPR-Cas, etc.) might be a solution to cope with the adverse effects of environmental stressors.

References

Akcan-Das, Z., Dimlioglu, G., Bor, M., and Ozdemir, F. (2016). Zinc induced activation of GABA-shunt in tobacco (*Nicotiana tabaccum* L.). *Environmental and Experimental Botany* 122: 78–84.

Akçay, N., Bor, M., Karabudak, T. et al. (2012). Contribution of gamma amino butyric acid (GABA) to salt stress responses of *Nicotiana sylvestris* CMSII mutant and wild type plants. *Journal of Plant Physiology* 169: 452–458.

Allan, W.L., Simpson, J.P., Clark, S.M., and Shelp, B.J. (2008). γ-Hydroxybutyrate accumulation in *Arabidopsis* and tobacco plants is a general response to abiotic stress: putative regulation by redox balance and glyoxylate reductase isoforms. *Journal of Experimental Botany* 59 (9): 2555–2564.

Barbosa, J.M., Singh, N.K., Cherry, J.H., and Locy, R.D. (2010). Nitrate uptake and utilization is modulated by exogenous γ-aminobutyric acid in *Arabidopsis thaliana* seedlings. *Plant Physiology and Biochemistry* 48: 443–450.

Bigeard, J., Colcombet, J., and Hirt, H. (2015). Signaling mechanisms in pattern-triggered immunity (PTI). *Molecular Plant* 8: 521–559.

Bor, M. and Turkan, I. (2019). Is there a room for GABA in ROS and RNS signalling? *Environmental and Experimental Botany* 161: 67–73.

Bor, M., Seckin, B., Ozgur, R. et al. (2009). Comparative effects of drought, salt, heavy metal and heat stresses on gamma aminobutryric acid levels of sesame (*Sesamum indicum* L.). *Acta Physiologiae Plantarum* 31: 655–659.

Bouché, N. and Fromm, H. (2004). GABA in plants: just a metabolite? *Trends in Plant Science* 9: 110–115.

Bouché, N., Fait, A., Bouchez, D. et al. (2003). Mitochondrial succinic semialdehyde dehydrogenase of the γ-aminobutyrate shunt is required to restrict levels of reactive oxygen intermediates in plants. *Proceedings of the National Academy of Sciences of the United States of America* 100: 6843–6848.

Cheng, B., Li, Z., Liang, L. et al. (2018). The γ-aminobutyric acid (GABA) alleviates salt stress damage during seeds germination of white clover associated with Na^+/K^+ transportation, dehydrins accumulation, and stress-related genes expression in white clover. *International Journal of Molecular Sciences* 19: 2520.

Copley, T.R., Aliferis, K.A., Kliebenstein, D.J., and Jabaji, S.H. (2017). An integrated RNAseq-^{1}H NMR metabolomics approach to understand soybean primary metabolism regulation in response to *Rhizoctonia foliar* blight disease. *BMC Plant Biology* 17: 84.

Cramer, G.R., Urano, K., Delrot, S. et al. (2011). Effects of abiotic stress on plants: a systems biology perspective. *BMC Plant Biology* 11: 163.

Dimlioglu, G., Das, Z.A., Bor, M. et al. (2015). The impact of GABA in harpin elicited biotic stress responses in *Nicotiana tabaccum*. *Journal of Plant Physiology* 188: 51–57.

Dover, S. and Halpern, Y. (1972). Control of the pathway of γ-aminobutyrate breakdown in *Escherichia coli* K-12. *Journal of Bacteriology* 110: 165–170.

Dubey, S., Misra, P., Dwivedi, S. et al. (2010). Transcriptomic and metabolomic shifts in rice roots in response to Cr (VI) stress. *BMC Genomics* 11: 648.

Fait, A., Yellin, A., and Fromm, H. (2005). GABA shunt deficiencies and accumulation of reactive oxygen intermediates: insight from *Arabidopsis* mutants. *FEBS Letters* 579: 415–420.

Forlani, G., Funck, D., and Giberti, S. (2014). D1-pyrroline-5-carboxylate reductase from Arabidopsis thaliana: stimulation or inhibition by chloride ions and feedback regulationby proline depend on whether NADPH or NADH acts as co-substrate. *The New Phytologist* 202: 911–919.

Fromm, H. (2020). GABA Signaling in plants – targeting the missing pieces of the puzzle. *Journal of Experimental Botany* 71: 6238–6245.

Ghori, N.-H., Ghori, T., Hayat, M.Q. et al. (2019). Heavy metal stress and responses in plants. *International journal of Environmental Science and Technology* 16: 1807–1828.

Golnari, S., Vafaee, Y., Nazari, F., and Ghader, N. (2021). Gamma-aminobutyric acid (GABA) and salinity impacts antioxidative response and expression of stress-related genes in strawberry cv. Aromas. *Brazilian Journal of Botany* 44: 639–651.

Gong Z, Xiong, L., and Shi, H. (2017). Plant abiotic stress response and nutrient use efficiency. *Science China Life Sciences* 63: 635–674.

Gramazio, P., Takayama, M., and Ezura, H. (2020). Challenges and prospects of new plant breeding techniques for GABA improvement in crops: tomato as an example. *Frontiers in Plant Science* 11: 577980.

Guo, X., Liu, D., and Chong, K. (2018). Cold Signaling in plants: insights into mechanisms and regulation. *Journal of Integrative Plant Biology* 60: 745–756.

Hasanuzzaman, M., Nahar, K., Alam, M.M. et al. (2013). Physiological, biochemical, and molecular mechanisms of heat stress tolerance in plants. *International Journal of Molecular Sciences* 14 (5): 9643–9684.

Hatmi, S., Gruau, C., Trotel-Aziz, P. et al. (2015). Drought stress tolerance in grapevine involves activation of polyamine oxidation contributing to improved immune response and low susceptibility to *Botrytis cinerea*. *Journal of Experimental Botany* 66: 775–787.

Haudecoeur, E., Planamente, S., Cirou, A. et al. (2009). Proline antagonizes GABA-induced quenching of quorum-sensing in *Agrobacterium tumefaciens*. *Proceedings of the National Academy of Sciences of the United States of America* 106: 14587–14592.

Hayat, F., Khan, U., Li, J. et al. (2023). Aminobutyric acid (GABA): a key player in alleviating abiotic stress resistance in horticultural crops: current insights and future directions. *Horticulture* 9: 647.

Hayes, S., Schachtschabel, J., Mishkind, M. et al. (2021). Hot topic: thermosensing in plants. *Plant, Cell & Environment* 44: 2018–2033.

Hildebrandt, T.M., Nunes Nesi, A., Araújo, W.L., and Braun, H.P. (2015). Amino acid catabolism in plants. *Molecular Plant* 8: 1563–1579.

Imran, A., Zia, R., Nawaz, M.S. et al. (2021). Plant survival under drought stress: implications, adaptive responses, and integrated rhizosphere management strategy for stress mitigation. *Microbiological Research* 242: 126626.

Iqbal, Z., Iqbal, M.S., Hashem, A. et al. (2021). Plant defense responses to biotic stress and its interplay with fluctuating dark/light conditions. *Frontiers in Plant Science* 12: 631810.

Jain, P., Dubey, H., Singh, P.K. et al. (2019). Deciphering signalling network in broad spectrum near isogenic lines of rice resistant to *Magnaporthe oryzae*. *Scientific Reports* 9: 16939.

Jalil, U., Ansari, M.I., Jalil, S.U., and Ansari, S.A. (2021). GABA shunt: a key-player in mitigation of ROS during stress. *Plant Growth Regulation* 94: 131–149.

Jones, J.D. and Dangl, J.L. (2006). The plant immune system. *Nature* 444: 323–329.

Kalhor, M.S., Aliniaeifard, S., Seif, M. et al. (2018). Enhanced salt tolerance and photosynthetic performance: implication of ɤ-amino butyric acid application in salt-exposed lettuce (*Lactuca sativa* L.) plants. *Plant Physiology and Biochemistry* 130: 157–172.

Karabudak, T., Bor, M., Özdemir, F., and Türkan, İ. (2014). Glycine betaine protects tomato (*Solanum lycopersicum*) plants at low temperature by inducing fatty acid desaturase7 and lipoxygenase gene expression. *Molecular Biology Reports* 41: 1401–1410.

Kaspal, M., Kanapaddalagamage, M.H., and Ramesh, S.A. (2021). Emerging roles of γ aminobutyric acid (GABA) gated channels in plant stress tolerance. *Plants (Basel)* 10: 2178.

Khan, M.I.R., Jalil, S.U., Chopra, P. et al. (2021). Role of GABA in plant growth, development and senescence. *Plant Gene* 26: 2352-4073.

Kidokoro, S., Yoneda, K., Takasaki, H. et al. (2017). Different cold-signaling pathways function in the responses to rapid and gradual decreases in temperature. *The Plant Cell* 29 (4): 760–774.

Kopecká, R., Kameniarová, M., Cerný, M. et al. (2023). Abiotic stress in crop production. *International Journal of Molecular Sciences* 24: 6603.

Kumar, S., Beena, A.S., Awana, M., and Singh, A. (2017). Physiological, biochemical, epigenetic and molecular analyses of wheat (*Triticum aestivum*) genotypes with contrasting salt tolerance. *Frontiers in Plant Science* 8: 1151.

Lancien, M. and Roberts, M.R. (2006). Regulation of *Arabidopsis thaliana* 14-3-3 gene expression by gamma-aminobutyric acid. *Plant, Cell & Environment* 29: 1430–1436.

Li, Z., Cheng, B., Liang, L. et al. (2018). The γ-aminobutyric acid (GABA) alleviates salt stress damage during seeds germination of white clover associated with Na^+/K^+ transportation, dehydrins accumulation, and stress-related genes expression in white clover. *International Journal of Molecular Sciences* 19: 2520.

Li, Y., Li, J.Q., Xue, S.Y. et al. (2020). Effects of Cu, Zn and Cd on larval development and respiratory metabolism of pacific abalone (Haliotis discus hannai). *Aquaculture Research* 51: 4255–4266.

Li, L., Dou, N., Zhang, H., and Wu, C. (2021). The versatile GABA in plants. *Plant Signaling & Behavior* 16 (3): 1862565.

Liu, Q., Piao, S., Chen, A. et al. (2019). Plant phenology and global climate change: current progresses and challenges. *Global Change Biology* 25: 1922–1940.

Liu, C., Wang, H., Zhang, X. et al. (2021). Activation of the aba signal pathway mediated by GABA improves the drought resistance of apple seedlings. *International Journal of Molecular Sciences* 22: 12676.

Ma, Y., Wang, P., Wang, M. et al. (2019). GABA mediates phenolic compounds accumulation and the antioxidant system enhancement in germinated hulless barley under NaCl stress. *Food Chemistry* 270: 593–601.

Ma, X., Gu, J., Luo, Q. et al. (2022). The synergistic benefits of β-aminobutyric acid and γ-aminobutyrate on salt and drought tolerance in cassava. *Plant Biotechnology Reports* 16: 29–41.

MacGregor, K.B., Shelp, B.J., Peiris, S., and Bown, A.W. (2003). Overexpression of glutamate decarboxylase in transgenic tobacco plants deters feeding by phytophagous insect larvae. *Journal of Chemical Ecology* 29: 2177–2182.

Majumdar, R., Minocha, R., Lebar, M.D. et al. (2019). Contribution of maize polyamine and amino acid metabolism toward resistance against *Aspergillus flavus* infection and aflatoxin production. *Frontiers in Plant Science* 10: 692.

Malabarba, J., Reichelt, M., and Pasquali, G. (2019). Tendril coiling in grapevine: jasmonates and a new role for GABA? *Journal of Plant Growth Regulation* 38: 39–45.

Maruyama, K., Takeda, M., Kidokoro, S. et al. (2009). Metabolic pathways involved in cold acclimation identified by integrated analysis of metabolites and transcripts regulated by DREB1A and DREB2A. *Plant Physiology* 150: 1972–1980.

Mazzucotelli, E., Tartari, A., Cattivelli, L., and Forlani, G. (2006). Metabolism of γ-aminobutyric acid during cold acclimation and freezing and its relationship to frost tolerance in barley and wheat. *Journal of Experimental Botany* 57: 3755–3766.

Mekonnen, D.W., Flügge, U.I., and Ludewig, F. (2016). Gamma-aminobutyric acid depletion affects stomata closure and drought tolerance of *Arabidopsis thaliana*. *Plant Science* 245: 25–34.

Meyerhoff, O., Müller, K., Roelfsema, M.R. et al. (2005). AtGLR3.4, a glutamate receptor channel-like gene is sensitive to touch and cold. *Planta* 222: 418–427.

Miyashita, Y. and Good, A.G. (2008). Glutamate deamination by glutamate dehydrogenase plays a central role in amino acid catabolism in plants. *Plant Signaling & Behavior* 3: 842–843.

Munns, R. and Tester, M. (2008). Mechanisms of salinity tolerance. *Annual Review of Plant Biology* 59: 651–681.

Nayyar, H., Kaur, R., Kaur, S., and Singh, R. (2014). γ-Aminobutyric acid (GABA) imparts partial protection from heat stress injury to rice seedlings by improving leaf turgor and upregulating osmoprotectants and antioxidants. *Journal of Plant Growth Regulation* 33: 408–419.

O'Leary, B.M., Neale, H.C., Geilfus, C.M. et al. (2016). Early changes in apoplast composition associated with defence and disease in interactions between *Phaseolus vulgaris* and the halo blight pathogen *Pseudomonas syringae* Pv. Phaseolicola. *Plant, Cell & Environment* 39: 2172–2184.

Palanivelu, R., Brass, L., Edlund, F., and Preuss, D. (2003). Pollen tube growth and guidance is regulated by POP2, an *Arabidopsis* gene that controls GABA levels. *Cell* 114 (1): 47–59.

Park, D.H., Mirabella, R., Bronstein, P.A. et al. (2010). Mutations gamma aminobutyric acid (GABA) transaminase genes in plants or *Pseudomonas syringae* reduce bacterial virulence. *The Plant Journal* 64: 318–330.

Pelvan, A., Bor, M., Yolcu, S. et al. (2021). Day and night fluctuations in GABA biosynthesis contribute to drought responses in *Nicotiana tabacum* L. *Plant Signaling & Behavior* 16: 1899672.

Priya, M., Sharma, L., Kaur, R. et al. (2019). GABA (γ-aminobutyric acid), as a thermo-protectant, to improve the reproductive function of heat stressed mung bean plants. *Scientific Reports* 9: 7788.

Ramesh, S.A., Tyerman, S.D., Xu, B. et al. (2015). GABA signalling modulates plant growth by directly regulating the activity of plant-specific anion transporters. *Nature Communications* 6: 7879.

Ramesh, S.A., Tyerman, S.D., Gilliham, M., and Xu, B. (2017). γ-Aminobutyric acid (GABA) signalling in plants. *Cellular and Molecular Life Sciences* 74: 1577–1603.

Ramos-Ruiz, R., Martinez, F., Knauf-Beiter, G., and Tejada Moral, M. (2019). The effects of GABA in plants. *Cogent Food & Agriculture* 5 (1): 1670553.

Ramputh, A. and Shelp, A.W. (1996). Rapid y-aminobutyric acid synthesis and the inhibition of the growth and development of oblique banded leaf roller larvae. *Plant Physiology* 111: 1349–1352.

Reddy, A.S., Ali, G.S., Celesnik, H., and Day, I.S. (2011). Coping with stresses: roles of calcium- and calcium/calmodulin-regulated gene expression. *The Plant Cell* 23: 2010–2032.

Renault, H., Roussel, V., El Amrani, A. et al. (2010). The *Arabidopsis* pop2-1 mutant reveals the involvement of GABA transaminase in salt stress tolerance. *BMC Plant Biology* 10: 20.

Rico, A. and Preston, G.M. (2008). *Pseudomonas syringae* pv. tomato DC3000 uses constitutive and apoplast-induced nutrient assimilation pathways to catabolize nutrients that are abundant in the tomato apoplast. *Molecular Plant-Microbe Interactions* 21 (2): 269–282.

Scholz, S.S., Reichelt, M., Mekonnen, D.W. et al. (2015). Insect herbivory-elicited GABA accumulation in plants is a wound-induced, direct, systemic, and jasmonate-independent defense response. *Frontiers in Plant Science* 6: 1128.

Schumann, G.L. and D'Arcy, C.J. (2010). *Essential Plant Pathology*. St. Paul, MI: American Phytopathological Society (APS Press).

Seifi, H.S., Curvers, K., De Vleesschauwer, D. et al. (2013). Concurrent over activation of the cytosolic glutamine synthetase and the GABA shunt in the ABA-deficient sitiens mutant of tomato leads to resistance against *Botrytis cinerea*. *The New Phytologist* 2013 (199): 490–504.

Sekmen, H., Ozgur, R., Uzilday, B., and Turkan, I. (2014). Reactive oxygen species scavenging capacities of cotton (*Gossypium hirsutum*) cultivars under combined drought and heat induced oxidative stress. *Environmental and Experimental Botany* 99: 141–149.

Shelp, B.J., Bown, A.W., and McLean, M.D. (1999). Metabolism and functions of gamma-aminobutyric acid. *Trends in Plant Science* 4: 446–452.

Shelp, B.J., Mullen, R.T., and Waller, J.C. (2012). Compartmentation of GABA metabolism raises intriguing questions. *Trends in Plant Science* 17 (2): 57–59.

Shelp, B.J., Bown, A.W., and Zarei, A. (2017). 4-Aminobutyrate (GABA): a metabolite and signal with practical significance. *Botany* 95: 1015–1032.

Shelp, B.J., Aghdam, M.S., and Flaherty, E.J. (2021). γ Aminobutyrate (GABA) regulated plant defense: mechanisms and opportunities. *Plants* 10: 1939.

Shi, S.Q., Shi, Z., Jiang, Z.P. et al. (2010). Effects of exogenous GABA on gene expression of *Caragana intermedia* roots under NaCl stress: regulatory roles for H_2O_2 and ethylene production. *Plant, Cell & Environment* 33: 149–162.

Signorelli, S., Dans, P.D., Coitiño, E.L. et al. (2015). Connecting proline and γ aminobutyric acid in stressed plants through non-enzymatic reactions. *PLoS One* 10 (3): e0115349.

Simpson, J.P., Allan, W.L., Clark, S.M., and Shelp, B.J. (2008). Gamma-hydroxybutyrate accumulation in *Arabidopsis* and tobacco plants is a general response to abiotic stress: putative regulation by redox balance and glyoxylate reductase isoforms. *Journal of Experimental Botany* 59: 2555–2564.

Solomon, P.S. and Oliver, R.P. (2002). Evidence that gamma-aminobutyric acid is a major nitrogen source during *Cladosporium fulvum* infection of tomato. *Planta* 214: 414–420.

Song, Y., Ligang, C., Liping, Z., and Diqiu, Y. (2010). Overexpression of OsWRKY72 gene interferes in the abscisic acid signal and auxin transport pathway of *Arabidopsis*. *Journal of Biosciences* 35: 459–471.

Steward, F., Thompson, J., and Dent, C. (1949). γ-Aminobutyric acid, a constituent of the potato tuber. *Science* 110: 439–440.

Takayama, M. and Ezura, H. (2015). How and why does tomato accumulate a large amount of GABA in the fruit? *Frontiers in Plant Science* 6: 612.

Tarkowski, L.P., Signorelli, S., and Hofte, M. (2020). γ-Aminobutyric acid and related amino acids in plant immune responses: emerging mechanisms of action. *Plant, Cell & Environment* 43 (5): 1103–1116.

Toubiana, D., Sade, N., Liu, L. et al. (2020). Correlation-based network analysis combined with machine learning techniques highlight the role of the GABA shunt in Brachypodium sylvaticum freezing tolerance. *Scientific Reports* 10: 4489.

Vijayakumari, K. and Puthur, J.T. (2016). γ-Aminobutyric acid (GABA) priming enhances the osmotic stress tolerance in *Piper nigrum* Linn. Plants subjected to PEG-induced stress. *Plant Growth Regulation* 78: 57–67.

Vo, K.T.X., Rahman, M.M., Rahman, M.M. et al. (2021). Proteomics and metabolomics studies on the biotic stress responses of rice: an update. *Rice* 14: 30.

Walters DR ,Newton AC, Fitt BD, Atkins SD, , Daniell TJ. (2010) Pathogenesis, parasitism and mutualism in the trophic space of microbe-plant interactions. *Trends in Microbiology* 18:365–373.

Wang, G., Kong, J., Cui, D. et al. (2019). Resistance against *Ralstonia solanacearum* in tomato depends on the methionine cycle and the γ-aminobutyric acid metabolic pathway. *The Plant Journal* 97: 1032–1047.

Ward, J.L., Forcat, S., Beckmann, M. et al. (2010). The metabolic transition during disease following infection of *Arabidopsis thaliana* by *Pseudomonas syringae* pv. tomato. *The Plant Journal* 63 (3): 443–457.

Weiland, M., Mancuso, S., and Baluska, F. (2016). Signalling via glutamate and GLRs in *Arabidopsis thaliana*. *Functional Plant Biology* 43 (1): 1–25.

Wu, C., Zhou, S., Zhang, Q. et al. (2006). Molecular cloning and differential expression of a c-aminobutyrate transaminase gene, OsGABA-T, in rice (*Oryza sativa*) leaves infected with blast fungus. *Journal of Plant Research* 119: 663–669.

Wu, X., Jia, Q., Ji, S. et al. (2020). Gamma-aminobutyric acid (GABA) alleviates salt damage in tomato by modulating Na^+ uptake, the GAD. Gene, amino acid synthesis and reactive oxygen species metabolism. *BMC Plant Biology* 20: 465.

Wu, Q., Su, N., Huang, X. et al. (2021). Hypoxia-induced increase in GABA content is essential for restoration of membrane potential and preventing ROS-induced disturbance to ion homeostasis. *Plant Communications* 2 (3): 100188.

Xiang, L., Hu, L., Xu, W. et al. (2016). Exogenous γ-aminobutyric acid improves the structure and function of photosystem II in muskmelon seedlings exposed to salinity-alkalinity stress. *PLoS One* 11 (10): e0164847.

Xu, J., Liu, T., Qu, F. et al. (2021). Nitric oxide mediates γ-aminobutyric acid-enhanced muskmelon tolerance to salinity–alkalinity stress conditions. *Scientia Horticulturae* 286: 110229.

Yang, J., Sun, C., Zhang, Y. et al. (2017). Induced resistance in tomato fruit by γ-aminobutyric acid for the control of alternaria rot caused by *Alternaria alternata*. *Food Chemistry* 221 (4): 1014–1020.

Yang, S., Gong, Z., Xiong, L. et al. (2020). Plant abiotic stress response and nutrient use efficiency. *Science China Life Sciences* 63: 635–674.

Yin, Y.G., Tominaga, T., Iijima, Y. et al. (2010). Metabolic alterations in organic acids and gamma-aminobutyric acid in developing tomato (*Solanum lycopersicum* L.) fruits. *Plant & Cell Physiology* 51: 1300–1314.

Yong, B., Xie, H., Li, Z. et al. (2017). Exogenous application of GABA improves PEG-induced drought tolerance positively associated with GABA-shunt, polyamines, and proline metabolism in white clover. *Frontiers in Physiology* 8: 1107.

Zhang, J., Zhang, Y., Du, Y. et al. (2011). Dynamic metabonomic responses of tobacco (*Nicotiana tabacum*) plants to salt stress. *Journal of Proteome Research* 10 (4): 1904–1914.

Zhang, M., Liu, Z., Fan, Y. et al. (2022). Characterization of GABA-transaminase gene from mulberry (*Morus multicaulis*) and its role in salt stress tolerance. *Genes* 13: 501.

Zhou, M., Hassan, M.J., Peng, Y. et al. (2021). γ-Aminobutyric acid (GABA) priming improves seed germination and seedling stress tolerance associated with enhanced antioxidant metabolism, DREB expression, and dehydrin accumulation in white clover under water stress. *Frontiers in Plant Science* 12: 776939.

Zhu, J.K. (2016). Abiotic stress signaling and responses in plants. *Cell* 167: 313–324.

10

GABA Priming Induced Modulations in the Redox Homeostasis of Plants under Osmotic Stress

Kakkuzhiyulla Parambath Raj Aswathi[1], *Kolothodi Chandran Jisha*[2], *Mathew Veena*[1], *Akhila Sen*[3], *Nair Gopalakrishnan Sarath*[3], *and Jos Thomas Puthur*[1]

[1] *Plant Physiology and Biochemistry Division, Department of Botany, University of Calicut, Malappuram, Kerala, 673635, India*
[2] *Department of Botany, MES Asmabi College, Thrissur, Kerala, 680671, India*
[3] *Department of Botany, Mar Athanasius College, Kothamangalam, Kerala, 686666, India*

Introduction

Numerous environmental challenges, such as salinity, cold, heat, and drought, have a negative effect on plants by creating water-deficit condition and, subsequently, osmotic stress inside the plant cell. A number of signaling cascades get initiated in response to this osmotic stress so as to maintain the osmoticum inside the cells. Many physiological changes, including reduced photosynthesis, increased reactive oxygen species (ROS) production, stomatal closure, and suppression of cell elongation, occur during a stressed state. Antioxidant machinery operates efficiently with increased ROS scavenging activity, and this could aid in cellular redox equilibrium maintenance (Darko et al. 2019). Together with this, plants adopt a range of other adaptation strategies to respond to the stress, including osmotic adjustment through the higher production of compatible solutes, and hormonal and metabolic regulation (Zhu et al. 2021; Aswathi et al. 2022). Crop productivity is negatively impacted by these variations in the physiological and molecular responses of plants under stress (Chaudhry and Sidhu 2022).

Various strategies are adopted to increase plants' tolerance toward osmotic stress, including conventional breeding, mutation breeding, genetic engineering, etc. However, these techniques have certain limitations like high cost, huge manpower requirements, biosafety issues, etc. Therefore, researchers are in search of alternatives, and one of the options is priming techniques to improve seed vigor, seedling establishment, and stress tolerance in plants. Priming involves the treatment of seeds/seedlings with natural or synthetic compounds that elicit certain physiological and molecular responses in plants to make them better adapted to a stressed environment (Jisha et al. 2013). Various priming agents are used from time to time, including water; osmotic agents such as mannitol, polyethylene glycol; hormones such as abscisic acid, salicylic acid; amino acids such as proline,

GABA in Plants: Biosynthesis, Plant Development, and Food Security, First Edition. Edited by Samiksha Singh, Durgesh Kumar Tripathi, and Vijay Pratap Singh.

γ-aminobutyric acid (GABA), etc. (Farooq et al. 2013; Vijayakumari et al. 2016). The application of GABA is found to be effective against osmotic stress in a variety of plant species.

GABA is a nonprotein amino acid found in organisms like bacteria, fungi, plants, and animals (Kinnersley and Turano 2000). It is a key metabolite, acts as a signal molecule, and is involved in stress tolerance response against a variety of biotic and abiotic stresses. At the onset of stress sensing, the synthesis of GABA is drastically elevated, and this serves as a cue for stress defense mechanism in plants (Li et al. 2018). Various studies have proved the potential of GABA against a variety of stresses, including heat, drought, salinity, etc. GABA application was found to increase the seed germination rate, provide better osmotic adjustment, improve photosynthetic efficiency, increase antioxidant activity, and upregulate stress-related genes, including *DREB* and *DHN* expression (Cheng et al. 2018: Shomali et al. 2021; Zhou et al. 2021).

GABA could activate the expression of stress-related genes, and GABA metabolism could control plant growth and development under stress via controlling carbon and nitrogen metabolisms (Deleu et al. 2013). Priming with GABA was found to be an efficient and cost-effective method for enhancing the stress tolerance of plants. Future research is expected to understand the precise mechanism of stress tolerance, which is accomplished through GABA priming. Research in this direction will be beneficial for addressing the food demand of the growing population under a challenging climate.

Role of GABA in Plants

GABA is a neurotransmitter in animals, and it plays a distinct role in signaling. It is the neurotransmitter in the mammalian central nervous system that mediates inhibitory synaptic currents via binding to receptors located in the pre- or postsynaptic membranes. GABA might function similarly as a signaling agent in plants. There are hints that plants have GABA receptors or that they use GABA as a modulatory ligand for glutamate receptors (Kinnersley and Turano 2000). GABA is reported to be transported by proline transporters in tomatoes (LeProT1) (Schwacke et al. 1999) and *Arabidopsis* (AtProT2) (Breitkreuz et al. 1999). Under conditions of water or salt stress, AtProT2 is substantially stimulated, which indicates its role in osmotic stress adaptation. Besides this, GABA has a role in nitrogen metabolism. The role of GABA both as a metabolite and as a signaling molecule needs to be explored more.

GABA Priming and Oxidative Stress Mitigation

Priming with GABA has stress-ameliorating potential in a variety of plants, viz. rice (Sheteiwy et al. 2019), pepper (Vijayakumari and Puthur 2016), white clover (Zhou et al. 2021), and lentil (Bhardwaj et al. 2021). GABA treatment modulates various morphological, physiological, and molecular responses in plants and makes the plants more adaptable to changing and adverse environmental conditions. The ability of plants to perform better overall due to the beneficial effects of GABA can help them tolerate osmotic stress. The mechanism of osmotic stress alleviation through GABA priming is depicted in Figure 10.1.

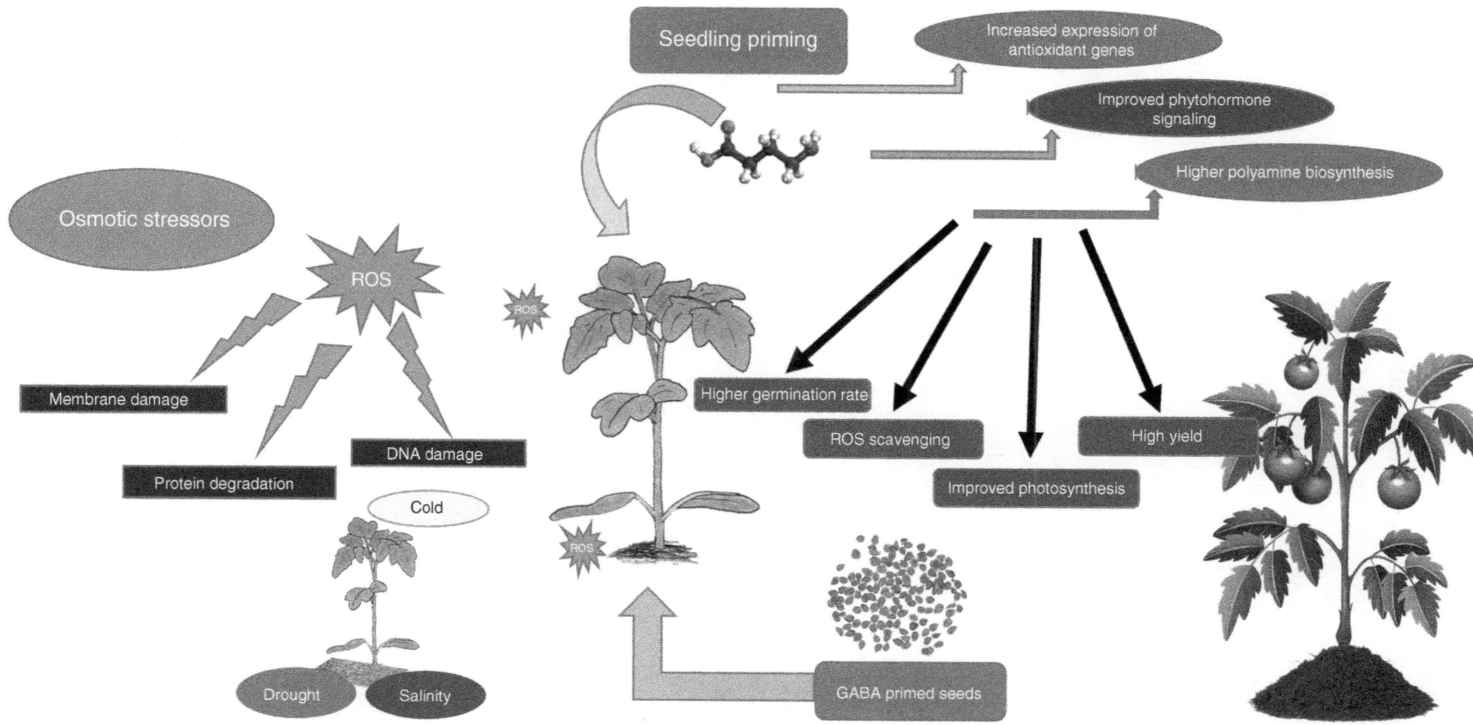

Figure 10.1 The mechanism of osmotic stress alleviation through GABA priming.

Morphological Response

Seed priming with GABA can alter the morphology of plants. It can enhance the growth of the plant by altering the root, shoot, flower, and fruit morphology. GABA priming can enhance root growth and root formation in plants. It can increase root elongation, lateral root growth, and also increase root hair formation. Zhou et al. (2021) noticed that during drought stress, the root length was more in GABA-primed plants when compared to nonprimed as well as stressed plants. In white clover, under salt stress, the seedlings that emerged from GABA-primed seeds exhibited increased root length compared to stressed and nonprimed plants (Cheng et al. 2018). The radicle length of citrus plants under salinity was also enhanced by priming the seeds with various concentrations of GABA (Ziogas et al. 2017).

The shoot length and branching can be enhanced under stressed conditions upon priming with GABA. Under drought stress, the seedling length and shoot weight per unit seedling length were enhanced in maize upon priming with GABA (Li et al. 2016a). The combined and individual treatment of salinity and osmotic stresses can negatively affect the seedling growth. However, priming with GABA increased the shoot length of rice plants under salinity stress, osmotic stress, and even combined stress (Sheteiwy et al. 2019). Li et al. (2022) proved that GABA priming in creeping bent grass enhanced the shoot growth under heat stress.

Priming plants with GABA can alter the leaf morphology under stress. It enhances photosynthesis by producing larger and thicker leaves. It can also positively alter the stomatal density, thus increasing photosynthesis further. Shomali et al. (2021) noticed that priming *Vicia faba* with GABA increased the area of the leaf, the length, width, and stomatal index were highest in primed plants when compared to the control plants.

GABA priming positively influences the yield of the plant. It can increase the number of flowers and fruits in plants by promoting floral induction and other parameters connected with flowering. It can fasten vegetative to reproductive transition. It can also alter the flowering time, thus enhancing the fruit setting, which is majorly achieved by promoting pollen tube formation. Under heat stress, GABA treatment was suitable for ameliorating its effect by increasing the number of flowers, pollen germination, and increased seed setting in lentils (Bhardwaj et al. 2021). Figure 10.2 represents the overall enhancement in morphological parameters of GABA-primed plants.

Physiological Response: With Special Emphasis on ROS and Antioxidant Machinery

The external application of GABA produces several stimulatory effects in plants by acting as a growth-enhancing substance in several plants (Kinnersley and Lin 2000; Roberts 2007). There are several research works on GABA priming for enhanced growth and stress tolerance in plants, such as in the case of *Lolium perenne* (Krishnan et al. 2013), *Oryza sativa* (Nayyar et al. 2014; Sheteiwy et al. 2019), black pepper (Vijayakumari and Puthur 2016), musk melon (Jin et al. 2019), *Arabidopsis thaliana* (Janse van Rensburg and Van den Ende 2020), tomato (Abd Elbar et al. 2021), and white clover (Zhou et al. 2021).

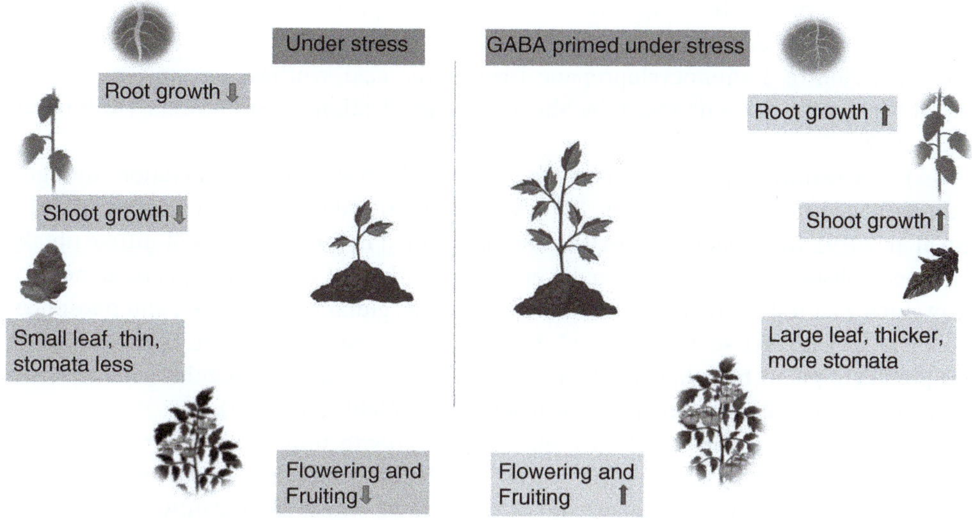

Figure 10.2 The morphological response of GABA-primed plants.

GABA is a vital nonproteinogenic amino acid in plants, which is seen in very low levels in the plant tissues and is engaged in many physiochemical functions that are essential for plant growth and development (Sita and Kumar 2020). The exogenous application of GABA alleviates the impacts of many abiotic stresses like drought (Abd El-Gawad et al. 2020), chilling (Malekzadeh et al. 2014; Wang et al. 2014), salinity (Wu et al. 2020), heavy metals (Seifikalhor et al. 2020), low light, and nitrogen starvation (Kinnersley and Lin 2000). These alleviating effects are achieved through regulation of the osmotic pressure, pH of the cytosol, C and N metabolism, and scavenging of ROS (Gilliham and Tyerman 2016). Moreover, the exogenous GABA application also enhanced the photosynthetic capacity of plants by lowering the ROS level, regulating the antioxidant enzyme activities, and maintaining membrane integrity (Aghdam et al. 2015). In tomato, several research works have already proved that GABA application can mitigate chilling injuries (Aghdam et al. 2012; Malekzadeh et al. 2014).

GABA regulates various physiological processes in plants by acting as a signaling molecule. Priming with GABA showed many enhanced physiological or biochemical defense responses in plants exposed to stress (Tian et al. 2005; Malekzadeh et al. 2012, 2014; Krishnan et al. 2013; Nayyar et al. 2014; Vijayakumari and Puthur 2016). According to Bouché and Fromm (2004), GABA regulates carbon: nitrogen (C : N) ratios due to its association with the tricarboxylic acid cycle. In plants, GABA is mostly produced from glutamate (Yang et al. 2015) and is concerned with physiological processes like nitrogen metabolism (Bouché and Fromm 2004), signaling, growth, and development (Fait et al. 2008; Renault et al. 2011), and also in stress alleviation (Kinnersley and Turano 2000; Podlesakova et al. 2019). GABA accumulation occurs during exposure to several types of stresses, which gets further enhanced through calcium signaling or through cytosolic pH acidification, which further causes the induction of the GABA biosynthetic pathway (Kinnersley and Turano 2000). Various enzymes are reported to get upregulated in plants under GABA priming. In *Brassica napus*, the mRNA expression for BnNRT2 nitrate transporter in roots gets enhanced as a result of GABA treatments (Beuve et al. 2004). Similarly, the expression of 1-aminocyclopropane-1-carboxylic acid synthase enzyme of ethylene metabolism was also enhanced by the GABA application in sunflowers (Kathiresan et al. 1997).

GABA is mainly important during environmental stresses like N starvation, drought, salinity, and low light (Barbosa et al. 2010; Vijayakumari and Puthur 2016; Wang et al. 2017; Jin et al. 2019). In *Arabidopsis*, GABA priming resulted in the regulation of nitrate uptake and metabolism by activating the major enzymes involved in nitrogen and carbon metabolism, such as nitrate reductase, glutamine synthetase, glutamate synthase, and phosphoenol pyruvate carboxylase (Barbosa et al. 2010). During stressed conditions, GABA promotes the growth of plants by enhancing their antioxidant capacity. It also regulates the reallocation and recycling of nitrogen during leaf senescence (Jalil et al. 2017).

The positive effects of GABA priming in plant growth under various environmental stresses were studied in several plant species. In rice, growth enhancement and heat stress mitigation were observed as a result of GABA priming, which significantly enhanced the photochemical efficiency, relative leaf water content, stomatal conductance, osmolytes accumulation, and activities of enzymatic and nonenzymatic antioxidants (Nayyar et al. 2014). In musk melon also, an enhancement in growth parameters like fresh weight, dry weight, and leaf area was observed after foliar pretreatment with GABA, along with mitigation of the growth limitation and membrane lipid peroxidation triggered by salinity-alkalinity stress (Jin et al. 2019). Similarly, under drought stress, GABA-primed ryegrass plants (*L. perenne*) exhibited higher relative water content (RWC), lower wilting, canopy temperature depression, and turf quality than the untreated plants under drought stress (Krishnan et al. 2013). According to Xiang et al. (2016), priming with GABA mitigated stress-induced photosynthetic reduction and alleviated the stress-induced injuries to the chloroplast. In musk melon seedlings, under unstressed conditions, GABA priming significantly enhanced the Chl *a* content and decreased the Chl *b* content when compared to control plants. But under salinity-alkalinity stress, the Chl *a* and Chl *b* content were increased sharply, and this was due to the higher content of Chl precursors (Pchl, URO III, ALA, PBG, Mg-proto IX, and Proto IX). Compared to the application of salinity-alkalinity stress alone, there was a reduction in the content of all these Chl precursors except ALA content (an increase of 57.8%) upon the GABA application under stress (Jin et al. 2019).

The effects of GABA priming in black pepper under PEG-induced osmotic stress were studied, and it was found that when the GABA-primed black pepper plants were grown hydroponically and exposed to PEG stress, those plants exhibited higher osmotic stress tolerance in terms of lower rates of wilting, lipid peroxidation, and better leaf RWC (Vijayakumari and Puthur 2016). Similarly, in tomatoes, the influence of GABA priming under chilling stress for 5 successive days during the flowering stage was investigated, and it was found that the applied GABA differentially influenced photosynthetic pigment composition by reducing the chlorophyll *a*/*b* ratio and enhancing the anthocyanin content. Moreover, in the GABA-treated tomato plants, root tissues also showed an enhanced thermotolerance and the applied GABA significantly reduced the chilling-induced oxidative stress by protecting cell membrane integrity and reducing malondialdehyde (MDA) content and H_2O_2 (Abd Elbar et al. 2021).

In rice seedlings (Sheteiwy et al. 2019), the photosynthetic system and the water relation parameters were improved by GABA under salinity and osmotic stress. In rice, GABA priming significantly enhanced the sugar, protein, and starch content of seedlings, along with a reduction in Na^+ concentrations, proline, free radical, and malonaldehyde content. According to Ma et al. (2019), under salinity stress, GABA priming induces phenolic compound enrichment in plants. Similarly, GABA priming also enhanced the ROS scavenging mechanisms (Renault et al. 2010) and thus regulates the antioxidant enzyme activities (Barbosa et al. 2010).

ROS like superoxide radical ($O_2^{\cdot -}$), hydrogen peroxide (H_2O_2), hydroxyl radical ($^{\cdot}OH$), singlet oxygen ($^1O_2$), peroxy radical ($ROO^{\cdot}$), and alkoxyl radicals ($RO^{\cdot}$) are produced in a threshold concentration in the plant cell under normal environmental conditions. However, several environmental conditions trigger excessive production of ROS (Maurya 2020). These ROS damage the molecular and cellular machinery by oxidizing the biomolecules, thereby leading to plant death (Shah et al. 2019; Bhuyan et al. 2020). To cope with these changes, plants strongly regulate the production of ROS through enzymatic and nonenzymatic antioxidants. The enzymatic antioxidant includes enzymes such as peroxidase (POD), superoxide dismutase (SOD), glutathione reductase (GR), catalase (CAT), and ascorbate peroxidase (APX) and nonenzymatic antioxidants include vitamins, flavonoids, and carotenoids (Kumar et al. 2017; Hasanuzzaman et al. 2020; Wongshaya et al. 2020).

GABA accumulation is the key mechanism for regulating redox homeostasis in many organisms (Allan et al. 2008; Tang et al. 2018). In maize, GABA treatment significantly enhanced the activity of SOD, PODs, and CATs (Tian et al. 2005). Similarly, GABA regulates the expression of genes involved in ethylene biosynthesis, polyamine (PA) metabolism, and ROS generation in *Caragana intermedia* subjected to NaCl stress (Shi et al. 2010). According to Bouché et al. (2003), in *A. thaliana* also, GABA inhibits the accumulation of ROS. Many research works have proved that the exogenous application of GABA improved the water stress tolerance in different plant species by regulating the antioxidant defense pathways (Li et al. 2018, 2019; Rezaei-Chiyaneh et al. 2018). In white clover, seeds primed with GABA exhibited significantly lower ROS levels and higher SOD, POD, CAT, APX, and MDHR (monodehydroascorbate reductase) activities than the seeds without GABA priming during germination under salt stress (Cheng et al. 2018). Ascorbate (ASA) and glutathione (GSH) are the two major nonenzymatic antioxidants, and the ratios of ASA/DHA and GSH/GSSG are also vital for oxidative homeostasis in plant cells (Gill and Tuteja 2010). Li et al. (2016b) stated that the GABA priming in creeping bent grass enhanced the contents of GSH, GSSG, ASA, and DHA and increased the ratios of ASA/DHA and GSH/GSSG in leaves under heat stress. Moreover, In *A. thaliana*, the GABA priming reduced H_2O_2 and $O_2^{\cdot -}$ accumulation in seedlings induced by high-temperature stress (Zhang et al. 2020).

In white clover, GABA priming significantly reduced the oxidative damage through acceleration in the activity of many antioxidant enzymes (SOD, POD, CAT, APX, DHAR, GR, and MDHR) and increasing the contents of nonenzymatic antioxidants (ASA, DAH, GSH, and GSSG), which were involved in ASA-GSH cycle during seed germination under water stress (Zhou et al. 2021). Similarly, in tomato, under chilling stress, the positive effect of GABA priming was due to the enhancement in activities of phenylalanine ammonia-lyase (PAL), CAT, SOD, and APX (Abd Elbar et al. 2021).

Tian et al. (2005) observed that in maize seeds, GABA priming reduced the effect of NaCl stress on seed germination by increasing the activity of antioxidant enzymes, such as SOD, PODs, and CATs. According to Yang et al. (2011), GABA priming protects peach fruits from the damaging effects of chilling injury by regulating its enzymatic antioxidant system, especially the activities of SOD, CAT, glutathione peroxidase (GPX), and glutathione S-transferase. Similarly, in wheat and tomato seedlings, the chilling stress damages were mitigated by GABA priming through increased activities of antioxidant enzymes, like CAT, SOD, etc. (Malekzadeh et al. 2012, 2014). GABA treatment alleviated the oxidative stress in the roots of barley seedlings by upregulating the activity of antioxidant enzymes (SOD, CAT, and POD) along with reduced MDA and ROS contents (Song et al. 2010).

The exogenous application of GABA relieved the tomato seedlings from chilling injury by enhancing the antioxidant enzyme activities and reducing ROS production (Malekzadeh et al. 2014). Moreover, the GABA priming alleviated the hypoxia damage in melon plants by triggering the PAs biosynthesis and conversion of free PAs to conjugated and insoluble bound forms as well as preventing PA degradation (Wang et al. 2014). The exogenous application of GABA protected maize seedlings from oxidative stress by inhibiting lipid peroxidation, reducing ROS levels, and triggering the activities of antioxidant enzymes (Salah et al. 2019). Several studies demonstrated that exogenous GABA stimulates the activities of various antioxidant enzymes and thereby reduces oxidative stress under different stresses (Nayyar et al. 2014; Li et al. 2016a). Among the different antioxidant enzymes, CAT, GR, and SOD were significantly activated by GABA application in maize seedlings. The exogenous application of GABA also regulates the expression of an NADPH-oxidase gene during salt stress in the roots of *Caragana intermedia* along with the reduction in H_2O_2 content due to the direct activation of CAT after the priming treatment (Shi et al. 2010).

Molecular Response

GABA functions as a signaling molecule that sets off a chain of events that upregulates genes linked to stress tolerance, especially those linked to antioxidant defense mechanisms. These genes produce proteins and enzymes that neutralize ROS and guard against oxidative injury. Cheng et al. (2018) conducted a study to ascertain the impact of GABA priming in white clover (*Trifolium repens* cv. Haifa) seeds during salt stress caused by 100 mM NaCl. Results proved that GABA enhanced the activities of antioxidant enzymes (SOD, GPOX, CAT, APX, DHAR, GR, and MDHR) and gene transcript levels of *Cu/ZnSOD*, *FeSOD*, *MnSOD*, *CAT*, *GPOX*, *APX*, *MDHR*, *GPX*, and *GST*, which was followed by a decrease in the accumulation of $O_2^{\cdot-}$, H_2O_2, and MDA during germination under salt stress. Similarly, the study by Shi et al. (2010) also revealed that exogenous GABA could negatively regulate many important genes involved in H_2O_2 synthesis, reducing the buildup of H_2O_2 in *Caragana intermedia* roots and raising the possibility that GABA could function as a signal molecule to control gene expression in response to salt stress.

GABA accumulation influenced the expression of glutamate decarboxylase (*GAD*), GABA aminotransferase (*GABA-T*), and succinic semialdehyde dehydrogenase (*SSADH*) involved in the GABA shunt route. Together with this, the augmented activity of antioxidant enzymes such as CAT, POD, GPX, and SOD provides stress tolerance (Kumar et al. 2017). GABA reduced oxidative injury caused by osmotic stress, salinity, and their combination in rice by modulating cellular and molecular signals. It was found that GABA-activated antioxidant enzymes could play critical roles in scavenging free radicals induced by H_2O_2 and O_2 and reducing lipid peroxidation by upregulating the genes involved in antioxidant enzymes (*APXa*, *CATa*, and *SOD1*) during rice seed germination and seedling growth under salinity and osmotic stress, as well as their combination (Sheteiwy et al. 2019).

Polyphenols are active components of tea plants that have high biological and antioxidant activity. Polyphenol content in plants can be considerably increased by stress or foreign chemicals. Ren et al. (2021) investigated whether exogenous GABA can boost polyphenol content and antioxidant activity in tea plants under heat stress. And they showed that most of the genes encoding the enzymes involved in flavonoid metabolism (*PAL*, *C4H*, *4CL*, *CHS*, *CHI*, *F3H*, *F3′H*, *F3′5′H*, *DFR*, *LAR*, *ANS*, *ANR*, and *FLS*) were upregulated. In addition, the activities of antioxidant enzymes were induced under heat-stress conditions. The accumulations of these phenolic substances could be correlated with the activities of related key enzymes and gene expression levels in flavonoids metabolism (Xiong et al. 2013).

During germination under water stress, as compared to the untreated seeds, the GABA-primed seeds showed higher expression of transcription factors (*DREB2*, *DREB3*, *DREB4*, and *DREB5*) and genes encoding DHNs (*SK2*, *Y2K*, *Y2SK*, and *Dehydrin b*) (Zhou et al. 2021). Likewise, GABA treatment improved the transcription level of *DREB*, *cAPX*, *MnSOD*, and *GST* genes, and as a result, the activity of enzymatic antioxidants was increased in strawberries. This higher expression of stress-related genes, particularly the antioxidants genes, imparted salinity tolerance to strawberries (Golnari et al. 2021).

Stress Signaling Cross-Talk

Plants are regularly exposed to various abiotic stressors individually and in combinations. So, plants should be examined for their tolerance to a variety of stresses in combination since combined stress necessitates the integration of defense or cross-talk response. According to the study of Sheteiwy et al. (2019), GABA-activated antioxidant enzymes that are essential for scavenging free radicals via activating the enzymatic antioxidants genes (*APXa*, *CATa*, and *SOD1*) and were found to provide protection against individual and/or combined effects of salinity and osmotic stresses in rice. Phytohormone signaling also has a role in cross-stress tolerance in plants. GABA was found to be involved in cross-talk between hormones and PAs and have the ability to regulate their levels under stressful conditions (Podlesakova et al. 2019). In *Citrus sinensis*, the levels of abscisic acid, salicylic acid, indole acetic acid, cinnamic acid, and trans-jasmonic acid were increased upon GABA treatment as compared to the control plants (Hijaz et al. 2018). Similar results of increased phytohormones in GABA-treated plants were reported in apple seedlings by Liu et al. (2021), wherein the expression level of ABA-related genes (*ABI1*, *ABI2*, *HAB1*, *ABF3*, *PYL4*, and *OST1*) altered under drought and this in turn leads to the closure of stomata, and thus the exogenous application of GABA increased tolerance against drought in apple seedlings. Exogenous GABA administration increased proline accumulation in bent grass and improved the osmotic adjustment under heat and drought stresses (Li et al. 2016b). Metabolism of PAs such as spermidine and putrescine are linked to GABA production (Shelp et al. 2012). GABA and BABA combined application also have a role in drought and salt stress tolerance in plants (Ma et al. 2022). Stress-related genes share a more or less closely associated expression pattern among various osmotic stressors, and hence, the alert for one stress can also be beneficial for the other stress, and it is because of the cross-talk response.

Conclusion and Future Prospects

Osmotic stress has a detrimental impact on plant growth and productivity. Augmented accumulation of ROS disrupts cell membranes and leads to cell death under extreme osmotic stress. Plants have developed ROS scavenging machinery to maintain cellular redox homeostasis by balancing the production and scavenging of ROS. Priming with GABA is highly advantageous for osmotic stress adaptation. GABA enhanced the photosynthetic pigments, shoot length, leaf area, root growth, and antioxidation potential in a variety of plant species. GABA priming was found to alleviate the stress effects by lowering ROS accumulation and lipid peroxidation. The physiological and molecular changes brought by GABA could help the plants for better stress adaptation. Further investigations are needed to understand the actual and detailed mechanism of osmotic stress tolerance mediated through GABA priming.

Acknowledgments

All the authors express sincere thanks to the University of Calicut for the support. ARKP and VM express thanks to the University Grant Commission (UGC) for providing funds in the form of SRF.

References

Abd Elbar, O.H., Elkelish, A., Niedbała, G. et al. (2021). Protective effect of γ-aminobutyric acid against chilling stress during reproductive stage in tomato plants through modulation of sugar metabolism, chloroplast integrity, and antioxidative defense systems. *Frontiers in Plant Science* 12: 663750.

Abd El-Gawad, H.G., Mukherjee, S., Farag, R. et al. (2020). Exogenous γ-aminobutyric acid (GABA)-induced signaling events and field performance associated with mitigation of drought stress in *Phaseolus vulgaris* L. *Plant Signaling & Behavior* 16: 1853384.

Aghdam, M.S., Asghari, M., Farmani, B. et al. (2012). Impact of postharvest brassinosteroids treatment on PAL activity in tomato fruit in response to chilling stress. *Scientia Horticulturae* 144: 116–120.

Aghdam, M.S., Naderi, R., Sarcheshmeh, M.A., and Babalar, M. (2015). Amelioration of postharvest chilling injury in anthurium cut flowers by γ-aminobutyric acid (GABA) treatments. *Postharvest Biology and Technology* 110: 70–76.

Allan, W.L., Simpson, J.P., Clark, S.M., and Shelp, B.J. (2008). γ-Hydroxybutyrate accumulation in *Arabidopsis* and tobacco plants is a general response to abiotic stress: putative regulation by redox balance and glyoxylate reductase isoforms. *Journal of Experimental Botany* 59 (9): 2555–2564.

Aswathi, K.P.R., Kalaji, H.M., and Puthur, J.T. (2022). Seed priming of plants aiding in drought stress tolerance and faster recovery: a review. *Plant Growth Regulation* 97: 235–253.

Barbosa, J.M., Singh, N.K., Cherry, J.H., and Locy, R.D. (2010). Nitrate uptake and utilization is modulated by exogenous γ-aminobutyric acid in *Arabidopsis thaliana* seedlings. *Plant Physiology and Biochemistry* 48: 443–450.

Beuve, N., Rispail, N., Laine, P. et al. (2004). Putative role of γ-aminobutyric acid (GABA) as a long-distance signal in up-regulation of nitrate uptake in *Brassica napus* L. *Plant, Cell & Environment* 27: 1035–1046.

Bhardwaj, A., Sita, K., Sehgal, A. et al. (2021). Heat priming of lentil (*Lens culinaris* Medik.) seeds and foliar treatment with γ-aminobutyric acid (GABA), confers protection to reproductive function and yield traits under high-temperature stress environments. *International Journal of Molecular Sciences* 22: 5825.

Bhuyan, M.B., Hasanuzzaman, M., Parvin, K. et al. (2020). Nitric oxide and hydrogen sulfide: two intimate collaborators regulating plant defense against abiotic stress. *Plant Growth Regulation* 90: 409–424.

Bouché, N., Fait, A., Bouchez, D. et al. (2003). Mitochondrial succinic-semialdehyde dehydrogenase of the γ-aminobutyrate shunt is required to restrict levels of reactive oxygen intermediates in plants. *Proceedings of the National Academy of Sciences of the United States of America* 100: 6843–6848.

Bouché, N. and Fromm, H. (2004). GABA in plants: just a metabolite. *Trends in Plant Science* 9: 110–115.

Breitkreuz, K.E., Shelp, B.J., Fischer, W.N. et al. (1999). Identification and characterization of GABA, proline and quaternary ammonium compound transporters from *Arabidopsis thaliana*. *FEBS Letters* 450 (3): 280–284.

Chaudhry, S. and Sidhu, G.P.S. (2022). Climate change regulated abiotic stress mechanisms in plants: a comprehensive review. *Plant Cell Reports* 41 (1): 1–31.

Cheng, B., Li, Z., Liang, L. et al. (2018). The γ-aminobutyric acid (GABA) alleviates salt stress damage during seeds germination of white clover associated with Na^+/K^+ transportation, dehydrins accumulation, and stress-related genes expression in white clover. *The International Journal of Molecular Sciences* 19: 2520–2528.

Darko, E., Végh, B., Khalil, R. et al. (2019). Metabolic responses of wheat seedlings to osmotic stress induced by various osmolytes under iso-osmotic conditions. *PLoS One* 14 (12): e0226151.

Deleu, C., Faes, P., Niogret, M.F., and Bouchereau, A. (2013). Effects of the inhibitor of the γ-aminobutyrate-transaminase, vinyl-γ-aminobutyrate, on development and nitrogen metabolism in Brassica napus seedlings. *Plant Physiology and Biochemistry* 64: 60–69.

Fait, A., Fromm, H., Walter, D. et al. (2008). Highway or byway: the metabolic role of the GABA shunt in plants. *Trends in Plant Science* 13: 14–19.

Farooq, M., Irfan, M., Aziz, T. et al. (2013). Seed priming with ascorbic acid improves drought resistance of wheat. *Journal of Agronomy and Crop Science* 199 (1): 12–22.

Gill, S.S. and Tuteja, N. (2010). Reactive oxygen species and antioxidant machinery in abiotic stress tolerance in crop plants. *Plant Physiology and Biochemistry* 48: 909–930.

Gilliham, M. and Tyerman, S.D. (2016). Linking metabolism to membrane signaling: the GABA–malate connection. *Trends in Plant Science* 21: 295–301.

Golnari, S., Vafaee, Y., Nazari, F. et al. (2021). Gamma-aminobutyric acid (GABA) and salinity impacts antioxidative response and expression of stress-related genes in strawberry cv. Aromas. *The Brazilian Journal of Botany* 44: 639–651.

Hasanuzzaman, M., Bhuyan, M.H.M., Zulfiqar, F. et al. (2020). Reactive oxygen species and antioxidant defense in plants under abiotic stress: revisiting the crucial role of a universal defense regulator. *Antioxidants* 9 (8): 681.

Hijaz, F., Nehela, Y., and Killiny, N. (2018). Application of gamma-aminobutyric acid increased the level of phytohormones in *Citrus sinensis*. *Planta* 248: 909–918.

Jalil, S.U., Ahmad, I., and Ansari, M.I. (2017). Functional loss of GABA transaminase (GABA-T) expressed early leaf senescence under various stress conditions in *Arabidopsis thaliana*. *Current Plant Biology* 9–10: 11–22.

Janse van Rensburg, H.C. and Van den Ende, W. (2020). Priming with γ-aminobutyric acid against *Botrytis cinerea* reshuffles metabolism and reactive oxygen species: dissecting signalling and metabolism. *Antioxidants (Basel)* 9 (12): 1174.

Jin, X., Liu, T., Xu, J. et al. (2019). Exogenous GABA enhances muskmelon tolerance to salinity-alkalinity stress by regulating redox balance and chlorophyll biosynthesis. *BMC Plant Biology* 19: 48.

Jisha, K.C., Vijayakumari, K., and Puthur, J.T. (2013). Seed priming for abiotic stress tolerance: an overview. *Acta Physiologiae Plantarum* 35 (5): 1381–1396.

Kathiresan, A., Tung, P., Chinnappa, C.C., and Reid, D.M. (1997). Gamma-aminobutyric acid stimulates ethylene biosynthesis in sunflower. *Plant Physiology* 115: 129–135.

Kinnersley, A.M. and Lin, F. (2000). Receptor modifiers indicate that 4-aminobutyric acid (GABA) is a potential modulator of ion transport in plants. *Plant Growth Regulation* 32: 65–76.

Kinnersley, A.M. and Turano, F.J. (2000). Gamma aminobutyric acid (GABA) and plant responses to stress. *CRC Critical Reviews in Plant Sciences* 19: 479–509.

Krishnan, S., Laskowski, K., Shukla, V., and Emily, B.M. (2013). Mitigation of drought stress damage by exogenous application of a non-protein amino acid γ-aminobutyric acid on perennial ryegrass. *American Society for Horticultural Science* 138: 358–366.

Kumar, N., Dubey, A.K., Upadhyay, A.K. et al. (2017). GABA accretion reduces Lsi-1 and Lsi-2 gene expressions and modulates physiological responses in *Oryza sativa* to provide tolerance towards arsenic. *Scientific Reports* 7 (1): 8786.

Liu, C., Wang, H., Zhang, X. et al. (2021). Activation of the ABA signal pathway mediated by GABA improves the drought resistance of apple seedlings. *International Journal of Molecular Sciences* 22 (23): 12676.

Li, W., Liu, J., Ashraf, U. et al. (2016a). Exogenous γ-aminobutyric acid (GABA) application improved early growth, net photosynthesis, and associated physio-biochemical events in maize. *Frontiers in Plant Science* 7: 919.

Li, Z., Peng, Y., and Huang, B. (2016b). Physiological effects of γ-aminobutyric acid application on improving heat and drought tolerance in creeping bentgrass. *The Journal of the American Society for Horticultural Science* 141: 76–84.

Li, Z., Zhu, Y., He, X. et al. (2018). The hydrogen sulfide, a downstream signaling molecule of hydrogen peroxide and nitric oxide, involves spermidine-regulated transcription factors and antioxidant defense in white clover in response to dehydration. *Environmental and Experimental Botany* 161: 255–264.

Li, Z., Yong, B., Cheng, B. et al. (2019). Nitric oxide, γ-aminobutyric acid, and mannose pretreatment influence metabolic profiles in white clover under water stress. *Journal of Integrative Plant Biology* 61: 1255–1273.

Li, Z., Burgess, P., Peng, Y., and Huang, B. (2022). Regulation of nutrient accumulation by γ-aminobutyric acid associated with GABA priming-enhanced heat tolerance in creeping bentgrass. *Grassland Research* 2: 1–8.

Ma, Y., Wang, P., Wang, M. et al. (2019). GABA mediates phenolic compounds accumulation and the antioxidant system enhancement in germinated hulless barley under NaCl stress. *Food Chemistry* 270: 593–601.

Ma, X., Gu, J., Luo, Q. et al. (2022). The synergistic benefits of β-aminobutyric acid and γ -aminobutyrate on salt and drought tolerance in cassava. *Plant Biotechnology Reports* 16: 29–41.

Malekzadeh, P., Khara, J., and Heidari, R. (2012). Effect of exogenous gamma-aminobutyric acid on physiological tolerance of wheat seedlings exposed to chilling stress. *Iranian Journal of Plant Physiology* 3: 611–617.

Malekzadeh, P., Khara, J., and Heydari, R. (2014). Alleviating effects of exogenous gamma-aminobutiric acid on tomato seedling under chilling stress. *Physiology and Molecular Biology of Plants* 20: 133–137.

Maurya, A.K. (2020). Oxidative stress in crop plants. In: *Agronomic Crops: Stress Responses and Tolerance* (ed. M. Hasanuzzaman), 349–380. Singapore: Springer.

Nayyar, H., Kaur, R., Kaur, S., and Singh, R. (2014). γ-Aminobutyric acid (GABA) imparts partial protection from heat stress injury to rice seedlings by improving leaf turgor and upregulating osmoprotectants and antioxidants. *Journal of Plant Growth Regulation* 33: 408–419.

Podlesakova, K., Ugena, L., Spichal, L. et al. (2019). Phytohormones and polyamines regulate plant stress responses by altering GABA pathway. *New Biotechnology* 48: 53–65.

Ren, T., Zheng, P., Zhang, K. et al. (2021). Effects of GABA on the polyphenol accumulation and antioxidant activities in tea plants (*Camellia sinensis* L.) under heat-stress conditions. *Plant Physiology and Biochemistry* 159: 363–371.

Renault, H., Roussel, V., Amrani, A.E. et al. (2010). The *Arabidopsis* pop2-1 mutant reveals the involvement of GABA transaminase in salt stress tolerance. *BMC Plant Biology* 10: 20.

Renault, H., El Amrani, A., Palanivelu, R. et al. (2011). GABA accumulation causes cell elongation defects and a decrease in expression of genes encoding secreted and cell wall-related proteins in *Arabidopsis thaliana*. *Plant & Cell Physiology* 52: 894–908.

Rezaei-Chiyaneh, E., Seyyedi, S.M., Ebrahimian, E. et al. (2018). Exogenous application of gamma-aminobutyric acid (GABA) alleviates the effect of water deficit stress in black cumin (*Nigella sativa* L.). *Industrial Crops and Products* 112: 741–748.

Roberts, M.R. (2007). Does GABA act as a signal in plants? Hints from molecular studies: hints from molecular studies. *Plant Signaling & Behavior* 2: 408–409.

Salah, A., Zhan, M., Cao, C.G. et al. (2019). Gamma-aminobutyric acid promotes chloroplast ultrastructure, antioxidant capacity, and growth of waterlogged maize seedlings. *Scientific Reports* 9: 484.

Schwacke, R., Grallath, S., Breitkreuz, K.E. et al. (1999). LeProT1, a transporter for proline, glycine betaine, and γ-amino butyric acid in tomato pollen. *The Plant Cell* 11 (3): 377–391.

Seifikalhor, M., Aliniaeifard, S., Bernard, F. et al. (2020). γ-Aminobutyric acid confers cadmium tolerance in maize plants by concerted regulation of polyamine metabolism and antioxidant defense systems. *Scientific Reports* 10: 1–18.

Shah, K., Chaturvedi, V., and Gupta, S. (2019). Climate change and abiotic stress-induced oxidative burst in rice. In: *Advances in Rice Research for Abiotic Stress Tolerance* (ed. M. Hasanuzzaman, M. Fujita, K. Nahar, and J.K. Biswas), 505–535. Cambridge, MA: Wood Head Publishing.

Shelp, B.J., Bozzo, G.G., Trobacher, C.P. et al. (2012). Hypothesis/review: contribution of putrescine to 4-aminobutyrate (GABA) production in response to abiotic stress. *Plant Science* 193: 130–135.

Sheteiwy, M.S., Shao, H., Qi, W. et al. (2019). GABA-alleviated oxidative injury induced by salinity, osmotic stress and their combination by regulating cellular and molecular signals in rice. *International Journal of Molecular Sciences* 20: 5709.

Shi, S.Q., Shi, Z., Jiang, Z.P. et al. (2010). Effects of exogenous GABA on gene expression of *Caragana intermedia* roots under NaCl stress: regulatory roles for H_2O_2 and ethylene production. *Plant, Cell & Environment* 33: 149–162.

Shomali, A., Aliniaeifard, S., Didaran, F. et al. (2021). Synergistic effects of melatonin and gamma-aminobutyric acid on protection of photosynthesis system in response to multiple abiotic stressors. *Cells* 10 (7): 1631.

Sita, K. and Kumar, V. (2020). Role of gamma amino butyric acid (GABA) against abiotic stress tolerance in legumes: a review. *Plant Physiology Reports* 25: 1–10.

Song, H., Xu, X., Wang, H. et al. (2010). Exogenous gamma-aminobutyric acid alleviates oxidative damage caused by aluminum and proton stresses on barley seedlings. *Journal of the Science of Food and Agriculture* 90: 1410–1416.

Tang, X., Yu, R., Zhou, Q. et al. (2018). Protective effects of γ-aminobutyric acid against H_2O_2-induced oxidative stress in RIN-m5F pancreatic cells. *Nutrition and Metabolism* 15: 1–9.

Tian, X.L., Wu, X.L., Li, Y., and Zhang, S.Q. (2005). The effect of gamma aminobutyric acid in superoxide dismutase, peroxidase and catalase activity response to salt stress in maize seedling. *Shi Yan Sheng Wu Xue Bao* 38: 75–79.

Vijayakumari, K., Jisha, K.C., and Puthur, J.T. (2016). GABA/BABA priming: a means for enhancing abiotic stress tolerance potential of plants with less energy investments on defence cache. *Acta Physiologiae Plantarum* 38: 1–14.

Vijayakumari, K. and Puthur, J.T. (2016). γ-Aminobutyric acid (GABA) priming enhances the osmotic stress tolerance in *Piper nigrum* Linn. Plants subjected to PEG-induced stress. *Plant Growth Regulation* 78: 57–67.

Wang, C., Fan, L., Gao, H. et al. (2014). Polyamine biosynthesis and degradation are modulated by exogenous gamma-aminobutyric acid in root-zone hypoxia-stressed melon roots. *Plant Physiology and Biochemistry* 82: 17–26.

Wang, Y., Gu, W., Meng, Y. et al. (2017). γ-Aminobutyric acid imparts partial protection from salt stress injury to maize seedlings by improving photosynthesis and up regulating osmoprotectants and antioxidants. *Scientific Reports* 7: 1–13.

Wongshaya, P., Chayjarung, P., Tothong, C. et al. (2020). Effect of light and mechanical stress in combination with chemical elicitors on the production of stilbene compounds and defensive responses in peanut hairy root culture. *Plant Physiology and Biochemistry* 157: 93–104.

Wu, X., Jia, Q., Ji, S. et al. (2020). Gamma-aminobutyric acid (GABA) alleviates salt damage in tomato by modulating Na^+ uptake, the GAD gene, amino acid synthesis and reactive oxygen species metabolism. *BMC Plant Biology* 20: 1–21.

Xiang, L.X., Hu, L.P., Xu, W.N. et al. (2016). Exogenous γ-aminobutyric acid improves the structure and function of photosystem II in muskmelon seedlings exposed to salinity-alkalinity stress. *PLoS One* 11 (10): e0164847.

Xiong, L., Li, J., Li, Y. et al. (2013). Dynamic changes in catechin levels and catechin biosynthesis-related gene expression in albino tea plants (*Camellia sinensis* L.). *Plant Physiology and Biochemistry* 71: 132–143.

Yang, A., Cao, S., Yang, Z. et al. (2011). γ-Aminobutyric acid treatment reduces chilling injury and activates the defence response of peach fruit. *Food Chemistry* 129: 1619–1622.

Yang, R., Yin, Y., and Gu, Z. (2015). Polyamine degradation pathway regulating growth and GABA accumulation in germinating fava bean under hypoxia-NaCl stress. *Journal of Agricultural Science and Technology* 17: 311–320.

Zhang, W., Sun, L., Yang, X. et al. (2020). Inhibition of NADPH oxidase within midbrain periaqueductal gray decreases pain sensitivity in Parkinson's disease via GABAergic signaling pathway. *Physiological Research* 69 (4): 711–720.

Zhou, M., Hassan, M.J., Peng, Y. et al. (2021). γ-Aminobutyric acid (GABA) priming improves seed germination and seedling stress tolerance associated with enhanced antioxidant metabolism, *DREB* expression, and dehydrin accumulation in white clover under water stress. *Frontiers in Plant Science* 12: 776939.

Zhu, D., Luo, F., Zou, R. et al. (2021). Integrated physiological and chloroplast proteome analysis of wheat seedling leaves under salt and osmotic stresses. *Journal of Proteomics* 234: 104097.

Ziogas, V., Tanou, G., Belghazi, M. et al. (2017). Characterization of β-amino-and γ-amino butyric acid-induced citrus seeds germination under salinity using nano LC–MS/MS analysis. *Plant Cell Reports* 36: 787–789.

11

Gamma-Aminobutyric acid-Mediated Heavy Metal Stress Tolerance in Plants

Srijita Ghosh[1] and Aryadeep Roychoudhury[2]

[1] *Post Graduate Department of Botany, Scottish Church College, Kolkata, West Bengal, 700006, India*
[2] *Discipline of Life Sciences, School of Sciences, Indira Gandhi National Open University, New Delhi, 110068, India*

Introduction

According to the information from September 2021, the world population was estimated to be around 7.9 billion, and by 2050, it may rise up to 9.6 billion. It is important to acknowledge that food security is a complex issue that is influenced by various external factors such as climate changes, agriculture practices, economic conditions, and political stability. There have been many localized food shortages and famines throughout history, but there is no such global food crisis at present. However, it is worth noting that some regions still face significant challenges in terms of food security. Even before the worldwide pandemic Covid-19, 135 million people across the globe were struggling with severe food security, and it has increased many folds in the current scenario (Sita and Kumar 2020).

Gamma-aminobutyric acid (GABA) is a nonprotein amino acid that acts as a primary inhibitory neurotransmitter in the central nervous system of mammals, including humans, and helps to regulate and reduce the activity of neurons in the brain and spinal cord. GABA plays a crucial role in modulating neuronal excitability and maintaining a balance between excitation and inhibition in the brain. It functions by binding to specific receptors, known as GABA receptors, located on the surface of neurons. When GABA molecules bind to these receptors, they open ion channels, allowing negatively charged chloride ions to enter the neuron. This influx of chloride ions hyperpolarizes the neuron, making it less likely to fire an action potential and transmit signals to other neurons. Consequently, GABA inhibits the activity of neurons, helping to calm and stabilize brain functions (Ochoa-de la Paz et al., 2021).

Due to its inhibitory nature, GABA is involved in various physiological processes, including the regulation of muscle tone, motor control, anxiety, stress response, sleep, and the overall balance of neuronal activity. Imbalances in GABA levels or functioning have been associated with various neurological and psychiatric disorders, such as epilepsy, anxiety disorders, insomnia, and mood disorders (Boyle et al. 1993).

GABA in Plants: Biosynthesis, Plant Development, and Food Security, First Edition. Edited by Samiksha Singh, Durgesh Kumar Tripathi, and Vijay Pratap Singh.

GABAergic drugs, such as benzodiazepines and barbiturates, target GABA receptors to enhance the inhibitory effects of GABA, resulting in sedative, anxiolytic (antianxiety), and anticonvulsant properties. These drugs are commonly used for the treatment of anxiety, insomnia, seizures, and muscle spasms (Smith et al. 2007).

In higher plants, GABA plays a multifaceted role in abiotic stress tolerance. Its ability to chelate metals, scavenge reactive oxygen species (ROS), regulate gene expression, and modulate ion homeostasis contributes to the ability of the plants to cope with various abiotic stresses and maintain cellular integrity in the presence of excessive stress conditions. GABA contributes to abiotic stress tolerance in various ways:

Chelation and sequestration: GABA can act as a chelating agent, binding to toxic heavy metal ions and forming stable complexes. This helps to reduce the bioavailability and toxicity of metals in plant cells. GABA can also sequester metal ions into vacuoles, isolating them from sensitive cellular compartments (Viehweger 2014).

Antioxidant activity: Various abiotic stresses such as drought, extreme temperature, excess salt, and heavy metal toxicity can induce oxidative stress in plants by generating ROS. GABA has been shown to have antioxidant properties and can help alleviate oxidative damage caused by abiotic stress. It can scavenge ROS and enhance the activity of antioxidant enzymes, thereby protecting plant cells from oxidative stress.

Regulation of gene expression: GABA can modulate the expression of genes involved in heavy metal tolerance. It has been found to up regulate the expression of genes encoding metal transporters, such as metallothioneins, which are involved in metal detoxification and sequestration. GABA can also regulate the expression of genes related to antioxidant defense systems, leading to enhanced metal tolerance (Ruttkay-Nedecky et al. 2013).

Modulation of ion homeostasis: GABA can influence ion homeostasis in plants, including the regulation of ion fluxes, as well as affect the uptake, transport, and distribution of ions (Wu et al. 2020).

Some of the important functions of GABA in plants are listed in Table 11.1.

Table 11.1 Role of GABA in plants.

Stress tolerance	GABA acts as a signaling molecule in plants, helping them cope with various abiotic stresses such as drought, salinity, extreme temperature, and oxidative stress. GABA accumulation in plant tissues during stress helps to regulate the expression of stress-responsive genes, enhance antioxidant defense systems, and maintain cellular homeostasis.
Osmotic regulation	GABA can act as an osmolyte, a molecule that helps plants regulate osmotic balance and protect cells from water stress. It helps maintain cell turgor under drought conditions, allowing plants to maintain water uptake and survive in a water-limited environment.
Antioxidant defense	GABA plays a crucial role in the antioxidant defense system of plants. It helps scavenge ROS and protect plant cells from oxidative damage caused by environmental stressors. GABA can enhance the activity of antioxidant enzymes such as catalase (CAT), superoxide dismutase (SOD), and peroxidases.

Table 11.1 (Continued)

Stress tolerance	GABA acts as a signaling molecule in plants, helping them cope with various abiotic stresses such as drought, salinity, extreme temperature, and oxidative stress. GABA accumulation in plant tissues during stress helps to regulate the expression of stress-responsive genes, enhance antioxidant defense systems, and maintain cellular homeostasis.
Modulation of growth and development	GABA influences plant growth and developmental processes, including seed germination, root development, flowering, and fruit ripening. It regulates hormone signaling pathways, such as auxin and gibberellins, and can modulate cell division, elongation, and differentiation.
Photosynthesis enhancement	GABA has been found to improve photosynthetic efficiency in plants by enhancing the activity of key enzymes involved in carbon fixation and reducing photorespiration. It helps optimize the use of light energy and enhances carbon assimilation, contributing to plant growth and productivity.
Nutrient uptake and transport	GABA can influence nutrient uptake and transport in plants. It has been reported to enhance the uptake of essential nutrients, such as nitrogen and potassium, and improve their allocation within plant tissues. GABA also affects ion transport across cell membranes, helping plants maintain ion homeostasis.
Indirect defense against pathogens and herbivores	GABA can act as a signaling molecule to activate plant defense responses against pathogens and herbivores. It induces the production of defense-related compounds, such as phytoalexins, antimicrobial peptides, and secondary metabolites, helping plants fend off microbial infections and herbivore attacks. These functions of GABA in plants contribute to their stress tolerance, growth, and overall fitness. Understanding the role of GABA in plant physiology can aid in developing strategies to enhance crop productivity, improve plant resilience to environmental stresses, and optimize agricultural practices.

Health Benefits of GABA

GABA is not only important in plant physiology but also has potential health benefits for humans. Some of the benefits in humans include anxiety and stress reduction by acting as an inhibitory neurotransmitter helping in promoting relaxation; improved sleep by calming the mind; mood enhancement by inhibiting the activity of certain brain regions associated with anxiety and depression; blood pressure regulation by promoting vasodilation and relaxing blood vessels, thereby reducing strain on the cardiovascular system; pain relief and modulation; and muscle relaxation by reducing muscle tension. It is important to note that while GABA supplementation is available, it can be challenging for GABA to cross the blood–brain barrier, limiting its direct effects on the brain.

Biosynthesis of GABA

GABA is generally synthesized by the GABA shunt metabolic pathway, which was first reported in higher plants. The GABA shunt, consisting of mainly three reactions catalyzed by the cytosolic enzymes glutamate decarboxylase (GAD, E.C.4.1.1.15) and succinic semialdehyde dehydrogenase (SSADH) (E.C.1.2.2.16), and the mitochondrial enzymes GABA transaminase (E.C.2.6.1.19), is essential for maintaining the optimal GABA levels in plants (Steward 1949). Alternatively, it may be synthesized by polyamine degradation or by a nonenzymatic reaction from proline during oxidative stress (Yang et al. 2015). The initial step of GABA biosynthesis involves the decarboxylation of glutamate, an abundant amino acid, by the enzyme GAD (nonreversible). This step converts glutamate into GABA and releases carbon dioxide (CO_2) as a byproduct. Glutamate, the precursor for GABA synthesis, is derived from various sources, such as de novo synthesis from amino acids such as glutamine and α-ketoglutarate (Rashmi et al. 2018). GAD requires the coenzyme pyridoxal 5′-phosphate (PLP), which is derived from vitamin B6. PLP serves as a cofactor for the decarboxylation reaction and facilitates the conversion of glutamate into GABA. The GABA shunt and the TCA cycle are connected by GABA permease (GABA-P) located in the mitochondrial membrane, which allows cytosolic GABA uptake into the mitochondria where finally, energy and carbon skeletons are provided by the TCA cycle (Michaeli et al. 2011).

After GABA is synthesized, it can be further metabolized by the enzyme GABA transaminase (GABA-T). GABA-T converts GABA into succinic semialdehyde (SSA) by transferring an amino group from GABA to α-ketoglutarate (Bown and Shelp 1997). SSA is subsequently converted into succinic acid by the enzyme succinic semialdehyde dehydrogenase (SSADH) (Ramos-Ruiz et al. 2019). Rarely, however, in plants as well as animals and *Escherichia coli*, SSA may be converted to γ-hydroxybutyrate (γHB) by succinic semialdehyde reductase (SSR) (Rashmi et al. 2018). Generally, GABA-T, as amino acceptor, requires pyruvate and α-ketoglutarate and ultimately produces alanine and glutamate. In plants, however, pyruvate is preferred over α-ketoglutarate, resulting in the phenomenon of glutamate recovery for GAD response in plants. Succinate synthesized either enters the TCA cycle as an electron donor in the electron transport chain (ETC) or, after release from SSADH, enters the TCA cycle (Akihiro et al. 2008) (Figure 11.1). Besides GAD and GABA-T, other enzymes involved in GABA biosynthesis pathways have been identified in various plant species, although their numbers may vary (Shelp et al. 2012b; Trobacher et al. 2013). GAD in plants contains a calmodulin-binding domain, which enables its activation by the Ca^{2+}/CaM complex and exhibits different pH optima (Priya et al. 2019). Finally, gamma-amino butyraldehyde, a precursor of GABA, is produced by the combined activities of 4-amino butyraldehyde dehydrogenase and diamine oxidase (DAO, E.C. 1.4.3.6) (Akihiro et al. 2008; Fait et al. 2008).

GABA Transport in Plants

The transport of GABA in plants involves multiple mechanisms to ensure its distribution and utilization within different tissues and cellular compartments. Plants can take up GABA from the external environment, such as the soil or surrounding water. GABA uptake is facilitated by specific transporters present in the plasma membrane of plant cells. These transporters, known as GABA permeases or GABA transporters (GATs), are responsible for GABA uptake and play a crucial role in regulating GABA levels in different plant tissues. GATs play essential roles in the accumulation and metabolism of GABA within plant cells. These transporters are membrane-bound proteins that facilitate the transport of GABA

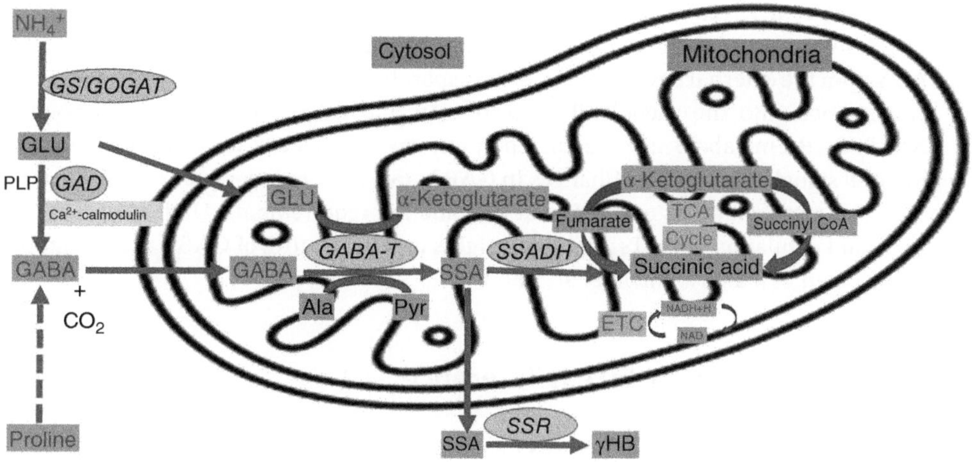

Figure 11.1 GABA biosynthesis in plants is mediated by GABA shunt. *Source:* Adapted from Ramos-Ruiz et al. (2019). Abbreviations: GAD, glutamate decarboxylase; GABA-P, GABA permease; GABA-T, GABA transaminase; Glu, glutamate; Ala, alanine; Pyr, pyruvate; SSADH, succinic semialdehyde dehydrogenase; ETC, electron transport chain; SSR, succinic semialdehyde reductase; SSA, succinic semialdehyde; γBH, γ-hydroxybutyric acid; GS/GOGAT, glutamine–synthetase/glutamate–synthase (AT).

Figure 11.2 Classes of GABA transporters based on substrate specificity.

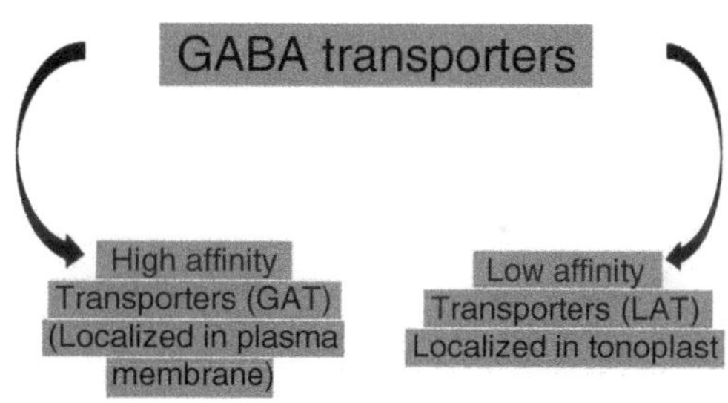

across cellular membranes. There are two main classes of GATs based on their substrate specificity: high-affinity transporters (GATs) and low-affinity transporters (LATs) (Ansari et al. 2021) (Figure 11.2). Long-distance transport of GABA occurs through the phloem, the vascular tissue responsible for transporting nutrients and signaling molecules throughout the plant. GABA is loaded into the phloem and transported to different plant organs, where it can fulfill various physiological roles (Scimemi 2014).

In *Arabidopsis thaliana*, GAT1 and GAT2 are examples of high-affinity GATs, while LAT1 is a low-affinity GAT. These transporters are expressed in various plant tissues, including roots, leaves, and flowers (Meyer et al. 2006). GABA is synthesized in the cytosol, and its accumulation in the vacuole relies on the activity of GATs. The subcellular localization of GATs can impact the accumulation of GABA within different organelles. The high-affinity GAT (GAT1) is localized in the plasma membrane, where it facilitates GABA uptake in root cells, particularly under stress conditions. On the other hand, the low-affinity GAT (LAT1) is localized in the tonoplast, which is responsible for sequestering GABA into the vacuole (Shelp et al. 2012a). The activity and regulation of GATs are crucial for maintaining proper GABA levels and their distribution within plant cells. They contribute to the uptake of GABA from the external environment, its transport between cellular compartments, and its sequestration into the vacuole. The expression and activity of GATs and enzymes involved in GABA metabolism are regulated by environmental cues, developmental stages, and internal signaling pathways. Changes in GABA transport and metabolism can occur in response to stress conditions or specific developmental requirements. Understanding the functions and regulation of GATs provides insights into the control of GABA accumulation and metabolism in plants (Minelli et al. 1995).

Role of GABA in Abiotic Stress Tolerance

Accumulation of GABA, as well as stress tolerance responses under various abiotic stress conditions, including drought, salinity, extreme temperature, and heavy metal toxicity, have been studied in this review (Figure 11.3).

GABA and Drought Stress

Drought stress poses a significant challenge for plants worldwide, impacting their growth and development. During drought conditions, plants experience water scarcity, leading to various physiological and biochemical changes (Waseem et al. 2021). One of the consequences of drought stress is the overproduction of ROS within plant cells (Hasan et al. 2019).

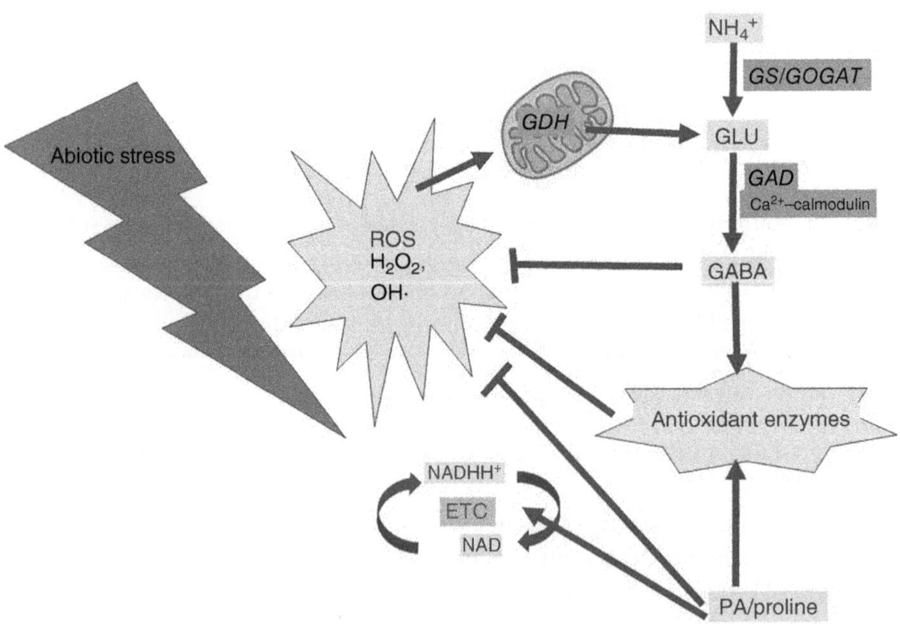

Figure 11.3 Production of ROS in plant cells is due to the exposure to abiotic stresses.

ROS, such as superoxide radicals (O^{2-}), hydrogen peroxide (H_2O_2), and hydroxyl radicals (OH^-), are highly reactive molecules that can cause damage to cellular components, including proteins, lipids, and DNA. ROS can initiate oxidative stress, leading to oxidative damage and disruption of cellular processes (Hasan et al. 2020b).

In drought-stressed plants, factors such as reduced water availability, increased temperature, and imbalances in cellular metabolism can lead to the accumulation of ROS. These ROS can be generated through processes such as photosynthesis, respiration, and ETC in mitochondria and chloroplasts. In addition, drought-induced cellular dehydration can further accelerate ROS production (Hasan et al. 2020a). The overproduction of ROS can have detrimental effects on plant cells by causing oxidative damage to biomolecules and disrupting cellular homeostasis. This oxidative stress can lead to membrane lipid peroxidation, protein oxidation, enzyme inactivation, and DNA damage. Consequently, the overall growth and development of plants are negatively affected under drought stress (Suzuki and Mittler 2006; Zhang and Becker 2015).

However, emerging evidence suggests that stress-induced ROS may have signaling functions in plants. ROS, including hydrogen peroxide (H_2O_2), have been found to induce the synthesis of proline, alanine (Ala), glutamate (Glu), and GABA under stress conditions (Bouche and Fromm 2004). These metabolites are associated with the GABA shunt, a pathway that accumulates in response to ROS generation. Certain ROS, like H_2O_2, have been proposed to activate signal transduction pathways within plant cells (Tripathy and Oelmüller 2012). It has been suggested that the decrease in reduced form of nicotinamide adenine dinucleotide phosphate (NADPH) levels in the nucleus due to H_2O_2 acts as a protective mechanism against H_2O_2-mediated toxicity. For instance, H_2O_2 reduction in wheat plants is crucial for regulating redox reactions and driving the expression of specific genes during seed growth (Khan et al. 2017; Hasan et al. 2018). Furthermore, ROS can promote the activity of mitochondrial glutamate dehydrogenase (GDH), which provides Glu as a precursor for the production of ornithine and GABA. This highlights the interconnectedness between ROS, Glu metabolism, and GABA production in response to drought stress. Plants have evolved various defense mechanisms to counteract the damaging effects of ROS. They possess an array of antioxidant systems, including enzymatic antioxidants such as SOD, CAT, peroxidases (POX), and nonenzymatic antioxidants such as ascorbate (vitamin C), glutathione (GSH), carotenoids and flavonoids (Pulido et al. 2009). These antioxidants help scavenge and neutralize ROS, minimizing their detrimental effects on cellular components. Understanding the impact of ROS production and the antioxidant defense system is essential for developing strategies to enhance drought tolerance. By improving the antioxidant capacity of the plants and minimizing ROS-mediated damage, scientists aim to enhance plant resilience to drought stress and improve crop productivity in a water-limited environment.

Overall, the role of ROS in plant biology is complex. While excessive ROS can be damaging, they can also act as signaling molecules in stress responses (Skopelitis et al. 2006). Understanding the interplay between ROS, metabolic pathways, and signaling networks is essential for deciphering plant responses to drought stress and developing strategies to enhance drought tolerance in plants.

Several studies have demonstrated the protective role of GABA in various plant species. In *Phaseolus vulgaris* L. (common bean), the application of GABA to the leaves increased fresh shoot weight, dry shoot weight, and leaf area. GABA improved drought tolerance in these plants through a process called osmotic adjustment, which involves the accumulation of compatible solutes. This adjustment helps plants maintain their water balance and withstand drought stress (Abd El-Gawad et al. 2020). Exogenous GABA application is seen to mitigate drought-induced leaf damage in plants. It is found to increase relative water content while decreasing electrolyte leakage and lipid peroxidation, which are indicators of cell membrane damage and oxidative stress. These effects contribute to the protection of plant tissues under drought conditions (Yong et al. 2017). Moreover, GABA treatment enhances drought resistance in plants by promoting the accumulation of amino acids, organic acids, and other osmotic compounds associated with secondary metabolism. These compounds help plants cope with water deficit and maintain cellular functions (Li et al. 2016). In terms of GABA metabolism, exogenous GABA treatment was found to increase the activity of GABA transaminase and α-ketoglutarate dehydrogenase in white clover leaves. However, the activity of GAD, an enzyme involved in GABA synthesis, was reduced under both control and drought conditions. This led to an increase in the levels of endogenous glutamate (Glu) and GABA. These changes in GABA metabolism contribute to the overall response of plants to drought stress (Yong et al. 2017).

GABA functions as a downstream signaling molecule of stress-related transcription factors like WRKY, MYB, and bZIP, which are known to confer drought tolerance in plants. During drought stress, GABA has been shown to increase the activity of Ca^{2+}-dependent protein kinase 26 (CDPK26) and mitogen-activated protein kinase 1 (MAPK1) by threefold and fivefold, respectively. These kinases play important roles in signal transduction pathways associated with drought stress responses in plants (Li et al. 2018).

ROS are produced in various plant cell organelles, including plasma membranes, peroxisomes, chloroplasts, and mitochondria, both under normal and stressful conditions. Chloroplasts and peroxisomes are the major sources of ROS generation under normal light conditions (Gill and Tuteja 2010). Excessive ROS production is associated with oxidative damage in plants (Guler and Pehlivan 2016; Alharbi et al. 2021), and its levels are regulated by factors such as genotype, growth phase, and environmental conditions like drought (Hasan et al. 2021a, b). Plants respond to ROS by increasing their antioxidant defenses (Slabbert and Krüger 2014). These defenses involve various compounds, such as carotenoids, proline, anthocyanins, ascorbate, tocopherols, flavonols, amino acids, as well as enzymes like SOD, CAT, POX, ascorbate peroxidase (APX), monodehydroascorbate reductase (MDHAR), and dehydroascorbate reductase (DHAR). When ROS production increases under stress, the antioxidant enzymes play a crucial role in protecting the integrity of cell membranes. Ascorbate, a nonenzymatic antioxidant, is particularly important in scavenging ROS and influencing key processes in plants during both stress and nonstress situations (Capell et al. 2004). The application of exogenous GABA has been shown to enhance the activities of CAT, POD, APX, MDHAR, and glutathione reductase (GR) enzymes while reducing H_2O_2 production, leading to greater drought tolerance in plants (Razik et al. 2020). However, in perennial ryegrass, GABA application did not significantly influence the activities of SOD and CAT under well-watered or drought conditions (Krishnan et al. 2013). Overall, such findings suggest that GABA improves tolerance to oxidative stress caused by abiotic stressors like drought by activating antioxidant pathways.

Stomatal guard cells, due to their important functions in plant cell signaling and their ability to respond to multiple stimuli, have become a prominent model system for studying various pathways that regulate plant tolerance to biotic and abiotic stresses (Murata et al. 2015; Yao et al. 2021). The effects of GABA on stomata have been observed in different plants, including dicot and monocot crops. Recent discoveries have shown that GABA and the ALMT1 protein are involved in plant development through the binding of GABA to the ALMT1 transporter (Renault et al. 2011; Ramesh et al. 2015). It has been suggested that this interaction may alter stomatal movement. Other proteins in the ALMT family, such as ALMT6, ALMT9, and ALMT12, are located in the vacuolar and plasma membranes and play a role in regulating stomatal movement by facilitating the influx and efflux of malate (Mekonnen et al. 2016). Meyer et al. (2010) proposed the existence of a cofactor (cytosolic or vacuolar) that binds to and regulates the action of the ALMT6 protein, as removing calcium ions (Ca^{2+}) from the activated protein had no effect on its activity (Meyer et al. 2011).

Recently, Xu et al. (2021a, b) reported GABA signaling pathway in plants, where cytosolic GABA signals produced by GAD2 affect stomatal opening, water-use efficiency (WUE), and drought tolerance by negatively regulating the activity of ALMT9. ALMT9 is a key component involved in transmitting GABA signals in cells under well-watered and drought conditions. Although the specific effect of GABA on stomatal closure during drought is unclear, it is suggested that GABA-mediated regulation of stomata may be influenced by light, dark, and low-intermediate signal concentrations.

GABA and Polyamines in Drought Stress

Polyamines (PAs), including putrescine (Put), spermidine (Spd), and spermine (Spm), are found in most eukaryotic cells and exist in free, conjugated or bound forms in higher plants (Gil-Amado and Gomez-Jimenez 2012; Hu et al. 2015). Under stressful conditions, plants often exhibit an increase in PA biosynthesis and catabolism (Mattoo et al. 2010; Wang et al. 2014a, b). PAs are involved in various physiological processes and are considered as signaling molecules that play important roles in developmental stages, such as embryonic differentiation, cell death responses, fruit ripening and seed germination (Kim et al. 2013; Silveira et al. 2013). PA biosynthesis can occur through the decarboxylation of ornithine or arginine mediated by ornithine decarboxylase (ODC) or arginine decarboxylase (ADC), respectively. The significance of GABA in plant tolerance and its relationship with PAs have been extensively studied (Hu et al. 2015). PA catabolism can generate GABA through the action of copper amine oxidase (CuAO) or polyamine oxidase (PAO), establishing a direct link between GABA and PAs. Yang et al. (2013) demonstrated that under hypoxic stress, PA degradation contributed to approximately 30% of the GABA content in *Vicia faba* (Yang et al. 2013). Three enzymes [ADC, ODC, and S-adenosylmethionine decarboxylase (SAMDC)] involved in PA production and two enzymes (CuAO and PAO) involved in PA breakdown were activated under drought conditions (Al-Quraan 2015; Yong et al. 2017). Wang et al. (2014a, b) suggested that increased endogenous GABA acts as negative feedback for PA breakdown in drought-stressed conditions. The findings indicated that GABA enhanced PA synthesis while inhibiting PA catabolism, leading to an overall increase in PA content in GABA-treated plants under drought conditions, and hence stress tolerance.

In soybean plants subjected to drought stress, GABA accumulation is achieved by increasing the activity of an enzyme called diamine oxidase (DAO), which promotes the breakdown of PAs. This suggests that PAs regulate the accumulation of GABA, which, in turn, contributes to drought tolerance responses in soybean plants (Xing et al. 2007). Similarly, in muskmelon plants exposed to salt stress, degradation of PAs and endogenous accumulation of GABA were observed. In addition, when exogenous GABA is applied to these plants, it blocks the degradation of PAs, leading to the accumulation of Put, Spm, and Spd, suggesting that GABA may act as a modulator of plant tolerance by influencing the metabolism of PAs in response to salt stress (Xiang et al. 2016). The interplay between PAs and GABA suggests a regulatory mechanism that helps plants cope with environmental challenges, although further research is needed to fully understand the underlying mechanism.

GABA and Salt Stress

Salinity, caused by high levels of salt in the soil, is a significant problem that restricts the growth and development of various plants (Machado and Serralheiro 2017; Naz et al. 2022). According to the Food and Agriculture Organization (FAO), salt stress has already damaged around 45 million hectares of irrigated land worldwide, which accounts for 19.5% of the total irrigated area (Parihar et al. 2015). Salt stress negatively affects plants by impairing their ability to absorb water, causing cell damage, stunting leaf growth, reducing available photosynthetic assimilates, and delaying germination (Stepien and Klobus 2005; Singh 2022). One of the consequences of salt stress is the overproduction of ROS, which can result in lipid peroxidation and damage to biomolecules (Barbosa et al. 2010; Ramzan et al. 2023).

Studies have shown that GABA can play a role in enhancing plant resilience to salt stress. GABA administration during salt stress can modify the activities of antioxidant enzymes involved in nitrogen metabolic pathways and affect the signaling of the nitrate uptake system (Beuve et al. 2004; Wu et al. 2020). Exogenous application of GABA has been found to improve the ability of tomato and cucumber seedlings to withstand salt stress by reducing sodium ion accumulation, increasing amino acid content, and enhancing antioxidant metabolism (Yin et al. 2010).

GABA accumulation rapidly increases in tomato and tea plants when anabolic metabolism is activated by salt stress induction. GABA helps reduce salt damage during seed germination by increasing starch catabolism, utilizing sugar and amino acids for growth maintenance, increasing Na^+/K^+ transportation for osmotic adjustment, and enhancing antioxidant defense potential (Cheng et al. 2018). Exogenous GABA treatment has been shown to inhibit H_2O_2 generation and reduce oxidative damage in salt-stressed roots of *Caragana intermedia* by regulating the expression of key genes involved in H_2O_2 and peroxidase production (Xiang et al. 2016).

GABA treatment has also been found to positively impact muskmelon and apple seedlings under saline conditions. It improves the structure and functions of Photosystem II, enhances nitrogen absorption and assimilation, promotes growth, scavenges ROS activities, and improves photosynthetic properties and chlorophyll levels (Chen et al. 2018). GABA application increases nitric oxide (NO) levels in muskmelon under salinity–alkalinity stress, which helps regulate Na^+/K^+ balance, enhance antioxidation, and maintain the stability of cell membranes (Shi et al. 2010; Li et al. 2020b). It has also been found to reduce salt damage and increase plant resistance to salt stress.

GABA supplementation, along with other approaches, has the potential to enhance the resilience of crops to saline conditions, which is crucial for addressing the increasing threat of soil salinity to global food security (Xu et al. 2021a, b; Zhang et al. 2022). Further research is needed to fully understand the mechanisms involved and to explore the applicability of GABA supplementation in different plant species and under various environmental conditions.

GABA and Heat Stress

Global warming has led to a significant increase in temperature, resulting in heat stress as a major constraint on crop growth and yield (Apel and Hirt 2004). Heat stress negatively affects various aspects of plant physiology, including growth, mineral nutrient content, and yield. Plants may exhibit symptoms such as oxidative damage, chlorophyll degradation, and photoinhibition under heat-stress conditions. GABA has been studied for its regulatory function in enhancing heat tolerance in plants (Jespersen et al. 2016).

External application of GABA can protect seedlings from heat stress by enhancing their antioxidant defense systems. For example, in creeping bent grass foliage, GABA has been found to regulate osmotic potential, metabolic homeostasis, and the TCA cycle, thereby promoting heat tolerance (Jespersen et al. 2016). GABA application also up regulates genes involved in heat shock factor pathways and phenylpropanoid biosynthesis, improving heat tolerance in perennial creeping bent grass. Transcript levels of heat shock proteins, heat shock factors, and APX 3 have been observed to increase under heat stress with exogenous GABA treatment (Ren et al. 2021).

Foliar treatment with GABA has been shown to alleviate the harmful effects of heat stress in creeping bent grass by enhancing the expression of heat-induced heat shock proteins and heat shock factors (Jespersen et al. 2016). GABA treatment induces changes in proteomic profiles, leading to increased accumulation of sugars and amino acids, which are crucial for energy supply and pathways involved in heat tolerance. GABA administration also enhances endogenous levels of GABA, glutamic acid, and threonine, regulating the GABA shunt and oxaloacetate pathway and improving heat tolerance (Liu et al. 2019; Chen et al. 2022).

In tea seedlings, exogenous GABA application has been found to contribute to the accumulation of polyphenols, up regulate genes related to flavonoid metabolism, and improve the antioxidant system under heat-stress conditions. GABA enhances the adaptability of roots to heat stress by boosting antioxidant capacity, vitality, and osmotic adjustment. It also regulates metabolites, improving antioxidant capacity, energy metabolism, cellular structures, and osmotic balance in roots (Nayyar et al. 2014; Li et al. 2016).

Overall, exogenous administration of GABA can enhance plant growth and survival in the face of heat stress by ameliorating antioxidant capacity and reducing oxidative damage (Li et al. 2019b; Zeng et al. 2021). It offers promising potential for improving heat tolerance in crops under the challenges posed by global warming. Further research is needed to explore the mechanisms involved and to evaluate the efficacy of GABA application in different plant species and environmental conditions (Li et al. 2020a; Rossi et al. 2021).

GABA and Cold Stress

Low-temperature stress poses a significant challenge to plant growth and development, resulting in various physiological, biochemical, and molecular changes (Raza et al. 2022). Cold stress can lead to detrimental effects such as leaf wilting, chlorophyll loss, reduced photosynthesis, impaired cell membrane fluidity, decreased enzyme activity, metabolic disruption, stunted growth (Ding et al. 2019; Hayat et al. 2021) as well as increased production of ROS, which are highly active and toxic, causing oxidative damage to plant structures (Megha et al. 2018).

Several studies have highlighted the beneficial effects of GABA treatment in mitigating the adverse effects of cold stress on different plant species. GABA has been shown to alleviate chilling injury in tomato seedlings by modulating antioxidant enzyme activity and scavenging ROS (Rawat et al. 2021). In tea plants, GABA enhances low-temperature tolerance and modulates various physiological and biochemical processes, including antioxidant activity, chlorophyll fluorescence, and membrane stability (Malekzadeh et al. 2014).

Exogenous GABA treatment has also been found to be beneficial during the postharvest storage of various fruits and vegetables. For example, GABA can regulate weight loss, chilling injury, and cell death in zucchini fruits stored at low temperature (Zhu et al. 2019). Preharvest application of GABA and Spm effectively prevents chilling damage and delays senescence in plants by increasing proline content, boosting antioxidant capacity, scavenging ROS, and improving cell membrane integrity (Madebo et al. 2021). In pear fruits, GABA treatment protects against peel browning during extended storage at low temperature by reducing browning indices, ROS levels, and malondialdehyde (MDA) content while increasing antioxidant enzyme activity. GABA application also increases GABA shunt activity, promotes glycine betaine accumulation, and enhances ATP production in cut flowers. In addition, GABA treatment alleviates membrane damage, elevates antioxidant capacity, and reduces chilling injury in banana peels (Jiao 2022).

In peach fruit, GABA treatment reduces chilling injury and weight loss while maintaining firmness and total soluble solid content. This is achieved by suppressing ROS production, increasing the activity and gene expression of antioxidant enzymes (Wang et al. 2014; Mohammadi et al. 2021). GABA application also increases the NADPH/NADP ratio and up regulates the expression and activity of glucose-6-phosphate dehydrogenase (G6PDH) (Li et al. 2019b).

Overall, GABA application shows promise in mitigating chilling injury in plants by regulating various physiological, biochemical, and molecular mechanisms. It helps maintain membrane integrity, enhances antioxidant capacity, and reduces oxidative damage caused by low-temperature stress (Soleimani Aghdam et al. 2016).

GABA and Heavy Metal Stress

Heavy metal pollution is a significant environmental issue caused by both natural processes and human activities such as urbanization, industrialization, and mining. This pollution can have detrimental effects on ecosystems and plant productivity worldwide. Soil contamination by heavy metals, including zinc (Zn), nickel (Ni), iron (Fe), chromium (Cr), cobalt (Co), lead (Pb), cadmium (Cd), mercury (Hg), and arsenic (As), resulting from agricultural practices, is a particular concern due to its potential threat to human health and ecological systems (Alengebawy et al. 2021).

One of the consequences of heavy metal accumulation in plants is the disruption of various biochemical, physiological, and morphological functions, leading to decreased crop productivity. However, there is promising research indicating that GABA, a natural compound, can play a significant role in mitigating heavy metal toxicity in plants (Naveedullah et al. 2013). GABA is considered environment friendly and can be mass-produced, making it highly applicable in various fields (Shahid et al. 2015).

In maize, GABA acts as a chelating agent, binding to Cd ions and forming stable complexes. This helps sequester Cd away from the plant root system, making it less available for uptake. GABA interacts with receptors, such as the aluminum-activated malate transporter (ALMT), triggering signaling pathways that affect Cd uptake or distribution (Seifikalhor et al. 2020). GABA influences the osmotic pressure and cytosolic pH within plant cells. By doing so, it could impact the absorption of Cd ions. Changes in osmotic pressure and pH might affect the availability of Cd for uptake by the plant roots in garden cress (Gill et al. 2012). GABA could regulate enzymatic activities involved in Cd uptake, transport, or sequestration within the plant. By inhibiting certain enzymatic processes, GABA can limit the transfer of Cd within the plant, particularly from roots to shoots, as seen in tobacco (Allan et al. 2008).

Application of GABA enhanced plant ability to cope with Cr-stress by promoting better water uptake and retention. External GABA application modulates stomatal conductance, water transport pathways, and osmotic adjustment mechanisms, ultimately contributing to improved water status in stressed plants (Nayyar et al. 2014). By regulating stomatal aperture, chlorophyll content, and electron transport, GABA optimizes the photosynthetic processes (Song et al. 2010) by up regulating the enzymes of antioxidant defense and glyoxylase system in Cr-stressed mustard seedlings (Mahmud et al. 2017).

In Cd-stressed microalga *Monoraphidium,* exogenous GABA application resulted in improved cell growth, lipid biosynthesis, and Cd-tolerance (Zhao et al. 2020). Several studies have explored the potential of GABA in detoxifying heavy metals in plants. GABA treatment led to a reduction in the expression of Lsi-1, an influx transporter responsible for facilitating the accumulation of As(III) from the solution into root cells, and Lsi-2, an efflux transporter that regulates the movement of As(III) and silicon (Si) from root cells to the apoplast in response to As(III) treatments in rice seedlings. GABA signaling influenced the gene expression of these transporters in a way that decreased their activity and abundance (Kumar et al. 2017).

In *Malus huphehensis* (a species of apple), exogenous GABA treatment activates the GABA shunt pathway, resulting in increased levels of malate, citric acid, and sucrose, as well as enhanced enzyme activity (Mahmud et al. 2017; Li et al. 2021). These effects help alleviate biomass reduction, root growth inhibition, and oxidative stress caused by alkaline stress. Again, exogenous application of GABA has shown effectiveness in mitigating Cd toxicity in apple seedlings. It reduces Cd content in plants and decreases the expression of genes related to Cd uptake and transport (Bor et al. 2009; Li et al. 2022). Exogenous application of GABA to Cr-stressed plants improves tolerance against heavy metal by reducing metal uptake and increasing water and chlorophyll content through the up regulation of

ROS scavenging enzymes (Hasanuzzaman and Fujita 2012). In rice, GABA accumulation enhanced the activities of antioxidant enzymes (MDHAR, and POD) against excessive As stress, which coincided with the reduced levels of H_2O_2 and its toxicity (Kumar et al. 2019). Elevated activities of antioxidant enzymes were reported with GABA treatment in Al-induced stress in barley seedlings (Song et al. 2010). Exogenous GABA application lowered Cr-accumulation in *Brassica juncea* (Mahmud et al. 2017), As-accumulation in rice (Kumar et al. 2019), tomato and brinjal (Suhel et al. 2023), and Cd-accumulation in maize, perlite (Seifikalhor et al. 2020) and apple (Li et al. 2022). These findings highlight the potential of GABA to mitigate the adverse effects of heavy metal pollution and promote growth in plants (Kumar et al. 2019).

Overall, GABA shows promise as a natural and mass-producible compound to alleviate the negative impacts of heavy metal pollution on plants. Its application may help detoxify heavy metals, enhance plant growth, and mitigate the potential threats posed to human health through the food chain.

Conclusion

The protective role of GABA in stress tolerance of plants has been extensively studied, and the findings have been promising. Exogenous application of GABA has been shown to improve the tolerance of plants to various abiotic stress factors, including drought, high salinity and extreme temperature. GABA also plays a crucial role in stomatal movement in plants, particularly under drought stress. It is believed to be involved in the closure and opening of stomata. This suggests that GABA can be a valuable tool in improving the yield and quality of plants. In stressed plants, GABA enhances the activities of the GABA shunt, which leads to increased photosynthetic efficiency, activation of antioxidative enzymes, reduced production of ROS, MDA, and H_2O_2, and protection of the cell membrane. However, the significance of the GABA shunt and its metabolites in abiotic stress responses still requires further investigation.

Overall, it is proposed that manipulating GABA levels could be a viable approach for improving plant stress resistance and enhancing physiological regulation under adverse environmental conditions. This comprehensive understanding of GABA synthesis and its stress-mitigating role can contribute to increasing the quantity and quality of agricultural products, support the achievement of sustainable development goals and provide insights into future research directions in this field.

Acknowledgments

Financial assistance from the Science and Engineering Research Board (SERB), Government of India through the grant EMR/2016/004799, and the Department of Higher Education, Science and Technology and Biotechnology, Government of West Bengal (DHESTBT) through the grant 264(Sanc.)/ST/P/S&T/1G-80/2017 is gratefully acknowledged.

References

Abd El-Gawad, H.G., Mukherjee, S., Farag, R. et al. (2020). Exogenous γ-aminobutyric acid (GABA)-induced signaling events and field performance associated with mitigation of drought stress in *Phaseolus vulgaris* L. *Plant Signaling & Behavior* 16: 1853384.

Akihiro, T., Koike, S., Tani, R. et al. (2008). Biochemical mechanism on GABA accumulation during fruit development in tomato. *Plant & Cell Physiology* 49: 1378–1389.

Alengebawy, A., Abdelkhalek, S.T., Qureshi, S.R., and Wang, M.Q. (2021). Heavy metals and pesticides toxicity in agricultural soil and plants: ecological risks and human health implications. *Toxics* 9: 42.

Alharbi, B.M., Elhakem, A.H., Alnusairi, G.S.H. et al. (2021). Exogenous application of melatonin alleviates salt stress-induced decline in growth and photosynthesis in *Glycine max* (L.) seedlings by improving mineral uptake, antioxidant and glyoxalase system. *Plant, Soil and Environment* 67: 208–220.

Allan, W.L., Simpson, J.P., Clark, S.M., and Shelp, B.J. (2008). γ-Hydroxybutyrate accumulation in *Arabidopsis* and tobacco plants is a general response to abiotic stress: putative regulation by redox balance and glyoxylate reductase isoforms. *Journal of Experimental Botany* 59: 2555–2564.

Al-Quraan, N.A. (2015). GABA shunt deficiencies and accumulation of reactive oxygen species under UV treatments: insight from *Arabidopsis thaliana* calmodulin mutants. *Acta Physiologia Plantarum* 37: 86.

Ansari, M.I., Jalil, S.U., Ansari, S.A., and Hasanuzzaman, M. (2021). GABA shunt: a key-player in mitigation of ROS during stress. *Plant Growth Regulation* 94: 131–149.

Apel, K. and Hirt, H. (2004). Reactive oxygen species: metabolism, oxidative stress, and signal transduction. *Annual Review of Plant Biology* 55: 373–399.

Barbosa, J.M., Singh, N.K., Cherry, J.H., and Locy, R.D. (2010). Nitrate uptake and utilization is modulated by exogenous-aminobutyric acid in *Arabidopsis thaliana* seedlings. *Plant Physiology and Biochemistry* 48: 443–450.

Beuve, N., Rispail, N., Laine, P. et al. (2004). Putative role of γ-aminobutyric acid (GABA) as a long-distance signal in up-regulation of nitrate uptake in *Brassica napus* L. *Plant, Cell & Environment* 27: 1035–1046.

Bor, M., Seckin, B., Ozgur, R. et al. (2009). Comparative effects of drought, salt, heavy metal and heat stresses on gamma-aminobutryric acid levels of sesame (*Sesamum indicum* L.). *Acta Physiologia Plantarum* 31: 655–659.

Bouche, N. and Fromm, H. (2004). GABA in plants: just a metabolite? *Trends in Plant Science* 9: 110–115.

Bown, A.W. and Shelp, B.J. (1997). The metabolism and functions of [gamma]-aminobutyric acid. *Plant Physiology* 115: 1–15.

Boyle, A.E., Segal, R., Smith, B.R., and Amit, Z. (1993). Bidirectional effects of GABAergic agonists and antagonists on the maintenance of voluntary ethanol intake in rats. *Pharmacology Biochemistry and Behavior* 46: 179–182.

Capell, T., Bassie, L., and Christou, P. (2004). Modulation of the polyamine biosynthetic pathway in transgenic rice confers tolerance to drought stress. *Proceedings of the National Academy of Sciences of the United States of America* 101: 9909–9914.

Chen, H., Liu, T., Xiang, L. et al. (2018). GABA enhances muskmelon chloroplast antioxidants to defense salinity-alkalinity stress. *Russian Journal of Plant Physiology* 65: 674–679.

Chen, X., Li, N., Liu, C. et al. (2022). Exogenous GABA improves the resistance of apple seedlings to Long-term drought stress by enhancing GABA shunt and secondary cell wall biosynthesis. *Tree Physiology* 42: 2563–2577.

Cheng, B., Li, Z., Liang, L. et al. (2018). The γ-aminobutyric acid (GABA) alleviates salt stress damage during seeds germination of white clover associated with Na^+/K^+ transportation, dehydrins accumulation, and stress-related genes expression in white clover. *International Journal of Molecular Sciences* 19: 2520.

Ding, Y., Shi, Y., and Yang, S. (2019). Advances and challenges in uncovering cold tolerance regulatory mechanisms in plants. *The New Phytologist* 222: 1690–1704.

Fait, A., Fromm, H., Walter, D. et al. (2008). Highway or byway: the metabolic role of the GABA shunt in plants. *Trends in Plant Science* 13: 14–19.

Gil-Amado, J.A. and Gomez-Jimenez, M.C. (2012). Regulation of polyamine metabolism and biosynthetic gene expression during olive mature-fruit abscission. *Planta* 235: 1221–1237.

Gill, S.S. and Tuteja, N. (2010). Reactive oxygen species and antioxidant machinery in abiotic stress tolerance in crop plants. *Plant Physiology and Biochemistry* 48: 909–930.

Gill, S.S., Khan, N.A., and Tuteja, N. (2012). Cadmium at high dose perturbs growth, photosynthesis and nitrogen metabolism while at low dose it up regulates sulfur assimilation and antioxidant machinery in garden cress (*Lepidium sativum* L.). *Plant Science* 182: 112–120.

Guler, N.S. and Pehlivan, N. (2016). Exogenous low-dose hydrogen peroxide enhances drought tolerance of soybean (*Glycine max* L.) through inducing antioxidant system. *Acta Biologica Hungarica* 67: 169–183.

Hasan, M.M., Hajar, A.S., Alharby, H.F., and Hakeem, K.R. (2018). Effects of magnetized water on phenolic compounds, lipid peroxidation and antioxidant activity of *Moringa* species under drought stress. *Journal of Animal and Plant Sciences* 28: 803–810.

Hasan, M.M., Alharby, H.F., Hajar, A.S. et al. (2019). The effect of magnetized water on the growth and physiological conditions of *Moringa* species under drought stress. *Polish Journal of Environmental Studies* 28: 1145–1155.

Hasan, M.M., Alharby, H.F., Uddin, M.N. et al. (2020a). Magnetized water confers drought stress tolerance in *Moringa* biotype via modulation of growth, gas exchange, lipid peroxidation and antioxidant activity. *Polish Journal of Environmental Studies* 29: 1625–1636.

Hasan, M.M., Ali, M.A., Soliman, M.H. et al. (2020b). Insights into 28-homobrassinolide (HBR) – mediated redox homeostasis, AsA–GSH cycle, and methylglyoxal detoxification in soybean under drought-induced oxidative stress. *Journal of Plant Interactions* 15: 371–385.

Hasan, M.M., Alabdallah, N.M., Alharbi, B.M. et al. (2021a). GABA: A key player in drought stress resistance in plants. *International Journal of Molecular Sciences* 22: 10136.

Hasan, M.M., Skalicky, M., Jahan, M.S. et al. (2021b). Spermine: its emerging role in regulating drought stress responses in plants. *Cells* 10: 261.

Hasanuzzaman, M. and Fujita, M. (2012). Heavy metals in the environment: current status, toxic effects on plants and possible phytoremediation. In: *Phyto-Technologies: Remediation of*

Environmental Contaminants (ed. N.A. Anjum, M.A. Pereira, I. Ahmad, et al.), 7–73. Boca Raton, FL: CRC Press.

Hayat, F., Ma, C., Iqbal, S. et al. (2021). Rootstock-mediated transcriptional changes associated with cold tolerance in *Prunus mume* leaves. *Horticulturae* 7: 572.

Hu, X., Xu, Z., Xu, W. et al. (2015). Application of γ-aminobutyric acid demonstrates a protective role of polyamine and GABA metabolism in muskmelon seedlings under ca $(NO_3)_2$. *Plant Physiology and Biochemistry* 92: 1–10.

Jespersen, D., Zhang, J., and Huang, B. (2016). Chlorophyll loss associated with heat-induced senescence in bentgrass. *Plant Science* 249: 1–12.

Jiao, C. (2022). Aminobutyric acid boosts chilling tolerance by promoting the methionine sulfoxide reductase-thioredoxin reductase system in peach fruit. *Horticulture, Environment and Biotechnology* 63: 353–361.

Khan, A., Anwar, Y., Hasan, M. et al. (2017). Attenuation of drought stress in *Brassica* seedlings with exogenous application of Ca^{2+} and H_2O_2. *Plants* 6: 20.

Kim, N.H., Kim, B.S., and Hwang, B.K. (2013). Pepper arginine decarboxylase is required for polyamine and γ-aminobutyric acid signaling in cell death and defense response. *Plant Physiology* 162: 2067–2083.

Krishnan, S., Laskowski, K., Shukla, V., and Merewitz, E.B. (2013). Mitigation of drought stress damage by exogenous application of a non-protein amino acid γ-aminobutyric acid on perennial ryegrass. *Journal of the American Society for Horticultural Science* 138: 358–366.

Kumar, N., Dubey, A.K., Upadhyay, A.K. et al. (2017). GABA accretion reduces Lsi-1 and Lsi-2 gene expressions and modulates physiological responses in *Oryza sativa* to provide tolerance towards arsenic. *Scientific Reports* 7: 8786.

Kumar, N., Gautam, A., Dubey, A.K. et al. (2019). GABA mediated reduction of arsenite toxicity in Rice seedling through modulation of fatty acids, stress responsive amino acids and polyamines biosynthesis. *Ecotoxicology and Environmental Safety* 173: 15–27.

Li, Z., Yu, J., Peng, Y., and Huang, B. (2016). Metabolic pathways regulated by γ-aminobutyric acid (GABA) contributing to heat tolerance in creeping bentgrass (*Agrostis stolonifera*). *Scientific Reports* 6: 30338.

Li, Z., Peng, Y., and Huang, B. (2018). Alteration of transcripts of stress-protective genes and transcriptional factors by γ-aminobutyric acid (GABA) associated with improved heat and drought tolerance in creeping Bentgrass (*Agrostis stolonifera*). *International Journal of Molecular Sciences* 19: 1623.

Li, J., Zhou, X., Wei, B. et al. (2019a). GABA application improves the mitochondrial antioxidant system and reduces peel browning in 'Nanguo' pears after removal from cold storage. *Food Chemistry* 297: 124903.

Li, Z., Cheng, B., Zeng, W. et al. (2019b). The transcriptional and post-transcriptional regulation in perennial creeping bentgrass in response to γ-aminobutyric acid (GABA) and heat stress. *Environmental and Experimental Botany* 162: 515–524.

Li, Y., Liu, B., Peng, Y. et al. (2020a). Exogenous GABA alleviates alkaline stress in *Malus hupehensis* by regulating the accumulation of organic acids. *Scientia Horticulturae* 261: 108982.

Li, Z., Cheng, B., Zeng, W. et al. (2020b). Proteomic and metabolomic profilings reveal crucial functions of γ-aminobutyric acid in regulating ionic, water, and metabolic homeostasis in creeping bentgrass under salt stress. *Journal of Proteome Research* 19: 769–780.

Li, L., Dou, N., Zhang, H., and Wu, C. (2021). The versatile GABA in plants. *Plant Signaling & Behavior* 16: 1862565.

Li, Y., Li, Y., Cui, Y. et al. (2022). GABA-mediated inhibition of cadmium uptake and accumulation in apples. *SSRN Electronic Journal* 300: 118867.

Li, Z., Zhou, M., Zeng, W. et al. (2023). Root metabolites remodeling regulated by γ-aminobutyric acid (GABA) improves adaptability to high temperature in creeping bentgrass. *Plant and Soil* 1–15.

Liu, T., Liu, Z., Li, Z. et al. (2019). Regulation of heat shock factor pathways by γ-aminobutyric acid (GABA) associated with thermotolerance of creeping Bentgrass. *International Journal of Molecular Sciences* 20: 4713.

Machado, R.M.A. and Serralheiro, R.P. (2017). Soil salinity: effect on vegetable crop growth. Management practices to prevent and mitigate soil salinization. *Horticulturae* 3: 30.

Madebo, M.P., Hu, S., Zheng, Y., and Jin, P. (2021). Mechanisms of chilling tolerance in melatonin treated postharvest fruits and vegetables: a review. *Journal of Future Foods* 1: 156–167.

Mahmud, J.A.L., Hasanuzzaman, M., Nahar, K. et al. (2017). Aminobutyric acid (GABA) confers chromium stress tolerance in *Brassica juncea* L. by modulating the antioxidant defense and glyoxalase systems. *Ecotoxicology* 26: 675–690.

Malekzadeh, P., Khara, J., and Heydari, R. (2014). Alleviating effects of exogenous gamma-aminobutyric acid on tomato seedling under chilling stress. *Physiology and Molecular Biology of Plants* 20: 133–137.

Mattoo, A.K., Minocha, S.C., Minocha, R., and Handa, A.K. (2010). Polyamines and cellular metabolism in plants: transgenic approaches reveal different responses to diamine putrescine versus higher polyamines spermidine and spermine. *Amino Acids* 38: 405–413.

Megha, S., Basu, U., and Kav, N.N.V. (2018). Regulation of low temperature stress in plants by microRNAs. *Plant, Cell & Environment* 41: 1–15.

Mekonnen, D.W., Flügge, U.I., and Ludewig, F. (2016). Gamma-aminobutyric acid depletion affects stomata closure and drought tolerance of *Arabidopsis thaliana*. *Plant Science* 245: 25–34.

Meyer, A., Eskandari, S., Grallath, S., and Rentsch, D. (2006). AtGAT1, a high affinity transporter for γ-aminobutyric acid in *Arabidopsis thaliana*. *The Journal of Biological Chemistry* 281: 7197–7204.

Meyer, S., Mumm, P., Imes, D. et al. (2010). AtALMT12 represents an R-type anion channel required for stomata movement in *Arabidopsis* guard cells. *The Plant Journal* 63: 1054–1062.

Meyer, S., Scholz-Starke, J., De Angeli, A. et al. (2011). Malate transport by the vacuolar AtALMT6 channel in guard cells is subject to multiple regulation. *The Plant Journal* 67: 247–257.

Michaeli, S., Fait, A., Lagor, K. et al. (2011). A mitochondrial GABA permease connects the GABA shunt and the TCA cycle, and is essential for normal carbon metabolism. *The Plant Journal* 67: 485–498.

Minelli, A., Brecha, N.C., Karschin, C. et al. (1995). GAT-1, a high-affinity GABA plasma membrane transporter, is localized to neurons and astroglia in the cerebral cortex. *The Journal of Neuroscience* 15: 7734–7746.

Mohammadi, M., Aelaei, M., and Saidi, M. (2021). Pre-harvest spray of GABA and spermine delays postharvest senescence and alleviates chilling injury of gerbera cut flowers during cold storage. *Scientific Reports* 11: 14166.

Murata, Y., Mori, I.C., and Munemasa, S. (2015). Diverse stomatal signaling and the signal integration mechanism. *Annual Review of Plant Biology* 66: 369–392.

Naveedullah, Hashmi, M.Z., Yu, C. et al. (2013). Risk assessment of heavy metals pollution in agricultural soils of siling reservoir watershed in Zhejiang Province, China. *BioMed Research International* 2013: 590306.

Nayyar, H., Kaur, R., Kaur, S., and Singh, R. (2014). γ-Aminobutyric acid (GABA) imparts partial protection from heat stress injury to rice seedlings by improving leaf turgor and upregulating osmoprotectants and antioxidants. *Journal of Plant Growth Regulation* 33: 408–419.

Naz, S., Mushtaq, A., Ali, S. et al. (2022). Foliar application of ascorbic acid enhances growth and yield of lettuce (*Lactuca sativa*) under saline conditions by improving antioxidant defence mechanism. *Functional Plant Biology* .

Ochoa-de la Paz, L.D., Gulias-Cañizo, R., Ruíz-Leyja, E.D.A. et al. (2021). The role of GABA neurotransmitter in the human central nervous system, physiology, and pathophysiology. *Revista Mexicana de Neurociencia* 22 (2): 67–76.

Parihar, P., Singh, S., Singh, R. et al. (2015). Effect of salinity stress on plants and its tolerance strategies: a review. *Environmental Science and Pollution Research* 22: 4056–4075.

Priya, M., Sharma, L., Kaur, R. et al. (2019). GABA (γ-aminobutyric acid), as a thermo-protectant, to improve the reproductive function of heat-stressed mungbean plants. *Scientific Reports* 9: 7788.

Pulido, P., Domínguez, F., and Cejudo, F.J. (2009). A hydrogen peroxide detoxification system in the nucleus of wheat seed cells: protection or signaling role? *Plant Signaling & Behavior* 4: 23–25.

Ramesh, S.A., Tyerman, S.D., Xu, B. et al. (2015). GABA signaling modulates plant growth by directly regulating the activity of plant-specification transporters. *Nature Communications* 6: 7879.

Ramos-Ruiz, R., Martinez, F., and Knauf-Beiter, G. (2019). The effects of GABA in plants. *Cogent Food & Agriculture* 5: 1670553.

Ramzan, M., Shah, A.A., Ahmed, M.Z. et al. (2023). Exogenous application of glutathione and gamma amino-butyric acid alleviates salt stress through improvement in antioxidative defense system and modulation of CaXTHs stress-related genes. *South African Journal of Botany* 157: 266–273.

Rashmi, D., Zanan, R., John, S. et al. (2018). γ-Aminobutyric acid (GABA): biosynthesis, role, commercial production, and applications. *Studies in Natural Products Chemistry* 57: 413–452.

Rawat, N., Singla-Pareek, S.L., and Pareek, A. (2021). Membrane dynamics during individual and combined abiotic stresses in plants and tools to study the same. *Physiologia Plantarum* 171: 653–676.

Raza, A., Su, W., Jia, Z. et al. (2022). Mechanistic insights into trehalose-mediated cold stress tolerance in rapeseed (*Brassica napus* L.) seedlings. *Frontiers in Plant Science* 13: 857980.

Razik, E.S.A., Alharbi, B.M., Pirzadah, T.B. et al. (2020). γ-Aminobutyric acid (GABA) mitigates drought and heat stress in sunflower (*Helianthus annuus* L.) by regulating its physiological, biochemical and molecular pathways. *Physiologia Plantarum* 172: 505–527.

Ren, T., Zheng, P., Zhang, K. et al. (2021). Effects of GABA on the polyphenol accumulation and antioxidant activities in tea plants (*Camellia sinensis* L.) under heat-stress conditions. *Plant Physiology and Biochemistry* 159: 363–371.

Renault, H., El Amrani, A., Palanivelu, R. et al. (2011). GABA accumulation causes cell elongation defects and a decrease in expression of genes encoding secreted and cell wall-related proteins in *Arabidopsis thaliana*. *Plant & Cell Physiology* 52: 894–908.

Rossi, S., Chapman, C., Yuan, B., and Huang, B. (2021). Improved heat tolerance in creeping bentgrass by γ-aminobutyric acid, proline, and inorganic nitrogen associated with differential regulation of amino acid metabolism. *Plant Growth Regulation* 93: 231–242.

Ruttkay-Nedecky, B., Nejdl, L., Gumulec, J. et al. (2013). The role of metallothionein in oxidative stress. *International Journal of Molecular Sciences* 14 (3): 6044–6066.

Scimemi, A. (2014). Plasticity of GABA transporters: an unconventional route to shape inhibitory synaptic transmission. *Frontiers in Cellular Neuroscience* 8: 128.

Seifikalhor, M., Aliniaeifard, S., Bernard, F. et al. (2020). Aminobutyric acid confers cadmium tolerance in maize plants by concerted regulation of polyamine metabolism and antioxidant defense systems. *Scientific Reports* 10: 3356.

Shahid, M., Khalid, S., Abbas, G. et al. (2015). Heavy metal stress and crop productivity. In: *Crop Production and Global Environmental Issues*, 1–25. Berlin/Heidelberg, Germany: Springer ISBN 9783319231624.

Shelp, B.J., Bozzo, G.G., Trobacher, C.P. et al. (2012a). Hypothesis/review: contribution of putrescine to 4-aminobutyrate (GABA) production in response to abiotic stress. *Plant Science* 193–194: 130–135.

Shelp, B.J., Mullen, R.T., and Waller, J.C. (2012b). Compartmentation of GABA metabolism raises intriguing questions. *Trends in Plant Science* 17: 57–59.

Shi, S.Q.; Shi, Z.; Jiang, Z.P.; Qi, L.W.; Sun, X.M.; Li, C.X.; Liu, J.F.; Xiao,W.F.; Zhang, S.G. Effects of exogenous GABA on gene expression of *Caragana intermedia* roots under NaCl stress: regulatory roles for H_2O_2 and ethylene production. *Plant, Cell & Environment* 2010, 33, 149–162.

Silveira, V., de Vita, A.M., Macedo, A.F. et al. (2013). Morphological and polyamine content changes in embryogenic and non-embryogenic callus of sugarcane. *Plant Cell, Tissue and Organ Culture* 114: 351–364.

Singh, D. (2022). Juggling with reactive oxygen species and antioxidant defense system – a coping mechanism under salt stress. *Plant Stress* 5: 100093.

Sita, K. and Kumar, V. (2020). Role of gamma amino butyric acid (GABA) against abiotic stress tolerance in legumes: a review. *Plant Physiology Reports* 25 (4): 654–663. *International Journal of Molecular Sciences* 2021, 22, 10136.

Skopelitis, D.S., Paranychianakis, N.V., Paschalidis, K.A. et al. (2006). Abiotic stress generates ROS that signal expression of anionic glutamate dehydrogenases to form glutamate for proline synthesis in tobacco and grapevine. *Plant Cell* 18: 2767–2781.

Slabbert, M.M. and Krüger, G.H.J. (2014). Antioxidant enzyme activity, proline accumulation, leaf area and cell membrane stability in water stressed *Amaranthus* leaves. *South African Journal of Botany* 95: 123–128.

Smith, S.S., Shen, H., Gong, Q.H., and Zhou, X. (2007). Neurosteroid regulation of GABA(A) receptors: focus on the alpha4 and delta subunits. *Pharmacology & Therapeutics* 116: 58–76.

Soleimani Aghdam, M., Naderi, R., Jannatizadeh, A. et al. (2016). Enhancement of postharvest chilling tolerance of *Anthurium* cut flowers by γ-aminobutyric acid (GABA) treatments. *Scientia Horticulturae* 198: 52–60.

Song, H., Xu, X., Wang, H. et al. (2010). Exogenous γ-aminobutyric acid alleviates oxidative damage caused by aluminium and proton stresses on barley seedlings. *Journal of the Science of Food and Agriculture* 90: 1410–1416.

Stepien, P. and Klobus, G. (2005). Antioxidant defense in the leaves of C3 and C4 plants under salinity stress. *Physiologia Plantarum* 125: 31–40.

Steward, F.C. (1949). γ-Aminobutyric acid: a constituent of the potato tuber? *Science* 110: 439–440.

Suhel, M., Husain, T., Prasad, S.M., and Singh, V.P. (2023). GABA requires nitric oxide for alleviating arsenate stress in tomato and brinjal seedlings. *Journal of Plant Growth Regulation* 42: 670–683.

Suzuki, N. and Mittler, R. (2006). Reactive oxygen species and temperature stresses: a delicate balance between signaling and destruction. *Physiologia Plantarum* 126: 45–51.

Tripathy, B.C. and Oelmüller, R. (2012). Reactive oxygen species generation and signaling in plants. *Plant Signaling & Behavior* 7: 1621–1633.

Trobacher, C.P., Clark, S.M., Bozzo, G.G. et al. (2013). Catabolism of GABA in apple fruit: subcellular localization and biochemical characterization of two c-aminobutyrate transaminases. *Postharvest Biology and Technology* 75: 106–113.

Viehweger, K. (2014). How plants cope with heavy metals. *Botanical Studies* 55: 35.

Wang, C., Fan, L., Gao, H. et al. (2014a). Polyamine biosynthesis and degradation are modulated by exogenous gamma-aminobutyric acid in root-zone hypoxia-stressed melon roots. *Plant Physiology and Biochemistry* 82: 17–26.

Wang, Y., Luo, Z., Huang, X. et al. (2014b). Effect of exogenous -aminobutyric acid (GABA) treatment on chilling injury and antioxidant capacity in banana peel. *Scientia Horticulturae* 168: 132–137.

Waseem, M., Nie, N.F., Yao, G.Q. et al. (2021). Dew absorption by leaf trichomes in *Caragana korshinskii*: an alternative water acquisition strategy to withstand drought in arid environment. *Physiologia Plantarum* 172: 528–539.

Wu, X., Jia, Q., Ji, S. et al. (2020). Gamma-aminobutyric acid (GABA) alleviates salt damage in tomato by modulating Na^+ uptake, the GAD. Gene, amino acid synthesis and reactive oxygen species metabolism. *BMC Plant Biology* 20: 465.

Xiang, L., Hu, L., Xu, W. et al. (2016). Exogenous γ-aminobutyric acid improves the structure and function of photosystem II in muskmelon seedlings exposed to salinity-alkalinity stress. *PLoS One* 11: e0164847.

Xing, S.G., Jun, Y.B., Hau, Z.W., and Liang, L.Y. (2007). Higher accumulation of γ-aminobutyric acid induced by salt stress through stimulating the activity of diamine oxidases in *Glycine max* (L.) Merr. Roots. *Plant Physiology and Biochemistry* 45: 560–566.

Xu, B., Long, Y., Feng, X. et al. (2021a). GABA signalling modulates stomatal opening to enhance plant water use efficiency and drought resilience. *Nature Communications* 12: 1952.

Xu, J., Liu, T., Qu, F. et al. (2021b). Nitric oxide mediates γ-aminobutyric acid-enhanced muskmelon tolerance to salinity–alkalinity stress conditions. *Scientia Horticulturae* 286: 110229.

Yang, R., Guo, Q., and Gu, Z. (2013). GABA shunt and polyamine degradation pathway on -aminobutyric acid accumulation in germinating fava bean (*Vicia faba* L.) under hypoxia. *Food Chemistry* 136: 152–159.

Yang, R., Yin, Y., and Gu, Z. (2015). Polyamine degradation pathway regulating growth and GABA accumulation in germinating fava bean under hypoxia-NaCl stress. *Journal of Agricultural Science and Technology* 17: 311–320.

Yao, G.Q., Nie, Z.F., Turner, N.C. et al. (2021). Combined high leaf hydraulic safety and efficiency provides drought tolerance in *Caragana* species adapted to low mean annual precipitation. *The New Phytologist* 229: 230–244.

Yin, Y.G., Tominaga, T., Iijima, Y. et al. (2010). Metabolic alterations in organic acids and γ-aminobutyric acid in developing tomato (*Solanum lycopersicum* L.) fruits. *Plant & Cell Physiology* 51: 1300–1314.

Yong, B., Xie, H., Li, Z. et al. (2017). Exogenous application of GABA improves PEG-induced drought tolerance positively associated with GABA-shunt, polyamines, and proline metabolism in white clover. *Frontiers in Physiology* 8: 1107.

Zeng, W., Hassan, M.J., Kang, D. et al. (2021). Photosynthetic maintenance and heat shock protein accumulation relating to γ-aminobutyric acid (GABA)-regulated heat tolerance in creeping bentgrass (*Agrostis stolonifera*). *South African Journal of Botany* 141: 405–413.

Zhang, L. and Becker, D.F. (2015). Connecting proline metabolism and signaling pathways in plant senescence. *Frontiers in Plant Science* 6: 552.

Zhang, M., Liu, Z., Fan, Y. et al. (2022). Characterization of GABA-transaminase gene from mulberry (*Morus multicaulis*) and its role in salt stress tolerance. *Genes* 13: 501.

Zhao, Y., Song, X., Zhong, D. et al. (2020). γ-Aminobutyric acid (GABA) regulates lipid production and cadmium uptake by *Monoraphidium* sp. QLY-1 under cadmium stress. *Bioresource Technology* 297: 122500.

Zhu, X., Liao, J., Xia, X. et al. (2019). Physiological and ITRAQ-based proteomic analyses reveal the function of exogenous γ-aminobutyric acid (GABA) in improving tea plant (*Camellia sinensis* L.) tolerance at cold temperature. *BMC Plant Biology* 19: 43.

12

GABA and Heat Stress

Zeba Azim[1], Shubhra Khare[2], Narsingh Bahadur Singh[1], Km Niharika[3], Ajey Singh[3], Ravi Kumar Yadav[4], and Nimisha Amist[5]

[1] *Plant Physiology Laboratory, Department of Botany, University of Allahabad, Prayagraj, Uttar Pradesh, 211002, India*
[2] *DepartmentofApplied Sciences and Humanities, Invertis University, Bareilly, Uttar Pradesh, 243123, India*
[3] *Department of Botany, University of Lucknow, Lucknow, Uttar Pradesh, 226007, India*
[4] *Department of Botany, Kashi Naresh Government P. G. College, Bhadohi, Uttar Pradesh, 221304, India*
[5] *Department of Botany, Ewing Christian College, Prayagraj, Uttar Pradesh, 211003, India*

Introduction

Worldwide, the population is increasing, and it is predicted that there will be more than 9 billion people by 2050. People face many challenges among which increasing global food production will require more agricultural land to attain food security. To achieve sustainable intensification goals, i.e., zero hunger, food security is also a high-priority concern (UN 2019). To withstand environmental challenges, there is an imperative necessitate for collaborative research to develop crops under present climatic variations (Gomez-Zavaglia et al. 2020; Malhi et al. 2021; Qin et al. 2022).

In plants, several physiological processes that deal via a diverse assortment of signal systems and regulatory mechanisms areaffected negatively byenvironmentalstresses (heat,drought,cold,heavymetals,andsalt),resultinginadeclineincropproductivity(Hayat et al. 2022; Mushtaq et al.2022; Zhao et al. 2022). Climate change and allied factors such as biotic and abiotic stresses hinder plant development processes, thereby affecting crop yield (Jalil and Ansari 2020a, 2020b). The abiotic stresses (soil alkalinity, water logging cold, salinity, metal/metalloid toxicity, drought, heat, and nutrient deficiency) are amplified by the increasing unpredictability of climate change. Under stern environments, reactive oxygen species (ROS) are formed due to excessive accumulation of oxygen radicals (hydroxyl radical, –OH, singlet oxygen, $_1O^2$; hydrogen peroxide, H_2O_2), and superoxide anion, $O^{2^{\cdot}-}$) can lead to cellular damage, apoptosis thereby declining crop yield (Hasanuzzaman et al. 2020). Plants develop a remarkable surfeit of direct and indirect stress-specific responses and adaptation strategies, including cellular, physiological, biochemical, and molecular processes for scavenging ROS to alleviate oxidative stresses

GABA in Plants: Biosynthesis, Plant Development, and Food Security, First Edition. Edited by Samiksha Singh, Durgesh Kumar Tripathi, and Vijay Pratap Singh.

(Beebe et al. 2011). There is considerable interest in boosting heat stress tolerance by exogenous application of natural compounds and biomolecules with shielding impact in heat-sensitive plants (Godoy et al. 2021). Heat stress hampers the growth and metabolism of plants through notable changes that have been reported in morphology, physiology, biochemistry, and molecular level, resulting in a reduction in crop yield (Kaushal et al. 2016). At the time of vegetative growth, heat stress causes a decrease in photosynthetic pigments, scorching, and necrosis, therefore retarding plant growth, but at the reproductive stage, heat stress showsa more damaging effect in plants. It impairsthe growth and function of reproductive parts and hampers fertilization via damaging pollen development, germination, and tube growth (Zinn et al. 2010; Hatfield and Prueger 2015; Kaur et al. 2015; Sita et al. 2017).

Gamma-aminobutyric acid (GABA), a ubiquitous, four-carbon, nonprotein amino acid, is present in both prokaryotes and eukaryotes, and the enzymes implicated in GABA metabolism are highly conserved. GABA forms an essential part of a group offree amino acids. GABA acts as a metabolite and signaling molecule in both biotic and abiotic stress responses, and via upregulating antioxidant defense systems, it promotes plant growth and mitigates stress (Hasanuzzaman et al. 2020; Xu et al. 2021). At first, the molecule GABA was reported in potato (*Solanum tuberosum*) (Thompson et al. 1953) and afterward found in mammalian brains (Awapara et al. 1950). In stress conditions, GABA actsas an endogenous signaling molecule that rapidly accumulates in plant tissue and affects plant metabolism (Roberts 2007). In leaf senescence, during abiotic stress, GABA plays a significant role in the recycling and reallocation of nitrogen (Jalil et al. 2017) and is also involved in the plant'sdefense mechanism toward biotic stress caused by insects and necrotrophic fungi (Bown and Shelp 2016). During seed germination, the GABA concentration increased significantly (Kim et al. 2013). Increased level of GABA concentration is also reported in response to various stresses, and its level depends upon the growth and metabolism of plants (Shelp et al. 2017). In response to abiotic stress, the GABA concentration increased many folds, whether in normal conditions. GABA is present in the range between 0.03 and 2 mol/g (FW) in plant tissue (Bhattacharya et al. 2018). The function of GABA in respiratory metabolism is significant and also involved in the activity of evolutionarilyconserved enzymes in the tricarboxylic acid (TCA) cycle. During stress conditions, GABA accumulates and binds to the transporter of aluminum-activated malate directly, and thereafter, the guard cell rectifying K+ (GORK) channel (Gilliham andTyerman2016). GABA is a regulatory signaling molecule that maintains cellular processes by linking and changing the protein receptors'activities (Ramesh et al. 2015). GABA is involved in regulating cytosolic pH, osmoregulation, buffering mechanisms in C and N metabolisms, and defensive function in oxidative stress (Bouche and Fromm 2004). The expression of GABA shunt-associated genes regulates abiotic stresses in *Arabidopsis* plants. GABA gets accumulated through the activation of calcium-induced enzyme glutamate decarboxylase(GAD) under heat stress (Locy et al. 2000). GABA is a vital amino acid in pollen fertility and also plays a significant role in fertilization and postpollination fertilization (Biancucci et al. 2015). The present study provides an outline of the obtainable knowledge of the significant role played by GABA-triggered heat responses and the potential application of GABA in heat-stressed plants.

GABA-Biosynthesis and Transport/Pathways in Plant

GABA shunt is a metabolic pathway by which GABA is synthesized in vivo. It is firstobserved in plants, and its functions for synthesis and maintaining GABA at appropriate levels within the cells. GABA metabolism in GABA shunt occurs in the cell cytosol (GABA synthesis) and mitochondria (GABA degradation). Cytosolic enzyme GAD, mitochondrial enzymes GABA-T, and succinic semialdehyde (SSA) dehydrogenase are significant enzymes that catalyzeits main reactions. GABA-T depends on either pyruvate (GABA-TP) or α-ketoglutarate (GABA-TK) as amino acceptors, with the products alanine or glycine. GABA-T favors 2-oxoglutarate despitepyruvate in plants and through GAD reaction recuperating glutamate. Therefore, SSADH is produced from succinate as an electron donor and may enter the TCA cycle or the electron transport chain in mitochondria (Akihiro et al. 2008). The GABA shunt pathway involves three reactions: (1) glutamate converted into GABA via the action of GAD in the cytosol (Breitkreuz et al.1999), (2) GABA converted into SSA in mitochondria that converts to succinate concurrently with NADH production by NAD^{+}-dependent enzyme SSADH, and (3) succinic semialdehyde (SSA) into γ-hydroxybutyric acid (GHB) via SSA reductase (SSR) enzyme (Breitkreuz et al. 2003; Hoover et al. 2007; Simpson et al. 2008) and then transfer to the TCA cycle. Under oxidative stress, GABA is mainly formed from glutamate by cytosolic GAD through an irreversible reaction or alternatively through proline (nonenzymatic reaction) or through polyamine degradation (Yang et al.2015b). Under oxidative stress, through nonenzymatic reactions of spermidine and putrescine (polyamines) along with proline synthesize GABA (Shelp et al. 2012). It is reported that under stressed conditions, the activity of GAD and DAO is highly elevated for the synthesis of GABA (Bouché et al. 2003; Bouche and Fromm 2004).

In the GABA transportation, it is moved to the mitochondria of the cell, where it is catabolized to SSA by GABA-T, and then SSA can bereduced to GHB by the cytosolic γ-hydroxybutyrate dehydrogenase (GHBDH) or oxidized by the mitochondrial succinic SSADH to produce succinate (Hoover et al. 2007; Simpson et al. 2008). Bouche and Fromm (2004) reported that in plants, the enhancement of GABA is the common response with restriction of glutamine synthesis, an increase in protein degradation, and reduction of protein synthesis. GABA transport across the subcellular membrane is mediated by the family of membrane-bound transporter protein, and based on substrate specificity, it is divided into two classes: (1) high-affinity GABA transporters-GAT1and GAT2 (2) low-affinity GABA transporters-LATs. GABA transporters are expressed in various plant tissues, including roots, leaves, and flowers, and in *Arabidopsis thaliana*, both types of GABA transporters are reported. GABA is synthesized in the cytosol and accumulated in the vacuole, and subcellular localization of GABA transporters can affect the accumulation of GABA in different organelles (Meyer et al. 2006).

GABA Morphological and Physiological Functions within Plants

Both exogenously applied GABA and endogenous GABA act in asimilar manner and stimulate GABA levels in plant cells under both stress and nonstress conditions (Li et al. 2017;Cekic 2018). GABA is anon-protein four-carbon amino acid and is absorbed directly by plant systems (Michaeli and Fromm 2015;Gilliham and Tyerman 2016). GABA can affect boththe morphology and physiology of plants. GABA significantly improved root and shoot length inmaize seedlings. Depending on the seedling's variety, it also enhances biomass. The application ofGABA positively influenced the morphological growth of the seedlings via upregulating thedifferent physiological and biochemical mechanisms (Li et al.2016b). Ziogas et al. (2017) reported that prior exposure to high concentrations of NaCl,higher germination rate, and longer radicles enhanced the adaptation of the GABA-treated citrusseeds in saline conditions. Malabarba et al. (2019) demonstratedthat the treatment of isolated grapevine tendrilspositively affected the tendril coiling.The amount of GABA present in the grapevine tendril is five times more with respect tograpevine bunches and branches that contain GABA in equivalent amounts under hypoxia incornseedlings (Wang et al. 2017) and in melon the exogenous GABA alleviated growth of retardedroots and restored biochemical functions (Wang et al. 2014). Malekzadeh et al. (2014) reported that chilling injury could be compensated by GABAtreatments in tomato seedlings by enhancement of the antioxidant enzymeactivities.

The application of GABA stimulates root growth both in primary and adventitious roots. GABA is involved in increasing seed germination in Chinese wild rice (Chu et al. 2019), barley (Sheng et al. 2018), wheat (Zhao et al. 2017; Al-Quraan et al.2019), adlay (Xu et al. 2017), barnyard millet (Sharma et al. 2016), and oats (Xu et al. 2010). During seed germination, GABA promotes the degradation of seed starch in a dose-dependent manner by stimulating the expression of the α-amylase gene (Sheng et al.2018).

Michaeli and Fromm (2015) reported that the metabolic pathway of GABA is divided between cytosol and the mitochondrial matrix, and therefore, GABA metabolites have the possibility to increase photosynthesis through the mitochondrial matrix. Enhancements in the photosynthesis parameters were reported in GABA-treated muskmelon seedlings. Carbon and nitrogen are the major essential elements for plant growth and productivity. Through the GABA shunt, carbon skeletons enter the Krebs Cycle (TCA cycle) in plants and function as a nitrogen storage metabolite. Assimilation of carbon from glutamate and the carbon flux through the TCA cycle, GABA shunt appears to be a part of the metabolic pathways and maintainsequilibrium between carbon and nitrogen (Michaeli and Fromm 2015). *A. thaliana*growswell in GABA-mediated artificial medium, and the growth depends on the NO_3 level in the substrate and enzymes involved in carbon and nitrogen metabolisms (Barbosa et al. 2010).

GABA and Abiotic Stress

GABA has been studied under various abiotic stress conditions, including drought, salinity, extremetemperature, and heavy metals. In stress conditions, the accumulation and metabolism of GABA play a positive role and enhancethe abiotic stress tolerance capacity of plants (Figure12.1). This enhancement may be due to significant regulation of GABA shunt and associated pathways. The endogenous GABA levels were increased by the exogenous application of GABA (Bao et al. 2015). The exogenous application of GABA efficiently alleviated germination or growth inhibition in many crop plants such as *Cucumis melo* L., *Oryza sativa*, *Lens culinaris* Medik, and *Triticum aestivum* under unfavorable

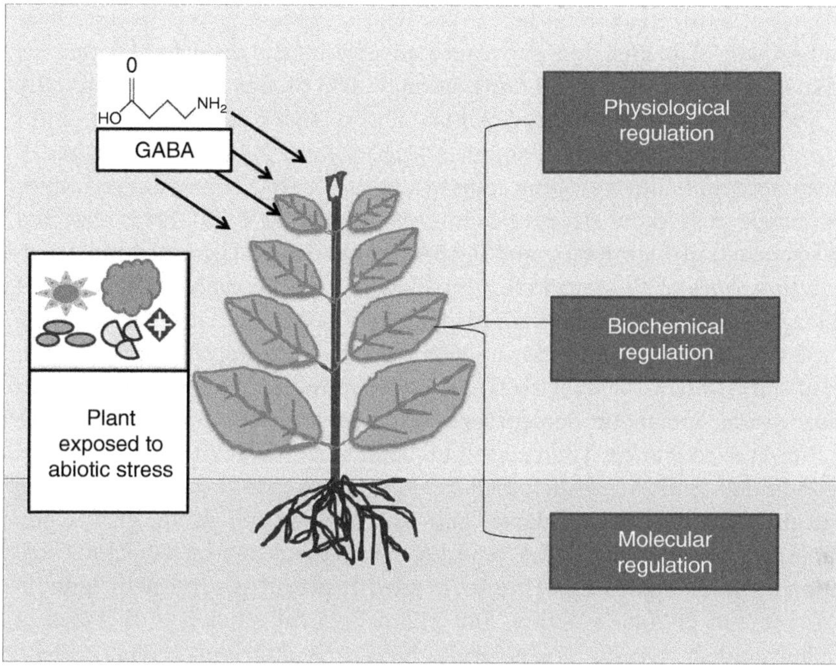

Figure 12.1 A diagrammatic representation reveals the potential role of GABA in improving heat stress resilience through physiological (enhancingroot and shoot growth, photosynthesis, soluble sugar accumulation, nutrient uptake, reducing stomatal opening and closing), biochemical(enhancing antioxidant system, phenolics, osmolytes accumulation, ascorbate glutathione metabolism, and molecular (interaction with transcription factors, phytohormones, stress response gene) regulation in plants.

environmental conditions such as light or oxygen stress (Ramos-Ruiz et al. 2019), salt (Kumar et al. 2017), drought (Li et al. 2016a), extreme temperatures (Nayyar et al. 2014). Ansari et al. (2014) reported that abiotic stresses cause precocious leaf senescence, resulting in the breakdown of macromolecules, which are further remobilized to the growing plant parts. During stress conditions, exogenously applied GABA enhanced photosynthetic mechanism and antioxidative enzymemachinery, subsequently reducing malondialdehyde (MDA) content and ROS accumulation (Jia et al. 2017), and ultimately maintainingthe membrane integrity (Aghdam et al. 2015).

GABA and Heat Stress

Global warming has emerged worldwide as a significantconstraint and may cause considerable economic loss in the coming years (Apeland Hirt2004). Heat stress is also a key issue for plants and other photosynthetic organisms that can limit plant growth by altering phenology, physiology, biochemical, and gene expression (Prasad et al. 2017). Heat stress has been increasing appreciably,negatively affecting plant growth, mineral nutrientcontent, and production. Furthermore, various symptoms, such as ultrastructural alterations, photoinhibition, and chlorophyll degradation displayed by heat-stressed plants (Jespersenet al.2016).Heat stress accelerates leaf senescence, causes water loss, inhibits normal growth of plants, and may cause adverse effects on plants. It may be related to oxidative damage to cell membranes through ROS production that eventually encumbers the productivity of heat-stressed plants (Liu and Huang 2000). A GABA shunt is required to limit ROS intermediates. It supplies succinate and NADH to maintain cellular oxidative balance and allows evading steps of the Krebs cycle concerning enzymes sensitive to oxidative stress under stressful conditions (Fait et al. 2005). Itwas observed that *GAD* was expressed at high levels, and GABA-T and SSADH decreased rapidly in heat-dried immature seeds of *Glycine max*L. Merrill at 40 °C, as a result, GABA content in the treated seeds increased to more than fivefold(Takahashi et al. 2013). Priya et al. (2019a) observed GABA accumulation during heat stress in *Arabidopsis* plants through calcium-induced activation of the enzyme GAD. GABA induces alterations in the antioxidant defense system, metabolic homeostasis, and heat shock factor pathway, ultimately enhancing heat tolerance. The exogenous application of GABA rice seedlings under stress conditionssignificantly increased growth and survival rates by improving upregulating osmoprotectants, antioxidants, and leaf turgor (Ren et al. 2021; Chen et al. 2022) (Table12.1). Nayyar et al. (2014) reported that in the heat-stressed plant, exogenous application of GABA significantly improves growth as well as survival by improving chlorophyll content, cellular viability, and photochemical efficiency of leaves of heat-stressed plantsand in *Agrostis stolonifera* by bolstering their antioxidant defense

Table 12.1 GABA modifies and develops various mechanisms for improving heat stress in crop plants.

Crop	Abiotic stress	Concentration of GABA	Mechanism of tolerance	References
Creeping bentgrass (*Agrostis stolonifera*)	Heat stress	0.5 mM	Increase antioxidant metabolism, reduces leaf senescence, improving osmotic adjustment, balance between photosynthesis and transpiration, and accumulates amino acids, carbohydrates, organic acids, and alcohol.	Li et al. (2016b)
Creeping bentgrass (*A. stolonifera*)	Heat stress	0.5 μmol/l	Biosynthesis of phenylpropanoids and theenhancement of HSF pathways, upregulation of genes such as CAD_3, ACS, AER, andCA_1.	Li et al. (2019)
Creeping bentgrass (*A. stolonifera*)	Heat stress	0.5 mmol/l	Higher relative water content, improved photosynthesis, and cell membrane stability, upregulate heat shock factor HSFs (HSFA-6a, HSFA-2c, and HSFB-2b), heat shock proteins (HSP17.8, HSP26.7, HSP70, and HSP90.1-b1), and ascorbate peroxidase $_3$	Liu et al. (2019)
Creeping bentgrass (*A. stolonifera*)	Heat stress	0.5 mmol/l	Maintain chlorophyll content, enhancing antioxidant capacity.	Zeng et al. (2021)

Table 12.1 (Continued)

Crop	Abiotic stress	Concentration of GABA	Mechanism of tolerance	References
Creeping bentgrass (*A. stolonifera*)	Heat stress	0.5 mM	Regulate amino acid metabolism.	Rossi et al. (2021)
Creeping bentgrass (*A. stolonifera*)	Heat stress	—	Enhanced various antioxidant enzyme functions, root vigor, osmoregulation capabilities of root system.	Li et al. (2023)
Tea (*Camellia sinensis*)	Heat stress	0.5 mM	Polyphenol accumulation and antioxidant system upregulation.	Ren et al. (2021)
Rice (*Oryza sativa*)	Heat stress	—	Improving leaf turgor and upregulating osmoprotectants and antioxidants.	Nayyar et al. (2014)
Mungbeans (*Vigna radiata*)	Heat stress	1 mM	Improving the reproductive function of heat stressed plants	Priya et al. (2019b)

systems regulating osmotic potential, metabolic homeostasis, and the TCA cycle (Li et al. 2016). The endogenous content of GABA-induced heat shock protein in *Sesamum indicum* and cowpea (Bor et al. 2009). Similarly, various parameters such as morphological, physiological, and biochemical were also enhanced with the application of GABA under elevated temperature stress in legumes (Ali et al. 2019). GABA levelsincreased in ripening grapes (Sweetman et al. 2014) and grape berries (Ayenew et al. 2015).

Priya et al. (2019b) reported that GABA supplementation is beneficial for the developmental and functional aspects of flowers in terms of reproductive function. A study on creeping bent grass demonstrated that GABA induces heat tolerance by improving HSF pathways and phenylpropanoid biosynthesis by inducing AER, ACS, CA1, WUE, and CAD_3 genes, whichhelp to reduce water loss and increase water balance, thereby enhancing its heat tolerance (Li et al. 2019). In Lichen, applying GABA decreases the function of GAD and GDH, preventing chlorophyll degradation and lipid peroxidation during heat stress (Cekic 2018). Exogenous application of GABA in heat-stressed plants enhanced the accumulation of sugars(glucose, sucrose, maltose, fructose, andgalactose), organic acids (malic acid, aconitic acid,oxalic acid, succinic acid, and threonic acid), amino acids (aspartic acid, glutamic acid, serine,threonine, valine, andalanine) and sugar alcohols (mannitol and myoinositol) (Li et al. 2016a, 2016c).

Conclusion

Under field conditions, plants must tolerate anample array of abiotic and biotic stresses. Many crop improvement programs strive to develop stress-tolerant cultivars to prevent productivity losses. In recent years, GABA metabolism has garnered significant attention because it often accumulates in response to a variety of abiotic and biotic stresses. Through the application of exogenous GABA, the levels of endogenous GABA areenhanced in plants, thereby improvingheat stress tolerance via positive regulation of the GABA shunt and associated pathways. In plants, GABA application increases the GABA shunt activity, thereby enhancingphotosynthetic efficiency, antioxidative enzymes, maintaining water balance, and decline in MDA content oxidative stress marker, ROS, and finally, the membrane integrity was retained in GABA-treated plants at high temperature. The studies suggest that GABA could be a precious tool in improving crop yields in heat-stressed plants. However, further research is needed to fully understand the molecular mechanisms and complete signaling pathways underlying the protective and sound effects of GABA in heat stress tolerance and to develop advanced applications in agriculture.

References

Aghdam, M.S., Naderi, R., Sarcheshmeh, M.A.A., and Babalar, M. (2015). Amelioration of postharvest chilling injury in anthurium cut flowers by γ-aminobutyric acid (GABA) treatments. *Postharvest Biology and Technology* 110: 70–76.

Akihiro, T., Koike, S., Tani, R. et al. (2008). Biochemical mechanism on GABA accumulation during fruit development in tomato. *Plant and Cell Physiology* 49 (9): 1378–1389.

Ali, S., Rizwan, M., Arif, M.S. et al. (2019). Approaches in enhancing thermotolerance in plants: an updated review. *Journal of Plant Growth Regulation* 39: 456–480.

AL-Quraan, N.A., AL-Ajlouni, Z.I., and Obedat, D.I. (2019). The GABA shunt pathway in germinating seeds of wheat (*Triticum aestivum* L.) and barley (*Hordeum vulgare* L.) under salt stress. *Seed Science Research* 29 (4): 250–260.

Ansari, M.I., Hasan, S., and Jalil, S.U. (2014). Leaf senescence and GABA shunt. *Bioinformation* 10: 730–732.

Apel, K. and Hirt, H. (2004). Reactive oxygen species: metabolism, oxidative stress, and signal transduction. *Annual Review of Plant Biology* 55: 373–399.

Awapara, J., Landua, A.J., Fuerst, R., and Seale, B. (1950). Free gamma-amino butyric acid in brain. *The Journal of Biological Chemistry* 187: 35–39.

Ayenew, B., Degu, A., Manela, N. et al. (2015). Metabolite profiling and transcript analysis reveal specificities in the response of a berry derived cell culture to abiotic stresses. *Frontiers in Plant Science* 6: 728.

Bao, H., Chen, X., Lv, S. et al. (2015). Virus-induced gene silencing reveals control of reactive oxygen species accumulation and salt tolerance in tomato by caminobutyric acid metabolic pathway. *Plant, Cell and Environment* 38: 600–613.

Barbosa, J.M., Singh, N.K., Cherry, J.H., and Locy, R.D. (2010). Nitrate uptake and utilization is modulated by exogenous γ-amino-butyric acid in Arabidopsis thaliana seedlings. *Plant Physiology and Biochemistry* 48: 443–450.

Beebe, S., Ramirez, J., Jarvis, A. et al. (2011). Genetic improvement of common beans and the challenges of climate change. In: *Crop Adaptation to Climate Change* (ed. S.S. Yadav, R.J. Redden, J.L. Hatfield, et al.), 356–369. New York: John Wiley & Sons.

Bhattacharya, S., Khatri, A., Swanger, S.A. et al. (2018). Triheteromeric GluN1/GluN2A/GluN2C NMDARs with unique single-channel properties are the dominant receptor population in cerebellar granule cells. *Neuron* 99: 315–328.

Biancucci, M., Mattioli, R., Forlani, G. et al. (2015). Role of proline and GABA in sexual reproduction of angiosperms. *Frontiers in Plant Science* 6: 680.

Bor, M., Seckin, B., Ozgur, R. et al. (2009). Comparative effects of drought, salt, heavy metal and heat stresses on gamma-aminobutryric acid levels of sesame (*Sesamum indicum* L.). *ActaPhysiologiaePlantarum* 31: 655–659.

Bouche, N. and Fromm, H. (2004). GABA in plants: just a metabolite? *Trends in Plant Science* 9 (3): 110–115.

Bouché, N., Fait, A., Bouchez, D. et al. (2003). Mitochondrial succinic-semialdehyde dehydrogenase of the γ-aminobutyrate shunt is required to restrict levels of reactive oxygen intermediates in plants. *Proceedings of the National Academy of Sciences of the United States of America* 100 (11): 6843–6848.

Bown, A.W. and Shelp, B.J. (2016). Plant GABA: not just a metabolite. *Trends in Plant Science* 21: 811–813.

Breitkreuz, K.E., Shelp, B.J., Fischer, W.N., and Rentsch, D. (1999). Identification and characterization of GABA, proline and quaternary ammonium compound transporters from *Arabidopsis thaliana*. *FEBS Letters* 450: 280–284.

Breitkreuz, K.E., Allan, W.L., Van Cauwenberghe, O.R. et al. (2003). A novel gamma-hydroxybutyrate dehydrogenase: identification and expression of an *Arabidopsis* cDNA and potential role under oxygen deficiency. *The Journal of Biological Chemistry* 278: 41552–41556.

Cekic, F.Ö. (2018). Exogenous GABA stimulates endogenous GABA and phenolic acid contents in tomato plants under salt stress. *Celal Bayar University Journal of Science* 14: 61–64.

Chen, X., Li, N., Liu, C. et al. (2022). Exogenous GABA improves the resistance of apple seedlings to long-term drought stress by enhancing GABA shunt and secondary cell wall biosynthesis. *Tree Physiology* 42 (12): 2563–2577.

Chu, C., Yan, N., Du, Y. et al. (2019). iTRAQ-based proteomic analysis reveals the accumulation of bioactive compounds in Chinese wild rice (*Zizania latifolia*) during germination. *Food Chemistry* 289: 635–644.

Fait, A., Yellin, A., and Fromm, H. (2005). GABA shunt deficiencies and accumulation of reactive oxygen intermediates: insight from *Arabidopsis* mutants. *FEBS Letters* 579 (2): 415–420.

Gilliham, M. and Tyerman, S.D. (2016). Linking metabolism to membrane signaling: the GABA–malate connection. *Trends in Plant Science* 21 (4): 295–301.

Godoy, F., Olivos-Hernández, K., Stange, C., and Handford, M. (2021). Abiotic stress in crop species: improving tolerance by applying plant metabolites. *Plants* 10 (2): 186.

Gomez-Zavaglia, A., Mejuto, J.C., and Simal-Gandara, J. (2020). Mitigation of emerging implications of climate change on food production systems. *Food Research International* 134: 109256.

Hasanuzzaman, M., Bhuyan, M.B., Zulfiqar, F. et al. (2020). Reactive oxygen species and antioxidant defense in plants under abiotic stress: revisiting the crucial role of a universal defense regulator. *Antioxidants* 9 (8): 681.

Hatfield, J.L. and Prueger, J.H. (2015). Temperature extremes: effect on plant growth and development. *Weather and Climate Extremes* 10: 4–10.

Hayat, F., Sun, Z., Ni, Z. et al. (2022). Exogenous melatonin improves cold tolerance of strawberry (*Fragaria* × *ananassa*Duch.) through modulation of DREB/CBF-COR pathway and antioxidant defense system. *Horticulturae* 8 (3): 194.

Hoover, G.J., Van Cauwenberghe, O.R., Breitkreuz, K.E. et al. (2007). Characteristics of an *Arabidopsis* glyoxylate reductase: general biochemical properties and substrate specificity for the recombinant protein, and developmental expression and implications for glyoxylate and succinic semialdehyde metabolism in planta. *Canadian Journal of Botany* 85: 883–895.

Jalil, S.U. and Ansari, M.I. (2020a). Stress implication and crop productivity. In: *Plant Ecophysiology and Adaptation under Climate Change: Mechanism and Perspectives* (ed. M. Hasanuzzaman), 73–86. Singapore: Springer.

Jalil, S.U. and Ansari, M.I. (2020b). Physiological role of gamma aminobutyric acid in salt stress tolerance. In: *Salt and Drought Stress Tolerance in Plants, Signaling and Communication in Plants* (ed. M. Hasanuzzaman and M. Tanveer), 337–350. Cham, Switzerland: Springer.

Jalil, S.U., Ahmad, I., and Ansari, M.I. (2017). Functional loss of GABA transaminase (GABA-T) expressed early leaf senescence under various stress conditions in *Arabidopsis thaliana*. *Current Plant Biology* 9–10: 11–22.

Jespersen, D., Zhang, J., and Huang, B. (2016). Chlorophyll loss associated with heat-induced senescence in bentgrass. *Plant Science* 249: 1–12.

Jia, Y., Zou, D., Wang, J. et al. (2017). Effects of c-aminobutyric acid, glutamic acid, and calcium chloride on rice (*Oryza sativa* L.) under cold stress during the early vegetative stage. *Journal of Plant Growth Regulation* 36: 240–253.

Kaur, R., Bains, T.S., Bindumadhava, H., and Nayyar, H. (2015). Responses of mungbean (*Vignaradiata* L.) genotypes to heat stress: effects on reproductive biology, leaf function and yield traits. *ScientiaHorticulturae* 197: 527–541.

Kaushal, N., Bhandari, K., Siddique, K.H., and Nayyar, H. (2016). Food crops face rising temperatures: an overview of responses, adaptive mechanisms, and approaches to improve heat tolerance. *Cogent Food & Agriculture* 2 (1): 1134380.

Kim, N.H., Kim, B.S., and Hwang, B.K. (2013). Pepper arginine decarboxylase is required for polyamine and γ-aminobutyric acid signaling in cell death and defense response. *Plant Physiology* 162 (4): 2067–2083.

Kumar, S., Jangde, S., Priya, T., and Yashu, B.R. (2017). Effect of GABA on morphology, yield and yield attributes in black gram (*Vigna mungo*[L.] Hepper) under salt stress condition. *International Journal of Pure Applied Biosciences* 5: 1223–1228.

Li, M.F., Guo, S.J., Yang, X.H. et al. (2016a). Exogenous gamma-aminobutyric acid increases salt tolerance of wheat by improving photosynthesis and enhancing activities of antioxidant enzymes. *Biologia Plantarum* 60 (1): 123–131.

Li, W., Lin, J., Ashraf, U. et al. (2016b). Exogenous γ-aminobutyric acid (GABA) application improved early growth, net photosynthesis, and associated physio-biochemical events in maize. *Frontiers in Plant Science* 7: 1–13.

Li, Z., Yu, J., Peng, Y., and Huang, B. (2016c). Metabolic pathways regulated by γ-aminobutyric acid (GABA) contributing to heat tolerance in creeping bentgrass (*Agrostisstolonifera*). *Scientific Reports* 6 (1): 30338.

Li, Y., Fan, Y., Ma, Y. et al. (2017). Effects of exogenous γ-aminobutyric acid (GABA) on photosynthesis and antioxidant system in pepper (*Capsicumannuum* L.) seedlings under low light stress. *Journal of Plant Growth Regulation* 36: 436–449.

Li, Z., Cheng, B., Zeng, W. et al. (2019). The transcriptional and post-transcriptional regulation in perennial creeping bentgrass in response to γ-aminobutyric acid (GABA) and heat stress. *Environmental and Experimental Botany* 162: 515–524.

Li, Z., Zhou, M., Zeng, W. et al. (2023). Root metabolites remodeling regulated by γ-aminobutyric acid (GABA) improves adaptability to high temperature in creeping bentgrass. *Plant and Soil* 500: 181.

Liu, X. and Huang, B. (2000). Heat stress injury in relation to membrane lipid peroxidation in creeping bent grass. *Crop Science* 40: 503–510.

Liu, T., Liu, Z., Li, Z. et al. (2019). Regulation of heat shock factor pathways by γ-aminobutyric acid (GABA) associated with thermotolerance of creeping bentgrass. *International Journal of Molecular Sciences* 20: 4713.

Locy, R.D., Wu, S.J., Bisnette, J. et al. (2000). The regulation of GABA accumulation by heat stress in *Arabidopsis*. In: *Plant Tolerance to Abiotic Stresses in Agriculture: Role of Genetic Engineering* (ed. J.H. Cherry, R.D. Locy, and A. Rychter), 39–52. Dordrecht, The Netherlands: Springer.

Malabarba, J., Reichelt, M., Pasquali, G., and Mithöfer, A. (2019). Tendril coiling in grapevine: jasmonates and a new role for GABA? *Journal of Plant Growth Regulation* 38: 39–45.

Malekzadeh, P., Khara, J., and Heydari, R. (2014). Alleviating effects of exogenous γ-aminobutyric acid on tomato seedlings under chilling stress. *Physiology and Molecular Biology of Plants* 20: 133–137.

Malhi, G.S., Kaur, M., and Kaushik, P. (2021). Impact of climate change on agriculture and its mitigation strategies: a review. *Sustainability* 13 (3): 1318.

Meyer, A., Eskandari, S., Grallath, S., and Rentsch, D. (2006). AtGAT1, a high affinity transporter for γ-aminobutyric acid in *Arabidopsis thaliana*. *Journal of Biological Chemistry* 281 (11): 7197–7204.

Michaeli, S. and Fromm, H. (2015). Closing the loop on the GABA shunt in plants: are GABA metabolism and signaling entwined? *Frontiers in Plant Science* 6: 419.

Mushtaq, N., Iqbal, S., Hayat, F. et al. (2022). Melatonin in micro-tom tomato: improved drought tolerance via the regulation of the photosynthetic apparatus, membrane stability, osmoprotectants, and root system. *Life* 12 (11): 1922.

Nayyar, H., Kaur, R., Kaur, S., and Singh, R. (2014). γ-Aminobutyric acid (GABA) imparts partial protection from heat stress injury to rice seedlings by improving leaf turgor and upregulating osmoprotectants and antioxidants. *Journal of Plant Growth Regulation* 33: 408–419.

Prasad, P.V., Bheemanahalli, R., and Jagadish, S.K. (2017). Field crops and the fear of heat stress – opportunities, challenges and future directions. *Field Crops Research* 200: 114–121.

Priya, M., Sharma, L., Kaur, R. et al. (2019a). GABA (γ-aminobutyric acid), as a thermo-protectant, to improve the reproductive function of heatstressedmungbean plants. *Scientific Reports* 9: 7788.

Priya, M., Sharma, L., Kaur, R. et al. (2019b). GABA (c-aminobutyric acid), as a thermo-protectant, to improve the reproductive function of heat-stressed mungbean plants. *Scientific Reports* 9: 1–14.

Qin, X., Zhang, K., Fan, Y. et al. (2022). The bacterial MtrABtwo-component system regulates the cell wall homeostasis responding to environmental alkaline stress. *Microbiology Spectrum* 10 (5): e02311–e02322.

Ramesh SA, Tyerman SD, Xu B, Bose J, Kaur S, Conn V, Domingos P, Ullah S, Wege S, Shabala S, et al. GABA signaling modulates plant growth by directly regulating the activity of plant-specific anion transporters. *Nat Commun.* 2015;6:7879. doi:https://doi.org/10.1038/ncomms8879.

Ramos-Ruiz, R., Martinez, F., and Knauf-Beiter, G. (2019). The effects of GABA in plants. *Cogent Food & Agriculture* 5: 1670553.

Ren, T., Zheng, P., Zhang, K. et al. (2021). Effects of GABA on the polyphenol accumulation and antioxidant activities in tea plants (*Camellia sinensis* L.) under heat-stress conditions. *Plant Physiology and Biochemistry* 159: 363–371.

Roberts, M.R. (2007). Does GABA act as a signal in plants? Hints from molecular studies: hints from molecular studies. *Plant Signaling & Behavior* 2: 408–409.

Rossi, S., Chapman, C., Yuan, B., and Huang, B. (2021). Improved heat tolerance in creeping bentgrass by γ-aminobutyric acid, proline, and inorganic nitrogen associated with differential regulation of amino acid metabolism. *Plant Growth Regulation* 93: 231–242.

Sharma, S., Saxena, D.C., and Riar, C.S. (2016). Analysing the effect of germination on phenolics, dietary fibres, minerals and γ-amino butyric acid contents of barnyard millet (*Echinochloafrumentaceae*). *Food Bioscience* 13: 60–68.

Shelp, B.J., Bozzo, G.G., Trobacher, C.P. et al. (2012). Hypothesis/review: contribution of putrescine to 4-aminobutyrate (GABA) production in response to abiotic stress. *Plant Science* 193–194: 130–135.

Shelp, B.J., Bown, A.W., and Zarei, A. (2017). 4-Aminobutyrate (GABA): a metabolite and signal with practical significance. *Botany* 95: 1015–1032.

Sheng, Y., Xiao, H., Guo, C. et al. (2018). Effects of exogenous gamma-aminobutyric acid on α-amylase activity in the aleurone of barley seeds. *Plant Physiology and Biochemistry* 127: 39–46.

Simpson, J.P., Di Leo, R., Dhanoa, P.K. et al. (2008). Identification and characterization of a plastid localized *Arabidopsis* glyoxylate reductase isoform: comparison with a cytosolic isoform and implications for cellular redox homeostasis and aldehydedetoxification. *Journal of Experimental Botany* 59: 2545–2554.

Sita, K., Sehgal, A., Kumar, J. et al. (2017). Identification of high-temperature tolerant lentil (*Lens culinaris* Medik.) genotypes through leaf and pollen traits. *Frontiers in Plant Science* 8: 744.

Sweetman, C., Sadras, V.O., Hancock, R.D. et al. (2014). Metabolic effects of elevated temperature on organic acid degradation in ripening *Vitis vinifera* fruit. *Journal of Experimental Botany* 65 (20): 5975–5988.

Takahashi, Y., Sasanuma, T., and Abe, T. (2013). Accumulation of gamma-aminobutyrate (GABA) caused by heat-drying and expression of related genes in immature vegetable soybean (edamame). *Breeding Science* 63 (2): 205–210.

Thompson, J.F., Pollard, J.K., and Steward, F.C. (1953). Investigations of nitrogen compounds and nitrogen metabolism in plants. III. Gamma-aminobutyric acid in plants, with special reference to the potato tuber and a new procedure for isolating amino acids other than alpha-amino acids. *Plant Physiology* 28: 401–414.

UN (2019). The Sustainable Development Goals Report 2019. https://unstats.un.org/sdgs/report/2019/The-Sustainable-Development-Goals-Report-2019.pdf. Retrieved 7 July 2020.

Wang, C.-Y., Li, J.R., Xia, Q.P. et al. (2014). Influence of exogenous γ-aminobutyric acid (GABA) on GABA metabolism and amino acid contents in roots of melon seedling under hypoxia stress. *The Journal of Applied Ecology* 25: 2011–2018.

Wang, Y., Gu, W., Meng, Y. et al. (2017). Υ- aminobutyric acid imparts partial protection from salt stress injury to maize seedlings by improving photosynthesis and upregulating osmoprotectants and antioxidants. *Scientific Reports* 7: 1–13.

Xu, J.G., Hu, Q.P., Duan, J.L., and Tian, C.R. (2010). Dynamic changes in γ-aminobutyric acid and glutamate decarboxylase activity in oats (*Avena nuda* L.) during steeping and germination. *Journal of Agricultural and Food Chemistry* 58 (17): 9759–9763.

Xu, L., Chen, L., Ali, B. et al. (2017). Impact of germination on nutritional and physicochemical properties of adlay seed (*Coixlachryma-jobi* L.). *Food Chemistry* 229: 312–318.

Xu, B., Sai, N., and Gilliham, M. (2021). The emerging role of GABA as a transport regulator and physiological signal. *Plant Physiology* 187 (4): 2005–2016.

Yang, R., Yin, Y., and Gu, Z. (2015). Polyamine degradation pathway regulating growth and GABA accumulation in germinating fava bean under hypoxia-NaCl stress. *Journal of Agricultural Science and Technology* 17: 311–320.

Zeng, W., Hassan, M.J., Kang, D. et al. (2021). Photosynthetic maintenance and heat shock protein accumulation relating to γ-aminobutyric acid (GABA)-regulated heat tolerance in creeping bentgrass (*Agrostis stolonifera*). *South African Journal of Botany* 141: 405–413.

Zhao, G.C., Xie, M.X., Wang, Y.C., and Li, J.Y. (2017). Molecular mechanisms underlying γ-aminobutyric acid (GABA) accumulation in giant embryo rice seeds. *Journal of Agricultural and Food Chemistry* 65 (24): 4883–4889.

Zhao, Z.Y., Wang, P.Y., Xiong, X.B. et al. (2022). Environmental risk of multi-year polythene film mulching and its green solution in arid irrigation region. *Journal of Hazardous Materials* 435: 128981.

Zinn, K.E., Tunc-Ozdemir, M., and Harper, J.F. (2010). Temperature stress and plant sexual reproduction: uncovering the weakest links. *Journal of Experimental Botany* 61 (7): 1959–1968.

Ziogas, V., Tanou, G., Belghazi, M. et al. (2017). Characterization of β-amino-and γ-amino butyric acid-induced citrus seeds germination under salinity using nanoLC–MS/MS analysis. *Plant Cell Reports* 36: 787–789.

13

GABA and Oxidative Stress and the Regulation of Antioxidants

Somayeh Rastegar[1] *and Pegah Sayyad-Amin*[2]

[1] *Department of Horticultural Sciences, Faculty of Agriculture and Natural Resources, University of Hormozgan, Bandar Abbas, 7916193145, Iran*
[2] *Department of Horticultural Science and Landscape, Ferdowsi University of Mashhad, Mashhad, 9177948974, Iran*

Introduction

Reactive oxygen species (ROS) are inadvertent outcomes of oxygen metabolism that can trigger oxidative stress when their generation surpasses the capability of antioxidants to counteract them. Oxidative stress stems from an asymmetry between ROS creation and removal, leading to detrimental consequences and functional disturbances within diverse cellular compartments. The outcome of a plant is contingent upon the intricate balance between ROS synthesis and the efficacy of antioxidant mechanisms. Plants react to alterations in their surroundings and stressors by producing elevated quantities of ROS (Das and Roychoudhury 2014). External stressors like salinity, ultraviolet (UV) radiation, heavy metals, temperature variations, drought, air pollutants, herbicides, and nutrient deficiencies can instigate the accumulation of reactive oxygen species (ROS) in plants. These stressors disrupt the equilibrium between ROS generation and elimination mechanisms, resulting in oxidative stress. For instance, heightened levels of salinity can disrupt ion balance in plant cells, triggering the generation of ROS. UV radiation exposure directly induces ROS production within plant tissues. Heavy metal stressors can elicit ROS formation through actions like inhibiting antioxidant enzyme function or disrupting electron transport processes. Extreme temperature fluctuations can impact the ROS production balance, such as through increased electron leakage and the efficiency of antioxidant defenses (Hasanuzzaman et al. 2020). In addition, both drought stress and inadequate nutrient supply can promote the accumulation of ROS in plants. To mitigate oxidative stress, plants have developed an intricate antioxidant defense mechanism. This system incorporates enzymatic antioxidants, including superoxide dismutase (SOD), catalase (CAT), peroxidases, along with nonenzymatic antioxidants like ascorbic acid (ASH) (vitamin C), glutathione (GHS), tocopherols (vitamin E), and carotenoids. These antioxidants work in conjunction to eliminate and counteract ROS, thereby shielding plant cells from oxidative injury (Sharma et al. 2012; Sachdev et al. 2021).

GABA in Plants: Biosynthesis, Plant Development, and Food Security, First Edition. Edited by Samiksha Singh, Durgesh Kumar Tripathi, and Vijay Pratap Singh.

Types and Characteristics of ROS

ROS comprise a group of highly reactive molecules or ions containing oxygen and possessing unpaired electrons. The four primary types of ROS include the superoxide anion ($O_2^{\bullet-}$), hydrogen peroxide (H_2O_2), hydroxyl radical ($OH^{\bullet}$), and singlet oxygen (O_2). These ROS species play crucial roles in plant physiology, both beneficial and detrimental (Bhattacharjee 2019; Savchenko and Tikhonov 2021).

1) **Superoxide anion ($O_2^{\bullet-}$):** This is the simplest and most abundant form of ROS. It is generated through the one-electron reduction of molecular oxygen (O_2). Sources of superoxide anion in plants include photosystem I, NADPH oxidase, xanthine oxidase, and the mitochondrial electron transport chain. Although relatively stable, superoxide anion can interact with other ROS or metal ions, leading to the formation of more harmful species such as H_2O_2 or $OH^{\bullet-}$ (Das and Roychoudhury 2014).
2) **Hydrogen peroxide (H_2O_2):** H_2O_2 is formed through the two-electron reduction of O_2 or the dismutation of superoxide anion ($O_2^{\bullet-}$). Plants produce H_2O_2 through enzymatic and nonenzymatic reactions involving SOD, glycolate oxidase, peroxisomal β-oxidation, and cell wall peroxidases. Hydrogen peroxide exhibits moderate reactivity and stability. It serves as a signaling molecule and participates in redox-regulated processes. However, excessive levels of H_2O_2 can cause oxidative damage by reacting with metal ions or generating $OH^{\bullet-}$ (Černý et al. 2018).
3) **Hydroxyl radical ($OH^{\bullet}$):** The $OH^{\bullet}$- represents the most potent and detrimental ROS. These radicals are formed either by the one-electron reduction of H_2O_2 or through the Fenton reaction involving H_2O_2 and Fe^{2+}. Within plant systems, $OH^{\bullet-}$ can emerge from photochemical reactions, UV radiation exposure, ozone exposure, or through plasma membrane NADPH oxidase activity. Owing to their pronounced reactivity and limited stability, $OH^{\bullet-}$ promptly engage in reactions with biomolecules situated in proximity to their source of generation, culminating in destructive harm (Choudhary et al. 2020).
4) **Singlet oxygen (O_2):** Singlet oxygen constitutes an energized form of molecular oxygen characterized by a unique spin arrangement. Its formation can occur through the transfer of energy from triplet chlorophyll to O_2 or via the interaction between O_2 and carotenoid radicals. In plant systems, singlet oxygen production can arise from diverse sources, such as photosystem II activity, exposure to herbicides, UV radiation, or ozone exposure. Notably, singlet oxygen demonstrates a heightened reactivity coupled with low stability, leading to the oxidation of unsaturated fatty acids, amino acids, and nucleotides, resulting in oxidative harm. The impacts of ROS on plants are multifaceted, where an excessive accumulation of ROS can incite oxidative stress and inflict damage on cellular constituents. Nonetheless, ROS also play indispensable roles as signaling molecules that orchestrate various physiological processes and stress responses. Hence, sustaining a delicate equilibrium between ROS generation and removal processes is indispensable for facilitating optimal plant growth, developmental processes, and adaptation to environmental adversities (Savchenko and Tikhonov 2021).

ROS Generation in Plants under Normal and Stress Conditions

Plants are continuously confronted with a myriad of environmental stressors that can impact their viability, growth, and overall development. Some of these stressors have the capacity to incite oxidative stress within plant systems, a state characterized by an imbalance where the production of ROS surpasses the antioxidative defense mechanisms. ROS are oxygen-containing molecules or ions possessing one or more unpaired electrons, rendering them exceptionally reactive and unstable. This reactive nature enables ROS to instigate oxidative harm to lipids, proteins, DNA, and other vital biomolecules, resulting in compromised cellular functionalities and eventual cell demise. Nevertheless, ROS carry out a dual role as pivotal signaling molecules that regulate diverse physiological pathways and stress responses within plants (Juan et al. 2021).

ROS can be generated in plants under both normal and stress conditions by different sources and mechanisms. Under normal conditions, ROS are mainly produced as byproducts of photosynthesis and respiration, which are essential metabolic processes for plant energy production and carbon fixation. Photosynthesis involves the transfer of electrons from water to $NADP^+$ through two photosystems (PSI and PSII) in the thylakoid membrane of chloroplasts (Foyer 2018). During this process, some electrons can leak from the electron transport chain and reduce molecular oxygen (O_2) to form superoxide anion ($O_2^{\cdot -}$), which can further be converted to H_2O_2 by SOD. H_2O_2 can then be reduced to water by ascorbate peroxidase (APX) or CAT or react with metal ions or other ROS to form more damaging species, such as $OH^{\cdot}$ or singlet oxygen (O_2). Respiration involves the transfer of electrons from organic substrates to O_2 through a series of complexes in the inner membrane of mitochondria. During this process, some electrons can also leak from the electron transport chain and reduce O_2 to form $O_2^{\cdot -}$, which can be further converted to H_2O_2 by SOD. H_2O_2 can then be reduced to water by glutathione peroxidase (GPX) or peroxiredoxin (PRX) or react with metal ions or other ROS to form $OH^{\cdot}$ (Madkour 2019; Sachdev et al. 2021). Under stress conditions, ROS production can be enhanced by various abiotic and biotic factors, such as salt, drought, extreme temperatures, toxic metals/metalloids, flooding, pathogens, herbivores, and herbicides. These factors can affect plant water relations, ion homeostasis, membrane integrity, energy status, nutrient uptake, translocation, detoxification, and hormone synthesis. These effects can alter the balance between ROS generation and scavenging in different cellular compartments and tissues. For example, salt stress can increase the influx of Na^+ and Cl^- into the cytosol, which can disrupt the ionic balance and osmotic potential of the cell. This can lead to stomatal closure and reduced CO_2 assimilation, resulting in excess light energy and electron pressure in the chloroplasts. This can increase the leakage of electrons from PSI and PSII and enhance the formation of $O_2^{\cdot -}$, H_2O_2, $OH^{\cdot}$, and O_2. Drought stress can also cause stomatal closure and reduced CO_2 assimilation, as well as decrease the water potential and turgor pressure of the cell. This can impair the activity of photosynthetic enzymes and increase the production of ROS in the chloroplasts (Cruz de Carvalho 2008; García-Caparrós et al. 2021). Extreme temperatures can affect the fluidity and permeability of membranes, as well as the structure and function of proteins and enzymes. This can impair the efficiency of photosynthesis and respiration and increase the generation of ROS in both chloroplasts and mitochondria. Toxic metals/metalloids can interfere with plant nutrient uptake, translocation, and detoxification. They can also participate in redox reactions with ROS or other biomolecules and generate more ROS or free radicals. For example, Fe^{2+} can react with H_2O_2 to form $OH^{\cdot}$ by the Fenton reaction. Flooding stress can reduce the availability of O_2 for respiration and cause hypoxia or anoxia in plant tissues. This can lead to a shift from aerobic to anaerobic metabolism and increase the accumulation of ethanol, lactate, pyruvate, alanine, succinate, malate, and other organic acids. These metabolites can act as substrates for ROS production when O_2 is restored by reaeration. Generally, ROS generation in plants is a complex and dynamic process that involves various sources and mechanisms under both normal and stress conditions. Plants have evolved sophisticated systems to regulate the production and scavenging of ROS to maintain a balance between oxidative stress and redox signaling (Huang et al. 2019; Hasanuzzaman and Fujita 2022).

The Importance of ROS Compartmentation for Plant Stress Adaptation

The process of ROS compartmentalization plays a pivotal role in the adaptive responses of plants to stress conditions. This mechanism entails the segregation and movement of ROS within distinct cellular locales, including chloroplasts, mitochondria, peroxisomes, cytosol, nucleus, vacuole, and cell wall. By orchestrating ROS localization, plants can finely modulate the generation and elimination of ROS in a specific spatial and temporal manner. Moreover, ROS compartmentation facilitates synchronized intercellular and interorganelle communication. Various elements contribute to the effective compartmentation of ROS in plants. Physical barriers formed by membranes serve to restrict the diffusion of ROS, while the presence of antioxidant enzymes and molecules aids in the neutralization of ROS, thereby mitigating oxidative damage (Gautam et al. 2017). Furthermore, the functions of systems responsible for both generating and scavenging ROS, alongside the activation of genes responsive to ROS, are critical in preserving ROS homeostasis across diverse cellular compartments. As an illustration, chloroplasts are equipped with a dual membrane configuration that restricts ROS diffusion, in conjunction with a substantial presence of antioxidant enzymes and molecules aimed at counteracting ROS and averting lipid oxidation. The dynamic control of photosynthesis and respiration activities within chloroplasts enables the precise regulation of ROS production and detoxification processes. Notably, chloroplasts showcase distinct patterns of gene expression that can sense ROS signals and trigger protective responses (Palma and Corpas 2021; Rawat et al. 2021). The compartmentation of ROS is essential for plant stress adaptation as it enables plants to maintain a delicate equilibrium between oxidative stress and redox signaling. Oxidative stress occurs when the production of ROS surpasses the capacity of the antioxidant defense system, resulting in cellular damage and even cell death. On the other hand, redox signaling involves ROS acting as second messengers, triggering various signal transduction pathways and gene expression changes that mediate stress responses. For instance, H_2O_2, a type of ROS, can function as both a redox signal and an oxidative stressor. It activates mitogen-activated protein kinases (MAPKs), calcium channels, transcription factors (TFs), and stress tolerance-related genes (Petrov et al. 2015). Nevertheless, heightened concentrations of H_2O_2 can lead to detrimental outcomes such as lipid peroxidation (LPO), protein oxidation, DNA damage, and eventual cell demise. Hence, meticulous control over the quantity and specific sites of H_2O_2, along with other ROS, is imperative in maintaining the delicate equilibrium between oxidative stress and redox signaling in plants (Gautam et al. 2017).

Antioxidant Defense System in Plants

Nonenzymatic Antioxidants

Antioxidants can be broadly categorized into lipid-soluble and membrane-associated compounds or water-soluble reductants. Examples of lipid-soluble antioxidants include α-tocopherol (vitamin E) and β-carotene, while ascorbate (vitamin C) and GHS represent water-soluble antioxidants.

Ascorbic acid (AsA), also known as vitamin C, is an abundant and highly potent water-soluble antioxidant in plants. It serves as a crucial defense mechanism against ROS, effectively neutralizing their damaging effects. AsA is found in various plant tissues, but it is particularly abundant in meristematic regions, photosynthetic cells, and certain fruits. Its presence allows it to play a vital role in protecting these vital plant structures (Venkatesh and Park 2014). This antioxidant functions effectively in neutralizing ROS by transferring electrons through enzymatic or nonenzymatic pathways. AsA has the capability to regenerate alpha-tocopherol from tocopheroxyl radicals, thus supporting the antioxidant potential of vitamin E. Furthermore, it can directly eliminate highly reactive OH^- and oxygen radicals (O^2). In chloroplasts, ASH acts as a cofactor for violaxanthin de-epoxidase, aiding in the dissipation of excess excitation energy and preventing potential harm to the photosynthetic system. In addition, ASH plays a crucial role in maintaining the functionality of enzymes containing transition metal ions as prosthetic groups. By regulating the redox state of these metal ions, it ensures the optimal operation of these enzymes (Akram et al. 2017; Hossain et al. 2017). The ASH redox system comprises L-ASH, monodehydroascorbate (MDHA), and dehydroascorbate (DHA). Both MDHA and DHA, the oxidized forms of ASH, exhibit relatively low stability in water-based environments. Nevertheless, DHA can undergo chemical reduction back to ASH with the assistance of GSH, facilitating the restoration of active ASH molecules. Furthermore, ASH is vital for the regeneration of other antioxidants like vitamin E. Its presence significantly contributes to maintaining the overall antioxidant capacity of the plant. In summary, ASH, abundant in plants and possessing potent antioxidant attributes while participating in various physiological processes, serves as a pivotal component in the plant's defense against oxidative injury induced by ROS (Smirnoff and Wheeler 2000; Bilska et al. 2019).

Glutathione (GSH) serves as a crucial internal protective system in plants, shielding against oxidative harm. Predominantly in a reduced state, GSH is dispersed across various cellular compartments such as mitochondria, peroxisomes, cytosol, vacuoles, apoplasts, and the endoplasmic reticulum. Within plant cells, GSH is involved in a multitude of physiological functions. It contributes to signal transmission, regulates sulfate transport, assists in xenobiotic detoxification, aids in metabolite conjugation, and impacts the activation of stress-responsive genes. Apart from its cellular functions, GSH is indispensable for growth and developmental processes in plants. It is vital for cell aging, differentiation, enzyme regulation, and defense against pathogens. During cellular processes, GSH can serve as a substrate, leading to the creation of oxidized glutathione (GSSG), where two GSH molecules are linked by a disulfide bond. This equilibrium between GSH and GSSG maintains the cell's redox status, with GSH acting to preserve the cellular reduced state and counteract the inhibitory consequences of oxidative stress induced by ROS (Gill et al. 2013). GSH exhibits impressive scavenging capacities, efficiently counteracting ROS like O^2, H_2O_2, and highly reactive OH^-. It assumes a critical function in the antioxidant defense system by engaging in the replenishment of other water-soluble antioxidants, such as ascorbate (ASH) via the ASH–GSH cycle. Moreover, GSH contributes to the elimination of inhaled pollutants like ozone, NO_2, and other detrimental substances. Acting as a primary nonprotein thiol, it plays a key role in the recycling of antioxidants like vitamins C and E. Overall, GHS serves as a fundamental intracellular protective mechanism in plants, providing protection against oxidative harm. Its capacity to modulate redox equilibrium, scavenge ROS, and partake in antioxidant defense mechanisms underscores its importance in preserving cellular well-being and enhancing plant vitality (Szalai et al. 2009; Hameed et al. 2014).

Proline serves not only as an osmolyte but also as a nonenzymatic antioxidant pivotal in alleviating the detrimental impacts of ROS in microbes, animals, and plants. In situations of salt, drought, and metal-induced stress, proline accumulates due to heightened synthesis or diminished breakdown. This accumulation empowers proline to fulfill various roles in stress resilience. Functioning as a protein stabilizer, free proline safeguards proteins against structural deformation, preserving their integrity. In addition, proline acts as an osmotic stabilizer, aiding cells in upholding water equilibrium and preventing desiccation. It also serves as a metal binder, forming complexes with metals and lessening their harmful effects. Moreover, proline operates as a scavenger of OH- and oxygen radicals (O_2), highly reactive species capable of causing cellular harm. Proline's efficacy in scavenging $OH^{\cdot-}$ surpasses that of other compounds like mannitol, sorbitol, and myo-inositol. When confronted with ROS stemming from metal, salt, and dehydration stressors, proline effectively quenches these destructive radicals (Rejeb et al. 2014; Meena et al. 2019). Moreover, proline serves as a redox-signaling molecule, participating in redox-signaling pathways that regulate various cellular processes. Overall, proline's multifaceted role as an antioxidant, protein stabilizer, osmoprotectant, metal chelator, and scavenger of ROS makes it a crucial component in reducing the adverse effects of stress in microbes, animals, and plants.

Tocopherols, commonly known as vitamin E, demonstrate impressive capabilities in scavenging ROS and lipid radicals. In plants, these molecules are primarily located in the thylakoid membrane and exhibit a lipid-soluble characteristic. The plant system encompasses four distinct isomers of tocopherols: α, β, γ, and δ. Among these variants, α-tocopherol displays the highest antioxidant efficacy. This heightened effectiveness can be ascribed to the presence of three methyl groups in its molecular configuration, enhancing its ability to counteract detrimental oxidative agents. Moreover, tocopherols play a crucial role in stabilizing polyunsaturated fatty acids within lipid bilayers and low-density lipoprotein (LDL) (Munné-Bosch and Alegre 2002). Tocopherols, notably α-tocopherol, act as potent antioxidants in plants, shielding against oxidative harm and upholding the stability and functionality of lipid-rich structures by safeguarding against lipoxygenase-related damage. Furthermore, these compounds contribute to the overall maintenance of lipid stability and functionality in the plant system.

In all photosynthetic entities, β-carotene, zeaxanthin, carotenoids, and tocopherols play a crucial role in photoprotection. They disperse surplus excitation energy as heat, scavenge ROS, and suppress LPO. Among these components, β-carotene excels as a top-notch scavenger of singlet oxygen. Carotenoids, lipid-soluble pigments predominantly present in plants and microorganisms, hold substantial importance in plant physiology (Swapnil et al. 2021). Carotenoids are a diverse group of over 600 identified compounds in plants with several metabolic functions, including conferring tolerance to oxidative stress. Carotenoids fulfill three primary roles. At first, they act as additional pigments for light absorption, capturing light within the 400–550 nm range and transferring it to chlorophyll molecules, thereby enhancing photosynthetic efficiency. Second, carotenoids act as antioxidants by neutralizing triplet sensitizer (Chl_3), molecular oxygen (O_2), and other harmful free radicals generated during photosynthesis. This antioxidative capability aids in shielding the photosynthetic apparatus from oxidative harm. Third, carotenoids play a structural function by contributing to the stability of the thylakoid membrane and proteins within light-harvesting complexes. Moreover, they are critical for the construction of photosystem I (PSI), a pivotal part of the photosynthetic system. In essence, β-carotene, zeaxanthin, carotenoids, and tocopherols empower photosynthetic organisms to effectively capture light energy, combat oxidative stress, and uphold the integrity of the photosynthetic machinery (Pérez-Gálvez et al. 2020).

Flavonoids, predominantly located in flowers, leaves, and pollen, can also amass in various other plant regions like aboveground structures, leaf exteriors, and vacuoles in the form of glycosides. They encompass diverse categories such as flavonols, flavones, isoflavones, and anthocyanins. These substances serve numerous roles in plants, functioning as signaling agents in plant–microbe communications, shielding against detrimental UV radiation, imparting vivid colors to flowers, fruits, and seeds, promoting plant reproduction and pollen development, and serving as a mechanism to fend off pathogenic microbes, insects, and animals. Flavonoids are widely recognized as potent bioactive compounds produced by plants, surpassing the antioxidant properties of well-known antioxidants like ascorbate (ASH) and α-tocopherol (Swallah et al. 2020). Through their ability to scavenge and neutralize radicals, flavonoids assist in preventing cellular damage during environmental stress. By inhibiting LPO and lipoxygenases, flavonoids play a critical role, with their antioxidant efficacy influenced by the configuration and quantity of hydroxyl groups within their ring structures. Plants synthesizing flavonoids and other phenolic compounds exhibit enhanced tolerance to severe UV exposure owing to their ability to absorb UV radiation. Moreover, stressors like injuries, drought, metal toxicity, and nutrient scarcity can activate the expression of genes involved in flavonoid biosynthesis, resulting in elevated flavonoid levels as part of the plant's defense mechanism. Phenolic compounds, including flavonoids, demonstrate antioxidant characteristics by functioning as electron donors, hydrogen suppliers, and quenchers of singlet oxygen. Categorized into polyphenols and simple phenols, these compounds are present in both consumable and nonconsumable plants. In addition, phenolic compounds possess metal-binding capabilities, further diversifying their roles in plant physiology (Grace 2005; Olszowy 2019).

Enzymatic Antioxidants

Various antioxidant enzymes fulfill distinct functions. Specific enzymes such as SOD, CAT, and peroxidase are involved in eliminating ROS, whereas enzymes like GHS peroxidase eradicate internal LPO byproducts. GHS *S*-transferase holds significance in eliminating detrimental secondary oxidation radicals.

SOD is an important antioxidant enzyme that plays a crucial role in protecting cells from the harmful effects of ROS. In higher plants, SOD isozymes are in various cell organelles, such as mitochondria, peroxisomes, and chloroplasts, and they protect these organelles from oxidative damage caused by ROS. Salt stress has been observed to increase SOD activity in some plants (Saibi and Brini 2018). Drought and cadmium stress induce an increase in SOD activity in different plant species. For instance, water stress elevates SOD activity in *Trifolium repens*, while aluminum stress results in heightened SOD activity in the root tips of *Hordeum vulgare*. The joint impact of drought and low light also affects SOD activity in *Picea asperata*. Various studies have delved into the regulatory mechanisms of SOD genes in plants. For instance, a TF named CsPIF8 has been identified to upregulate SOD expression in Citrus, enhancing cold tolerance. Furthermore, a microRNA, csn-miR398a-3p-1, has been observed to directly cleave Cu/Zn–SOD mRNA in *Clonorchis sinensis*, thereby exerting negative regulation over its gene expression. Overall, SOD serves as a crucial enzyme in the primary defense against ROS. Its functioning is influenced by diverse environmental stresses and can be modulated by TFs and microRNAs in plants.

Catalase (CAT) is a tetrameric enzyme crucial for the rapid breakdown of H_2O_2. This enzyme is present in all aerobic organisms and possesses a unique capability to degrade H_2O_2 without the need for additional reducing agents. A single CAT molecule can convert millions of H_2O_2 molecules into water (H_2O) and oxygen (O_2) each minute. CAT shows a dual role depending on H_2O_2 concentration. At elevated substrate levels, CAT efficiently breaks down harmful H_2O_2 through a "catalytic" process where H_2O_2 serves as both a hydrogen donor and acceptor. The activity of CAT can be impacted by various environmental stressors. For instance, a decline in CAT activity was noted in *Anabaena doliolum* under NaCl and copper stress. In wheat plants, CAT activity increased under drought stress, particularly in susceptible varieties. Conversely, a study by (Sharma et al. 2012) reported a reduction in CAT activity in rice seedlings under drought stress.

Enhanced CAT activity levels were seen in *P. asperata* under drought stress and intense light exposure. UV-B stress also triggered an increase in CAT activity in *Cassia auriculata* seedlings. Conversely, the combined impact of salt and drought stress resulted in reduced CAT activity in *Glycyrrhiza uralensis* seedlings. The control of the CAT gene family is believed to occur through multiple avenues. TFs can govern the expression of CAT genes at the transcriptional stage, while alternative splicing and mRNA sponging can modulate gene expression posttranscriptionally. These regulatory processes contribute to the precise adjustment of CAT activity in reaction to diverse environmental cues.

Peroxidases constitute a group of enzymes with significant roles in diverse biological processes. Among them, guaiacol peroxidase (GPOX) and APX stand out as crucial enzymes. GPOX is an enzyme that employs aromatic electron donors like guaiacol and pyrogallol. It participates in the breakdown of indole-3-acetic acid (IAA) and the formation of lignin. The activity of GPOX varies across plant species and under distinct stress environments. Ascorbate peroxidases (APXs) are hem-containing proteins that use ascorbate to reduce H_2O_2. Plants possess multiple APX isoforms located in mitochondria, cytosol, chloroplasts, and peroxisomes. APXs act as scavengers of H_2O_2, and their activity typically increases under various stress conditions. For instance, plants exposed to cadmium stress, salt stress, and water stress exhibit enhanced APX activity (Caverzan et al. 2012).

Glutathione reductase (GR) is a widespread enzyme responsible for converting oxidized glutathione (GSSG) back to its reduced form, GSH. GHS plays a vital function in upholding cellular redox equilibrium and serves as an antioxidant. While primarily located in chloroplasts, GR is also distributed in the cytosol and mitochondria. The interplay between GR and GSH is integral to a plant's resilience against diverse stressors (Gill et al. 2013).

Monodehydroascorbate reductase (MDHAR) is an enzyme crucial for regenerating ascorbate (ASH). MDHAR utilizes NADH as an electron source and eliminates H_2O_2 within the mitochondria and peroxisomes. Alterations in MDHAR activity have been noted under stress conditions, such as reduced activity in Cd-exposed poplar hybrids and elevated activity in Cd-exposed *Pinus sylvestris*. Dehydroascorbate reductase (DHAR) has a role in the regulation and replenishment of ASH. Higher DHAR levels have been linked to stress resilience in plants subjected to aluminum stress, ozone toxicity, drought, and salt stress (Mehla et al. 2017).

Glutathione peroxidase (GPX) comprises a group of enzymes that safeguard plant cells from oxidative harm by diminishing H_2O_2, organic, and lipid hydroperoxides with the aid of GSH. Selenium, a component of GPXs, functions as a peroxide scavenger. GHS *S*-transferases (GSTs) represent an extensive group of enzymes engaged in diverse activities like herbicide neutralization, hormonal balance, and detoxification of toxic substances. GSTs contribute to plant reactions to both biotic and abiotic stressors (Gill and Tuteja 2010).

To conclude, peroxidases such as GPOX and APX, as well as enzymes like GR, MDHAR, DHAR, GPX, and GST, are pivotal in aiding plants in combating oxidative stress and adapting to diverse environmental challenges. These enzymes are integral in neutralizing ROS and upholding cellular redox equilibrium, thereby enhancing plant resilience and viability amidst stressful conditions.

Relationships of GABA Shunt and ROS during Stress Conditions

The GABA shunt serves as a significant metabolic pathway contributing to carbon/nitrogen metabolism, nitrogen assimilation, and the mitigation of ROS during adverse conditions. It hinders the Krebs cycle, disrupts respiration, and results in heightened ROS generation under stress. Studies have shown that the overexpression of genes related to the GABA shunt improves stress resistance, while knockout mutants of the pathway become more susceptible to oxidative stress. Furthermore, the GABA shunt contributes to plant development and stress tolerance by maintaining appropriate levels of other metabolites. Components of the GABA shunt, such as GAD and Ca^{2+}/CaM, play crucial roles in scavenging ROS and regulating GABA synthesis during stress signaling. Knockout mutants lacking SSADH, a key enzyme in the GABA shunt, exhibit the accumulation of GHB, which in turn promotes ROS synthesis (Kim et al. 2011).

In conditions of oxygen deficiency, succinic semialdehyde (SSA) is converted to γ-hydroxybutyrate (GHB) through the action of SSA reductase. GHB levels increase under hypoxia and various stress conditions such as UV light, salinity, drought, and heat stress. GABA transaminase (GABA-T) functions as a key regulator during stress, reducing oxidative damage and enhancing stress tolerance in plants. GABA and proline (Pro) serve as osmoprotectants, preserving cell integrity and reducing degradation during oxidative and osmotic stress. These compounds are synthesized from glutamic acid and could quench

ROS production. Exogenous application of GABA regulates free radicals and ethylene production, offering protection against salt stress. Pro overproduction enhances osmotic stress tolerance in transgenic crops, and other amino acids are also involved in the response to osmotic stress. Changes in osmotic pressure led to increased amino acid content through altered gene expression of enzymes involved in their metabolism (Hayat et al. 2023).

The GABA shunt encompasses distinct enzymes that can bypass particular stages of the mitochondrial tricarboxylic acid cycle (TCAC). This pathway is central in counteracting the impacts of stress-induced alterations in respiratory mechanisms and diminishing oxidative injury. Moreover, GABA, which accumulates during stress, has the capability to bind directly to the aluminum-activated malate transporter (ALMT) and the guard cell outward rectifying K^+ (GORK) channel, consequently bolstering stress resilience. In acidic environments, heavy metals stimulate the release of malate via the root ALMT (Gilliham and Tyerman 2016; Khan et al. 2021). This leads to the creation of complexes between heavy metals and malate that are challenging for the plant to uptake. GABA is accountable for augmenting the concentrations of nonenzymatic compounds (such as ASH, GSH, phenols) and modulating the activity and expression of diverse enzymes (e.g., APX, ascorbate oxidase, SOD, MDHAR, GHS peroxidase, GHS *S*-transferase, GHS reductase, CAT) involved in antioxidants and osmolytes regulation (Kumar et al. 2017). These compounds aid in diminishing the quantities of ROS and indicators of oxidative harm like malondialdehyde and protein carbonylation. These enzymes have a critical function in detoxifying ROS and forestalling LPO in varied cellular compartments. GABA also contributes to the restoration of ion balance and membrane integrity by reducing the functions of NADPH oxidase, lipoxygenase, and polyphenol oxidase. Furthermore, GABA is engaged in signaling pathways, either as a signal itself or as a stimulus for subsequent processes. It can induce ROS production, trigger the activation of antioxidant enzymes, interact with plant hormones, and instigate protective responses against fungi and insects. The rise in intracellular Ca^{2+} levels due to stress stimulates the activity of glutamate decarboxylase (GAD), which transforms glutamate into GABA. Subsequently, GABA modulates the operation of ALMTs, potentially setting off downstream signaling cascades (Shelp et al. 2021; Guo et al. 2023). Nonetheless, further exploration is necessary to unravel the intricate dynamics of these relationships. Elevated concentrations of intracellular GABA have the potential to stimulate the activation of genes linked to signaling and metabolism, thereby augmenting plant development, vitality, and resilience to stress. These favorable outcomes may have repercussions on the caliber and quantity of crop output. External administration of GABA can elevate the internal GABA levels, leading to advantageous impacts on plant maturation and proliferation. GABA showcases its antioxidant characteristics by neutralizing ROS and inhibiting LPO. It accomplishes this by contributing hydrogen atoms or electrons to ROS and by bolstering the synthesis or performance of other antioxidants, such as ascorbate, GHS, tocopherols, and carotenoids (Sita and Kumar 2020). Apart from its role as an antioxidant, GABA serves as a signaling agent that regulates diverse signal transduction pathways and gene expressions associated with plant stress reactions. It interacts with plant hormones such as abscisic acid (ABA) (Xu et al. 2021a), ethylene (ET) (Kathiresan et al. 1997), salicylic acid (SA), jasmonic acid (JA), gibberellins (GA) (Xie et al. 2020), cytokinins (CK), and brassinosteroids (BR) influence numerous facets of plant growth and development. Moreover, GABA promotes communication between distinct stress reactions and triggers the activation of stress-responsive genes. Given its beneficial effects on plants and its capacity to enhance stress resilience, GABA could serve as a method for enhancing the quality of agricultural yields. Nevertheless, additional investigation is necessary, especially across diverse crop varieties and practical agricultural settings, to comprehensively grasp the potential advantages of GABA (Podlešáková et al. 2019; Xu et al. 2021b) (Mishra et al. 2023).

The Response of GABA under Abiotic Stress Conditions

Synthesis of Ascorbic Acid (AsA)

AsA, a water-soluble antioxidant, is crucial for neutralizing ROS and recycling other antioxidants. It is produced from glucose via two primary routes: the L-galactose pathway and the L-gulose pathway. The key enzymes involved in AsA biosynthesis include GDP-L-galactose phosphorylase (GGP), L-galactono-1,4-lactone dehydrogenase (GLDH), GDP-D-mannose-3′,5′-epimerase (GME), GDP-L-gulose phosphorylase (GULP), and L-gulono-1,4-lactone dehydrogenase (GULDH) (Smirnoff 2011). Research has demonstrated that a majority of genes involved in the L-galactose pathway were upregulated following GABA treatment. This upregulation is believed to be partly responsible for the concurrent elevation in AsA levels (Zhou et al. 2022). GABA has been found to induce the synthesis and accumulation of AsA in plants under various stress conditions, such as drought, salinity, temperature extremes, heavy metals, and pathogens (Hasan et al. 2021). Studies have demonstrated that exogenous application of GABA increases AsA content in wheat, rice, maize, tomato, and tobacco plants under different stressors. The molecular mechanisms underlying GABA's regulation of AsA synthesis are not fully understood, but some pathways have been proposed. One possible pathway is that GABA acts as a signaling molecule, modulating the expression and activity of enzymes involved in AsA biosynthesis through specific TFs or binding motifs. The application of GABA has been shown to reduce the production of H_2O_2 by increasing the activity of enzymes such as APX MDHAR, and DHAR. These enzymes play a crucial role in the antioxidant defense system of plants (Abdel Razik et al. 2021). Research has demonstrated that kiwifruit treated with GABA exhibited improved tolerance to chilling stress and better maintenance of quality when compared to the control group. The application of GABA resulted in heightened levels of AsA in kiwifruit, which served to safeguard cell membranes and mitigate the harm induced by oxidative stress triggered by chilling conditions. Analysis of gene expression patterns revealed an increase in the expression of genes involved in AsA synthesis and regeneration, along with a reduction in the expression of catabolic genes post-GABA treatment, leading to an increase in AsA levels. Furthermore, TFs associated with AsA production, such as bHLHs and HZ1, were augmented by the application of GABA (Liu et al. 2023). Aghdam et al. (2016) demonstrated that GABA is effective in preserving the levels of ASH, total phenols, flavonoids, and other compounds in peach fruit, while also exhibiting the capacity to scavenge free radicals.

In research, treating peach fruit with GABA was found to enhance the expression of genes responsible for GSH biosynthesis, resulting in increased GSH levels. The AsA–GSH cycle, which involves enzymes like GPX, MDHAR, DHAR, GR, and APX, is pivotal for scavenging H_2O_2 in plants. GABA treatment upregulated the expression of genes related to the AsA–GSH cycle, sustaining elevated AsA and GSH levels throughout cold storage and mitigating chilling-induced damage in peach fruit. TFs such as ERFs have been identified to regulate AsA metabolic pathways in various plant species. The upregulation of PpERF genes in the peach group treated with GABA indicates their potential role in modulating the metabolic pathways of AsA and GSH (Zhou et al. 2022). Additional investigation is warranted to clarify the specific mechanisms through which GABA governs the AsA. Unraveling these mechanisms can offer valuable understanding for manipulating AsA levels in plants, potentially paving the way for devising strategies to augment antioxidant capability and stress resilience in crops.

Synthesis of Phenolic Compounds

Plants are renowned for being rich sources of exogenous antioxidants, which encompass a diverse array of phytochemicals with profound effects on human health. Among these phytochemicals, phenolic compounds, terpenoids, and peptides stand out as key contributors to the antioxidant activity found in plants. However, it is the phenolic compounds that reign supreme as the most abundant and diverse group of plant antioxidants. Within the realm of phenolic compounds, flavonoids emerge as a notable subclass with an extensive repertoire of biological activities and health benefits for humans (Kaurinovic and Vastag 2019). These compounds exhibit anti-inflammatory, anti-allergic, anti-cancer, anti-diabetic, and neuroprotective effects, to name just a few. Another vital subclass of phenolic compounds is phenolic acids, known for their antioxidant properties. Phenolic acids can act as potent scavengers of free radicals by donating hydrogen atoms or electrons, and they also possess the ability to form complexes with metal ions. Lignans, on the other hand, represent a distinctive type of phenolic compound primarily found in seeds, grains, and nuts. These compounds showcase a remarkable range of effects, including estrogenic and antiestrogenic properties, as well as antioxidant and anticancer activities (Kaurinovic and Vastag 2019).

Under environmental stresses, such as salinity, plants respond by stimulating the synthesis and accumulation of phenolic compounds. This adaptive mechanism serves to protect plants from damage caused by ROS generated because of the stress conditions. By scavenging free radicals, interrupting radical chain reactions, and decomposing peroxides, phenolics play a crucial role in defending plants against the detrimental effects of ROS. In recent studies, the application of exogenous GABA (gamma-aminobutyric acid) has emerged as a promising approach to enhance the content of phenolic compounds in plants. GABA exerts its influence on phenolic synthesis by regulating the expression of genes and the activity of enzymes involved in phenylpropanoid metabolism. This regulatory role enables GABA to modulate the production of phenolic compounds in plants, thereby contributing to their antioxidant capacity. Moreover, it has been observed that low concentrations of NaCl and nitrogen (N) treatment can further stimulate the synthesis of phenolic compounds in plants (Wang et al. 2021). Interestingly, the presence of GABA enhances plant growth and triggers an increased demand for phenolic compounds, leading to the synthesis of larger quantities of these beneficial molecules. By delving deeper into the intricate interactions between GABA, environmental stressors, and phenolic synthesis, researchers can unravel the underlying mechanisms and potentially unlock new strategies to optimize phenolic compound production in plants. Such endeavors hold immense promise not only for enhancing the nutritional value of plants but also for harnessing their bioactive compounds for human health benefits (Xie et al. 2021). A research investigation examined the impact of externally applied GABA on internal GABA levels and phenolic compounds in tomato plants exposed to salt stress. The outcomes indicated that the addition of GABA resulted in elevated levels of gallic acid and coumaric acid, along with a modest rise in endogenous GABA concentrations. These results imply that exogenous GABA has the potential to bolster the plant's stress response by augmenting specific phenolic compounds and GABA levels in the presence of salt stress (Çekiç 2018).

The mechanism by which GABA operates in plants to combat various stressors, such as salt stress and drought stress, differs (Zhao et al. 2023). The impact of GABA on the accumulation of phenolic compounds in germinated barley exposed to NaCl stress varied depending on the concentration used. External application of GABA influenced phenolic synthesis by modulating gene expression levels and enzyme activity associated with phenylpropanoid metabolism (Ma et al. 2019). Furthermore, GABA serves as a signaling agent to mitigate oxidative harm and control the production of phenolic compounds in soybean sprouts when subjected to NaCl stress. Application of GABA led to a rise in the levels of overall phenolic compounds, phenolic acids, and isoflavones in the sprouts. Moreover, the function and gene expression of crucial enzymes related to phenolic compound synthesis, including phenylalanine ammonia lyase, cinnamic acid 4-hydroxylase, and 4-coumarate coenzyme A ligase, were augmented by GABA (Zhao et al. 2021).

Conclusion

Plant metabolic pathways produce ROS. These ROS are typically balanced by their production and removal under normal circumstances. However, adverse environmental conditions can upset this equilibrium, causing an upsurge in ROS levels. The excessive ROS generation leads to harmful effects and oxidative harm in plant cells. In response to this, plants exhibit stress-specific reactions and possess a sophisticated antioxidant system to scavenge ROS. This antioxidant defense system comprises both enzymatic elements like GHS peroxidase, CAT, SOD and GHS reductase, as well as nonenzymatic components such as, carotenoids, α-tocopherols, proline, GHS and flavonoids. The GABA shunt, a signaling and metabolic pathway, plays a crucial role in regulating GABA content and protecting plants from oxidative damage during stress. It is actively involved in oxidative stress tolerance. Although research has primarily focused on the role of the GABA shunt in abiotic stress, further investigations are needed to understand its function and the impact of its metabolites. Subsequent research should investigate the physiological reasons for the elevated GABA levels during stress, such as the activation of GAD, inhibition of catabolic processes, or increased activity within the GABA shunt pathway. This information will advance our comprehension of how GABA impacts plant growth and the initiation of defense responses. Moreover, identifying surface-bound GABA receptors and other elements engaged in GABA-regulated gene control is essential for leveraging the capabilities of this system to enhance crop yield and quality under various environmental stresses.

References

Abdel Razik, E.S., Alharbi, B.M., Pirzadah, T.B. et al. (2021). γ-Aminobutyric acid (GABA) mitigates drought and heat stress in sunflower (*Helianthus annuus* L.) by regulating its physiological, biochemical and molecular pathways. *Physiologia Plantarum* 172: 505–527.

Aghdam, M.S., Razavi, F., and Karamneghad, F. (2016). Maintaining the postharvest nutritional quality of peach fruits by γ-aminobutyric acid. *Iranian Journal of Plant Physiology* 5: 1457–1463.

Akram, N.A., Shafiq, F., and Ashraf, M. (2017). Ascorbic acid – a potential oxidant scavenger and its role in plant development and abiotic stress tolerance. *Frontiers in Plant Science* 8: 613.

Bhattacharjee, S. (2019). *Reactive Oxygen Species in Plant Biology*. New Delhi, India: Springer.

Bilska, K., Wojciechowska, N., Alipour, S., and Kalemba, E.M. (2019). Ascorbic acid – the little-known antioxidant in woody plants. *Antioxidants* 8: 645.

Caverzan, A., Passaia, G., Rosa, S.B. et al. (2012). Plant responses to stresses: role of ascorbate peroxidase in the antioxidant protection. *Genetics and Molecular Biology* 35: 1011–1019.

Çekiç, F. (2018). Exogenous GABA stimulates endogenous GABA and phenolic acid contents in tomato plants under salt stress. *Celal Bayar University Journal of Science* 14: 61–64.

Černý, M., Habánová, H., Berka, M., Luklová, M., Brzobohatý, B., 2018. Hydrogen peroxide: its role in plant biology and crosstalk with signalling networks. *International Journal of Molecular Sciences* https://doi.org/10.3390/ijms19092812.

Choudhary, A., Kumar, A., and Kaur, N. (2020). ROS and oxidative burst: roots in plant development. *Plant Diversity* 42: 33–43. https://doi.org/10.1016/j.pld.2019.10.002.

Cruz de Carvalho, M.H. (2008). Drought stress and reactive oxygen species: production, scavenging and signaling. *Plant Signaling & Behavior* 3: 156–165.

Das, K. and Roychoudhury, A. (2014). Reactive oxygen species (ROS) and response of antioxidants as ROS-scavengers during environmental stress in plants. *Frontiers in Environmental Science* 2: 53.

Foyer, C.H. (2018). Reactive oxygen species, oxidative signaling and the regulation of photosynthesis. *Environmental and Experimental Botany* 154: 134–142. https://doi.org/10.1016/j.envexpbot.2018.05.003.

García-Caparrós, P., De Filippis, L., Gul, A., Hasanuzzaman, M., Ozturk, M., Altay, V., Lao, M.T., 2021. Oxidative stress and antioxidant metabolism under adverse environmental conditions: a review. *The Botanical Review* 87, 421–466. https://doi.org/10.1007/s12229-020-09231-1.

Gautam, V., Kaur, R., Kohli, S.K. et al. (2017). ROS compartmentalization in plant cells under abiotic stress condition. In: *Reactive Oxygen Species and Antioxidant Systems in Plants: Role and Regulation Under Abiotic Stress* (ed. M.I.R. Khan and N.A. Khan), 89–114. Singapore: Springer https://doi.org/10.1007/978-981-10-5254-5_4.

Gill, S.S. and Tuteja, N. (2010). Reactive oxygen species and antioxidant machinery in abiotic stress tolerance in crop plants. *Plant Physiology and Biochemistry* 48: 909–930.

Gill, S.S., Anjum, N.A., Hasanuzzaman, M. et al. (2013). Glutathione and glutathione reductase: a boon in disguise for plant abiotic stress defense operations. *Plant Physiology and Biochemistry* 70: 204–212.

Gilliham, M. and Tyerman, S.D. (2016). Linking metabolism to membrane signaling: the GABA–malate connection. *Trends in Plant Science* 21: 295–301.

Grace, S.C. (2005). Phenolics as antioxidants. In: *Antioxidants and Reactive Oxygen Species in Plants* (ed. N. Smirnoff), 141–168. Oxford: Blackwell Publishing.

Guo, Z., Gong, J., Luo, S., Zuo, Y., Shen, Y., 2023. Role of gamma-aminobutyric acid in plant defense response. *Metabolites* https://doi.org/10.3390/metabo13060741

Hameed, A., Sharma, I., Kumar, A. et al. (2014). Glutathione metabolism in plants under environmental stress. In: *Oxidative Damage to Plants* (ed. P. Ahmad), 183–200. Amsterdam, the Netherlands: Elsevier.

Hasan, M.M., Alabdallah, N.M., Alharbi, B.M. et al. (2021). GABA: a key player in drought stress resistance in plants. *International Journal of Molecular Sciences* 22: 10136.

Hasanuzzaman, M., Fujita, M., 2022. Plant oxidative stress: biology, physiology and mitigation. *Plants*. https://doi.org/10.3390/plants11091185

Hasanuzzaman, M., Bhuyan, M.H.M.B., Zulfiqar, F. et al. (2020). Reactive oxygen species and antioxidant defense in plants under abiotic stress: revisiting the crucial role of a universal defense regulator. *Antioxidants* 9: 681.

Hayat, F., Khan, U., Li, J., Ahmed, N., Khanum, F., Iqbal, S., Altaf, M.A., Ahmad, J., Javed, H.U., Peng, Y., Ma, X., Tu, P., Chen, J., Shahid, M.A., 2023. Gamma-aminobutyric acid (GABA): a key player in alleviating abiotic stress resistance in horticultural crops: current insights and future directions. *Horticulturae* https://doi.org/10.3390/horticulturae9060647

Hossain, M.A., Munné-Bosch, S., Burritt, D.J. et al. (2017). *Ascorbic Acid in Plant Growth, Development and Stress Tolerance*. Cham, Switzerland: Springer.

Huang, H., Ullah, F., Zhou, D.-X. et al. (2019). Mechanisms of ROS regulation of plant development and stress responses. *Frontiers in Plant Science* 10: 800.

Juan, C.A., Pérez de la Lastra, J.M., Plou, F.J., and Pérez-Lebeña, E. (2021). The chemistry of reactive oxygen species (ROS) revisited: outlining their role in biological macromolecules (DNA, lipids and proteins) and induced pathologies. *International Journal of Molecular Sciences* 22: 4642.

Kathiresan, A., Tung, P., Chinnappa, C.C., and Reid, D.M. (1997). [Gamma]-aminobutyric acid stimulates ethylene biosynthesis in sunflower. *Plant Physiology* 115: 129–135.

Kaurinovic, B. and Vastag, D. (2019). *Flavonoids and Phenolic Acids as Potential Natural Antioxidants*. London: IntechOpen.

Khan, M.I.R., Jalil, S.U., Chopra, P. et al. (2021). Role of GABA in plant growth, development and senescence. *Plant Gene* 26: 100283.

Kim, K.-J., Pearl, P.L., Jensen, K. et al. (2011). Succinic semialdehyde dehydrogenase: biochemical–molecular–clinical disease mechanisms, redox regulation, and functional significance. *Antioxidants & Redox Signaling* 15: 691–718.

Kumar, N., Dubey, A.K., Upadhyay, A.K. et al. (2017). GABA accretion reduces Lsi-1 and Lsi-2 gene expressions and modulates physiological responses in *Oryza sativa* to provide tolerance towards arsenic. *Scientific Reports* 7: 8786.

Liu, Q., Li, X., Jin, S. et al. (2023). γ-Aminobutyric acid treatment induced chilling tolerance in postharvest kiwifruit (*Actinidia chinensis* cv. Hongyang) via regulating ascorbic acid metabolism. *Food Chemistry* 404: 134661.

Ma, Y., Wang, P., Wang, M. et al. (2019). GABA mediates phenolic compounds accumulation and the antioxidant system enhancement in germinated hulless barley under NaCl stress. *Food Chemistry* 270: 593–601.

Madkour, L.H. (2019). Function of reactive oxygen species (ROS) inside the living organisms and sources of oxidants. *Pharmaceutical Sciences & Analytical Research Journal* 2: 180023.

Meena, M., Divyanshu, K., Kumar, S. et al. (2019). Regulation of l-proline biosynthesis, signal transduction, transport, accumulation and its vital role in plants during variable environmental conditions. *Heliyon* 5: e02952.

Mehla, N., Sindhi, V., Josula, D. et al. (2017). An introduction to antioxidants and their roles in plant stress tolerance. In: *Reactive Oxygen Species and Antioxidant Systems in Plants: Role and Regulation Under Abiotic Stress* (ed. M.I.R. Khan and N.A. Khan), 1–23. Singapore: Springer.

Mishra, V., Gahlowt, P., Singh, S. et al. (2023). GABA: a key player of abiotic stress regulation. *Plant Signaling & Behavior* 18: 2163343.

Munné-Bosch, S. and Alegre, L. (2002). The function of tocopherols and tocotrienols in plants. *Critical Reviews in Plant Sciences* 21: 31–57.

Olszowy, M. (2019). What is responsible for antioxidant properties of polyphenolic compounds from plants? *Plant Physiology and Biochemistry* 144: 135–143.

Palma, J.M. and Corpas, F.J. (2021). Subcellular compartmentalization of plant antioxidants and ROS generating systems. *Frontiers in Plant Science* 12: 643239.

Pérez-Gálvez, A., Viera, I., and Roca, M. (2020). Carotenoids and chlorophylls as antioxidants. *Antioxidants* 9: 505.

Petrov, V., Hille, J., Mueller-Roeber, B., and Gechev, T.S. (2015). ROS-mediated abiotic stress-induced programmed cell death in plants. *Frontiers in Plant Science* 6: 69.

Podlešáková, K., Ugena, L., Spíchal, L. et al. (2019). Phytohormones and polyamines regulate plant stress responses by altering GABA pathway. *New Biotechnology* 48: 53–65.

Rawat, N., Singla-Pareek, S.L., and Pareek, A. (2021). Membrane dynamics during individual and combined abiotic stresses in plants and tools to study the same. *Physiologia Plantarum* 171: 653–676.

Rejeb, K.B., Abdelly, C., and Savouré, A. (2014). How reactive oxygen species and proline face stress together. *Plant Physiology and Biochemistry* 80: 278–284.

Sachdev, S., Ansari, S.A., Ansari, M.I. et al. (2021). Abiotic stress and reactive oxygen species: generation, signaling, and defense mechanisms. *Antioxidants* 10: 277.

Saibi, W. and Brini, F. (2018). Superoxide dismutase (SOD) and abiotic stress tolerance in plants: an overview. In: *Superoxide Dismutase: Structure, Synthesis and Applications* (ed. S. Magliozzi), 101–142. Hauppauuge, NY: Nova Science Publishers.

Savchenko, T. and Tikhonov, K. (2021). Oxidative stress-induced alteration of plant central metabolism. *Life* 11: 304.

Sharma, P., Jha, A.B., Dubey, R.S., Pessarakli, M., 2012. Reactive oxygen species, oxidative damage, and antioxidative defense mechanism in plants under stressful conditions. *Journal of Botany* 2012, 34–59. https://doi.org/10.1155/2012/217037.

Shelp, B.J., Aghdam, M.S., and Flaherty, E.J. (2021). γ-Aminobutyrate (GABA) regulated plant defense: mechanisms and opportunities. *Plants* 10: 1939.

Sita, K. and Kumar, V. (2020). Role of gamma amino butyric acid (GABA) against abiotic stress tolerance in legumes: a review. *Plant Physiology Reports* 25: 654–663.

Smirnoff, N. (2011). Vitamin C: the metabolism and functions of ascorbic acid in plants. In: *Advances in Botanical Research* (ed. F. Rebeille and R. Douce), 107–177. London: Elsevier.

Smirnoff, N. and Wheeler, G.L. (2000). Ascorbic acid in plants: biosynthesis and function. *Critical Reviews in Plant Sciences* 19: 267–290.

Swallah, M.S., Sun, H., Affoh, R. et al. (2020). Antioxidant potential overviews of secondary metabolites (polyphenols) in fruits. *International Journal of Food Science* 2020: 9081686.

Swapnil, P., Meena, M., Singh, S.K. et al. (2021). Vital roles of carotenoids in plants and humans to deteriorate stress with its structure, biosynthesis, metabolic engineering and functional aspects. *Current Plant Biology* 26: 100203. https://doi.org/10.1016/j.cpb.2021.100203.

Szalai, G., Kellős, T., Galiba, G., and Kocsy, G. (2009). Glutathione as an antioxidant and regulatory molecule in plants under abiotic stress conditions. *Journal of Plant Growth Regulation* 28: 66–80.

Venkatesh, J. and Park, S.W. (2014). Role of l-ascorbate in alleviating abiotic stresses in crop plants. *Botanical Studies* 55: 1–19.

Wang, M., Zhu, Y., Wang, P., Gu, Z., Yang, R., 2021. Effect of γ-aminobutyric acid on Phenolics metabolism in barley seedlings under low NaCl treatment. *Antioxidants* https://doi.org/10.3390/antiox10091421.

Xie, T., Ji, J., Chen, W. et al. (2020). GABA negatively regulates adventitious root development in poplar. *Journal of Experimental Botany* 71: 1459–1474.

Xie, C., Wang, P., Sun, M. et al. (2021). Nitric oxide mediates γ-aminobutyric acid signaling to regulate phenolic compounds biosynthesis in soybean sprouts under NaCl stress. *Food Bioscience* 44: 101356.

Xu, B., Long, Y., Feng, X. et al. (2021a). GABA signalling modulates stomatal opening to enhance plant water use efficiency and drought resilience. *Nature Communications* 12: 1952.

Xu, B., Sai, N., Gilliham, M., 2021b. The emerging role of GABA as a transport regulator and physiological signal. *Plant Physiology* 187, 2005–2016. https://doi.org/10.1093/plphys/kiab347.

Zhao, Y., Xie, C., Wang, P., Gu, Z., Yang, R., 2021. GABA regulates Phenolics accumulation in soybean sprouts under NaCl stress. *Antioxidants*. https://doi.org/10.3390/antiox10060990.

Zhao, Q., Ma, Y., Huang, X. et al. (2023). GABA application enhances drought stress tolerance in wheat seedlings (*Triticum aestivum* L.). *Plants* 12: 2495.

Zhou, C., Dong, W., Jin, S. et al. (2022). γ-Aminobutyric acid treatment induced chilling tolerance in postharvest peach fruit by upregulating ascorbic acid and glutathione contents at the molecular level. *Frontiers in Plant Science* 13: 1059979.

14

GABA in Relation to Cold and Chilling Stress

Somayeh Rastegar[1] *and Emad Hamdy Khedr*[2]

[1] *Department of Horticultural Sciences, Faculty of Agriculture and Natural Resources, University of Hormozgan, Bandar Abbas, 7916193145, Iran*
[2] *Department of Pomology, Faculty of Agriculture, Cairo University, Giza, 12613, Egypt*

Introduction

Low temperature is a prominent abiotic stress factor that places limitations on the growth, productivity, and geographical range of agricultural crops. It has the potential to cause significant crop losses (Janská et al. 2010). Cold stress affects approximately 57% of the world's land areas and 26% of rural areas (Cramer et al. 2011). Plants that are sensitive to cold can experience damage when exposed to temperatures around 15 °C, while tolerant plants can endure temperatures just below 5 °C (Rao 2004). Typically, tropical plants are more susceptible to cold stress, whereas subtropical, temperate, and arctic plants, as well as mosses, lichens, and specific tree species, exhibit greater tolerance (Ahmad and Prasad 2011). The impact of cold on plants is determined by several factors, including the rate of temperature decrease, the duration of exposure, and other related stresses (Buchanan et al. 2015). Depending on a species' tolerance, damage from low temperatures, often referred to as "chilling injury" can manifest within a few hours or days, indicating the species' level of sensitivity. In cases of abrupt cold exposure, injury can occur within minutes, which is termed "cold shock" (Phadtare et al. 1999).

Low temperatures trigger a complex cascade of physiological, biochemical, and molecular responses in plants. Some of the negative effects of cold stress are leaf curling, chlorophyll degradation, impaired photosynthesis, reduced cell membrane fluidity, decreased enzyme activity, metabolic imbalance, and growth retardation (Hussain et al. 2018). Low temperatures also result in the accumulation of highly reactive and harmful oxygen species (ROS), such as hydrogen peroxide (H_2O_2), superoxide radicals (O^{2-}), and hydroxyl radicals (OH^-). ROS are produced by membrane-bound electron transport and various metabolic pathways. ROS damage plant membranes and affect transport processes and membrane-bound enzyme activity. ROS also harms macromolecules such as lipids, proteins, and

GABA in Plants: Biosynthesis, Plant Development, and Food Security, First Edition. Edited by Samiksha Singh, Durgesh Kumar Tripathi, and Vijay Pratap Singh.

nucleic acids, leading to oxidative damage. Therefore, preserving the structure and stability of cell membranes is essential for optimal plant growth and development (Ding et al. 2019).

Chilling stress causes rapid and coordinated changes in the thermodynamic microenvironment of each plant cell, affecting the molecular composition of all organelles. Enzymatic reactions are slowed down due to reduced diffusion rates of substrates. This affects photosynthesis performance, including carbon fixation, stomatal conductance, electron transport, and the functions of PSI and PSII (Liu et al. 2019b). As a result, ROS are generated (Kohli et al. 2019), causing immediate damage to membranes and disrupting transport processes and membrane-bound enzyme activity. Moreover, chilling stress leads to structural damage, especially in chloroplasts, which are among the most sensitive organelles. Symptoms of chilling stress include chloroplast swelling and disorganization, thylakoid expansion, and an increase in plastoglobule numbers (El-Dakak et al. 2023). Other symptoms include grana disorganization, changes in thylakoid and chloroplast membranes accompanied by lipid droplet accumulation, and stroma darkening (Bhattacharya 2022). However, chilling stress-resistant plants show a decrease in the size and number of starch grains, and more compact grana disks are present. Structural and physiological changes are closely related to the accumulation of ROS in the chloroplast, which is considered the main site for ROS production under unfavorable stress conditions, ultimately damaging the photosynthetic apparatus. The severity of these injuries depends on the duration of exposure to damaging temperatures and the physiological age of the plant (Song et al. 2021; Bhattacharya 2022).

Membranes are the primary targets for CI development. Increased oxidation of membrane lipids leads to loss of membrane integrity, which has been observed in chilling-sensitive crops under low temperatures (Liang et al. 2020; Khedr and Khedr 2023a). Antioxidant enzymes play an important role in protecting cellular membranes and enhancing chilling tolerance. These enzymes include superoxide dismutase (SOD), catalase (CAT), ascorbate peroxidase (APX), glutathione peroxidase (GPX), glutathione S-transferase (GST), monodehydroreductase (MDHAR), and dehydroascorbate reductase (DHAR). They work together to scavenge ROS and reduce oxidative damage. The activity of these enzymes is often increased in response to biotic stress to balance the increased production of ROS (Liu et al. 2019b; Khedr and Khedr 2023b).

Plants have also evolved various mechanisms at the molecular level to cope with chilling stress. One such mechanism is the expression of cold-responsive genes, which are involved in stress sensing, signal transduction, and the activation of stress-responsive pathways. Transcription factors such as C-repeat-binding factors (CBFs) and dehydration-responsive element-binding (DREB) proteins play a key role in regulating the expression of downstream cold-responsive genes. These genes encode proteins involved in various processes, such as Osmo protection, detoxification, antioxidative defense, membrane stabilization, and repair mechanisms (Thomashow 2010; Ritonga and Chen 2020).

In addition to these physiological and molecular responses, plants can also acclimate to chilling stress through changes in their growth and development patterns. For example, plants may modify their phenology, such as changing their flowering time or postponing leaf senescence, to avoid periods of low temperatures (Ruelland and Collin 2012).

Chilling stress tolerance varies among different plant species and cultivars, with some naturally equipped to endure lower temperatures while others are more susceptible. Plant breeders and genetic engineers are actively engaged in enhancing chilling tolerance through selective breeding and genetic modification techniques. Chilling stress has diverse impacts, encompassing cellular, molecular, and physiological changes, as well as developmental adaptations. A comprehensive understanding of these responses and their underlying mechanisms is pivotal for devising strategies to bolster chilling tolerance in crops and mitigate the detrimental effects of low temperatures on agricultural productivity (Ding et al. 2019; Liang et al. 2020).

GABA, a four-carbon molecule, is present in various organisms, including autotrophs and heterotrophs. Initially discovered in potato tubers, GABA has been found in plants, animals, and microbial organisms (Ramesh et al. 2017). GABA is a naturally occurring active substance that plays an important role in different physiological activities of plants, including growth, development, signaling, and stress responses (Kaspal et al. 2021). Supplementing plants with exogenous GABA has been demonstrated to promote their growth. In addition, this effect is attained by elevating the levels of endogenous GABA, amino acids, and hormones and by upregulating key genes responsible for regulating phytohormones (Hijaz et al. 2018).

Understanding GABA's role in aiding plants in coping with cold stress is crucial for sustainable cultivation in a challenging environment. This chapter provides a comprehensive summary of current insights and research progress on GABA-induced responses to cold stress and their potential applications in plant management.

Plant Strategies to Overcome Cold Stress

When plants encounter chilling stress, they initiate a range of physiological and molecular responses to contend with adverse conditions. According to Satyakam et al. (2022), plants typically employ fundamental adaptation mechanisms, which include molecular, physiological, biochemical, and morphological adaptations (Figure 14.1), to adapt to cold conditions. Plants typically respond to chilling stress by employing one or more of the following strategies:

1) **Membrane Stability and Fluidity:** Chilling stress affects the fluidity and integrity of cellular membranes. To maintain membrane stability, plants undergo changes in lipid composition by increasing the proportion of unsaturated fatty acids and altering the lipid-to-protein ratio. This adjustment helps to maintain membrane fluidity and functionality under low temperatures (Liang et al. 2020).
2) **ROS Scavenging:** Chilling stress leads to the accumulation of ROS, which is highly reactive and can cause oxidative damage to cells. To counteract this, plants activate antioxidant defense mechanisms. They increase the production of enzymes such as superoxide dismutase (SOD), catalase (CAT), peroxidases, and glutathione reductase, which help scavenge ROS and protect cellular components from oxidative damage (Suzuki and Mittler 2006; Huang et al. 2019; Sachdev et al. 2021).
3) **Cold-Responsive Genes:** Plants activate specific genes in response to chilling stress. These cold-responsive genes are involved in various metabolic pathways and cellular processes. One group of genes that plays a crucial role in cold acclimation is the C-repeat-binding factor (CBF)/dehydration-responsive element-binding (DREB) gene

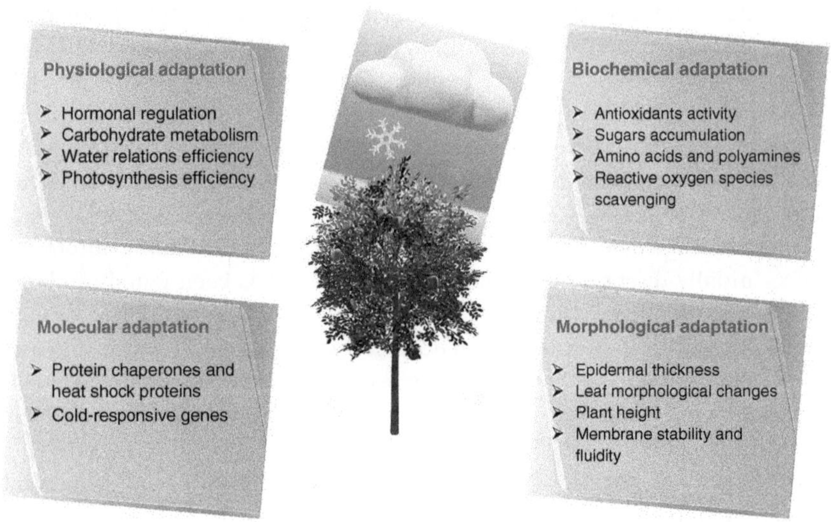

Figure 14.1 Schematic illustration of plant adaptation strategies to mitigate chilling stress. *Source:* Photo by E.H. Khedr and S. Rastegar.

family. These genes encode transcription factors that regulate the expression of downstream cold-responsive genes. The CBF/DREB pathway controls the synthesis of proteins involved in stress tolerance, including proteins for osmoprotection, cryoprotection, and detoxification (Chinnusamy et al. 2010).

4) **Accumulation of Compatible Solutes:** Compatible solutes, such as proline, sugars (e.g., sucrose), and polyols (e.g., sorbitol), accumulate in plant cells under chilling stress. These solutes act as osmoprotectants and help maintain cellular water balance, stabilize proteins, and scavenge ROS. They also contribute to the protection of membranes and cellular structures from chilling-induced damage (Aghaee et al. 2011; Ruelland and Collin 2012).
5) **Protein Chaperones and Heat Shock Proteins (HSPs):** Chilling stress can disrupt protein folding and lead to the accumulation of misfolded or denatured proteins. To address this, plants increase the production of protein chaperones and HSPs. These proteins assist in proper protein folding, refolding damaged proteins, and preventing protein aggregation under stress conditions (Aghdam et al. 2013).
6) **Calcium signaling:** Chilling stress triggers changes in intracellular calcium levels, which serve as important signaling molecules in plant cells. Calcium acts as a second messenger and regulates various physiological processes, including gene expression, enzyme activation, and ion channel activities. Calcium signaling is involved in the activation of cold-responsive genes and the modulation of cellular responses to chilling stress (Guo et al. 2018; Yuan et al. 2018; Aslam et al. 2022; Manasa et al. 2022).
7) **Hormonal Regulation:** Chilling stress also affects plant hormone levels and signaling. Abscisic acid (ABA), a stress-related hormone, is known to accumulate under chilling conditions. ABA regulates stomatal closure, leading to reduced water loss and improved water-use efficiency. In addition, other hormones like ethylene, jasmonic acid, and salicylic acid are involved in signaling pathways that regulate cold acclimation and defense responses (Ashraf and Rahman 2018).

These physiological and molecular responses work together to enhance chilling tolerance in plants. It's important to note that the specific responses can vary depending on the plant species, genotype, developmental stage, and duration of chilling stress. Researchers continue to study these responses to gain a deeper understanding of plant adaptation to chilling stress and to develop strategies to improve crop resilience under low-temperature conditions.

γ-Aminobutyric Acid (GABA)

In plants, GABA serves two major functions. First, it acts as a substrate for key pathways, playing a vital role in connecting primary and secondary metabolic pathways. Its functions include providing carbon skeletons and energy, regulating cytosolic pH, participating in nitrogen metabolism, and potentially affecting insects. Second, GABA is implicated in stress perception, as evidenced by studies linking GABA accumulation with physiological reactions in plants. It is involved in signaling pathways, serving as a signal molecule itself or triggering downstream signals related to ROS generation, antioxidant enzyme activity, interaction with plant hormones, and defense against fungi and insects (Guo et al. 2023). Moreover, GABA has been shown to modulate the growth and development of plants under non-stress conditions. It promotes germination, influences root and above-ground growth patterns, modulates photosynthetic capacity, and plays a role in nitrogen metabolism. Increased intracellular GABA levels can stimulate the expression of genes associated with signaling and metabolism, thus regulating plant growth, vigor, and stress tolerance (Ramos-Ruiz et al. 2019). These positive effects can ultimately impact the quality and quantity of crop yield. The exogenous application of GABA can enhance endogenous GABA levels, providing beneficial effects on plant development and growth like naturally formed GABA. Given its positive impact on plants and its ability to induce stress tolerance, GABA holds promise as a tool for improving agricultural crop quality (Srivastava et al. 2021). Overall, GABA plays a significant role not only in normal plant growth and development but also in response to various abiotic stresses. With the anticipated increase in adverse growing conditions due to global climate change, the importance of GABA in plant adaptation is underscored.

GABA Biosynthesis in Plants

Glutamate Decarboxylation: Chemically, the GABA is abbreviated as GABA with the chemical formula $C_4H_9NO_2$, lying in the category of amino acid but not considered a protein. The versatile GABA is predominantly indulged in growth and development via the GABA shunt, a sidestep of a tricarboxylic acid (TCA) cycle. In plants, GABA is derived from glutamate by irreversible decarboxylation, catalyzed through glutamic acid decarboxylase (GAD) in the cytosolic region. Consequently, GABA is translocated into the mitochondrial matrix and exposed to transamination to succinic semialdehyde (SSA) through GABA transaminase (GABA-T/POP2). Subsequently, SSA is oxidized to succinate through the activity of a dehydrogenase enzyme SSADH (NAD^+-dependent succinate semialdehyde dehydrogenase), and successively succinate feeds into the TCA cycle. Ultimately, glutamate and its carbon skeleton enter the TCA cycle through this GABA shunt and recycle (Che-Othman et al. 2020; Li et al. 2021).

Polyamine (PA) Degradation: Alternatively, biosynthesis of GABA is mediated by the PA metabolic pathway, in which putrescine is obtained from arginine through multistep routes. In the presence of oxygen, PA oxidase converts putrescene into either 4-amino butyraldehyde or spermidine but degrades to 4-amino butyraldehyde, and in turn, oxidized to GABA through NAD^+-dependent 4-amino-butyraldehyde dehydrogenase (Yuan et al. 2023).

Proline Degradation: A third source of GABA synthesis in plants is proline, another proteinogenic amino acid that is involved in osmotic adjustment and stress tolerance. Proline can be nonenzymatically converted to GABA under oxidative stress conditions, such as high light intensity, drought, and salinity. This pathway is also known as the proline oxidation pathway or the nonenzymatic pathway (Forlani et al. 2019).

Response Strategies of GABA in Cold Stress Tolerance

The different roles of GABA in plants might be functionally linked and difficult to be separated from each other. Studies have demonstrated that GABA can provide greater stress resistance in plants by regulating the expression of genes related to signaling, hormone biosynthesis, transcriptional regulation, the production of ROS, and PA metabolism. Furthermore, GABA plays a vital role in modulating the antioxidant system during plant growth and the transcription of genes that encode antioxidant enzymes, thereby mitigating plant oxidative damage (Hasanuzzaman et al. 2020). The application of GABA is also significant in respiratory metabolism and changes the activity of many enzymes from the TCA cycle. Table 14.1 presents the role of exogenous GABA treatments in alleviating chilling stress in various crops.

Mitigating ROS Generation and Improving Antioxidant Systems During Cold Stress in Plants

ROS are produced as byproducts of aerobic metabolism in plants and are generated in various cellular compartments, such as chloroplasts, mitochondria, apoplasts, and peroxisomes (Gill and Tuteja 2010). Recent evidence suggests that ROS in plants has a dual role, acting as both harmful byproducts and vital signaling molecules involved in plant growth and development. ROS plays a crucial role in plant responses to abiotic and biotic stresses, including drought, salinity, metal toxicity, heat shock, and cold stress. Precise regulation of ROS levels is crucial for maintaining cellular balance and optimizing plant stress adaptation (Hasanuzzaman et al. 2020). GABA plays a central role in the plant defense system and is actively involved in mitigating ROS production in response to various stresses, including cold exposure. It serves as a key molecule in ROS regulation and contributes to the plant's ability to cope with low temperatures and oxidative stress (Ansari et al. 2021). The GABA shunt pathway provides an efficient bypass for two enzymes in the TCA cycle that are known to be susceptible to oxidative stress. By diverting metabolites through the GABA shunt, plants can circumvent the potential damage caused by oxidative stress on these enzymes, ensuring the continued operation of the TCA cycle and maintaining metabolic homeostasis (Janse van Rensburg and Van den Ende 2020). Furthermore, various

Table 14.1 Role of exogenous GABA treatments in alleviating chilling stress in various crops.

Plant	Mechanism of action and main results	Reference
Tomato (*Lycopersicon esculentum*)	At GABA concentrations of 100, 250, 500, and 750 μmol/l, the results demonstrated a significant reduction in antioxidant enzyme activity, electrolyte leakage, MDA, and proline content.	Malekzadeh et al. (2014)
Rice (*Oryza sativa* L.)	At the seedling stage of rice, the optimal concentrations of culture components for enhancing cold tolerance were determined to be 2.21 mg/ml of glutamate, 2.94 mM of calcium chloride, and 2.68 mM of γ-aminobutyric acid.	Jia et al. (2016)
Wheat (*Triticum aestivum*)	Enhancing cold acclimation and frost tolerance through GABA accumulation	Mazzucotelli et al. (2006)
Citrus	Exogenous GABA is quickly metabolized to succinic acid and fed into the plant TCA cycle	Hijaz and Killiny (2019).
Tea (*Camellia sinensis*)	When tea plants were supplied with 5.0 mM GABA and compared to those without GABA supplementation, it was observed that endogenous GABA underwent degradation, leading to the promotion of the synthesis of other stress response substances. This process also provided feedback to the TCA (tricarboxylic acid) cycle. Notably, at low temperatures, a comparison between low and optimum temperature conditions revealed upregulated levels of other amino acids and differentially abundant proteins within the TCA cycle	Zhu et al. (2019)
Maize (*Zea mays*)	Application of 50 mg/l GABA improved early growth, net photosynthesis, and associated physiological and biochemical processes in maize.	Li et al. (2016)

components related to GABA have been shown to have crucial functions in scavenging and detoxifying ROS during stress conditions. These components actively contribute to the plant's antioxidant defense system, helping to neutralize ROS and minimize oxidative damage. Their involvement highlights the significance of GABA-related mechanisms in maintaining cellular redox balance and protecting plants from the harmful effects of ROS accumulation during stress (Khan et al. 2021). An enzyme called succinic semialdehyde dehydrogenase (SSADH), which is part of the GABA shunt pathway, plays a crucial role in maintaining the balance of ROS in plants. SSADH facilitates the conversion of succinic semialdehyde to succinate within the mitochondria. Previous research has demonstrated the indispensability of SSADH in regulating ROS levels and ensuring their homeostasis in plants (Bao et al. 2015). The application of GABA under low-temperature conditions has been observed to modulate the activities of oxidative enzymes, indicating a significant role for GABA in mitigating ROS during cold stress in plants. This suggests that GABA actively participates in the regulation of ROS levels, thereby contributing to the plant's ability to cope with the detrimental effects of cold stress.

The enzymes APX (ascorbate peroxidase), MDHAR (monodehydroascorbate reductase), DHAR (dehydroascorbate reductase), and GR (glutathione reductase) are key components of the ascorbate-glutathione cycle, which is involved in ROS detoxification. These enzymes work in conjunction with other antioxidant enzymes to facilitate the scavenging of ROS. Studies have shown that higher activities of APX, MDHAR, DHAR, and GR are associated with improved chilling tolerance in fruits, suggesting their involvement in mechanisms that enhance the plant's ability to cope with cold stress (Yang et al. 2011a). The application of exogenous GABA has been found to stimulate the activity of APX, MDHAR, and DHAR enzymes, thereby enhancing the ability of crops to withstand chilling stress. This suggests that GABA treatment may activate the antioxidant defense system, which plays a crucial role in scavenging excessive ROS and reducing oxidative damage caused by chilling stress in fruits. By improving the antioxidant capacity, GABA treatment can potentially enhance chilling tolerance in plants (Yang et al. 2011a; Li et al. 2017). Earlier studies have demonstrated that the application of GABA resulted in increased activity of SOD (superoxide dismutase), CAT (catalase), and APX (ascorbate peroxidase) in tomato seedlings subjected to chilling stress (Malekzadeh et al. 2014). Similarly, Cheng et al. (2018) reported that GABA-induced antioxidative defense in white clover involved the upregulation of various genes, including Cu/ZnSOD, MnSOD, FeSOD, GPOX, CAT, APX, MDHAR, GST, and GPX. These findings suggest that GABA treatment can enhance the antioxidant defense system by activating the expression of multiple genes involved in ROS scavenging and detoxification.

GABA Improves Nonenzyme Antioxidant System

Regulating Phenol Metabolism

Phenolic compounds function as antioxidants through their ability to react with different types of free radicals. The antioxidant mechanisms of action employed by PCs include hydrogen atom transfer, single electron transfer, sequential proton loss electron transfer, and chelation of transition metals. These diverse mechanisms allow phenolic compounds to effectively neutralize free radicals and reduce oxidative stress (Santos-Sánchez et al. 2019). Under chilling stress, the activity of two enzymes associated with phenol metabolism, polyphenol oxidase (PPO) and phenylalanine ammonia-lyase (PAL), was analyzed. The presence of GABA treatment was found to further increase PAL activity. Conversely, GABA treatment led to a decrease in PPO activity, indicating an elevation in antioxidative defense (Abd Elbar et al. 2021). PAL is a key enzyme involved in the phenylpropanoid pathway, which is responsible for the biosynthesis of various phenolic compounds. Phenolic compounds play crucial roles in plant defense against stress, including their antioxidant and antimicrobial properties. PAL catalyzes the conversion of phenylalanine to cinnamic acid, which is a crucial step in the phenylpropanoid pathway (Zhang and Liu 2015).

The increased activity of PAL observed in GABA-treated plants suggests that GABA treatment enhanced the phenylpropanoid pathway and the synthesis of phenolic compounds. This indicates that GABA may play a role in stimulating the production of phenolic compounds under chilling stress conditions (Xie et al. 2019). The upregulation of PAL activity by GABA treatment suggests that GABA may function as a signaling molecule or modulator of gene expression, triggering the activation of PAL and subsequent phenolic compound biosynthesis. The increased activity of PAL may contribute to the enhanced antioxidant defenses and stress tolerance observed in GABA-treated tomato plants (Malekzadeh et al. 2014). These findings align with previous studies demonstrating

GABA- mediated upregulation of PAL activity in banana, supporting the role of GABA in phenol metabolism and stress response (Wang et al. 2014). Further research is needed to investigate the specific molecular mechanisms by which GABA regulates PAL activity and its impact on phenolic compound accumulation. Understanding these mechanisms will provide valuable insights into the role of GABA in enhancing stress tolerance and its potential application in improving plant resilience to chilling stress.

In addition, GABA has been studied for its potential role in mitigating chilling injury in fruits. Chilling injury refers to the physiological damage that occurs in certain fruits when exposed to temperatures below their optimal range but above freezing. The downregulation of POX and PPO activities by GABA treatment indicates the involvement of GABA in phenol metabolism and suggests its potential as a regulator of phenolic compounds in tomato plants under chilling stress (Abd Elbar et al. 2021).

Regulating Ascorbic Acid Metabolism

Ascorbic acid (AsA), also known as vitamin C, plays a crucial role as an antioxidant in plant stress responses. Its ability to act as an electron donor allows it to directly scavenge ROS, thereby mitigating the damage caused by environmental stress (Smirnoff 2011). AsA is widely recognized as one of the most important antioxidants in plants, contributing to the defense against oxidative stress and the protection of cell membranes (Akram et al. 2017). Under stress conditions, plants often experience the overproduction of ROS. AsA acts as a potent antioxidant and functions as a scavenger of ROS. It helps neutralize ROS and protects plant cells from oxidative damage caused by stress. AsA is a universal nonenzymatic antioxidant in plants. It works in conjunction with other antioxidants to maintain the redox balance within plant cells. This helps prevent oxidative stress and maintains the overall health and vitality of plants (Smirnoff 2011).

The exogenous application of AsA has been shown to enhance stress tolerance in plants. It can improve plant growth, photosynthetic efficiency, transpiration rate, and the activity of antioxidant enzymes. In addition, AsA protects lipids and proteins from oxidative damage caused by various abiotic stresses such as salinity and drought (Athar and Ashraf 2006; Latif et al. 2016).

According to previous reports, the L-galactose pathway has been identified as the primary pathway for AsA biosynthesis in plants (Wang et al. 2013). Several enzymes, including GDP-D-mannose pyrophosphorylase (GMP), GDP-L-galactose phosphorylase (GGP), and L-galactono-1,4-lactone dehydrogenase (GalLDH), have been identified as key factors involved in AsA synthesis. These enzymes play crucial roles in catalyzing the sequential reactions leading to the biosynthesis of AsA (Hasanuzzaman et al. 2019; Liao et al. 2023).

It was observed that treatment with GABA resulted in the upregulation of gene expression in the L-galactose pathway, including genes such as AcPGIs, AcPMM, and AcGGP2. Furthermore, the transcript levels of three transcription factors, namely, bHLH1, bHLH4, and HZ1, were found to be consistent with the expression patterns of key genes in the L-galactose pathway. Interestingly, GABA treatment also led to an upregulation of the transcriptional levels of these three transcription factors. These findings suggest that the enhancement of chilling tolerance in kiwifruit induced by GABA treatment may be associated with the regulation of AsA content at the transcriptional level. The upregulation of genes and transcription factors involved in the L-galactose pathway indicates that GABA might modulate AsA biosynthesis by influencing gene expression, thereby contributing to improved chilling tolerance in kiwifruit (Liu et al. 2023).

Promoting Polyamine Synthesis

PAs, such as putrescine, spermidine, and spermine, are involved in various biological processes and are present in higher plants. Under cold stress conditions, plants employ a defense mechanism by producing and accumulating compatible solutes, including PAs. PAs play a role in mediating low-temperature tolerance by contributing to the production and accumulation of compatible solutes (Shao et al. 2022). The alteration in PAs during the initial phase of stress is recognized as a signaling event that initiates subsequent signal transduction processes. This includes the activation of transcription factors and stress-responsive pathways, such as the ICE-CBF-COR transcriptional cascade, as well as mechanisms involved in ROS scavenging and the antioxidant defense system (Wei et al. 2021; Hwarari et al. 2022).

PAs play a significant role in plant responses to abiotic stresses, including low-temperature stress, and their association with cold tolerance has been well established. A growing body of evidence indicates that the application of exogenous PAs can enhance tolerance to low temperatures in various plant species. For instance, studies have demonstrated that priming with spermidine in rice enhances PA metabolism, thereby improving tolerance to chilling stress (Sheteiwy et al. 2017). In addition, the spray application of putrescine has been shown to effectively reduce chilling injury in stored peach fruit, irrespective of the applied dose or timing of application. Conversely, the foliar spraying of winter oilseed rape seedlings with spermidine, spermine, and putrescine is believed to induce a defensive response that enhances cold resistance (Jankovska-Bortkevič 2020).

GABA and PAs are interconnected in the plant metabolic network, and they can influence each other's levels and functions. For example, GABA can stimulate the activity of arginine decarboxylase (ADC) and ornithine decarboxylase (ODC), leading to increased PA synthesis. Conversely, PAs can inhibit the activity of glutamate decarboxylase (GAD), leading to decreased GABA synthesis. Moreover, GABA and PAs can share common intermediates and cofactors, such as pyridoxal phosphate (PLP) and *S*-adenosylmethionine (SAM). The interaction between GABA and PAs can modulate the plant response to cold stress in several ways. First, GABA and PAs can act as osmolytes that help maintain the osmotic balance and water potential of plant cells under cold stress. This can prevent membrane dehydration and shrinkage that can compromise membrane fluidity. Second, GABA and PAs can function as antioxidants that scavenge ROS that are generated under cold stress. ROS can cause oxidative damage to membrane lipids, proteins, and nucleic acids, leading to membrane peroxidation and disruption. Third, GABA and PAs can act as signaling molecules that regulate the expression of genes involved in cold tolerance, such as CBFs and DREB proteins. These transcription factors control the expression of downstream cold-responsive genes that encode proteins involved in various processes, such as Osmo protection, detoxification, antioxidative defense, membrane stabilization, and repair mechanisms (Podlešáková et al. 2019; Xu et al. 2022).

GABA can promote the synthesis of PAs by increasing the expression of enzymes involved in PAs biosynthesis. PAs are organic compounds that help protect plant cells under stress conditions. GABA treatment increases the expression of key enzymes involved in PAs synthesis, such as arginine decarboxylase (ADC) and ornithine decarboxylase (ODC), leading to increased PAs content in plants.

Protecting Chloroplast Integrity

The structure of mesophyll cells, including chloroplasts, is crucial for the photosynthesis process and has a significant impact on the plant's photosynthetic assimilation capacity. Under stress conditions, the morphology and ultrastructure of chloroplasts are disrupted, leading to a decrease in photosynthetic efficiency, dry matter accumulation, and crop yield loss. Chloroplasts in plants subjected to abiotic stress are known to be primary sites for generating ROS, which can contribute to damage. Chilling stress affects chloroplast ultrastructure, including abnormal shape, disintegration or disappearance of the chloroplast envelope, thylakoid dilation, and accumulation of lipid droplets (Liu et al. 2018). Chilling stress induces significant ultrastructural damage, particularly to chloroplasts, which are considered the most severely affected organelles. The observable symptoms include chloroplast swelling and disorganization, as well as thylakoid dilation, ultimately resulting in an increased number of plastoglobules (Karim et al. 2014) (Zbierzak et al. 2013). Chilling stress manifests in various symptoms, such as disorganization of the grana, alterations in the thylakoid and chloroplast membrane structures, accompanied by the accumulation of lipid droplets and a darkening of the stroma. These structural and physiological changes are closely associated with the buildup of ROS within the chloroplast, which is recognized as the primary site for ROS generation under unfavorable stress conditions. The increased ROS levels contribute to the impairment of the photosynthetic apparatus, leading to further damage (Partelli et al. 2011). During stress conditions, the morphology and ultrastructure of chloroplasts undergo direct disturbances, resulting in a substantial reduction in photosynthetic efficiency. This, in turn, leads to the accumulation of dry matter and ultimately results in a loss of crop yield. The compromised state of chloroplasts hampers their ability to carry out photosynthesis effectively, which is crucial for the synthesis of carbohydrates and other essential compounds necessary for plant growth and development. As a consequence, the overall productivity and yield of crops are negatively impacted under stress conditions (Shao et al. 2014). Previous studies have provided confirmation that chloroplasts in plants experiencing abiotic stress serve as the primary sites for the generation of ROS. Abiotic stress conditions, such as high light intensity, drought, salinity, or extreme temperatures, can disrupt the balance between ROS production and scavenging mechanisms within chloroplasts. As a result, the excessive generation of ROS within chloroplasts can cause oxidative damage to various cellular components, including lipids, proteins, and nucleic acids. This oxidative stress can have detrimental effects on plant physiology and overall growth. Hence, understanding the role of chloroplasts as the main contributors to ROS production under abiotic stress is crucial for comprehending the mechanisms by which plants respond and adapt to adverse environmental conditions (Gill and Tuteja 2010). Therefore, the exogenous application of GABA has been shown to enable plants to enhance their photosynthetic capacity under stress conditions. This effect is achieved by GABA's ability to modulate the level of ROS within the plant. GABA treatment can alter the activities of antioxidant enzymes, such as superoxide dismutase (SOD), catalase (CAT), and peroxidase (POD), which are responsible for scavenging ROS and reducing oxidative damage.

By enhancing the activity of these antioxidant enzymes, GABA helps to maintain the balance between ROS production and scavenging, thereby reducing oxidative stress and preserving the integrity of cellular membranes. This, in turn, allows for improved photosynthetic efficiency and overall plant performance under stress conditions (Aghdam et al. 2015). The application of GABA has been found to improve the photosynthesis process by alleviating chlorophyll degradation. GABA application has been shown to effectively reduce chilling injury in plants by offering protection to thylakoid membranes and chloroplast envelopes. This protection helps to maintain the structural integrity of these components and prevent swelling, which is a common symptom of chilling stress. By mitigating these detrimental effects of chilling stress, GABA application contributes to the preservation of the photosynthetic apparatus and enhances photosynthetic efficiency in plants. (Zbierzak et al. 2013; Karim et al. 2014). When GABA is applied externally, it helps maintain the stability of membranes by reducing the formation of lipid peroxidation products. This is significant because it prevents damage to the membranes. In addition, it leads to a notable increase in the activities of important antioxidant enzymes such as SOD, CAT, APX, and PAL. These enzymes play a vital role in creating a defense system against chilling stress and protecting the membranes from damage. This enhancement of antioxidant enzyme activities is crucial for mitigating the harmful effects of chilling stress and ensuring the overall health and performance of plants in adverse conditions (Abd Elbar et al. 2021). Exogenous application of GABA can protect the integrity of chloroplasts in plants under chilling stress. It prevents the degradation of chlorophyll and carotenoids, maintains the structure and function of thylakoid membranes (which are essential for photosynthesis), and improves the efficiency of photosynthesis. This protection helps prevent the loss of photosynthetic capacity and the disruption of pigment composition in plants. In addition, GABA treatment in chilling-stressed tomato leaves leads to changes in nonreducing sugar levels and alters the shape of starch grains within the chloroplasts (Abd Elbar et al. 2021). Examination of the chloroplasts using transmission electron microscopy reveals that chilling stress causes clustered accumulation of swollen and round chloroplasts with indistinct thylakoid membranes. These chloroplasts are filled with large-sized spherical starch grains. However, GABA pretreatment normalizes the structure of the chloroplasts, and the starch grains appear to be ellipsoidal in shape (Abd Elbar et al. 2021). GABA application can reduce chilling injury by protecting thylakoid membranes and chloroplast envelopes and reducing swelling. This protective effect helps improve the photosynthesis process by alleviating chlorophyll degradation. GABA's action includes preserving the integrity of thylakoid membranes and chloroplast envelopes, which in turn helps maintain chlorophyll content and photosynthetic activities. Overall, GABA treatment can protect chloroplast integrity under chilling stress. It prevents chlorophyll degradation, maintains thylakoid membrane structure, and improves photosynthetic efficiency. GABA's protective effects on chloroplast structure and function contribute to the overall preservation of photosynthetic capacity and pigment composition in plants exposed to chilling stress (Xiang et al. 2016).

Maintaining Higher ATP Content and Energy Charge

Adenosine triphosphate (ATP) is the main source of energy to maintain cell metabolism. On the other hand, it has been reported that membrane damage could be a consequence of

ATP depletion, as evidenced by increased electrolyte leakage under stress. Studies showed that the shortage of ATP may cause fruit browning and chilling symptoms, while the high level of ATP can alleviate the chilling in horticultural crops (Liu et al. 2019a). It was found that exogenous GABA could reduce the chilling symptoms of fruit and pericarp and improve the cold tolerance of fruit (Wang et al. 2014; Malekzadeh et al. 2017). Zhao et al. (2019) studied the cold tolerance of nectarine fruit stored at different low temperatures and found that it was closely related to energy metabolism. In addition, exogenous GABA alleviates the chilling of peach fruit by regulating energy metabolism and GABA metabolism (Shang et al. 2011; Yang et al. 2011b). There may be some relationship between energy metabolism and GABA metabolism in reducing the chilling of fruit.

GABA has been recognized as an important signaling molecule and regulator of stress responses in plants. Studies have shown that GABA plays a crucial role in maintaining cellular energy homeostasis, particularly by influencing ATP content and energy charge under cold and chilling stress conditions (Yang et al. 2011a).

Cold and chilling stress disrupt cellular metabolism and energy production, leading to a decline in ATP levels, which is a critical energy currency in cells (Lyons and Breidenbach 1990). However, exogenous application or endogenous accumulation of GABA has been found to mitigate the adverse effects of chilling stress on ATP content and energy charge in plants.

GABA acts as a modulator of energy metabolism by regulating key enzymes involved in ATP synthesis, such as ATP synthase, and enzymes involved in glycolysis and the TCA cycle. GABA enhances the activity of ATP synthase, the enzyme responsible for ATP synthesis in mitochondria, thereby promoting ATP production and maintaining higher ATP content under chilling stress conditions (Aghdam et al. 2016).

Moreover, GABA has been shown to influence glycolytic and TCA cycle enzymes, leading to increased substrate availability and enhanced energy production. GABA treatment or GABA accumulation in plants under chilling stress has been associated with upregulated expression and activity of enzymes involved in glycolysis, such as phosphofructokinase and pyruvate kinase, and TCA cycle enzymes, such as isocitrate dehydrogenase and succinate dehydrogenase. These metabolic adjustments facilitated by GABA contribute to the maintenance of higher ATP levels and energy charges in plants exposed to chilling stress. The maintenance of higher ATP content and energy charge through GABA-mediated regulation is crucial for sustaining essential cellular processes and promoting stress tolerance (Palma et al. 2019). ATP serves as a source of energy for cellular activities, including ion transport, protein synthesis, and signal transduction. By preserving ATP levels, GABA helps to sustain these vital processes and prevent energy depletion, which could otherwise lead to cellular dysfunction and damage under chilling stress conditions (Aghdam et al. 2016).

In summary, the application of GABA has emerged as a promising approach to mitigate the detrimental effects of chilling stress on plant energy metabolism. By enhancing ATP synthesis and maintaining higher ATP content and energy charge, GABA contributes to the overall stress tolerance of plants exposed to cold and chilling stress. Further research is needed to elucidate the precise molecular mechanisms underlying GABA-mediated regulation of energy metabolism and to explore its potential application in improving crop resilience to chilling stress.

Future Perspectives, Challenges, and Conclusion

Future Perspectives

Future research should focus on unraveling the complex signaling pathways involved in GABA responses in plants. This includes identifying the components responsible for perceiving GABA levels, understanding the downstream components of GABA signal transduction pathways, and investigating the interactions between GABA and other signaling molecules, such as nitric oxide and hydrogen. By gaining a better understanding of these pathways, researchers can uncover the precise mechanisms by which GABA influences plant growth and stress responses. Exploring the interactions between GABA and other key metabolites or signaling molecules will provide further insights into the role of GABA in plant stress tolerance. For example, understanding how GABA interacts with PAs and other metabolites during low-temperature stress can help in developing strategies to enhance cold tolerance in plants. Research should also investigate the crosstalk between GABA and hormones, such as ABA, to determine how these molecules cooperate or antagonize each other in stress responses. Studying GABA in various plant species, including crop plants and horticultural plants, is essential to gain a comprehensive understanding of its roles and potential applications. Different plant species may have distinct GABA signaling pathways and responses to stress. Comparative studies will help identify conserved mechanisms and species-specific adaptations, leading to the development of tailored strategies for enhancing stress tolerance and productivity in different crops.

Challenges

Deciphering the complex regulatory networks associated with GABA metabolism and signaling in plants is a significant challenge. GABA is involved in multiple physiological processes, and its interactions with other molecules and pathways are intricate. Untangling these networks and understanding the precise roles of GABA in different stress responses will require sophisticated experimental approaches and integrative systems biology approaches. Studying the dynamics of key metabolites, including GABA, during stress responses remains a challenge. Metabolite levels can fluctuate rapidly in response to environmental cues, making it difficult to capture their precise dynamics. Advanced metabolomics technologies and high-resolution sampling methods will be crucial for monitoring metabolite changes with high temporal and spatial resolution. While the protective effects of exogenous GABA application in stress tolerance have shown promise, translating these findings into practical applications for horticulture and crop production is a challenge. Further research is needed to optimize the application methods, dosage, and timing of GABA treatments to maximize their benefits. In addition, the feasibility, cost-effectiveness, and scalability of GABA-based interventions in agricultural practices need to be evaluated.

Conclusion

Recent research on GABA in plants has revealed its pivotal roles in stress tolerance and growth regulation. While significant progress has been made, there remains a need to delve deeper into the underlying molecular mechanisms and the intricate interactions between GABA and other components within plant signaling networks. Future research should prioritize untangling these complex signaling pathways, exploring GABA's interactions with other molecules, and conducting comparative studies across diverse plant species. Overcoming the challenges posed by intricate regulatory networks and metabolite dynamics is paramount for advancing our comprehension of GABA's significance in plants. Ultimately, this knowledge can be translated into practical applications aimed at boosting stress tolerance and enhancing crop productivity in agricultural systems.

References

Abd Elbar, O.H., Elkelish, A., Niedbała, G. et al. (2021). Protective effect of γ-aminobutyric acid against chilling stress during reproductive stage in tomato plants through modulation of sugar metabolism, chloroplast integrity, and antioxidative defense systems. *Frontiers in Plant Science* 12: 663750.

Aghaee, A., Moradi, F., Zare-Maivan, H. et al. (2011). Physiological responses of two rice (*Oryza sativa* L.) genotypes to chilling stress at seedling stage. *African Journal of Biotechnology* 10: 7617–7621.

Aghdam, M.S., Sevillano, L., Flores, F.B., and Bodbodak, S. (2013). Heat shock proteins as biochemical markers for postharvest chilling stress in fruits and vegetables. *Scientia Horticulturae* 160: 54–64.

Aghdam, M.S., Naderi, R., Sarcheshmeh, M.A.A., and Babalar, M. (2015). Amelioration of postharvest chilling injury in anthurium cut flowers by γ-aminobutyric acid (GABA) treatments. *Postharvest Biology and Technology* 110: 70–76.

Aghdam, M.S., Naderi, R., Jannatizadeh, A. et al. (2016). Impact of exogenous GABA treatments on endogenous GABA metabolism in anthurium cut flowers in response to postharvest chilling temperature. *Plant Physiology and Biochemistry* 106: 11–15. https://doi.org/10.1016/j.plaphy.2016.04.045.

Ahmad, P. and Prasad, M.N.V. (2011). *Environmental Adaptations and Stress Tolerance of Plants in the Era of Climate Change*. New York: Springer Science & Business Media.

Akram, N.A., Shafiq, F., and Ashraf, M. (2017). AsA – a potential oxidant scavenger and its role in plant development and abiotic stress tolerance. *Frontiers in Plant Science* 8: 613.

Ansari, M.I., Jalil, S.U., Ansari, S.A., and Hasanuzzaman, M. (2021). GABA shunt: a key-player in mitigation of ROS during stress. *Plant Growth Regulation* 94: 131–149.

Ashraf, M.A. and Rahman, A. (2018). Hormonal regulation of cold stress response. In: *Cold Tolerance in Plants: Physiological, Molecular and Genetic Perspectives* (ed. S.H. Wani and V. Herath), 65–88. Cham, Switzerland: Springer International Publishing.

Aslam, M., Fakher, B., Ashraf, M.A. et al. (2022). Plant low-temperature stress: signaling and response. *Agronomy* 12: 702.

Athar, R. and Ashraf, M. (2006). Interactive effect of foliarly applied ascorbic acid and salt stress on wheat (*Triticum aestivum* L.) at the seedling stage. *Pakistan Journal of Botany* 38: 1407–1414.

Bao, H., Chen, X., Lv, S. et al. (2015). Virus-induced gene silencing reveals control of reactive oxygen species accumulation and salt tolerance in tomato by γ-aminobutyric acid metabolic pathway. *Plant, Cell & Environment* 38: 600–613.

Bhattacharya, A. (2022). Lipid metabolism in plants under low-temperature stress: a review. In: *Physiological Processes in Plants Under Low Temperature Stress* (ed. A. Bhattacharya), 409–516. Singapore: Springer.

Buchanan, B.B., Gruissem, W., and Jones, R.L. (2015). *Biochemistry and Molecular Biology of Plants*. Oxford: John Wiley & Sons.

Cheng, B., Li, Z., Liang, L. et al. (2018). The γ-aminobutyric acid (GABA) alleviates salt stress damage during seeds germination of white clover associated with Na^+/K^+ transportation, dehydrins accumulation, and stress-related genes expression in white clover. *International Journal of Molecular Sciences* 19: 2520.

Che-Othman, M.H., Jacoby, R.P., Millar, A.H., and Taylor, N.L. (2020). Wheat mitochondrial respiration shifts from the tricarboxylic acid cycle to the GABA shunt under salt stress. *New Phytologist* 225: 1166–1180.

Chinnusamy, V., Zhu, J.-K., and Sunkar, R. (2010). Gene regulation during cold stress acclimation in plants. In: *Plant Stress Tolerance: Methods and Protocols* (ed. R. Sunkar), 39–55. Totowa, NJ: Humana Press.

Cramer, G.R., Urano, K., Delrot, S. et al. (2011). Effects of abiotic stress on plants: a systems biology perspective. *BMC Plant Biology* 11: 1–14.

Ding, Y., Shi, Y., and Yang, S. (2019). Advances and challenges in uncovering cold tolerance regulatory mechanisms in plants. *New Phytologist* 222: 1690–1704.

El-Dakak, R.A., Badr, R.H., Zeineldein, M.H. et al. (2023). Effect of chilling and salinity stress on photosynthetic performance and ultrastructure of chloroplast in faba beans (*Vicia faba* L.) leaves. *Rendiconti Lincei. Scienze Fisiche e Naturali* 34: 447–456. https://doi.org/10.1007/s12210-022-01131-3.

Forlani, G., Trovato, M., Funck, D., and Signorelli, S. (2019). Regulation of proline accumulation and its molecular and physiological functions in stress defence. In: *Osmoprotectant-Mediated Abiotic Stress Tolerance in Plants: Recent Advances and Future Perspectives* (ed. M.A. Hossain, V. Kumar, D.J. Burritt, et al.), 73–97. Cham, Switzerland: Springer International Publishing https://doi.org/10.1007/978-3-030-27423-8_3.

Gill, S.S. and Tuteja, N. (2010). Reactive oxygen species and antioxidant machinery in abiotic stress tolerance in crop plants. *Plant Physiology and Biochemistry* 48: 909–930.

Guo, X., Liu, D., and Chong, K. (2018). Cold signaling in plants: insights into mechanisms and regulation. *Journal of Integrative Plant Biology* 60: 745–756.

Guo, Z., Gong, J., Luo, S. et al. (2023). Role of gamma-aminobutyric acid in plant defense response. *Metabolites* https://doi.org/10.3390/metabo13060741.

Hasanuzzaman, M., Bhuyan, M.H.M.B., Anee, T.I. et al. (2019). Regulation of ascorbate-glutathione pathway in mitigating oxidative damage in plants under abiotic stress. *Antioxidants* https://doi.org/10.3390/antiox8090384.

Hasanuzzaman, M., Bhuyan, M.H.M.B., Parvin, K. et al. (2020). Regulation of ROS metabolism in plants under environmental stress: a review of recent experimental evidence. *International Journal of Molecular Sciences* 21: 8695.

Hijaz, F. and Killiny, N. (2019). Exogenous GABA is quickly metabolized to succinic acid and fed into the plant TCA cycle. *Plant Signaling & Behavior* 14 (3): e1573096. https://doi.org/10.1080/15592324.2019.1573096.

Hijaz, F., Nehela, Y., and Killiny, N. (2018). Application of gamma-aminobutyric acid increased the level of phytohormones in *Citrus sinensis*. *Planta* 248: 909–918.

Huang, H., Ullah, F., Zhou, D.-X. et al. (2019). Mechanisms of ROS regulation of plant development and stress responses. *Frontiers in Plant Science* 10: 800.

Hussain, H.A., Hussain, S., Khaliq, A. et al. (2018). Chilling and drought stresses in crop plants: implications, cross talk, and potential management opportunities. *Frontiers in Plant Science* https://doi.org/10.3389/fpls.2018.00393.

Hwarari, D., Guan, Y., Ahmad, B. et al. (2022). ICE-CBF-COR signaling cascade and its regulation in plants responding to cold stress. *International Journal of Molecular Sciences* https://doi.org/10.3390/ijms23031549.

Jankovska-Bortkevič, E. (2020). Winter oilseed rape cold resistance characteristics and their alteration after treatment with polyamines and proline. Doctoral dissertation. Nature Research Centre.

Janse van Rensburg, H.C. and Van den Ende, W. (2020). Priming with γ-aminobutyric acid against *Botrytis cinerea* reshuffles metabolism and reactive oxygen species: dissecting signalling and metabolism. *Antioxidants* 9: 1174.

Janská, A., Maršík, P., Zelenková, S., and Ovesná, J. (2010). Cold stress and acclimation – what is important for metabolic adjustment? *Plant Biology* 12: 395–405.

Jia, Y., Zou, D., Wang, J. et al. (2016). Effects of γ-aminobutyric acid, glutamic acid, and calcium chloride on rice (*Oryza sativa* L.) under cold stress during the early vegetative stage. *Journal of Plant Growth Regulation* 36 (1): 240–253. https://doi.org/10.1007/s00344-016-9634-x.

Karim, S., Alezzawi, M., Garcia-Petit, C. et al. (2014). A novel chloroplast localized Rab GTPase protein CPRabA5e is involved in stress, development, thylakoid biogenesis and vesicle transport in *Arabidopsis*. *Plant Molecular Biology* 84: 675–692.

Kaspal, M., Kanapaddalagamage, M.H., and Ramesh, S.A. (2021). Emerging roles of γ aminobutyric acid (GABA) gated channels in plant stress tolerance. *Plants* 10: 2178.

Khan, M.I.R., Jalil, S.U., Chopra, P. et al. (2021). Role of GABA in plant growth, development and senescence. *Plant Gene* 26: 100283.

Khedr, E.H. and Khedr, N. (2023a). Optimization of postharvest progesterone treatment to alleviate chilling injury in mango fruit, maintaining intracellular energy, cell wall stability, and antioxidant activity. *Postharvest Biology and Technology* 206: 112572. https://doi.org/10.1016/j.postharvbio.2023.112572.

Khedr, E.H. and Khedr, N. (2023b). Unraveling the mechanisms behind salt stress alleviation and enhanced fruit storability in pomegranate through triacontanol treatment. *Scientia Horticulturae* 112634. https://doi.org/10.1016/j.scienta.2023.112634.

Kohli, S.K., Khanna, K., Bhardwaj, R. et al. (2019). Assessment of subcellular ROS and NO metabolism in higher plants: multifunctional signaling molecules. *Antioxidants* 8 (12): 641.

Latif, M., Akram, N.A., and Ashraf, M. (2016). Regulation of some biochemical attributes in drought-stressed cauliflower (*Brassica oleracea* L.) by seed pre-treatment with ascorbic acid. *The Journal of Horticultural Science and Biotechnology* 91: 129–137.

Li, W., Liu, J., Ashraf, U. et al. (2016). Exogenous γ-aminobutyric acid (GABA) application improved early growth, net photosynthesis, and associated physio-biochemical events in maize. *Frontiers in Plant Science* 7: https://doi.org/10.3389/fpls.2016.00919.

Li, Y., Fan, Y., Ma, Y. et al. (2017). Effects of exogenous γ-aminobutyric acid (GABA) on photosynthesis and antioxidant system in pepper (*Capsicum annuum* L.) seedlings under low light stress. *Journal of Plant Growth Regulation* 36: 436–449.

Li, L., Dou, N., Zhang, H., and Wu, C. (2021). The versatile GABA in plants. *Plant Signaling & Behavior* 16: 1862565. https://doi.org/10.1080/15592324.2020.1862565.

Liang, S., Kuang, J., Ji, S. et al. (2020). The membrane lipid metabolism in horticultural products suffering chilling injury. *Food Quality and Safety* 4: 9–14.

Liao, G., Xu, Q., Allan, A.C., and Xu, X. (2023). l-Ascorbic acid metabolism and regulation in fruit crops. *Plant Physiology* 192: 1684.

Liu, X., Zhou, Y., Xiao, J., and Bao, F. (2018). Effects of chilling on the structure, function and development of chloroplasts. *Frontiers in Plant Science* https://doi.org/10.3389/fpls.2018.01715.

Liu, J., Li, F., Li, T. et al. (2019a). Fibroin treatment inhibits chilling injury of banana fruit via energy regulation. *Scientia Horticulturae* 248: 8–13. https://doi.org/10.1016/j.scienta.2018.12.052.

Liu, X., Xu, N., Wu, Y. et al. (2019b). Photosynthesis, chilling acclimation and the response of antioxidant enzymes to chilling stress in mulberry seedlings. *Journal of Forestry Research* 30: 2021–2029.

Liu, Q., Li, X., Jin, S. et al. (2023). γ-Aminobutyric acid treatment induced chilling tolerance in postharvest kiwifruit (*Actinidia chinensis* cv. Hongyang) via regulating ascorbic acid metabolism. *Food Chemistry* 404: 134661.

Lyons, J.M. and Breidenbach, R.W. (1990). Relation of chilling stress to respiration. *Chilling Injury of Horticultural Crops* 14: 223–234.

Malekzadeh, P., Khara, J., and Heydari, R. (2014). Alleviating effects of exogenous gamma-aminobutiric acid on tomato seedling under chilling stress. *Physiology and Molecular Biology of Plants* 20: 133–137.

Malekzadeh, P., Khosravi-Nejad, F., Hatamnia, A.A., and Sheikhakbari Mehr, R. (2017). Impact of postharvest exogenous γ-aminobutyric acid treatment on cucumber fruit in response to chilling tolerance. *Physiology and Molecular Biology of Plants* 23: 827–836. https://doi.org/10.1007/s12298-017-0475-2.

Manasa, S.L., Panigrahy, M., Panigrahi, K.C.S., and Rout, G.R. (2022). Overview of cold stress regulation in plants. *The Botanical Review* 88: 359–387. https://doi.org/10.1007/s12229-021-09267-x.

Mazzucotelli, E., Tartari, A., Cattivelli, L., and Forlani, G. (2006). Metabolism of γ-aminobutyric acid during cold acclimation and freezing and its relationship to frost tolerance in barley and wheat. *Journal of Experimental Botany* 57: 3755–3766.

Palma, F., Carvajal, F., Jiménez-Muñoz, R. et al. (2019). Exogenous γ-aminobutyric acid treatment improves the cold tolerance of zucchini fruit during postharvest storage. *Plant Physiology and Biochemistry* 136: 188–195.

Partelli, F.L., Batista-Santos, P., Scotti-Campos, P. et al. (2011). Characterization of the main lipid components of chloroplast membranes and cold induced changes in *Coffea* spp. *Environmental and Experimental Botany* 74: 194–204. https://doi.org/10.1016/j.envexpbot.2011.06.001.

Phadtare, S., Alsina, J., and Inouye, M. (1999). Cold-shock response and cold-shock proteins. *Current Opinion in Microbiology* 2: 175–180.

Podlešáková, K., Ugena, L., Spíchal, L. et al. (2019). Phytohormones and polyamines regulate plant stress responses by altering GABA pathway. *New Biotechnology* 48: 53–65.

Ramesh, S.A., Tyerman, S.D., Gilliham, M., and Xu, B. (2017). γ-Aminobutyric acid (GABA) signalling in plants. *Cellular and Molecular Life Sciences* 74: 1577–1603. https://doi.org/10.1007/s00018-016-2415-7.

Ramos-Ruiz, R., Martinez, F., and Knauf-Beiter, G. (2019). The effects of GABA in plants. *Cogent Food & Agriculture* 5: 1670553. https://doi.org/10.1080/23311932.2019.1670553.

Rao, N.K. (2004). Plant genetic resources: advancing conservation and use through biotechnology. *African Journal of Biotechnology* 3: 136–145.

Ritonga, F.N. and Chen, S. (2020). Physiological and molecular mechanism involved in cold stress tolerance in plants. *Plants* https://doi.org/10.3390/plants9050560.

Ruelland, E. and Collin, S. (2012). Chilling stress. In: *Plant Stress Physiology* (ed. S. Shabala), 94–117. Wallingford: CABI.

Sachdev, S., Ansari, S.A., Ansari, M.I. et al. (2021). Abiotic stress and reactive oxygen species: generation, signaling, and defense mechanisms. *Antioxidants* 10: 277.

Santos-Sánchez, N.F., Salas-Coronado, R., Villanueva-Cañongo, C., and Hernández-Carlos, B. (2019). Antioxidant compounds and their antioxidant mechanism. *Antioxidants* 10: 1–29.

Satyakam, Zinta, G., Singh, R.K., and Kumar, R. (2022). Cold adaptation strategies in plants – an emerging role of epigenetics and antifreeze proteins to engineer cold resilient plants. *Frontiers in Genetics* 13: https://doi.org/10.3389/fgene.2022.909007.

Shang, H., Cao, S., Yang, Z. et al. (2011). Effect of exogenous γ-aminobutyric acid treatment on proline accumulation and chilling injury in peach fruit after long-term cold storage. *Journal of Agricultural and Food Chemistry* 59: 1264.

Shao, Q., Wang, H., Guo, H. et al. (2014). Effects of shade treatments on photosynthetic characteristics, chloroplast ultrastructure, and physiology of *Anoectochilus roxburghii*. *PLoS One* 9: e85996.

Shao, J., Huang, K., Batool, M. et al. (2022). Versatile roles of polyamines in improving abiotic stress tolerance of plants. *Frontiers in Plant Science 13*: 1003155.

Sheteiwy, M., Shen, H., Xu, J. et al. (2017). Seed polyamines metabolism induced by seed priming with spermidine and 5-aminolevulinic acid for chilling tolerance improvement in rice (*Oryza sativa* L.) seedlings. *Environmental and Experimental Botany* 137: 58–72. https://doi.org/10.1016/j.envexpbot.2017.02.007.

Smirnoff, N. (2011). Vitamin C: the metabolism and functions of ascorbic acid in plants. In: *Advances in Botanical Research* (ed. F. Rebeille and R. Douce), 107–177. London: Elsevier.

Song, Y., Feng, L., Alyafei, M.A.M. et al. (2021). Function of chloroplasts in plant stress responses. *International Journal of Molecular Sciences* 22: 13464.

Srivastava, V., Mishra, S., Chowdhary, A.A. et al. (2021). The γ-aminobutyric acid (GABA) towards abiotic stress tolerance. In: *Compatible Solutes Engineering for Crop Plants Facing Climate Change* (ed. S.H. Wani, M.P. Gangola, and B.R. Ramadoss), 171–187. Cham, Switzerland: Springer International Publishing.

Suzuki, N. and Mittler, R. (2006). Reactive oxygen species and temperature stresses: a delicate balance between signaling and destruction. *Physiologia Plantarum* 126: 45–51.

Thomashow, M.F. (2010). Molecular basis of plant cold acclimation: insights gained from studying the CBF cold response pathway. *Plant Physiology* 154: 571–577. https://doi.org/10.1104/pp.110.161794.

Wang, J., Zhang, Z., and Huang, R. (2013). Regulation of ascorbic acid synthesis in plants. *Plant Signaling & Behavior* 8: e24536.

Wang, Y., Luo, Z., Huang, X. et al. (2014). Effect of exogenous γ-aminobutyric acid (GABA) treatment on chilling injury and antioxidant capacity in banana peel. *Scientia Horticulturae* 168: 132–137. https://doi.org/10.1016/j.scienta.2014.01.022.

Wei, X., Liu, S., Sun, C. et al. (2021). Convergence and divergence: signal perception and transduction mechanisms of cold stress in *Arabidopsis* and rice. *Plants* https://doi.org/10.3390/plants10091864.

Xiang, L., Hu, L., Xu, W. et al. (2016). Exogenous γ-aminobutyric acid improves the structure and function of photosystem II in muskmelon seedlings exposed to salinity-alkalinity stress. *PLoS One* 11: e0164847.

Xie, T., Ji, J., Chen, W. et al. (2019). γ-Aminobutyric acid is closely associated with accumulation of flavonoids. *Plant Signaling & Behavior* 14: 1604015.

Xu, M., Yang, Q., Bai, G. et al. (2022). Polyamine pathways interconnect with GABA metabolic processes to mediate the low-temperature response in plants. *Frontiers in Plant Science* 13: 1035414.

Yang, A., Cao, S., Yang, Z. et al. (2011a). γ-Aminobutyric acid treatment reduces chilling injury and activates the defence response of peach fruit. *Food Chemistry* 129: 1619–1622. https://doi.org/10.1016/j.foodchem.2011.06.018.

Yuan, P., Yang, T., and Poovaiah, B.W. (2018). Calcium signaling-mediated plant response to cold stress. *International Journal of Molecular Sciences* 19: 3896.

Yuan, D., Wu, X., Gong, B. et al. (2023). GABA metabolism, transport and their roles and mechanisms in the regulation of abiotic stress (hypoxia, salt, drought) resistance in plants. *Metabolites* https://doi.org/10.3390/metabo13030347.

Zbierzak, A.M., Porfirova, S., Griebel, T. et al. (2013). A TIR–NBS protein encoded by *Arabidopsis* chilling sensitive 1 (CHS 1) limits chloroplast damage and cell death at low temperature. *The Plant Journal* 75: 539–552.

Zhang, X. and Liu, C.-J. (2015). Multifaceted regulations of gateway enzyme phenylalanine ammonia-lyase in the biosynthesis of phenylpropanoids. *Molecular Plant* 8: 17–27.

Zhao, H., Jiao, W., Cui, K. et al. (2019). Near-freezing temperature storage enhances chilling tolerance in nectarine fruit through its regulation of soluble sugars and energy metabolism. *Food Chemistry* 289: 426–435. https://doi.org/10.1016/j.foodchem.2019.03.088.

Zhu, X., Liao, J., Xia, X. et al. (2019). Physiological and iTRAQ-based proteomic analyses reveal the function of exogenous γ-aminobutyric acid (GABA) in improving tea plant (*Camellia sinensis* L.) tolerance at cold temperature. *BMC Plant Biology* 19: 43. https://doi.org/10.1186/s12870-019-1646-9.

15

Role of GABA Under Bacterial Stress in Plants

Kuldeep Lahry[1], Akhilesh Kumar Pandey[2], and Sudhir Singh[3]

[1] *Institut Régional du Cancer de Montpellier (ICM), Université de Montpellier, Montpellier, 34095, France*
[2] *Department of Biotechnology, Faculty of Biosciences and Biotechnology, Invertis University, Bareilly, Uttar Pradesh, 243123, India*
[3] *Centre for Molecular Biology, Central University of Jammu, Bagla, Jammu and Kashmir, 181143, India*

Introduction

The rising impact of global climate change has introduced increased environmental complexities for plant ecosystems. Unlike mobile animals, plants, being sessile organisms, cannot relocate to more favorable growth conditions. Consequently, they must skillfully adapt to their surroundings, navigating through a spectrum of challenges, including both biotic and abiotic stresses such as drought, salt, and extreme temperatures, to ensure their survival and sustained growth (Saijo and Loo, 2020; Tarkowski et al. 2020; Wali et al. 2016). In response to these stressors, plants deploy various defense mechanisms. Central to these defense strategies are the primary metabolism, through which plants, under stress conditions, reduce photosynthesis and elevate respiration to efficiently mobilize carbon and nitrogen resources (Li et al. 2018). Additionally, the stress response in plants involves critical factors such as pH homeostasis, reactive oxygen species (ROS), and calcium ions (Ca^{2+}) acting as signaling molecules. Maintaining intracellular pH homeostasis, a fundamental cellular activity, plays a pivotal role in sustaining normal plant growth, regulating development, and orchestrating responses to environmental stress (Mittler et al. 2022; Yadav et al. 2022; Yuan et al. 2022). ROS, known for their oxidative properties, not only inflict cell damage that leads to alterations in plant morphology and structure but also function as regulators of defense-related gene transcription by interacting with transcription factors (Mittler et al. 2022; Singh et al. 2019). Furthermore, fluctuations in calcium ion (Ca^{+2}) concentration represent the initial phase in the plant stress response. Calcium signals activate calmodulin and other calmodulin receptors, subsequently modulating gene expression and initiating protein phosphorylation processes (Yadav et al. 2022; Yuan et al. 2022). These complex molecular events highlight the sophisticated strategies plants employ to navigate and respond to the dynamic challenges imposed by changing environmental conditions.

GABA in Plants: Biosynthesis, Plant Development, and Food Security, First Edition. Edited by Samiksha Singh, Durgesh Kumar Tripathi, and Vijay Pratap Singh.

Within the intricate framework of stress responses, γ-aminobutyric acid (GABA) assumes significance as a class of four-carbon nonprotein amino acids prevalent in a wide range of organisms, including animals, plants, fungi, and microorganisms (Ramos-Ruiz et al. 2018). It serves as a crucial component within the free amino acid pool and functions as a vital inhibitory neurotransmitter in animals, primarily accumulating in the central nervous system of mammals (Li et al. 2019a). GABA exists in unbound form and possesses an amino group on its γ-carbon rather than on α-carbon. GABA is zwitterionic at physiological pH values. It is highly soluble in water and structurally can attain multiple conformations in solution including a cyclic structure similar to proline (Christensen et al. 1994).

The discovery of GABA in plants traces back more than 70 years to its identification in potato tubers (*Solanum tuberosum*) (Steward et al. 1949). Since then, extensive research has delved into its physiological role (Adem et al. 2020; Bown and Shelp 1997, 2020; Bouché et al. 2003b; Fait et al. 2008; Fromm 2020; Gilliham and Tyerman, 2016; Kinnersley and Turano 2000; Li et al. 2021; Long et al. 2020; Ramesh et al. 2017; Shelp and Zarei 2017; Shelp et al. 1999, 2017). In the kingdom of plants, GABA is prevalent and documented to occur in various tissues and organs, with its role traditionally defined as a metabolite. However, over the past few decades, researches have shown a spectrum of GABA concentrations, ranging from micromolar (low) to millimolar (high) in diverse tissues, organs, and compartments (Binzel et al. 1987; Fougère et al. 1991; Handa et al. 1983; Kisaka et al. 1989; Narayan and Nair 1990; Pérez-Alfocea et al. 1994; Reggiani et al. 1988; Rhodes et al. 1986; Serraj et al. 1998; Stress' et al. 1986; Wallace2 et al. 1984). This variability suggests that GABA may function not only as a metabolite but also as a signaling molecule in plants, adding a new dimension to our understanding of its role in plant biology (Bouché and Fromm 2004; Bouché et al. 2003b; Fromm 2020; Li et al. 2021).

Endogenous GABA is primarily synthesized through a process called GABA shunt (Fait et al. 2008), bypassing two steps of the TCA cycle. This includes irreversible decarboxylation of glutamate (Glu) by glutamate decarboxylase (GAD) in the cytosol. Subsequently, GABA is transported to mitochondria by GABA permease (GABP) where it undergoes transamination to succinic semialdehyde (SSA) by GABA transaminase (GABA-T/POP2). SSA is further oxidized to succinate by NAD^+-dependent semialdehyde dehydrogenase (SSADH). Succinate enters into tricarboxylic acid (TCA) cycle to maintain C/N balance (Li et al. 2021). In succession, GABA can be synthesized from α-ketoglutarate by glutamate dehydrogenase (GDH) (Bouché and Fromm 2004). Additionally, GABA can be derived from polyamines, such as spermidine and putrescine, through the catalytic action of two enzymes – diamine oxidase (DO) and polyamine oxidase (PAO) (Shelp et al. 2012, 2021). GABA can also be synthesized by nonenzymatic oxidation of proline in oxidative stress condition (Signorelli et al. 2015). The catabolism of GABA leads directly to SSA, which, in subsequent steps, transforms into succinate or hydroxybutyrate (Shelp et al. 1999).

GABA plays a crucial role in responding to both biotic and abiotic stresses, contributing to the maintenance of carbon/nitrogen (C/N) balance, enhancement of plant photosynthesis, quenching of ROS, and the regulation of plant development. The roles for plant defense were attributed to the several primary metabolites including sucrose and glutamate (Seifi et al. 2013b; Tauzin and Giardina 2014). Among such metabolites, GABA stands out for its role in plant stress physiology (Bown and Shelp 2016; Mahmud et al. 2017; Wang et al. 2019a). GABA is actively involved in modulating tolerance to various abiotic stresses, including hypoxia (Brikis et al. 2018; Fan et al. 1997; Reggiani et al. 2000; Streeter and Thompson 1972; Yang et al. 2015), salinity (Akçay et al. 2012; Al-Quraan et al. 2013; Cheng et al. 2018; Che-Othman et al. 2020; Ji et al. 2020; Jin et al. 2019; Ma et al. 2019; Su et al. 2019; Wang et al. 2017; Wu et al. 2020b), low temperature (Aghdam et al. 2016; Malekzadeh et al. 2014; Mazzucotelli et al. 2006; Toubiana et al. 2020; Wang et al. 2014; Yang et al. 2016; Zhu et al. 2019), high temperature (Ayenew et al. 2015; Cherry et al. 2000; Li et al. 2016; Sweetman et al. 2014; Takahashi et al. 2013), heavy metal (Daş et al. 2016; Dubey et al. 2010; Kang et al. 2015; Kumar et al. 2017), ROS and drought (Bor et al. 2009; Filho et al. 2018; Li et al. 2019b; Pal et al. 2016; Raggi, 1994; Serraj et al. 1998; Thompson et al. 1966). For instance, GABA enhances resistance to salt stress by promoting the antioxidant defense system (Che-Othman et al. 2020; Jin et al. 2019; Wang et al. 2017) and augments drought stress resistance by regulating stomatal aperture (Mekonnen et al. 2016). Consequently, this regulatory role promotes plant growth, extends shelf-life and improves crop storage quality (Aghdam et al. 2022; Malekzadeh et al. 2017; Nazoori et al. 2020; Shang et al. 2011; Sheng et al. 2017; Wang et al. 2014). In muskmelon seedlings, GABA governs osmotic balance as well as influences the accumulation of H_2O_2 and the ascorbic acid-glutathione cycle for tolerance to saline–alkali stress (Jin et al. 2019).

Furthermore, studies indicate that exogenous GABA inhibits H_2O_2 production and regulates H_2O_2-producing enzyme gene expression in *Caragana intermedia* under salt stress (Shi et al. 2010).

The transmembrane transport of GABA within plants has become a focal point of research, exploring its movement from the apoplast to the cytoplasm and various organelles (Yuan et al. 2023). The signaling role of GABA gains further support from its accumulation in response to diverse abiotic and biotic stresses. Cell membranes with GABA binding sites, the presence of GABA-gated anion channels, and their regulation by agonists and antagonists of GABA receptors in animals underscore the evolving understanding of GABA's role in signaling and stress tolerance (Kaspal et al. 2021). The identification of potential GABA binding sites on ion channels implicated in plant stress emphasizes the need to reassess the multifaceted roles of GABA in signaling and stress response. The distinct functions of GABA in plants may be complexly interconnected and challenging to isolate from one another (Michaeli and Fromm 2015).

The detailed biochemical mode of action of GABA has been discussed (Michaeli et al. 2011; Shelp et al. 2017). The application of exogenous GABA has been shown to augment the activity of antioxidant enzymes and the glyoxalase system, crucial for detoxifying methylglyoxal (Abdel Razik et al. 2021). Additionally, supplementation with exogenous GABA has demonstrated the capacity to stimulate plant growth by elevating endogenous GABA levels, amino acids, and hormone concentrations through the upregulation of essential genes involved in phytohormones (Hijaz et al. 2018; Rossi et al. 2021; Salvatierra et al. 2016; Yong et al. 2017). These recent advancements are beginning to unveil the underlying mechanisms associated with GABA-induced resistance. GABA serves as a natural bioactive substance, playing a pivotal role in various physiological activities of plants, encompassing growth, development, signaling, and responses to stress (Li et al. 2021).

While the majority of studies on GABA in plants have traditionally focused on abiotic stress, there is a growing body of literature indicating the involvement of GABA metabolism and signaling in plant immunity against both pests and pathogens (Bown and

Shelp, 2016; Scholz et al. 2015; Seifi et al. 2013a; Seifikalhor et al. 2019; Tarkowski et al. 2019, 2020). This chapter discusses the involvement of GABA in biotic stress in plants focusing on bacterial infection, its molecular mechanisms with few examples discussed in detail.

GABA and Biotic Stress in Plants

The continuous threat posed by various pests and pathogens poses a persistent challenge to the growth and development of plants. A multitude of research supports the notion that plants respond to biotic stress by elevating GABA levels (Bown et al. 2006; Seifikalhor et al. 2019; Shelp et al. 2006, 2017; Tarkowski et al. 2020). For instance, rice cells exposed to a cell-wall elicitor from the rice blast fungus (*Magnaporthe grisea*) exhibited a significant increase in GABA levels (Takahashi et al. 2008). Similarly, increased GABA production was observed when *Jatropha curcas* L. stems infected with the Jatropha mosaic virus (JMV), tobacco leaves infiltrated with a hairpin elicitor, tomato leaves infected with fungus *Botrytis cinereal*, potato plants inoculated with potato virus Y, *Phaseolus vulgaris* leaves exposed to *P. syringae* pv. *phaseolicola* (Pph) 1302A, Arabidopsis infected with *Fusarium graminearum*, citrus fruits in response to the *Candidatus Liberibacter asiaticus* and its vector *Diaphorina citri* (Chen et al. 2018; Dimlioğlu et al. 2015; Kogovšek et al. 2016; Nehela and Killiny, 2019; O'Leary et al. 2016; Seifi et al. 2013a; Sidhu et al. 2010; Tarkowski et al. 2019; Zaini et al. 2018). GABA accumulation is involved in reducing the toxicity and biomass of *Lasiodiplodia theobromae* and its byproducts (Salvatore et al. 2020). Conversely, a comparative proteomics analysis revealed a substantial downregulation of GABA production in tomato stems inoculated with *Ralstonia solanacearum* isolates (Wang et al. 2018, 2019a). This finding aligns with transcriptome profiling, which indicates that the expression of GAD2 and SSADH1, two genes associated with GABA shunt, are downregulated following VIGS (Virus-Induced Gene Silencing) (Wang et al. 2018).

Moreover, GABA plays a role in plant defense mechanisms against insects that feed on plants (Bown et al. 2006; Huang et al. 2011; Mithöfer and Boland 2012). When plant tissues are wounded, cell disruption occurs due to insect herbivory. *Arabidopsis thaliana* mutants with reduced GABA levels (*gad*1/2 double mutant) and those enriched with GABA (*gad*1/2xpop2-5) exhibit rapid, systemic GABA production and accumulation independent of jasmonic acid (Scholz et al. 2015). On the contrary, *Clematis terniflora* D. C. responds to UVB rays and darkness by experiencing an excessive increase in jasmonic acid, resulting in a notable increase in GABA concentration (Yang et al. 2019). Given that GABA functions as an inhibitory neurotransmitter in the brain systems of invertebrates (Tarkowski et al. 2020), the enhanced levels of GABA in plants might serve as a defense mechanism against insect herbivores. Similarly, GABA contributes to plant defense mechanisms against harmful insects and shows the ability to confer resistance in treated crops against plant pathogenic fungi. Notably, GABA exhibits a measurable inhibitory impact on the larvae of *Choristoneura rosaceana* and *Spodoptera littoralis*, leading to delayed development and reduced growth rates. Similar outcomes were observed in GABA-enriched Arabidopsis mutants when infested by insects (Bown et al. 2002, 2006; Ramputh and Bown 1996; Scholz et al. 2015). In response to insect damage, localized increases in GABA were identified in tobacco and soybean leaves (Bown et al. 2002). The effectiveness of GABA against insects appears to stem from both its direct action and its role as a signaling molecule.

When exposed to *Botrytis cinerea*, *sitiens* tomato mutants exhibited a biphasic defensive response characterized by an early localized hypersensitivity response followed by the over-activation of the GABA shunt (Seifi et al. 2013a). Pre-inoculation administration of exogenous GABA increased antioxidant enzymes, such as catalase and peroxidase, rendering tomatoes and pears resistant to fungal infections (Fu et al. 2017; Yang et al. 2017; Yu et al. 2014) suggesting that GABA-induced resistance activates antioxidant enzymes, thereby restraining ROS induced plant cell death. Presently, there is no evidence indicating that organisms have developed tolerance or adaptation to GABA. Considering that exogenous and endogenous GABA undergo the same metabolic processes and naturally participate in plant metabolism, the development of resistance against insects or fungi is considered unlikely.

In response to stress, plants typically exhibit an immediate elevation in Ca^{2+} concentration resulting in accumulation of GABA as the Ca^{2+}/calmodulin complex activates GAD in the cytosol (Tian et al. 2019; Wang et al. 2019b; Wu et al. 2020a; Yu et al. 2019). Subsequently, GABA is either catabolized to uphold the C/N balance or serves to mitigate the accumulation of ROS by stimulating antioxidant enzymes in plants. Additionally, GABA may function as a signaling molecule, initiating the activation of other chemicals involved in plant stress responses. The well-documented phenomenon of GABA accumulation in plants under stress conditions underscores its role in enhancing the plant's ability to withstand such stressors. By establishing a connection between amino acid metabolism and TCA cycle, GABA plays a pivotal role in central C/N metabolism within the context of plant-microbe interactions. Accumulation of GABA by increased GAD activity facilitate its entry into TCA cycle sustain cell viability through the modification of C/N metabolism (Michaeli et al. 2011). Furthermore, GABA acts to restrict the levels of ROS that could lead to cell death through activation of antioxidant enzymes such as catalase, superoxide dismutase, and peroxidase (Bouché et al. 2003a; Fait et al. 2005; Ludewig et al. 2008).

GABA and Bacterial Stress Response in Plants

GABA is synthesized by numerous bacteria including *Pseudomonas aeruginosa*, lactic acid bacteria, and *Escherichia coli* (Han et al. 2008; Noe and Nickerson 1958; Richard and Foster 2003; Siragusa et al. 2007). GABA can act as a signal molecule among bacteria and to their host as they are known to release it. There are different binding sites for GABA in different bacteria, for instance, GabP permease in *E. coli*, *Bacillus subtilis*, *Corynebacterium glutamicum* (Brechtel and King 1998; Hu and King 1998; Zhao et al. 2012). In *Lactobacillus*, *Lactococcus lactis* and *E. coli*, GABA has been shown to provide resistance to acid stress and even for some species (e.g., *Shigella flexneri* and *Listeria monocytogenes*), it is essential to colonize the gastrointestinal tract (De Biase and Pennacchietti 2012; Richard and Foster 2004; Tramonti et al. 2006).

On the other hand, the infection of plants with bacteria can also trigger GABA accumulation in various plant species which is similar to the response to fungal infection at molecular level e.g., soybean (*Glycine max* L.), tomato (*Solanum lycopersicon* L.), and bean

(*Phaseolus vulgaris* L.) (Lang et al. 2016; O'Leary et al. 2016; Park et al. 2010; Wang et al. 2019a) (Table 15.1). Growing evidence suggests that GABA is involved in bacteria-plant signaling. In response to damage, plants increase GABA production, a process crucial for tumor development, when infected with *Agrobacterium tumefaciens*. GABA-mediated signaling has been found to influence the production of quorum-sensing molecules in *Pseudomonas aeruginosa*, while also reducing virulence in *A. tumefaciens* and *P. syringae* (Figure 15.1) (Chevrot et al. 2006; McCraw et al. 2016; Park et al. 2010). The complex interplay between bacterial pathogens and host plants through GABA-mediated signaling raises the intriguing possibility of unidentified mechanisms at play in microbial consortia like the human gut microbiome. An elevated endogenous GABA levels in plant tissues on bacterial pathogens often associated with the activation of defense systems suggesting that GABA

Table 15.1 Bacterial species that can induce alterations in host GABA metabolism.

Pathogen	Lifestyle	Host	Tissue	Effect	Reference
Agrobacterium tumefaciens	Biotroph	*Arabidopsis thaliana*	Crown gall tumor	Increased GABA levels	Lang et al. (2016)
Pseudomonas syringae	Hemi-biotroph	*Phaseolus vulgaris*	Leaf	Increased GABA levels	O'Leary et al. (2016)
Pseudomonas syringae	Hemi-biotroph	*Arabidopsis thaliana*	Leaf	Increased GABA levels	Park et al. (2010)
Ralstonia solanacearum	Hemi-biotroph	*Solanum lycopersicon*	Leaf	Repression of GABA synthesis, upregulation of GABA catabolism	Wang et al. (2019a)

Source: Adapted from Tarkowski et al. (2020).

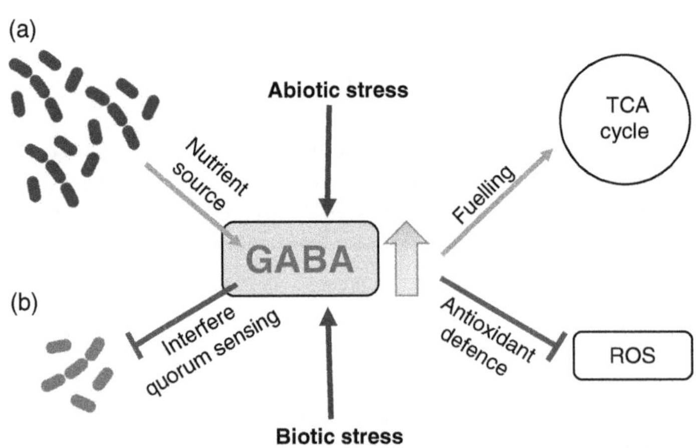

Figure 15.1 Different roles of GABA in plants against bacterial infection. GABA levels are increased in the plant in abiotic or biotic stress (including bacterial infection that increase Ca^{2+} influx through PTI and ETI). The accumulated GABA (a) can be utilized by some bacteria as nutrient source (b) interferes with quorum sensing (QS) in response to some bacterial pathogens. GABA helps in plant immunity by fueling TCA cycle and providing antioxidant defense against ROS produced.

may play a coordinating role in shaping the plant's response to microbial threats. Recent studies validate the intricate function of GABA in modulating plant responses to bacterial stress. The *Arabidopsis* pop2-1 mutant, for instance, exhibits swift GABA accumulation when exposed to various bacterial challenges (Palanivelu et al. 2003). Additionally, GABA plays a regulatory role in stomatal opening (Abdel Razik et al. 2021; Mekonnen et al. 2016) which can be an important strategy for plants to restrict bacterial and fungal entry. Considering its influence on transcriptional regulation and hormone production, GABA may serve a pivotal role in coordinating signaling pathways during interactions between microbes and plants. Understanding the interplay between GABA and other signaling molecules such as salicylic acid and jasmonic acid, can provide a more comprehensive understanding of the complex signaling networks activated by bacterial stress. Overall, role of GABA as a signaling molecule between bacteria and plants seems to extend beyond stress responses.

Molecular Basis of GABA Accumulation in Response to Bacterial Pathogens

Glu-dependent Accumulation of GABA

The microbial infection triggers the Pattern-Triggered Immunity (PTI) as typically the initial defense mechanism (primary defense) and the Effector-Triggered Immunity (ETI) as the secondary defense layer in plants (Jones and Dangl 2006). A crucial early response after pathogen recognition, occurring shortly after ligand-receptor interactions, involves an increased influx of calcium ions (Ca^{2+}), signifying the activation of both immune layers (Seybold et al. 2014). Ca^{2+} channel activation, particularly within the Cyclic Nucleotide Gated Channels (CNGC) family, is mediated by second messengers derived from intracellular domains of receptors or co-receptor proteins, such as cyclic GMP (cGMP) (Ma et al. 2012; Zhao et al. 2013). Released Ca^{2+} in the cytosol binds to numerous Ca^{2+} binding domains on proteins such as calmodulin (CaM), calcineurin B-like (CBLs) proteins, and Ca^{2+} dependent protein kinases (CDPKs). Subsequently, downstream activation of transcription factors that control essential immunological responses occurs, such as salicylic acid production and programmed cell death induction (de la Torre et al. 2013; Schulz et al. 2013; Tarkowski et al. 2020; Zhang et al. 2010).

However, most of the plant GADs possess a CaM-binding domain, enabling them to decarboxylate in response to elevated cytosolic Ca^{2+} levels when activated at neutral pH. This facilitates the regulation of GABA accumulation triggered by pathogens through Ca^{2+} influx as part of the plant's defense mechanism. Moreover, cytosolic acidification can also lead to GAD activation (Shelp et al. 2006; Snedden et al. 1995). The Glu-dependent GABA accumulation on bacterial infection has been shown in multiple studies e.g., when *Pseudomonas syringae* DC 3000 introduced into Arabidopsis leaves, GABA accumulation is associated with the induction of the GAD gene (Ward et al. 2010). Similarly, the study by Seifi et al. suggested that the primary source of cytosolic Glu for the GABA shunt in tomato *sitiens* mutants is the overactivation of the GS/GOGAT cycle (Seifi et al. 2013a). It is important to note that GS/GOGAT cycle may function as a detrimental electron drain in a condition of reduced photosynthetic activity which is common in various plant-pathogen interactions (Liu et al. 2010). The regulation of C/N ratio by the GS/GOGAT cycle is influenced by metabolic cues including changes in nitrogen availability and carbohydrate and amino acid levels (Ishiyama et al. 2004; Oliveira and Coruzzi 1999).

Glu-independent Accumulation of GABA

As mentioned earlier, plants also possess the capability to synthesize GABA through the breakdown of PA (Flores and Filner 1985; Shelp et al. 2012). PA are chemical molecules containing two or more amino groups with diverse roles in plant physiology, responding to various stressors, and aligning with the functions of GABA in plants (Aloisi et al. 2016; Bitrián et al. 2012; Galston and Sawhney 1990; Liu et al. 2015; Seifi and Shelp 2019). In rice cell suspensions treated with *P. oryzae* lysates, GABA levels increased in a GAD-independent manner (Forlani et al. 2014). Similarly, grape (*Vitis vinifera* L.) varieties exhibiting high PA catabolism displayed elevated GABA levels and enhanced resistance against *B. cinerea* (Hatmi et al. 2015). Comparative in vitro studies on susceptible and resistant maize lines indicated that increased PA catabolism correlated with maize (*Zea mays* L.) resistance to *Aspergillus flavus* infection (Majumdar et al. 2019).

The significance of PA-dependent GABA synthesis has been shown in bacterial infections. In pepper (*Capsicum annuum* L.), arginine decarboxylase (CaADC1) binds to the effector AvrBsT of the bacterial pathogen *Xanthomonas campestris*, triggering a hypersensitive response (HR) that inhibits spread of pathogen (Kim et al. 2013). CaADC1 produces arginine, contributing to PA biosynthesis. CaADC1 mutants, as reported by (Kim et al. 2013), exhibited reduced GABA levels and increased susceptibility to *X. campestris*. Additionally, GABA production under oxidative stress is suggested to follow a Pro-dependent pathway, involving both enzymatic and nonenzymatic processes (Signorelli et al. 2015), though its presence in plant-pathogen interactions remains unexplored.

The cellular location of GABA accumulation plays a pivotal role in shaping plant-pathogen interactions. The intracellular GABA accumulation actively drives metabolic pathways linked to plant defense. However, the extracellular accumulation may be beneficial for pathogens residing in the apoplast, such as *P. syringae* or *Cladosporium fulvum*, competing for GABA assimilation. Even the high level of GABA in apoplast has been shown to help *P. syringae* in nutritional adaptation (Figure 15.1) (O'Leary et al. 2016; Rico and Preston 2008; Solomon and Oliver 2002).

The Involvement of GABA in the Interaction of Microbes with Plants

In this section, we provide an in-depth analysis of two compelling examples that highlight the critical role of GABA in the interactions between bacteria and plant hosts.

Ralstonia solanacearum

Ralstonia solanacearum is a soil-borne, gram-negative betaproteobacterium recognized for causing substantial crop productivity losses globally. It serves as the causal agent of bacterial wilt disease affecting over 250 plant species and stands out as one of the most destructive plant pathogens (Mansfield et al. 2012). It infects crucial crops like tomatoes, potatoes, pepper, and bananas, demonstrating prolonged survival in the environment (Elphinstone 2005; Graham et al. 1979; Grey and Steck 2001). The infection process involves root penetration, colonization of xylem vessels, and obstruction of the plant's conducting system, leading to wilting and eventual death (Genin 2010; Xue et al. 2020).

During its multiplication, *R. solanacearum* secretes exopolysaccharides and the toxicity associated with the bacterium is associated with the secretion of the effector protein RipI *via* type-III secretion. RipI, a 47 kDa protein composed of 430 amino acids encoded by locus RSc0041 (Zhuo et al. 2020), enhances the activity of GAD by increasing its affinity with the plant protein CaM. Elevated activity of GAD leads to increased levels of GABA by converting more glutamate into GABA, promoting the extensive proliferation of *R. solanacearum* (Xian et al. 2020). Nonetheless, an excess of GABA is not beneficial for the plant until consumed by associated bacteria, as it can induce plant cell death. A knockout mutant in the GABA transaminase-encoding gene *gabT* is unable to utilize GABA as a nutrient, significantly compromising its ability to cause disease in plants. This emphasizes the crucial role of the *gabT* gene in mediating the pathogenicity of *R. solanacearum*, as demonstrated by experiments involving the inactivation and overexpression of *gabT* in tomato plants (Xian et al. 2021).

Pseudomonas syringae pv. *tabaci*

Pseudomonas species, gram-negative aerobic bacilli, encompass both plant pathogenic and plant growth-promoting bacteria, while also being recognized as animal pathogens. *Pseudomonas syringae* pv. tomato (Pto) DC3000 serves as a widely utilized model organism for investigating plant resistance to bacterial infections due to its well-established type-III effector system (Cunnac et al. 2009). The GABA-metabolizing enzyme, *gabT*, was initially identified in *P. fluorescens* (Scott and Jakoby 1959) with the GABA uptake system also characterized in *P. fluorescens* (Guthrie and Nicholson-Guthrie 1989).

Various chemoreceptors in different bacterial species of *Pseudomonas* respond to GABA. McpG acts as a chemoreceptor for GABA in *P. putida* (Reyes-Darias et al. 2015) and *P. syringae* pv. *tabaci* (Tumewu et al. 2020), PctC in *P. aeruginosa*, (Rico-Jiménez et al. 2013), and Psc in *P. syringaepv. actinidiae* (Ehrhardt et al. 2021; Mckellar et al. 2015). The virulence of *P. syringae* pv. tomato (Pst) was diminished in *gab*T triple mutant strain, which cannot degrade GABA. The virulence was further reduced when inoculated into the *Arabidopsis* pop2/gaba-t mutant, leading to elevated GABA levels (Park et al. 2010). Inoculation with the *P. syringae* pathogen in *Arabidopsis* was shown to elevate GABA levels, conferring resistance to pathogenicity. This increase in GABA was accompanied by increased expression of GAD. The *Arabidopsis thaliana* genome encodes five GAD genes (GAD1, GAD2, GAD3, GAD4, and GAD5). Notably, GAD1 and GAD4 transcript levels significantly increased under *Pseudomonas syringae* pv. tomato DC3000 infection. Furthermore, susceptibility to the pathogen increased in *gad*1/2/4 triple and *gad*1/2/4/5 quadruple mutants, underscoring the protective role of GABA against Pst infection (Deng et al. 2020).

Conclusions and Future Perspectives

Many bacteria produce GABA and release it, suggesting that bacteria can use GABA as signal molecule to communicate with other bacteria or host organisms. In contrast, the infection with bacteria has been shown to upregulate GABA expression in host plant and exogenous GABA provide antioxidant defense by scavenging ROS and increase in antioxidant enzymes in plants (Figure 15.1) (Abd Elbar et al. 2021; Cheng et al. 2018; Salah et al. 2019; Yang et al. 2017; Yu et al. 2014). The accumulated GABA can be converted to succinate in mitochondria and fueled to TCA cycle (Figure 15.1) (Fait et al. 2008; Michaeli et al. 2011). Increased GABA in host plant provide protection to against some bacteria by interfering with the quorum sensing (Figure 15.1) (Chevrot et al. 2006). Nonetheless, the accumulated GABA can be utilized by some bacteria as nutrient source (Figure 15.1) (O'Leary et al. 2016; Rico and Preston 2008).

While there is limited evidence of GABA directly scavenging ROS, research on crops like tomatoes and pears demonstrates that exogenous GABA treatments can effectively counteract biotic stress (Yang et al. 2017; Yu et al. 2014). This is attributed to the induction of transcript and protein activity levels of antioxidant enzymes, which, in turn, limit the HR. Coordination between the creation of ROS within the lesion and enhanced energy supply in the outer layer helps provide an efficient defense against pathogens that exploit necrotrophy (Govrin and Levine, 2000; Seifi et al. 2013a). In addition to the potential roles of GABA signaling and the significance of the GABA shunt, GABA may indirectly impact ROS homeostasis through PAs. PA catabolism can generate GABA and H_2O_2, with H_2O_2 acting as a signaling molecule and potentially causing harm to plant cell structures (Gupta et al. 2016). Research indicates that PA can influence the expression of the catalase (CAT) gene, suggesting potential signaling functions (Sung et al. 2011). PA's role in controlling ROS accumulation during plant-pathogen interactions has also been reported (Lou et al. 2016; Seifi and Shelp 2019) Additionally, the $NADH/NAD^+$ ratio, elevated during stresses like cold, anoxia, and drought, can preferentially catabolize GABA to gamma-hydroxybutyrate (GHB), offering potential benefits in counteracting excessive ROS formation under stress conditions (Allan et al. 2008; Kaplan et al. 2007).

The association between GABA and bacterial infections in plants offers a novel perspective on plant immunity. Previously recognized solely as a neurotransmitter, GABA now reveals its adaptive role in plant defense and signaling processes. In summary, this chapter describes the role of GABA in plant stress with a focus on bacterial infection and the molecular details of the process available. It is a molecule with shear importance for plants in both abiotic and biotic stresses. Beyond its physiological relevance, GABA emerges as a key player in plant defense mechanisms, exerting regulatory control over gene expression and signaling pathways. GABA functions both as a rapid-acting metabolite as well as a contributor to specialized and prolonged defense responses, suggesting potential applications in crop protection strategies. The implications of GABA in mediating interactions between microbes and plants, as well as its complex interaction with other signaling molecules, open exciting possibilities for further research. A thorough comprehensive understanding of GABA's role in plant-microbe's interactions can aid in designing more potent strategies against microbial infection in plants. There still remain questions to be explored such as – Can bacterial GABA help bacteria during infection and therefore some bacteria know how to utilize plant's GABA for their benefit? How are the levels of GABA sensed in plant? What are the signaling pathways regulated by GABA in plants and bacteria during infection? Is there direct evidence for involvement of endogenous GABA in plant during bacterial infection?

References

Abd Elbar, O.H., Elkelish, A., Niedbała, G. et al. (2021). Protective effect of γ-aminobutyric acid against chilling stress during reproductive stage in tomato plants through modulation of sugar metabolism, chloroplast integrity, and antioxidative defense systems. *Frontiers in Plant Science* 12: 663750. https://doi.org/10.3389/FPLS.2021.663750/BIBTEX.

Abdel Razik, E.S., Alharbi, B.M., Pirzadah, T.B. et al. (2021). γ-Aminobutyric acid (GABA) mitigates drought and heat stress in sunflower (*Helianthus annuus* L.) by regulating its physiological, biochemical and molecular pathways. *Physiologia Plantarum* 172 (2): 505–527. https://doi.org/10.1111/PPL.13216.

Adem, G.D., Chen, G., Shabala, L. et al. (2020). GORK channel: a master switch of plant metabolism? *Trends in Plant Science* 25 (5): 434–445. https://doi.org/10.1016/j.tplants.2019.12.012.

Aghdam, M.S., Naderi, R., Jannatizadeh, A. et al. (2016). Impact of exogenous GABA treatments on endogenous GABA metabolism in anthurium cut flowers in response to postharvest chilling temperature. *Plant Physiology and Biochemistry* 106: 11–15. https://doi.org/10.1016/J.PLAPHY.2016.04.045.

Aghdam, M.S., Flaherty, E.J., and Shelp, B.J. (2022). γ-Aminobutyrate improves the postharvest marketability of horticultural commodities: advances and prospects. *Frontiers in Plant Science* 13: 884572. https://doi.org/10.3389/FPLS.2022.884572/BIBTEX.

Akçay, N., Bor, M., Karabudak, T. et al. (2012). Contribution of gamma amino butyric acid (GABA) to salt stress responses of *Nicotiana sylvestris* CMSII mutant and wild type plants. *Journal of Plant Physiology* 169 (5): 452–458. https://doi.org/10.1016/J.JPLPH.2011.11.006.

Allan, W.L., Simpson, J.P., Clark, S.M., and Shelp, B.J. (2008). Gamma-hydroxybutyrate accumulation in *Arabidopsis* and tobacco plants is a general response to abiotic stress: putative regulation by redox balance and glyoxylate reductase isoforms. *Journal of Experimental Botany* 59 (9): 2555–2564. https://doi.org/10.1093/JXB/ERN122.

Aloisi, I., Cai, G., Serafini-Fracassini, D., and Del Duca, S. (2016). Polyamines in pollen: from microsporogenesis to fertilization. *Frontiers in Plant Science* 7: https://doi.org/10.3389/FPLS.2016.00155.

Al-Quraan, N.A., Sartawe, F.A., and Qaryouti, M.M. (2013). Characterization of γ-aminobutyric acid metabolism and oxidative damage in wheat (*Triticum aestivum* L.) seedlings under salt and osmotic stress. *Journal of Plant Physiology* 170 (11): 1003–1009. https://doi.org/10.1016/J.JPLPH.2013.02.010.

Ayenew, B., Degu, A., Manela, N. et al. (2015). Metabolite profiling and transcript analysis reveal specificities in the response of a berry derived cell culture to abiotic stresses. *Frontiers in Plant Science* 6: https://doi.org/10.3389/FPLS.2015.00728.

Binzel, M.L., Hasegawa, P.M., Rhodes, D. et al. (1987). Solute accumulation in tobacco cells adapted to NaCl. *Plant Physiology* 84 (4): 1408–1415. https://doi.org/10.1104/PP.84.4.1408.

Bitrián, M., Zarza, X., Altabella, T. et al. (2012). Polyamines under abiotic stress: metabolic crossroads and hormonal crosstalks in plants. *Metabolites* 2 (3): 516–528. https://doi.org/10.3390/METABO2030516.

Bor, M., Seckin, B., Ozgur, R. et al. (2009). Comparative effects of drought, salt, heavy metal and heat stresses on gamma-aminobutryric acid levels of sesame (*Sesamum indicum* L.). *Acta Physiologiae Plantarum* 31 (3): 655–659. https://doi.org/10.1007/S11738-008-0255-2/METRICS.

Bouché, N. and Fromm, H. (2004). GABA in plants: just a metabolite? *Trends in Plant Science* 9 (3): 110–115. https://doi.org/10.1016/J.TPLANTS.2004.01.006.

Bouché, N., Fait, A., Bouchez, D. et al. (2003a). Mitochondrial succinic-semialdehyde dehydrogenase of the gamma-aminobutyrate shunt is required to restrict levels of reactive oxygen intermediates in plants. *Proceedings of the National Academy of Sciences of the United States of America* 100 (11): 6843–6848. https://doi.org/10.1073/PNAS.1037532100.

Bouché, N., Lacombe, B., and Fromm, H. (2003b). GABA signaling: a conserved and ubiquitous mechanism. *Trends in Cell Biology* 13 (12): 607–610. https://doi.org/10.1016/j.tcb.2003.10.001.

Bown, A.W. and Shelp, B.J. (1997). The metabolism and functions of [gamma]-aminobutyric acid. *Plant Physiology* 115 (1): 1–5. https://doi.org/10.1104/PP.115.1.1.

Bown, A.W. and Shelp, B.J. (2016). Plant GABA: not just a metabolite. *Trends in Plant Science* 21 (10): 811–813. https://doi.org/10.1016/J.TPLANTS.2016.08.001.

Bown, A.W. and Shelp, B.J. (2020). Does the GABA shunt regulate cytosolic GABA? *Trends in Plant Science* 25 (5): 422–424. https://doi.org/10.1016/j.tplants.2020.03.001.

Bown, A.W., Hall, D.E., and MacGregor, K.B. (2002). Insect footsteps on leaves stimulate the accumulation of 4-aminobutyrate and can be visualized through increased chlorophyll fluorescence and superoxide production. *Plant Physiology* 129 (4): 1430–1434. https://doi.org/10.1104/PP.006114.

Bown, A.W., MacGregor, K.B., and Shelp, B.J. (2006). Gamma-aminobutyrate: defense against invertebrate pests? *Trends in Plant Science* 11 (9): 424–427. https://doi.org/10.1016/J.TPLANTS.2006.07.002.

Brechtel, C.E. and King, S.C. (1998). 4-Aminobutyrate (GABA) transporters from the amine-polyamine-choline superfamily: substrate specificity and ligand recognition profile of the 4-aminobutyrate permease from *Bacillus subtilis*. *The Biochemical Journal* 333 (Pt 3): 565–571. https://doi.org/10.1042/BJ3330565.

Brikis, C.J., Zarei, A., Chiu, G.Z. et al. (2018). Targeted quantitative profiling of metabolites and gene transcripts associated with 4-aminobutyrate (GABA) in apple fruit stored under multiple abiotic stresses. *Horticulture Research* 5 (1): https://doi.org/10.1038/S41438-018-0069-3.

Chen, F., Liu, C., Zhang, J. et al. (2018). Combined metabonomic and quantitative RT-PCR analyses revealed metabolic reprogramming associated with fusarium graminearum resistance in transgenic *Arabidopsis thaliana*. *Frontiers in Plant Science* 8: https://doi.org/10.3389/FPLS.2017.02177.

Cheng, B., Li, Z., Liang, L. et al. (2018). The γ-aminobutyric acid (GABA) alleviates salt stress damage during seeds germination of white clover associated with Na^+/K^+ transportation, dehydrins accumulation, and stress-related genes expression in white clover. *International Journal of Molecular Sciences* 19 (9): https://doi.org/10.3390/IJMS19092520.

Che-Othman, M.H., Jacoby, R.P., Millar, A.H., and Taylor, N.L. (2020). Wheat mitochondrial respiration shifts from the tricarboxylic acid cycle to the GABA shunt under salt stress. *The New Phytologist* 225 (3): 1166–1180. https://doi.org/10.1111/NPH.15713.

Cherry, J.H., Locy, R.D., and Rychter, A. (2000). *Plant Tolerance to Abiotic Stresses in Agriculture: Role of Genetic Engineering*, 376. Dordrecht, The Netherlands: Springer Retrieved from https://books.google.com/books/about/Plant_Tolerance_to_Abiotic_Stresses_in_A.html?id=clZtHnhrxy4C.

Chevrot, R., Rosen, R., Haudecoeur, E. et al. (2006). GABA controls the level of quorum-sensing signal in Agrobacterium tumefaciens. *Proceedings of the National Academy of Sciences of the United States of America* 103 (19): 7460–7464. https://doi.org/10.1073/PNAS.0600313103.

Christensen, H.N., Greene, A.A., Kakuda, D.K., and MacLeod, C.L. (1994). Special transport and neurological significance of two amino acids in a configuration conventionally designated as D. *Journal of Experimental Biology* 196 (1): 297–305. https://doi.org/10.1242/JEB.196.1.297.

Cunnac, S., Lindeberg, M., and Collmer, A. (2009). Pseudomonas syringae type III secretion system effectors: repertoires in search of functions. *Current Opinion in Microbiology* 12 (1): 53–60. https://doi.org/10.1016/J.MIB.2008.12.003.

Daş, Z.A., Dimlioğlu, G., Bor, M., and Özdemir, F. (2016). Zinc induced activation of GABA-shunt in tobacco (*Nicotiana tabaccum* L.). *Environmental and Experimental Botany* 122: 78–84. https://doi.org/10.1016/J.ENVEXPBOT.2015.09.006.

De Biase, D. and Pennacchietti, E. (2012). Glutamate decarboxylase-dependent acid resistance in orally acquired bacteria: function, distribution and biomedical implications of the gadBC operon. *Molecular Microbiology* 86 (4): 770–786. https://doi.org/10.1111/MMI.12020.

de la Torre, F., Gutiérrez-Beltrán, E., Pareja-Jaime, Y. et al. (2013). The tomato calcium sensor Cbl10 and its interacting protein kinase Cipk6 define a signaling pathway in plant immunity. *The Plant Cell* 25 (7): 2748–2764. https://doi.org/10.1105/TPC.113.113530.

Deng, X., Xu, X., Liu, Y. et al. (2020). Induction of γ-aminobutyric acid plays a positive role to *Arabidopsis* resistance against *Pseudomonas syringae*. *Journal of Integrative Plant Biology* 62 (11): 1797–1812. https://doi.org/10.1111/JIPB.12974/SUPPINFO.

Dimlioğlu, G., Daş, Z.A., Bor, M. et al. (2015). The impact of GABA in harpin-elicited biotic stress responses in *Nicotiana tabaccum*. *Journal of Plant Physiology* 188: 51–57. https://doi.org/10.1016/J.JPLPH.2015.08.005.

Dubey, S., Misra, P., Dwivedi, S. et al. (2010). Transcriptomic and metabolomic shifts in rice roots in response to Cr (VI) stress. *BMC Genomics* 11 (1): https://doi.org/10.1186/1471-2164-11-648.

Ehrhardt, M.K.G., Gerth, M.L., and Johnston, J.M. (2021). Structure of a double CACHE chemoreceptor ligand-binding domain from *Pseudomonas syringae* provides insights into the basis of proline recognition. *Biochemical and Biophysical Research Communications* 549: 194–199. https://doi.org/10.1016/J.BBRC.2021.02.090.

Elphinstone, J.G. (2005). The current bacterial wilt situation: a global overview. In: *Bacterial Wilt Disease and the Ralstonia solanacearum Complex* (ed. C. Allen, P. Piror, and A.C. Hayward), 9–28. St. Paul, MN: APS Press Retrieved from https://www.sciepub.com/reference/100376 (25 February 2024).

Fait, A., Yellin, A., and Fromm, H. (2005). GABA shunt deficiencies and accumulation of reactive oxygen intermediates: insight from *Arabidopsis* mutants. *FEBS Letters* 579 (2): 415–420. https://doi.org/10.1016/J.FEBSLET.2004.12.004.

Fait, A., Fromm, H., Walter, D. et al. (2008). Highway or byway: the metabolic role of the GABA shunt in plants. *Trends in Plant Science* 13 (1): 14–19. https://doi.org/10.1016/j.tplants.2007.10.005.

Fan, T.W.M., Higashi, R.M., Frenkiel, T.A., and Lane, A.N. (1997). Anaerobic nitrate and ammonium metabolism in flood-tolerant rice coleoptiles. *Journal of Experimental Botany* 48 (9): 1655–1666. https://doi.org/10.1093/JXB/48.9.1655.

Filho, E.G.A., Braga, L.N., Silva, L.M.A. et al. (2018). Physiological changes for drought resistance in different species of *Phyllanthus*. *Scientific Reports* 8 (1): https://doi.org/10.1038/S41598-018-33496-7.

Flores, H.E. and Filner, P. (1985). Polyamine catabolism in higher plants: characterization of pyrroline dehydrogenase. In: *Polyamines in Plants* (ed. A.W. Galston and T.A. Smith), 75–89. Dordrecht, The Netherlands: Springer https://doi.org/10.1007/978-94-009-5171-6_7.

Forlani, G., Bertazzini, M., and Giberti, S. (2014). Differential accumulation of γ-aminobutyric acid in elicited cells of two rice cultivars showing contrasting sensitivity to the blast pathogen. *Plant Biology* 16 (6): 1127–1132. https://doi.org/10.1111/PLB.12165.

Fougère, F., Le Rudulier, D., and Streeter, J.G. (1991). Effects of salt stress on amino acid, organic acid, and carbohydrate composition of roots, bacteroids, and cytosol of alfalfa (*Medicago sativa* L.). *Plant Physiology* 96 (4): 1228–1236. https://doi.org/10.1104/PP.96.4.1228.

Fromm, H. (2020). GABA signaling in plants: targeting the missing pieces of the puzzle. *Journal of Experimental Botany* 71 (20): 6238–6245. https://doi.org/10.1093/JXB/ERAA358.

Fu, D., Sun, Y., Yu, C. et al. (2017). Comparison of the effects of three types of aminobutyric acids on the control of *Penicillium* expansum infection in pear fruit. *Journal of the Science of Food and Agriculture* 97 (5): 1497–1501. https://doi.org/10.1002/JSFA.7891.

Galston, A.W. and Sawhney, R.K. (1990). Polyamines in plant physiology. *Plant Physiology* 94 (2): 406–410. https://doi.org/10.1104/PP.94.2.406.

Genin, S. (2010). Molecular traits controlling host range and adaptation to plants in *Ralstonia solanacearum*. *The New Phytologist* 187 (4): 920–928. https://doi.org/10.1111/J.1469-8137.2010.03397.X.

Gilliham, M. and Tyerman, S.D. (2016). Linking metabolism to membrane signaling: the GABA – malate connection. *Trends in Plant Science* 21 (4): 295–301. https://doi.org/10.1016/j.tplants.2015.11.011.

Govrin, E.M. and Levine, A. (2000). The hypersensitive response facilitates plant infection by the necrotrophic pathogen *Botrytis cinerea*. *Current Biology* 10 (13): 751–757. https://doi.org/10.1016/S0960-9822(00)00560-1.

Graham, R.L., Lawler, E.L., Lenstra, J.K., and Kan, A.H.G.R. (1979). Optimization and approximation in deterministic sequencing and scheduling: a survey. *Annals of Discrete Mathematics* 5 (C): 287–326. https://doi.org/10.1016/S0167-5060(08)70356-X.

Grey, B.E. and Steck, T.R. (2001). The viable but nonculturable state of *Ralstonia solanacearum* may be involved in long-term survival and plant infection. *Applied and Environmental Microbiology* 67 (9): 3866–3872. https://doi.org/10.1128/AEM.67.9.3866-3872.2001.

Gupta, K., Sengupta, A., Chakraborty, M., and Gupta, B. (2016). Hydrogen peroxide and polyamines act as double edged swords in plant abiotic stress responses. *Frontiers in Plant Science* 7 (September): https://doi.org/10.3389/FPLS.2016.01343.

Guthrie, G.D. and Nicholson-Guthrie, C.S. (1989). Gamma-aminobutyric acid uptake by a bacterial system with neurotransmitter binding characteristics. *Proceedings of the National Academy of Sciences of the United States of America* 86 (19): 7378–7381. https://doi.org/10.1073/PNAS.86.19.7378.

Han, T.C., Kwon, D.H., Hegazy, M., and Lu, C.D. (2008). Transcriptome analysis of agmatine and putrescine catabolism in *Pseudomonas aeruginosa* PAO1. *Journal of Bacteriology* 190 (6): 1966–1975. https://doi.org/10.1128/JB.01804-07.

Handa, S., Bressan, R.A., Handa, A.K. et al. (1983). Solutes contributing to osmotic adjustment in cultured plant cells adapted to water stress. *Plant Physiology* 73 (3): 834–843. https://doi.org/10.1104/PP.73.3.834.

Hatmi, S., Gruau, C., Trotel-Aziz, P. et al. (2015). Drought stress tolerance in grapevine involves activation of polyamine oxidation contributing to improved immune response and low susceptibility to *Botrytis cinerea*. *Journal of Experimental Botany* 66 (3): 775–787. https://doi.org/10.1093/JXB/ERU436.

Hijaz, F., Nehela, Y., and Killiny, N. (2018). Application of gamma-aminobutyric acid increased the level of phytohormones in *Citrus sinensis*. *Planta* 248 (4): 909–918. https://doi.org/10.1007/S00425-018-2947-1/METRICS.

Hu, L.A. and King, S.C. (1998). Membrane topology of the *Escherichia coli* gamma-aminobutyrate transporter: implications on the topography and mechanism of prokaryotic and eukaryotic transporters from the APC superfamily. *The Biochemical Journal* 336 (Pt 1): 69–76. https://doi.org/10.1042/BJ3360069.

Huang, T., Jander, G., and De Vos, M. (2011). Non-protein amino acids in plant defense against insect herbivores: representative cases and opportunities for further functional analysis. *Phytochemistry* 72 (13): 1531–1537. https://doi.org/10.1016/J.PHYTOCHEM.2011.03.019.

Ishiyama, K., Inoue, E., Watanabe-Takahashi, A. et al. (2004). Kinetic properties and ammonium-dependent regulation of cytosolic isoenzymes of glutamine synthetase in *Arabidopsis*. *The Journal of Biological Chemistry* 279 (16): 16598–16605. https://doi.org/10.1074/JBC.M313710200.

Ji, J., Shi, Z., Xie, T. et al. (2020). Responses of GABA shunt coupled with carbon and nitrogen metabolism in poplar under NaCl and CdCl2 stresses. *Ecotoxicology and Environmental Safety* 193: https://doi.org/10.1016/J.ECOENV.2020.110322.

Jin, X., Liu, T., Xu, J. et al. (2019). Exogenous GABA enhances muskmelon tolerance to salinity-alkalinity stress by regulating redox balance and chlorophyll biosynthesis. *BMC Plant Biology* 19 (1): https://doi.org/10.1186/S12870-019-1660-Y.

Jones, J.D.G. and Dangl, J.L. (2006). The plant immune system. *Nature* 444 (7117): 323–329. https://doi.org/10.1038/NATURE05286.

Kang, S.M., Radhakrishnan, R., You, Y.H. et al. (2015). Enterobacter asburiae KE17 association regulates physiological changes and mitigates the toxic effects of heavy metals in soybean. *Plant Biology* 17 (5): 1013–1022. https://doi.org/10.1111/PLB.12341.

Kaplan, F., Kopka, J., Sung, D.Y. et al. (2007). Transcript and metabolite profiling during cold acclimation of *Arabidopsis* reveals an intricate relationship of cold-regulated gene expression with modifications in metabolite content. *The Plant Journal: For Cell and Molecular Biology* 50 (6): 967–981. https://doi.org/10.1111/J.1365-313X.2007.03100.X.

Kaspal, M., Kanapaddalagamage, M.H., and Ramesh, S.A. (2021). Emerging roles of γ-aminobutyric acid (GABA) gated channels in plant stress tolerance. *Plants* 10 (10): https://doi.org/10.3390/PLANTS10102178.

Kim, N.H., Kim, B.S., and Hwang, B.K. (2013). Pepper arginine decarboxylase is required for polyamine and γ-aminobutyric acid signaling in cell death and defense response. *Plant Physiology* 162 (4): 2067–2083. https://doi.org/10.1104/PP.113.217372.

Kinnersley, A.M. and Turano, F.J. (2000). Gamma aminobutyric acid (GABA) and plant responses to stress. *Critical Reviews in Plant Sciences* 19 (6): 479–509. https://doi.org/10.1080/07352680091139277.

Kisaka, H., Kida, T., and Miwa, T. (1989). The metabolism and physiological roles of 4-aminobutyric acid. *Biochemistry (Life Science Advances)* 8 (3): 21–25. https://doi.org/10.5511/PLANTBIOTECHNOLOGY.23.267.

Kogovšek, P., Pompe-Novak, M., Petek, M. et al. (2016). Primary metabolism, phenylpropanoids and antioxidant pathways are regulated in potato as a response to potato virus Y infection. *PLoS One* 11 (1): https://doi.org/10.1371/JOURNAL.PONE.0146135.

Kumar, N., Dubey, A.K., Upadhyay, A.K. et al. (2017). GABA accretion reduces Lsi-1 and Lsi-2 gene expressions and modulates physiological responses in *Oryza sativa* to provide tolerance towards arsenic. *Scientific Reports* 7 (1): https://doi.org/10.1038/S41598-017-09428-2.

Lang, J., Gonzalez-Mula, A., Taconnat, L. et al. (2016). The plant GABA signaling downregulates horizontal transfer of the *Agrobacterium tumefaciens* virulence plasmid. *The New Phytologist* 210 (3): 974–983. https://doi.org/10.1111/NPH.13813.

Li, Z., Yu, J., Peng, Y., and Huang, B. (2016). Metabolic pathways regulated by γ-aminobutyric acid (GABA) contributing to heat tolerance in creeping bentgrass (*Agrostis stolonifera*). *Scientific Reports* 6: https://doi.org/10.1038/SREP30338.

Li, Z., Peng, Y., and Huang, B. (2018). Alteration of transcripts of stress-protective genes and transcriptional factors by γ-aminobutyric acid (GABA) associated with improved heat and drought tolerance in creeping bentgrass (*Agrostis stolonifera*). *International Journal of Molecular Sciences* 19 (6): https://doi.org/10.3390/IJMS19061623.

Li, K.X., He, M., Ye, W. et al. (2019a). TMEM16B regulates anxiety-related behavior and GABAergic neuronal signaling in the central lateral amygdala. *eLife* 8: https://doi.org/10.7554/ELIFE.47106.

Li, Z., Huang, T., Tang, M. et al. (2019b). iTRAQ-based proteomics reveals key role of γ-aminobutyric acid (GABA) in regulating drought tolerance in perennial creeping bentgrass (*Agrostis stolonifera*). *Plant Physiology and Biochemistry* 145: 216–226. https://doi.org/10.1016/J.PLAPHY.2019.10.018.

Li, L., Dou, N., Zhang, H., and Wu, C. (2021). The versatile GABA in plants. *Plant Signaling & Behavior* 16 (3): https://doi.org/10.1080/15592324.2020.1862565.

Liu, G., Ji, Y., Bhuiyan, N.H. et al. (2010). Amino acid homeostasis modulates salicylic acid-associated redox status and defense responses in *Arabidopsis*. *The Plant Cell* 22 (11): 3845–3863. https://doi.org/10.1105/TPC.110.079392.

Liu, J.H., Wang, W., Wu, H. et al. (2015). Polyamines function in stress tolerance: from synthesis to regulation. *Frontiers in Plant Science* 6 (October): https://doi.org/10.3389/FPLS.2015.00827.

Long, Y., Tyerman, S.D., and Gilliham, M. (2020). Cytosolic GABA inhibits anion transport by wheat ALMT1. *The New Phytologist* 225 (2): 671–678. https://doi.org/10.1111/NPH.16238.

Lou, Y.R., Bor, M., Yan, J. et al. (2016). *Arabidopsis* NATA1 acetylates putrescine and decreases defense-related hydrogen peroxide accumulation. *Plant Physiology* 171 (2): 1443–1455. https://doi.org/10.1104/PP.16.00446.

Ludewig, F., Hüser, A., Fromm, H. et al. (2008). Mutants of GABA transaminase (POP2) suppress the severe phenotype of succinic semialdehyde dehydrogenase (ssadh) mutants in *Arabidopsis*. *PLoS One* 3 (10): https://doi.org/10.1371/JOURNAL.PONE.0003383.

Ma, Y., Walker, R.K., Zhao, Y., and Berkowitz, G.A. (2012). Linking ligand perception by PEPR pattern recognition receptors to cytosolic Ca^{2+} elevation and downstream immune signaling in plants. *Proceedings of the National Academy of Sciences of the United States of America* 109 (48): 19852–19857. https://doi.org/10.1073/PNAS.1205448109/-/DCSUPPLEMENTAL/PNAS.201205448SI.PDF.

Ma, Y., Wang, P., Wang, M. et al. (2019). GABA mediates phenolic compounds accumulation and the antioxidant system enhancement in germinated hulless barley under NaCl stress. *Food Chemistry* 270: 593–601. https://doi.org/10.1016/J.FOODCHEM.2018.07.092.

Mahmud, J.A.L., Hasanuzzaman, M., Nahar, K. et al. (2017). γ-aminobutyric acid (GABA) confers chromium stress tolerance in *Brassica juncea* L. by modulating the antioxidant defense and glyoxalase systems. *Ecotoxicology* 26 (5): 675–690. https://doi.org/10.1007/S10646-017-1800-9.

Majumdar, R., Minocha, R., Lebar, M.D. et al. (2019). Contribution of maize polyamine and amino acid metabolism toward resistance against *Aspergillus flavus* infection and aflatoxin production. *Frontiers in Plant Science* 10: https://doi.org/10.3389/FPLS.2019.00692.

Malekzadeh, P., Khara, J., and Heydari, R. (2014). Alleviating effects of exogenous gamma-aminobutiric acid on tomato seedling under chilling stress. *Physiology and Molecular Biology of Plants: An International Journal of Functional Plant Biology* 20 (1): 133–137. https://doi.org/10.1007/S12298-013-0203-5.

Malekzadeh, P., Khosravi-Nejad, F., Hatamnia, A.A., and Sheikhakbari Mehr, R. (2017). Impact of postharvest exogenous γ-aminobutyric acid treatment on cucumber fruit in response to chilling tolerance. *Physiology and Molecular Biology of Plants* 23 (4): 827–836. https://doi.org/10.1007/S12298-017-0475-2/METRICS.

Mansfield, J., Genin, S., Magori, S. et al. (2012). Top 10 plant pathogenic bacteria in molecular plant pathology. *Molecular Plant Pathology* 13 (6): 614–629. https://doi.org/10.1111/J.1364-3703.2012.00804.X.

Mazzucotelli, E., Tartari, A., Cattivelli, L., and Forlani, G. (2006). Metabolism of gamma-aminobutyric acid during cold acclimation and freezing and its relationship to frost tolerance in barley and wheat. *Journal of Experimental Botany* 57 (14): 3755–3766. https://doi.org/10.1093/JXB/ERL141.

McCraw, S.L., Park, D.H., Jones, R. et al. (2016). GABA (γ-aminobutyric acid) uptake via the GABA permease GabP represses virulence gene expression in *Pseudomonas syringae* pv. tomato DC3000. *Molecular Plant-Microbe Interactions* 29 (12): 938–949. https://doi.org/10.1094/MPMI-08-16-0172-R/ASSET/IMAGES/LARGE/MPMI-08-16-0172-R_F5.JPEG.

Mckellar, J.L.O., Minnell, J.J., and Gerth, M.L. (2015). A high-throughput screen for ligand binding reveals the specificities of three amino acid chemoreceptors from *Pseudomonas syringae* pv. actinidiae. *Molecular Microbiology* 96 (4): 694–707. https://doi.org/10.1111/MMI.12964.

Mekonnen, D.W., Flügge, U.I., and Ludewig, F. (2016). Gamma-aminobutyric acid depletion affects stomata closure and drought tolerance of *Arabidopsis thaliana*. *Plant Science: An International Journal of Experimental Plant Biology* 245: 25–34. https://doi.org/10.1016/J.PLANTSCI.2016.01.005.

Michaeli, S. and Fromm, H. (2015). Closing the loop on the GABA shunt in plants: are GABA metabolism and signaling entwined? *Frontiers in Plant Science* 6 (June): https://doi.org/10.3389/FPLS.2015.00419.

Michaeli, S., Fait, A., Lagor, K. et al. (2011). A mitochondrial GABA permease connects the GABA shunt and the TCA cycle, and is essential for normal carbon metabolism. *The Plant Journal* 67 (3): 485–498. https://doi.org/10.1111/J.1365-313X.2011.04612.X.

Mithöfer, A. and Boland, W. (2012). Plant defense against herbivores: chemical aspects. *Annual Review of Plant Biology* 63: 431–450. https://doi.org/10.1146/ANNUREV-ARPLANT-042110-103854.

Mittler, R., Zandalinas, S.I., Fichman, Y., and Van Breusegem, F. (2022). Reactive oxygen species signalling in plant stress responses. *Nature Reviews. Molecular Cell Biology* 23 (10): 663–679. https://doi.org/10.1038/S41580-022-00499-2.

Narayan, V.S. and Nair, P.M. (1990). Metabolism, enzymology and possible roles of 4-aminobutyrate in higher plants. *Phytochemistry* 29 (2): 367–375. https://doi.org/10.1016/0031-9422(90)85081-P.

Nazoori, F., ZamaniBahramabadi, E., Mirdehghan, S.H., and Rafie, A. (2020). Extending the shelf life of pomegranate (*Punica granatum* L.) by GABA coating application. *Journal of Food Measurement and Characterization* 14 (5): 2760–2772. https://doi.org/10.1007/S11694-020-00521-1/METRICS.

Nehela, Y. and Killiny, N. (2019). "*Candidatus Liberibacter asiaticus*" and Its vector, diaphorina citri, augment the tricarboxylic acid cycle of their host via the γ-aminobutyric acid shunt and polyamines pathway. *Molecular Plant-Microbe Interactions* 32 (4): 413–427. https://doi.org/10.1094/MPMI-09-18-0238-R.

Noe, F.F. and Nickerson, W.J. (1958). Metabolism of 2-pyrrolidone and gamma-aminobutyric acid by *Pseudomonas aeruginosa*. *Journal of Bacteriology* 75 (6): 674–681. https://doi.org/10.1128/JB.75.6.674-681.1958.

O'Leary, B.M., Neale, H.C., Geilfus, C.M. et al. (2016). Early changes in apoplast composition associated with defence and disease in interactions between *Phaseolus vulgaris* and the halo blight pathogen *Pseudomonas syringae* Pv. phaseolicola. *Plant, Cell and Environment* 39 (10): 2172–2184. https://doi.org/10.1111/PCE.12770.

Oliveira, I.C. and Coruzzi, G.M. (1999). Carbon and amino acids reciprocally modulate the expression of glutamine synthetase in *Arabidopsis*. *Plant Physiology* 121 (1): 301–309. https://doi.org/10.1104/PP.121.1.301.

Pal, S., Zhao, J., Khan, A. et al. (2016). Paclobutrazol induces tolerance in tomato to deficit irrigation through diversified effects on plant morphology, physiology and metabolism. *Scientific Reports* 6: https://doi.org/10.1038/SREP39321.

Palanivelu, R., Brass, L., Edlund, A.F., and Preuss, D. (2003). Pollen tube growth and guidance is regulated by POP2, an *Arabidopsis* gene that controls GABA levels. *Cell* 114 (1): 47–59. https://doi.org/10.1016/S0092-8674(03)00479-3.

Park, D.H., Mirabella, R., Bronstein, P.A. et al. (2010). Mutations in γ-aminobutyric acid (GABA) transaminase genes in plants or *Pseudomonas syringae* reduce bacterial virulence. *The Plant Journal* 64 (2): 318–330. https://doi.org/10.1111/J.1365-313X.2010.04327.X.

Pérez-Alfocea, F., Santa-Cruz, A., Guerrier, G., and Bolarín, M.C. (1994). NaCl stress-induced organic solute changes on leaves and calli of *Lycopersicon esculentum*, *L. pennellii* and their interspecific hybrid. *Journal of Plant Physiology* 143 (1): 106–111. https://doi.org/10.1016/S0176-1617(11)82105-X.

E M Scott, W B Jakoby (1959). Soluble gamma-aminobutyric-glutamic transaminase from *Pseudomonas fluorescens*. https://pubmed.ncbi.nlm.nih.gov/13654294/.

Raggi, V. (1994). Changes in free amino acids and osmotic adjustment in leaves of water-stressed bean. *Physiologia Plantarum* 91 (3): 427–434. https://doi.org/10.1111/J.1399-3054.1994.TB02970.X.

Ramesh, S.A., Tyerman, S.D., Gilliham, M., and Xu, B. (2017). γ-Aminobutyric acid (GABA) signalling in plants. *Cellular and Molecular Life Sciences* 74 (9): 1577–1603. https://doi.org/10.1007/S00018-016-2415-7.

Ramos-Ruiz, R., Poirot, E., and Flores-Mosquera, M. (2018). GABA, a non-protein amino acid ubiquitous in food matrices. *Cogent Food & Agriculture* 4 (1): https://doi.org/10.1080/23311932.2018.1534323.

Ramputh, A.I. and Bown, A.W. (1996). Rapid [gamma]-aminobutyric acid synthesis and the inhibition of the growth and development of oblique-banded leaf-roller larvae. *Plant Physiology* 111 (4): 1349–1352. https://doi.org/10.1104/PP.111.4.1349.

Reggiani, R., Cantu, C.A., Brambilla, I., and Bertani, A. (1988). Accumulation and interconversion of amino acids in rice roots under anoxia. *Plant and Cell Physiology* 29 (6): 981–987. https://doi.org/10.1093/OXFORDJOURNALS.PCP.A077604.

Reggiani, R., Nebuloni, M., Mattana, M., and Brambilla, I. (2000). Anaerobic accumulation of amino acids in rice roots: role of the glutamine synthetase/glutamate synthase cycle. *Amino Acids* 18 (3): 207–217. https://doi.org/10.1007/S007260050018.

Reyes-Darias, J.A., García, V., Rico-Jiménez, M. et al. (2015). Specific gamma-aminobutyrate chemotaxis in pseudomonads with different lifestyle. *Molecular Microbiology* 97 (3): 488–501. https://doi.org/10.1111/MMI.13045.

Rhodes, D., Handa, S., and Bressan, R.A. (1986). Metabolic changes associated with adaptation of plant cells to water stress. *Plant Physiology* 82 (4): 890–903. https://doi.org/10.1104/PP.82.4.890.

Richard, H.T. and Foster, J.W. (2003). Acid resistance in *Escherichia coli*. *Advances in Applied Microbiology* 52: 167–186. https://doi.org/10.1016/S0065-2164(03)01007-4.

Richard, H. and Foster, J.W. (2004). *Escherichia coli* glutamate- and arginine-dependent acid resistance systems increase internal pH and reverse transmembrane potential. *Journal of Bacteriology* 186 (18): 6032. https://doi.org/10.1128/JB.186.18.6032-6041.2004.

Rico, A. and Preston, G.M. (2008). *Pseudomonas syringae* pv. tomato DC3000 uses constitutive and apoplast-induced nutrient assimilation pathways to catabolize nutrients that are abundant in the tomato apoplast. *Molecular Plant-Microbe Interactions* 21 (2): 269–282. https://doi.org/10.1094/MPMI-21-2-0269.

Rico-Jiménez, M., Muñoz-Martínez, F., García-Fontana, C. et al. (2013). Paralogous chemoreceptors mediate chemotaxis towards protein amino acids and the non-protein amino acid gamma-aminobutyrate (GABA). *Molecular Microbiology* 88 (6): 1230–1243. https://doi.org/10.1111/MMI.12255.

Rossi, S., Chapman, C., Yuan, B., and Huang, B. (2021). Improved heat tolerance in creeping bentgrass by γ-aminobutyric acid, proline, and inorganic nitrogen associated with differential regulation of amino acid metabolism. *Plant Growth Regulation* 93 (2): 231–242. https://doi.org/10.1007/S10725-020-00681-6/METRICS.

Saijo, Y. and Loo, E.P. (2020). Plant immunity in signal integration between biotic and abiotic stress responses. *The New Phytologist* 225 (1): 87–104. https://doi.org/10.1111/NPH.15989.

Salah, A., Zhan, M., Cao, C. et al. (2019). γ-Aminobutyric acid promotes chloroplast ultrastructure, antioxidant capacity, and growth of waterlogged maize seedlings. *Scientific Reports* 9 (1): https://doi.org/10.1038/S41598-018-36334-Y.

Salvatierra, A., Pimentel, P., Almada, R., and Hinrichsen, P. (2016). Exogenous GABA application transiently improves the tolerance to root hypoxia on a sensitive genotype of *Prunus rootstock. Environmental and Experimental Botany* 125: 52–66. https://doi.org/10.1016/J.ENVEXPBOT.2016.01.009.

Salvatore, M.M., Félix, C., Lima, F. et al. (2020). Effect of γ-aminobutyric acid (GABA) on the metabolome of two strains of lasiodiplodia theobromae isolated from grapevine. *Molecules* 25 (17): https://doi.org/10.3390/MOLECULES25173833.

Scholz, S.S., Reichelt, M., Mekonnen, D.W. et al. (2015). Insect herbivory-elicited GABA accumulation in plants is a wound-induced, direct, systemic, and jasmonate-independent defense response. *Frontiers in Plant Science* 6 (December): https://doi.org/10.3389/FPLS.2015.01128.

Schulz, P., Herde, M., and Romeis, T. (2013). Calcium-dependent protein kinases: hubs in plant stress signaling and development. *Plant Physiology* 163 (2): 523–530. https://doi.org/10.1104/PP.113.222539.

Seifi, H.S. and Shelp, B.J. (2019). Spermine differentially refines plant defense responses against biotic and abiotic stresses. *Frontiers in Plant Science* 10: https://doi.org/10.3389/FPLS.2019.00117.

Seifi, H.S., Curvers, K., De Vleesschauwer, D. et al. (2013a). Concurrent overactivation of the cytosolic glutamine synthetase and the GABA shunt in the ABA-deficient sitiens mutant of tomato leads to resistance against *Botrytis cinerea. The New Phytologist* 199 (2): 490–504. https://doi.org/10.1111/NPH.12283.

Seifi, H.S., Van Bockhaven, J., Angenon, G., and Höfte, M. (2013b). Glutamate metabolism in plant disease and defense: friend or foe? *Molecular Plant-Microbe Interactions* 26 (5): 475–485. https://doi.org/10.1094/MPMI-07-12-0176-CR.

Seifikalhor, M., Aliniaeifard, S., Hassani, B. et al. (2019). Diverse role of γ-aminobutyric acid in dynamic plant cell responses. *Plant Cell Reports* 38 (8): 847–867. https://doi.org/10.1007/S00299-019-02396-Z.

Serraj, R., Shelp, B.J., and Sinclair, T.R. (1998). Accumulation of γ-aminobutyric acid in nodulated soybean in response to drought stress. *Physiologia Plantarum* 102 (1): 79–86. https://doi.org/10.1034/J.1399-3054.1998.1020111.X.

Seybold, H., Trempel, F., Ranf, S. et al. (2014). Ca^{2+} signalling in plant immune response: from pattern recognition receptors to Ca^{2+} decoding mechanisms. *The New Phytologist* 204 (4): 782–790. https://doi.org/10.1111/NPH.13031.

Shang, H., Cao, S., Yang, Z. et al. (2011). Effect of exogenous γ-aminobutyric acid treatment on proline accumulation and chilling injury in peach fruit after long-term cold storage. *Journal of Agricultural and Food Chemistry* 59 (4): 1264–1268. https://doi.org/10.1021/JF104424Z.

Shelp, B.J. and Zarei, A. (2017). Subcellular compartmentation of 4-aminobutyrate (GABA) metabolism in arabidopsis: an update. *Plant Signaling & Behavior* 12 (5): e1322244. https://doi.org/10.1080/15592324.2017.1322244.

Shelp, B.J., Bown, A.W., and McLean, M.D. (1999). Metabolism and functions of gamma-aminobutyric acid. *Trends in Plant Science* 4 (11): 446–452. https://doi.org/10.1016/S1360-1385(99)01486-7.

Shelp, B.J., Bown, A.W., and Faure, D. (2006). Extracellular γ-aminobutyrate mediates communication between plants and other organisms. *Plant Physiology* 142 (4): 1350–1352. https://doi.org/10.1104/PP.106.088955.

Shelp, B.J., Bozzo, G.G., Trobacher, C.P. et al. (2012). Hypothesis/review: contribution of putrescine to 4-aminobutyrate (GABA) production in response to abiotic stress. *Plant Science: An International Journal of Experimental Plant Biology* 193–194: 130–135. https://doi.org/10.1016/J.PLANTSCI.2012.06.001.

Shelp, B.J., Bown, A.W., and Zarei, A. (2017). 4-Aminobutyrate (GABA): a metabolite and signal with practical significance. *Botany* 95 (11): 1015–1032. https://doi.org/10.1139/Cjb-2017-0135.

Shelp, B.J., Aghdam, M.S., and Flaherty, E.J. (2021). γ-Aminobutyrate (GABA) regulated plant defense: mechanisms and opportunities. *Plants* 10 (9): 1939. https://doi.org/10.3390/PLANTS10091939.

Sheng, L., Shen, D., Luo, Y. et al. (2017). Exogenous γ-aminobutyric acid treatment affects citrate and amino acid accumulation to improve fruit quality and storage performance of postharvest citrus fruit. *Food Chemistry* 216: 138–145. https://doi.org/10.1016/J.FOODCHEM.2016.08.024.

Shi, S.Q., Shi, Z., Jiang, Z.P. et al. (2010). Effects of exogenous GABA on gene expression of *Caragana intermedia* roots under NaCl stress: regulatory roles for H_2O_2 and ethylene production. *Plant, Cell & Environment* 33 (2): 149–162. https://doi.org/10.1111/J.1365-3040.2009.02065.X.

Sidhu, O.P., Annarao, S., Pathre, U. et al. (2010). Metabolic and histopathological alterations of Jatropha mosaic begomovirus-infected *Jatropha curcas* L. by HR-MAS NMR spectroscopy and magnetic resonance imaging. *Planta* 232 (1): 85–93. https://doi.org/10.1007/S00425-010-1159-0.

Signorelli, S., Dans, P.D., Coitiño, E.L. et al. (2015). Connecting proline and γ-aminobutyric acid in stressed plants through non-enzymatic reactions. *PLoS One* 10 (3): https://doi.org/10.1371/JOURNAL.PONE.0115349.

Singh, A., Kumar, A., Yadav, S., and Singh, I.K. (2019). Reactive oxygen species-mediated signaling during abiotic stress. *Plant Gene* 18: 100173. https://doi.org/10.1016/J.PLGENE.2019.100173.

Siragusa, S., De Angelis, M., Di Cagno, R. et al. (2007). Synthesis of γ-aminobutyric acid by lactic acid bacteria isolated from a variety of italian cheeses. *Applied and Environmental Microbiology* 73 (22): 7283. https://doi.org/10.1128/AEM.01064-07.

Snedden, W.A., Arazi, T., Fromm, H., and Shelp, B.J. (1995). Calcium/calmodulin activation of soybean glutamate decarboxylase. *Plant Physiology* 108 (2): 543–549. https://doi.org/10.1104/PP.108.2.543.

Solomon, P.S. and Oliver, R.P. (2002). Evidence that gamma-aminobutyric acid is a major nitrogen source during *Cladosporium fulvum* infection of tomato. *Planta* 214 (3): 414–420. https://doi.org/10.1007/S004250100632.

Steward, F.C., Thompson, J.F., and Dent, C.E. (1949). γ-Aminobutyric acid: a constituent of potato tubers? *Science* 110: 439–440. Retrieved from https://cir.nii.ac.jp/crid/1573387449104528512.

Streeter, J.G. and Thompson, J.F. (1972). Anaerobic accumulation of gamma-aminobutyric acid and alanine in radish leaves (*Raphanus sativus* L.). *Plant Physiology* 49 (4): 572–578. https://doi.org/10.1104/PP.49.4.572.

Stress', W., Rhodes, D., Handa, S., and Bressan, R.A. (1986). Metabolic changes associated with adaptation of plant cells to water stress. *Plant Physiology* 82 (4): 890–903. https://doi.org/10.1104/PP.82.4.890.

Su, N., Wu, Q., Chen, J. et al. (2019). GABA operates upstream of H^+-ATPase and improves salinity tolerance in *Arabidopsis* by enabling cytosolic K^+ retention and Na^+ exclusion. *Journal of Experimental Botany* 70 (21): 6349–6361. https://doi.org/10.1093/JXB/ERZ367.

Sung, M.S., Chow, T.J., and Lee, T.M. (2011). Polyamine acclimation alleviates hypersalinity-induced oxidative stress in a marine green macroalga, *Ulva fasciata*, by modulation of antioxidative enzyme gene expression1. *Journal of Phycology* 47 (3): 538–547. https://doi.org/10.1111/J.1529-8817.2011.00999.X.

Sweetman, C., Sadras, V.O., Hancock, R.D. et al. (2014). Metabolic effects of elevated temperature on organic acid degradation in ripening *Vitis vinifera* fruit. *Journal of Experimental Botany* 65 (20): 5975–5988. https://doi.org/10.1093/JXB/ERU343.

Takahashi, H., Matsumura, H., Kawai-Yamada, M., and Uchimiya, H. (2008). The cell death factor, cell wall elicitor of rice blast fungus (*Magnaporthe grisea*) causes metabolic alterations including GABA shunt in rice cultured cells. *Plant Signaling & Behavior* 3 (11): 945–953. https://doi.org/10.4161/PSB.6112.

Takahashi, Y., Sasanuma, T., and Abe, T. (2013). Accumulation of gamma-aminobutyrate (GABA) caused by heat-drying and expression of related genes in immature vegetable soybean (edamame). *Breeding Science* 63 (2): 205–210. https://doi.org/10.1270/JSBBS.63.205.

Tarkowski, Ł.P., de Poel, B., Van, H., and M., and Van den Ende, W. (2019). Sweet immunity: inulin boosts resistance of lettuce (*Lactuca sativa*) against grey mold (*Botrytis cinerea*) in an ethylene-dependent manner. *International Journal of Molecular Sciences* 20 (5): https://doi.org/10.3390/IJMS20051052.

Tarkowski, Ł.P., Signorelli, S., and Höfte, M. (2020). γ-Aminobutyric acid and related amino acids in plant immune responses: emerging mechanisms of action. *Plant, Cell & Environment* 43 (5): 1103–1116. https://doi.org/10.1111/PCE.13734.

Tauzin, A.S. and Giardina, T. (2014). Sucrose and invertases, a part of the plant defense response to the biotic stresses. *Frontiers in Plant Science* 5 (JUN): https://doi.org/10.3389/FPLS.2014.00293.

Thompson, J.F., Stewart, C.R., and Morris, C.J. (1966). Changes in amino acid content of excised leaves during incubation I. The effect of water content of leaves and atmospheric oxygen level. *Plant Physiology* 41 (10): 1578–1584. https://doi.org/10.1104/PP.41.10.1578.

Tian, W., Hou, C., Ren, Z. et al. (2019). A calmodulin-gated calcium channel links pathogen patterns to plant immunity. *Nature* 572 (7767): 131–135. https://doi.org/10.1038/s41586-019-1413-y.

Toubiana, D., Sade, N., Liu, L. et al. (2020). Correlation-based network analysis combined with machine learning techniques highlight the role of the GABA shunt in *Brachypodium sylvaticum* freezing tolerance. *Scientific Reports* 10 (1): https://doi.org/10.1038/S41598-020-61081-4.

Tramonti, A., De Canio, M., Delany, I. et al. (2006). Mechanisms of transcription activation exerted by GadX and GadW at the gadA and gadBC gene promoters of the glutamate-based acid resistance system in *Escherichia coli*. *Journal of Bacteriology* 188 (23): 8118–8127. https://doi.org/10.1128/JB.01044-06.

Tumewu, S.A., Matsui, H., Yamamoto, M. et al. (2020). Requirement of γ-aminobutyric acid chemotaxis for virulence of *Pseudomonas syringae* pv. tabaci 6605. *Microbes and Environment* 35 (4): https://doi.org/10.1264/jsme2.ME20114.

Wali, M., Gunsè, B., Llugany, M. et al. (2016). High salinity helps the halophyte *Sesuvium portulacastrum* in defense against Cd toxicity by maintaining redox balance and photosynthesis. *Planta* 244 (2): 333–346. https://doi.org/10.1007/S00425-016-2515-5.

Wallace2, W., Secor3, J., and Schrader, L. E. (1984). Rapid accumulation of γ-aminobutyric acid and alanine in soybean leaves in response to an abrupt transfer to lower temperature, darkness, or mechanical manipulation. *Plant Physiology* 75 (1): 170–175. https://doi.org/10.1104/PP.75.1.170.

Wang, Y., Luo, Z., Huang, X. et al. (2014). Effect of exogenous γ-aminobutyric acid (GABA) treatment on chilling injury and antioxidant capacity in banana peel. *Scientia Horticulturae* 168: 132–137. https://doi.org/10.1016/J.SCIENTA.2014.01.022.

Wang, Y., Gu, W., Meng, Y. et al. (2017). γ-Aminobutyric acid imparts partial protection from salt stress injury to maize seedlings by improving photosynthesis and upregulating osmoprotectants and antioxidants. *Scientific Reports* 7: https://doi.org/10.1038/SREP43609.

Wang, Q., Chen, D., Wu, M. et al. (2018). MFS transporters and GABA metabolism are involved in the self-defense against DON in *Fusarium graminearum*. *Frontiers in Plant Science* 9: https://doi.org/10.3389/FPLS.2018.00438.

Wang, G., Kong, J., Cui, D. et al. (2019a). Resistance against *Ralstonia solanacearum* in tomato depends on the methionine cycle and the γ-aminobutyric acid metabolic pathway. *The Plant Journal: For Cell and Molecular Biology* 97 (6): 1032–1047. https://doi.org/10.1111/TPJ.14175.

Wang, J., Liu, X., Zhang, A. et al. (2019b). A cyclic nucleotide-gated channel mediates cytoplasmic calcium elevation and disease resistance in rice. *Cell Research* 29 (10): 820–831. https://doi.org/10.1038/s41422-019-0219-7.

Ward, J.L., Forcat, S., Beckmann, M. et al. (2010). The metabolic transition during disease following infection of *Arabidopsis thaliana* by *Pseudomonas syringae* pv. tomato. *The Plant Journal: For Cell and Molecular Biology* 63 (3): 443–457. https://doi.org/10.1111/J.1365-313X.2010.04254.X.

Wu, F., Chi, Y., Jiang, Z. et al. (2020a). Hydrogen peroxide sensor HPCA1 is an LRR receptor kinase in *Arabidopsis*. *Nature* 578 (7796): 577–581. https://doi.org/10.1038/s41586-020-2032-3.

Wu, X., Jia, Q., Ji, S. et al. (2020b). Gamma-aminobutyric acid (GABA) alleviates salt damage in tomato by modulating Na^+ uptake, the GAD gene, amino acid synthesis and reactive oxygen species metabolism. *BMC Plant Biology* 20 (1): https://doi.org/10.1186/S12870-020-02669-W.

Xian, L., Yu, G., Wei, Y. et al. (2020). A bacterial effector protein hijacks plant metabolism to support pathogen nutrition. *Cell Host & Microbe* 28 (4): 548–557.e7. https://doi.org/10.1016/J.CHOM.2020.07.003.

Xian, L., Yu, G., and Macho, A.P. (2021). The GABA transaminase GabT is required for full virulence of *Ralstonia solanacearum* in tomato. *MicroPublication Biology* https://doi.org/10.17912/MICROPUB.BIOLOGY.000478.

Xue, H., Lozano-Durán, R., and Macho, A.P. (2020). Insights into the root invasion by the plant pathogenic bacterium *Ralstonia solanacearum*. *Plants* 9 (4): https://doi.org/10.3390/PLANTS9040516.

Yadav, M., Pandey, J., Chakraborty, A. et al. (2022). A comprehensive analysis of calmodulin-like proteins of glycine max indicates their role in calcium signaling and plant defense against insect attack. *Frontiers in Plant Science* 13: https://doi.org/10.3389/FPLS.2022.817950.

Yang, R., Guo, Y., Wang, S., and Gu, Z. (2015). Ca^{2+} and aminoguanidine on γ-aminobutyric acid accumulation in germinating soybean under hypoxia-NaCl stress. *Journal of Food and Drug Analysis* 23 (2): 287–293. https://doi.org/10.1016/J.JFDA.2014.07.004.

Yang, R., Feng, L., Wang, S. et al. (2016). Accumulation of γ-aminobutyric acid in soybean by hypoxia germination and freeze-thawing incubation. *Journal of the Science of Food and Agriculture* 96 (6): 2090–2096. https://doi.org/10.1002/JSFA.7323.

Yang, J., Sun, C., Zhang, Y. et al. (2017). Induced resistance in tomato fruit by γ-aminobutyric acid for the control of alternaria rot caused by *Alternaria alternata*. *Food Chemistry* 221: 1014–1020. https://doi.org/10.1016/J.FOODCHEM.2016.11.061.

Yang, B., Chen, M., Wang, T. et al. (2019). A metabolomic strategy revealed the role of JA and SA balance in *Clematis terniflora* DC. Response to UVB radiation and dark. *Physiologia Plantarum* 167 (2): 232–249. https://doi.org/10.1111/PPL.12883.

Yong, B., Xie, H., Li, Z. et al. (2017). Exogenous application of GABA improves PEG-induced drought tolerance positively associated with GABA-shunt, polyamines, and proline metabolism in white clover. *Frontiers in Physiology* 8 (December): 305904. https://doi.org/10.3389/FPHYS.2017.01107/BIBTEX.

Yu, C., Zeng, L., Sheng, K. et al. (2014). γ-Aminobutyric acid induces resistance against *Penicillium* expansum by priming of defence responses in pear fruit. *Food Chemistry* 159: 29–37. https://doi.org/10.1016/J.FOODCHEM.2014.03.011.

Yu, X., Xu, G., Li, B. et al. (2019). The receptor kinases BAK1/SERK4 regulate Ca^{2+} channel-mediated cellular homeostasis for cell death containment. *Current Biology* 29 (22): 3778–3790.e8. https://doi.org/10.1016/j.cub.2019.09.018.

Yuan, P., Tanaka, K., and Poovaiah, B.W. (2022). Calcium/calmodulin-mediated defense signaling: what is looming on the horizon for AtSR1/CAMTA3-mediated signaling in plant immunity. *Frontiers in Plant Science* 12: https://doi.org/10.3389/FPLS.2021.795353.

Yuan, D., Wu, X., Gong, B. et al. (2023). GABA metabolism, transport and their roles and mechanisms in the regulation of abiotic stress (hypoxia, salt, drought) resistance in plants. *Metabolites* 13 (3): https://doi.org/10.3390/METABO13030347.

Zaini, P.A., Nascimento, R., Gouran, H. et al. (2018). Molecular profiling of pierce's disease outlines the response circuitry of *Vitis vinifera* to *Xylella fastidiosa* infection. *Frontiers in Plant Science* 9: https://doi.org/10.3389/FPLS.2018.00771.

Zhang, Y., Xu, S., Ding, P. et al. (2010). Control of salicylic acid synthesis and systemic acquired resistance by two members of a plant-specific family of transcription factors. *Proceedings of the National Academy of Sciences of the United States of America* 107 (42): 18220–18225. https://doi.org/10.1073/PNAS.1005225107.

Zhao, Z., Ding, J.Y., Ma, W., and hua, Zhou, N. Y., and Liu, S. J. (2012). Identification and characterization of γ-aminobutyric acid uptake system GabPCg (NCgl0464) in *Corynebacterium glutamicum*. *Applied and Environmental Microbiology* 78 (8): 2596. https://doi.org/10.1128/AEM.07406-11.

Zhao, Y., Qi, Z., and Berkowitz, G.A. (2013). Teaching an old hormone new tricks: cytosolic Ca^{2+} elevation involvement in plant brassinosteroid signal transduction cascades. *Plant Physiology* 163 (2): 555–565. https://doi.org/10.1104/PP.112.213371.

Zhu, X., Liao, J., Xia, X. et al. (2019). Physiological and iTRAQ-based proteomic analyses reveal the function of exogenous γ-aminobutyric acid (GABA) in improving tea plant (*Camellia sinensis* L.) tolerance at cold temperature. *BMC Plant Biology* 19 (1): https://doi.org/10.1186/S12870-019-1646-9.

Zhuo, T., Wang, X., Chen, Z. et al. (2020). The *Ralstonia solanacearum* effector RipI induces a defence reaction by interacting with the bHLH93 transcription factor in *Nicotiana benthamiana*. *Molecular Plant Pathology* 21 (7): 999–1004. https://doi.org/10.1111/MPP.12937.

16

GABA-Mediated Salt Stress Tolerance Through Physiological and Molecular Mechanisms

Riya Johnson[1], *Joy Mulakkal Joel*[1], *Koravantakamparambil Sulaiman Anjitha*[1], *Louis Noble*[1], *Parammal Faseela*[2], *and Jos Thomas Puthur*[1]

[1] *Plant Physiology and Biochemistry Division, Department of Botany, University of Calicut, Malappuram, Kerala, 673635, India*
[2] *Department of Botany, Kahm Unity Women's College, Manjeri, Kerala, 676122, India*

Introduction

Salt stress conditions pose a significant challenge, as it is one of the damaging environmental stresses that curtail crop productivity and quality. Increased soil salinity causes cascades of reactions in plants at both the morphophysiological and molecular levels due to ionic toxicity, osmotic stress, and oxidative stress simultaneously (Hasanuzzaman and Fujita 2022). Notably, an excess of salt and chlorine ions can adversely affect plant root systems. Moreover, the presence of excessive sodium and chlorine ions hampers the retention of other essential ions, thereby impeding overall plant development and disrupting metabolic processes in response to salt stress. Additionally, salt stress leads to reduced water absorption capabilities in plants, contributing to dry conditions (Acharya et al. 2022). The detrimental effects of salt stress trigger the production of reactive oxygen species (ROS) within plants, further harming them through oxidative processes. It is crucial to recognize that salt stress adversely impacts all stages of plant development, with seed germination being particularly affected (Kumar et al. 2020).

In recent times, abiotic stressors have prompted researchers to explore newer techniques as a means to enhance global agricultural production. One noteworthy molecule in this context is gamma-aminobutyric acid (GABA), a nonprotein ubiquitous molecule first discovered in plants (potato tuber) and distributed throughout prokaryotic and eukaryotic organisms (Shelp et al. 1999). It is a crucial metabolite/endogenous signaling molecule in various regulatory systems in plants, and it accumulates in plant cells under adverse environmental conditions. Various biotic and abiotic stress conditions, including low temperature, radiation, low pH, hypoxia, darkness, and leaf senescence, trigger a rapid accumulation of GABA in plants (Ansari et al. 2014). Also, GABA plays a critical role in carbon and nitrogen metabolism, essential for plant development (Ansari et al. 2021). Under salt-induced desiccation, GABA functions as an active osmolyte without adverse

GABA in Plants: Biosynthesis, Plant Development, and Food Security, First Edition. Edited by Samiksha Singh, Durgesh Kumar Tripathi, and Vijay Pratap Singh.

effects and possesses free radical scavenging properties (Carillo 2018). The GABA-shunt pathway also plays a pivotal role in regulating GABA metabolism in cytosol by regulating glutamate dehydrogenase (GAD) through decarboxylating glutamate into GABA (Jalil and Ansari 2020).

Furthermore, the application of GABA has been found to promote plant growth, enhance antioxidant metabolism, and stimulate the production of antioxidant genes in a dose-dependent manner, thereby mitigating oxidative damage caused by stress in plants. GABA application also aids in regulating plant osmoregulation, thereby improving stress resistance. Specifically, exogenous GABA efficiently inhibits free radical production and reduces oxidative damage by modulating the expression of key genes related to free radical production and genes encoding antioxidant enzymes during salt stress (Li et al. 2017). Studies on white clover seeds treated with GABA have shown increased germination rates and reduced salt-related damage during germination (Cheng et al. 2018). While exogenous GABA administration has been shown to enhance abiotic stress tolerance in many plants, there remains a scarcity of information on how GABA regulates salt tolerance. This chapter explores the interconnectedness between osmoregulation, the antioxidant defense system, and important gene expression during salt stress, all of which are influenced by GABA-regulated salt tolerance.

Concept of Salt Stress to Plants

Various biotic and abiotic stresses adversely affect plant growth and development. Salinity is one of the major global problems that negatively affect crop productivity and global food security. The presence of excessive salt concentrations makes soil and water saline. Salt stress is typically brought on by high Na^+ and Cl^- levels. Salt stress lowers water potential, creates ion imbalances, disturbs ion homeostasis, and increases toxicity. Growth inhibition is directly connected to the osmotic potential of soil water or the concentration of soluble salts, as salt stress involves both osmotic and ionic stress. Reduced early development and plant productivity result from this altered water status. According to estimates, more than one billion hectares (ha) of the earth's land are thought to have been salinized. As the world's population grows, the food insecurity problem is made worse by the salinization of the world's shrinking agricultural area (Soltabayeva et al. 2021).

The impacts of stress on morphology can be seen in a variety of ways, including plant height, total biomass (fresh or dry weight), and other morphological markers. Increased salt concentration in the growth environment increases the effects of salinity stress on the development of plants. Germination is a useful indicator of salinity stress because salinity stress influences the pace at which seeds germinate. As one of the most important morphological processes, germination is a better stress indicator because stress can be detected at the early stages of the plant (Soltabayeva et al. 2021). At 100 mM NaCl, the germination rates of tomato and sunflower decreased by 71 and 62%, respectively. While the weight of tomato and sunflower decreased with the application of 50 mM NaCl, the weight of wheat, maize, rice, *Arabidopsis*, and ryegrass decreased at levels of 100–150 mM NaCl. In citrus and acacia, 100 mM NaCl generated biomass modifications during the early stages of growth (Pérez-Tornero et al. 2009; Ghaleb et al. 2010; Abbas et al. 2013).

Reduced photosynthesis due to salinity stress is caused by disruption of the photosynthetic machinery in plants. Increased soil NaCl levels cause osmotic stress, which limits cell growth rate and shrinks stomatal aperture, limiting photosynthetic efficiency. The first noticeable effect of salt stress on plants is the reduction in leaf area. In the initial stages, the smaller leaf surface area caused by the reduced cell extensibility is more significant than the lower photosynthetic rate per unit area (Shabala and Lew 2002). The drop in the regeneration capacity of ribulose bisphosphate (RuBP) and PSII sensitivity are further factors contributing to the reduction in photosynthetic activity. Salt stress interferes with the photosynthetic machinery by affecting the orientation of the chloroplast lamellar system and chloroplast integrity, resulting in reduced photosystem activity (Sharif et al. 2019). The decline in photosynthesis is linked to reduction in chlorophyll content and changes in the chlorophyll's ultrastructure. Chlorophyll content can be considered one of the physiological criteria that is a helpful predictor for choosing salt-tolerant cultivars. With an increase in salinity level, cotton cultivars' chlorophyll contents were found to significantly decrease. Inhibition of enzymes involved in chlorophyll production may be the cause of this decline (Zhang et al. 2014).

The osmotic effects of salt stress on a range of plant metabolic processes result in water deficit, which leads to oxidative stress due to the generation of ROS like superoxides and hydroxyl and peroxy radicals. ROS, which are by-products of ionic and hyperosmotic stressors, lead to membrane dysfunction and cell death. Membrane proteins, lipids, and nucleic acids are oxidatively damaged as a result of this rise in ROS levels in plant tissues (AbdElgawad et al. 2016). Plants create an effective system of enzymatic and nonenzymatic antioxidants to scavenge high ROS levels. Reduced ascorbate (ASC), glutathione (GSH), flavonoids, phenolics, and tocopherols are examples of nonenzymatic antioxidants. Enzymatic antioxidants involved in ROS detoxification include catalase (CAT), superoxide dismutase (SOD), peroxidase (POX), and enzymes of the ascorbate-glutathione cycle, like ASC peroxidase (APX), GSH reductase (GR), monodehydroascorbate dehydrogenase (MDHAR), and dehydroascorbate reductase (DHAR) (Gill and Tuteja 2010).

High salt uptake inhibits the uptake of other nutritional ions, particularly K^+, resulting in a K^+ deficit. Numerous plants respond to increased NaCl treatment by increasing Na^+ and Cl^- levels while decreasing Mg^{2+}, K^+, and Ca^{2+} levels (Isayenkov and Maathuis 2019). A mechanism utilized by plants to balance the osmoticum between the cytosol, vacuole, and external environment is the increase in the content of soluble organic molecules in the cytoplasm of plants exposed to salt stress. These substances also stabilize enzyme structures and protect enzymatic systems. Certain compounds, such as proline, glycine, β-alanine, glycine betaine, and complex sugars like raffinose, that plants frequently produce and/or accumulate in stressful environments can be used to modify the osmotic pressure (Rahneshan et al. 2018).

Salt Stress and Related Metabolic Changes

Different abiotic stressors trigger sensitive and dynamic responses from plant metabolism. Metabolic alterations are a key component of plant stress responses because they can occur quickly in response to stresses. Salinity harms plants at almost every stage of life, including seed germination, growth, and development. These effects are a result of salinity-induced activations of molecular networks that are involved in stress recognition, ionic balance, signal transduction, modulation of the expression of stress-related genes, protein, and consequently, metabolisms (Balasubramaniam et al. 2023). Metabolic pathways that produce metabolites are varied and complex in plants. Both primary and secondary metabolites aid in plant adaptability and are intermediates or by-products of cellular regulation mechanisms. Due to their ability to supply energy and chemical building blocks, primary metabolites such as amino acids, sugars, and lipids are essential for the growth and development of plants. Flavonoids, terpenoids, phenolics, saponins, alkaloids, cardiac glycosides, proanthocyanidins, and quinones are just a few of the specialized molecules known as secondary metabolites that have antioxidant activity (Arif et al. 2020). Stress-related increases in the production of secondary metabolites shield cellular structures from the oxidative damage and are essential for interacting with the environment by supplying resistance to biotic and abiotic stressors.

When plants sense salinity stress, they undergo metabolic reprogramming involving modifications in the production of primary and secondary metabolites for regulating osmotic homeostasis and signaling pathways. The accumulation of compatible solutes such as proline, hydroxyproline, sugars, polyamines, glycine betaine, and proteins from the late embryogenesis abundant (LEA) superfamily is an important strategy utilized by plants for maintaining a low intracellular osmotic potential during high salinity. Proline, one of these metabolites, dominates in osmotic adjustment during salt stress. Stimulation of the proline biosynthesis route and the inhibition of the proline catabolic route can result in proline accumulation under osmotic stress (Zhao et al. 2020).

Salinity causes disturbances in secondary metabolites with osmoregulatory functions that are frequently engaged in plant defense and stress adaptation. These substances also play a number of physiological roles, such as signal modulation, ROS scavenging, enzyme activation, and photoprotection. In general, the accumulation of phenolics and flavonoids was significantly influenced by the degree of salt stress (Arif et al. 2020). NaCl stress substantially raised the content of phenolics, as well as the transcription of phenylalanine ammonia-lyase (PAL) genes and their activity in *Lonicera japonica* (Yan et al. 2016). NaCl stress in *Salicornia brachiata* caused a clear and consistent upregulation of numerous flavonoids. There was a noticeable buildup of quercetin, kaempferol, and anthocyanin derivatives such as cyanidin, delphinidin, and hesperidin in the shoots and roots. Additionally, with higher NaCl dosages, *S. brachiata* had significantly higher amounts of flavones and peonidin. In salt-treated *S. brachiata* and *Sesuvium portulacastrum* plants, phenylpropanoids and lignans were upregulated. 1-*O*-feruloyl-D-glucose, 1-*O*-caffeoyl-D-glucose, and curcumin diglucoside were among the phenylpropanoid compounds that increased at the higher salt level. Lignans like secoisolariciresinol, sesamolin, pinoresinol, and matairesinol were also accumulated at maximum NaCl dosage (Benjamin et al. 2019).

GABA and Salinity Stress Tolerance

Under circumstances of abiotic stresses such as salinity stress, GABA tends to build up within the plant cells (Mekonnen et al. 2016; Ma et al. 2018). Earlier studies have suggested that the utilization of GABA enhances plants' ability to withstand salinity stress by promoting the activity of antioxidant enzymes and mitigating the accumulation of ROS within the cells (Jalil and Ansari 2020), regulating the C/N ratio, Na^+ and K^+ absorption, enhancing phenolic bioactive compounds, amino acids, and proline, and controlling cell osmosis (Cheng et al. 2018; Xu et al. 2019; Khan et al. 2021). In a study by Wang et al. (2017), GABA application under salinity treatment led to an increase in sugar content, proline accumulation, and seedling growth in maize, when compared to untreated plantlets (Wang et al. 2017). GABA metabolism within *Arabidopsis thaliana* was commonly upregulated in response to NaCl stress (Renault et al. 2010). Recent work on GABA administration in plants under salinity stress is represented in Table 16.1. The NaCl stress initiates the signal transduction pathway, resulting in an increase in the cytosolic calmodulin-dependent activity of the L-glutamate decarboxylase enzyme. Nonetheless, it appears that GABA is essential for plant development. Exogenous GABA has been shown to increase plants' adaptation toward stressed environmental circumstances by enhancing overall growth, photosynthetic efficiency, antioxidative defense mechanisms, and nitrogen metabolism (Vijayakumari et al. 2016; Ma et al. 2018; Salah et al. 2019). The functions of GABA vary across different species, depending on the specific cellular compartment where it is active. A schematic model representing the positive effect of GABA under salinity stress is depicted in Figure 16.1.

GABA Improves Photosynthesis and Chlorophyll Fluorescence Parameters Under Salt Stress

Photosynthetic apparatus and photosynthesis are the targets that are greatly damaged by salt stress. Salt stress inhibits photosynthetic capacity by decreasing CO_2 fixation, stomatal conductance, and transpiration in plants. GABA, a non-proteinogenic amino acid in plant cells, probably plays a dual role as both a signaling molecule and a metabolite when exposed to various biotic and abiotic stresses. GABA application has beneficial effects on various photosynthesis-associated functions in plants. Salinity stress negatively affects plant photosynthetic capacity, and GABA effectively improves salt stress tolerance. Recently, exogenous GABA reduced salinity-associated stress damages, resulting in improved growth and biomass by decreasing ROS concentration but enhanced photosynthetic pigments of mungbean plants (Ullah et al. 2023a).

Exogenous GABA application improved photosynthetic capacity, net photosynthetic rate, transpiration rate, and stomatal conductance under salt stress in maize seedlings, which ultimately resulted in enhancement in CO_2 utilization in maize seedling leaves (Wang et al. 2017). Jin et al. (2019) studied the interactions between cellular redox signaling, chlorophyll biosynthesis, and PSII photochemical efficiency involved in GABA-induced salinity stress tolerance in *Cucumis melon* and found that pretreatment with exogenous GABA mitigated salinity stress by excessive accumulation of Chl and its precursors, thus avoiding photooxidation injury. Furthermore, the genetic, physiological, and molecular mechanisms of GABA-mediated effects and the regulation of stress signaling by GABA in plants under high-salt environmental conditions have been described recently (Dabravolski and Isayenkov 2023). Chen et al. (2018) reported that exogenous GABA increases the tolerance of *Cucumis melon* L. seedlings under salinity-alkalinity stress via effects on the chloroplast antioxidant system.

Table 16.1 Recent works related to the effects of GABA in plants under salinity stress.

Plant species	Salt conc.	GABA conc.	Effect of salinity on plants	Effect of GABA	Reference
Cucumis melon	50 mM alkalic salt ($NaCl/Na_2SO_4/NaHCO_3/Na_2CO_3$ 1 : 9 : 9 : 1 M ratio) solution	50 mM	MDA level, relative electrical conductivity (REC), and the activities of ascorbate peroxidase (APX), ascorbate reductase (DHAR), and SOD all increased during salt stress. Reduced shoot fresh and dry weight and leaf area, reduced ascorbate and glutathione (AsA and GSH) contents, and reduced monodehydroascorbate reductase (MDAR) activities and glutathione reductase (GR) were observed.	Pretreatment with GABA significantly inhibited stress-induced changes in H_2O_2, AsA, and GSH content.	Jin et al. (2019)
Cyperus esculentus	100 and 200 mM NaCl	2 mM	Substantial increase in H_2O_2, MDA, soluble sugars, proteins, and superoxide dismutase. Considerable decrease in growth, activities of peroxidase (POD) and catalase (CAT), glutamine oxoglutarate aminotransferase (GOGAT), NO_3^- and NH_4^+ concentrations, and nitrate reductase (NR).	Exogenous GABA application lowers chlorophyll degradation and improves the photosynthetic efficiency and growth under salinity stress. Compared to untreated GABA peers, the treated GABA peers had considerably lower levels of Na^+ and Cl^- ions, MDA and H_2O_2, as well as higher levels of osmolytes, chlorophyll pigments, antioxidant enzymes, Mg^{2+}, and K^+, and N metabolism.	Ullah et al. (2023b)
Lactuca sativa	40 and 80 mM NaCl	25 μM	Salinity caused a significant decline in the germination percentage and maximum quantum yield of PS II (F_v/F_m). Elevated electrolyte leakage and proline levels were also noted in plants treated with higher concentrations of salt.	GABA considerably shortened the mean germination time in lettuce seeds exposed to salt, improved F_v/F_m, reduced the specific energy fluxes per reaction center, and increased the electron transport flux in the photosynthetic system of lettuce. Furthermore, GABA treatment helped to recover the decrease in non-photochemical quenching (NPQ) and quenching coefficients (qN, qP, qL) caused by salt stress and decreased proline content. Improvement in L-ascorbate peroxidase, superoxide dismutase, and catalase activities were also shown by GABA + NaCl supplemented plants.	Kalhor et al. (2018)

Lolium perenne	175 mM NaCl	1 mM	Under 175 mM NaCl, germination percentage, shoot weight, shoot carbon, and nitrogen levels fell in comparison to the non-stressed control. Salinity stress resulted in increased activity of shoot SOD and decreased CAT and APX activities.	GABA improved seed germination, growth of seedlings, and C and N levels and altered carbohydrate metabolism and antioxidation machinery to improve salinity tolerance.	Tang et al. (2022)
Oryza sativa	150 mM NaCl	0.5 mM	Considerable increase in proline, glutathione reductase (GR), and MDA.	GABA priming significantly improved photosynthetic system and water relations, increased GABA, glutathione reductase levels, and reduced Na^+, proline, free radical, and MDA concentrations. Furthermore, it improved K^+ concentration, antioxidant enzyme activities, phenolic metabolism, and their transcription levels in seedlings under stress.	Sheteiwy et al. (2019)
Triticum aestivum	100 and 200 mM NaCl	0.5 mM	Decline of pigment content, SOD and CAT activities, and increase in MDA and electrolytic activity demonstrated by increased membrane damage.	GABA + NaCl groups were found to have higher net photosynthetic rates, intercellular CO_2 concentrations, PS II efficiency, and NPQ than NaCl groups. GABA + NaCl treatment significantly outperformed the NaCl groups in terms of SOD and CAT activity.	Li et al. (2016)
Triticum aestivum	150 mM NaCl	50 mg/l	Salt stress decreased dry weight, relative water content, and altered the ionic balance in the stressed seedlings.	Exogenous GABA has the power to prevent salt ion transfer to leaves and maintain the overall wellness of leaves. Under salt-stressed circumstances, pretreatment with GABA boosted seedling biomass, K^+ content in leaves, but lowered Na^+ concentration in leaves and roots via enhancing Na^+ exclusion and K^+ retention.	Wang et al. (2019)
Triticum aestivum	100 mM NaCl	2.0 mM	Increased glucose content, nitric oxide (NO) production, and oxidative stress, together with decreased growth traits and photosynthetic efficiency of plants.	GABA application enhanced nitrogen and sulfur assimilation, growth, photosynthesis, and ion homeostasis under salt stress. Oxidative stress mitigation was attained through changes in the ascorbate-glutathione cycle and proline metabolism.	Khanna et al. (2021)

(*Continued*)

Table 16.1 (Continued)

Plant species	Salt conc.	GABA conc.	Effect of salinity on plants	Effect of GABA	Reference
Vigna radiata	50 and 100 mM NaCl : Na_2SO_4 solution (1 : 1 ratio)	1.5 mM	Mungbean plants treated with salinity promote antioxidant enzymes to eliminate ROS at the cost of decreased growth and biomass, since more energy is allocated to antioxidant defense mechanisms instead of organ development.	Exogenous GABA promoted growth and physiological metabolism, greatly reducing the harms brought on by salt stress. Increased K^+/Na^+ ratio, improved production of photosynthetic pigments, and decreased H_2O_2 and MDA levels boosted osmolyte accumulations, and increased N concentration and N-metabolizing enzyme activity were observed.	Ullah et al. (2023a)
Zea mays	150 mM NaCl	0.5 mM	Excessive levels of Na^+ change the amount of potassium and calcium levels, enzyme activity, and metabolic processes, leading to the production of ROS and inducing oxidative stress in maize.	In general, GABA activated enzymes involved in nitrogen metabolism and restored the essential TCA cycle substrates through the modulation of GABA-shunt metabolism, enhancing the balance of carbon and nitrogen metabolism in maize and increasing salt tolerance.	Wang et al. (2023)

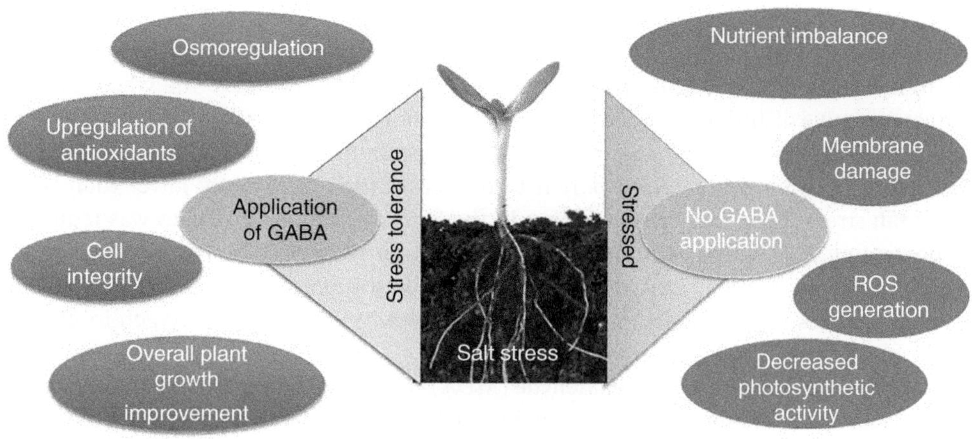

Figure 16.1 Schematic model representing the positive effect of GABA under salinity stress.

PSII is considered to be the primary site of the photosynthetic apparatus injury during salinity stress, due to the decreased excitation energy reaching PSII reaction centers (RCs) and changes in the pigment–protein complexes of thylakoid membranes. GABA protects PSII from salt stress damage by increasing the absorption of light energy and electron transfer in PSII and improves chlorophyll fluorescence parameters in plants. Based on the interpretation of the JIP-test and chlorophyll *a* fluorescence parameter, Xiang et al. (2016) concluded that exogenous GABA alleviated stress-related damage on the acceptor side of PSII in muskmelon seedlings. It also restored energy distribution, the reaction center status, and enhanced the ability of PSII to repair reaction centers in stressed seedlings. Thus, GABA may play a crucial role in protecting the chloroplast structure and function of PSII against the deleterious effects of salinity stress.

Moreover, chlorophyll fluorescence and photosynthesis improved under salinity stress in maize seedlings also (Wang et al. 2017). Additionally, GABA in combination with melatonin (*N*-acetyl-5-methoxytryptamine) effectively improved the salt stress tolerance of broad beans (*Vicia faba* L.) and increased the maximum quantum yield of PS II and minimized the negative effect of stress on the non-photochemical quenching and the energy fluxes of light absorption (Shomali et al. 2021). Even though salinity caused a significant decline in the maximum quantum yield of PSII (F_v/F_m) of lettuce plants, GABA application improved F_v/F_m. GABA decreased specific energy fluxes per reaction center (RC) for energy absorption and dissipation, while enhancing electron transport flux in the photosynthetic apparatus of lettuce plants. Moreover, the decline in non-photochemical quenching (NPQ) and quenching coefficients (qP, qL, qN) by salt stress were recovered by GABA application (Kalhor et al. 2018).

GABA Alleviates Oxidative Injury Induced by Salt Stress via Accumulation of the Osmolytes in Plants

Excessive ROS frequently accumulate under salinity stress, and the cellular ion homeostasis is an important adaptive trait of plants under salt stress. Salt stress decreases the stability and integrity of the biological membranes due to ROS accumulation, and exogenous application of GABA can reduce the inhibitory effects of ROS accumulation in plants. Li et al. (2016) investigated that GABA appeared to impart partial protection to the wheat plants against oxidative disturbance induced by salt. Exogenous GABA regulated the sodium ion flux in tomato plants by reducing the absorption of sodium ions. Due to the reduction of absorption, the sodium ions transported to the leaves are significantly lower than that under salt stress. Thus, Na^+ accumulation in the leaves and roots of tomato was reduced by preventing Na^+ influx in roots and transportation to leaves. Moreover, exogenous GABA under salt stress enhanced the reduction of both superoxide anion and hydrogen peroxide radicals in cells and subsequently decomposed into water and oxygen in plants (Wu et al. 2020; Aljuaid and Ashour 2022).

Electrolyte leakage and Malondialdehyde (MDA) content are often used to assess stress-induced damage in plants. GABA can trap the reactive intermediates during lipid peroxidation, and GABA reacts readily with MDA to form different products (Deng et al. 2010). The electrolyte conductivity and MDA as indexes of membrane injury were lowered by GABA treatments than in the plants under NaCl alone in wheat (Li et al. 2016). MDA content is regarded as an important indicator of oxidative stress in plants, and Wu et al. (2020) observed a significant decrease in MDA content following exogenous GABA treatment under salt stress in tomato plants. Foliar application of 1 mM GABA reduced MDA and H_2O_2 content in maize seedlings (Aljuaid and Ashour 2022).

GABA appears to impart partial protection to various abiotic stresses in most plants by increasing leaf turgor, increasing osmolytes, and reducing oxidative damage by stimulation of antioxidants. Al-Quraan and Al-Omari (2017) investigated the response of two lentil cultivars (*Lens culinaris* Medik.) to salt, osmotic, and oxidative stresses *via* the characterization of seed germination, accumulation of ROS, and GABA level. GABA content in lentil seedlings significantly reduced lipid peroxidation, and the results showed that GABA shunt is a key signaling and metabolic pathway that allows adaptation of lentil seedlings to salt, osmotic, and oxidative stresses. The metabolic and signaling role of GABA in plants against coping with various environmental stresses such as cold, heat, drought, salt, and heavy metal in legumes have been reviewed in detail (Sita and Kumar 2020).

GABA is an important component of the free amino acid pool in plant cells, mainly acts as an osmotic substance and plays an important role in imparting protection of plants to salt stress (Li et al. 2021). Exogenous GABA application in tomato leaves treated with NaCl exhibited increased accumulation of GABA and glutamate. Glutamate can maintain the flow of the GABA shunt, and the increase in glutamate content provides a large amount of raw material for the endogenous synthesis of GABA. This process can reduce the salt-induced damage in plant cells by consuming H^+ and thus reducing the degree of acidification of cells (Wu et al. 2020). They also found that the content of serine, proline, and alanine was increased, and thus the metabolic balance under salt stress with exogenous GABA was maintained in tomato plants. The accumulation of amino acids in leaves may help to reduce MDA content by inducing the formation of strong hydrogen-bonded water around protein and protecting the natural state of cellular protein polymers in plant cells.

GABA can alleviate salinity injury by accumulating osmolytes in plant cells and thus acts as an osmotic substance. Soluble sugars, proline, glycine betaine, and trehalose are among the major osmolytes in plants. Proline is the main nitrogen-containing osmolyte in plants, which can increase the cytoplasm water concentration and protect the metabolic center of the cytoplasm by preventing the vacuoles from absorbing water from the cytoplasm. Moreover, free amino acids are considered to be osmoregulation substances that protect plants from salt stress by reducing membrane permeability. Exogenous application of GABA has been shown to increase proline and soluble sugar content under salt stress conditions (Wang et al. 2017; Aljuaid and Ashour 2022). Exogenous GABA-induced increases in proline and soluble sugar contents might be indirectly associated with its participation in the citric acid cycle. The two intermediate products of the citric acid cycle – succinic acid and α-ketoglutarate – can be generated by the GABA oxidation reaction (Wang et al. 2017). Moreover, the findings of Sheteiwy et al. (2019) recommended that the high tolerance to osmotic and salinity stress in rice is closely associated with the capability of GABA priming to control ROS level by inducing antioxidant enzymes, secondary metabolism, and the expression level of *calcineurin B-like protein-interacting protein kinase* (CIPK) genes. The key physiological and molecular mechanisms associated with GABA under salinity stress are represented in Figure 16.2.

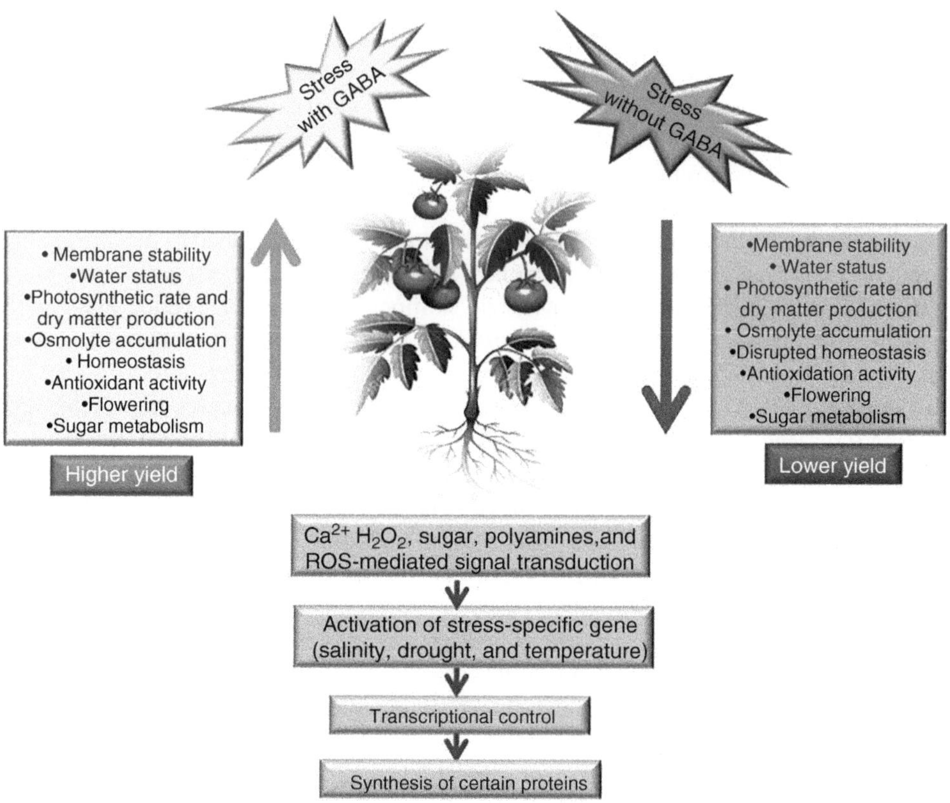

Figure 16.2 Physiological and molecular mechanisms associated with GABA exposure to salinity stress.

Molecular Changes Associated with GABA-Induced Salinity Stress Tolerance

GABA is a potential stress-responsive molecule that can alter certain stress-responsive genes to impart stress tolerance in plants. In the case of salinity stress, GABA plays a major role in maintaining seed viability. During salinity stress, glutamate decarboxylase (*GAD*) genes, which are responsible for the conversion of glutamate (Glu) to GABA, get activated, leading to the accumulation of GABA-shunt metabolites (GABA, glutamate, and alanine). The enhanced levels of GABA aid in respiratory mechanisms, maintaining nitrogen–carbon metabolism and synthesis of osmolytes for enhanced stress tolerance (Al-Quraan et al. 2019). *GAD* genes can also be activated by the exogenous application of GABA, which results in the endogenous accumulation of GABA by *GAD* gene-mediated conversion. It was observed that tomato plants subjected to salinity stress exhibit enhanced stress tolerance (Wu et al. 2020). Apart from this, exogenous application of GABA in muskmelon plants under salinity-alkalinity stress significantly upregulates the expression of *spermidine synthase (SPDS), spermine synthase (SPMS), arginine decarboxylase (ADC), and carbamoyl putrescine amido hydrolase (CPA)* genes, resulting in increased levels of polyamines such as putrescine, spermidine, and spermine, which aid in photosynthesis by maintaining chlorophyll synthesis and also inhibit ROS accumulation by altering redox homeostasis (Xu et al. 2019). Polyamine contributes to stress tolerance with the help of their oxidative product GABA, and this degradation is facilitated by the enzyme diamine oxidase (DAO). When *Glycine max* plants are subjected to salinity stress, *DAO* genes get upregulated, leading to a higher accumulation of GABA (Xing et al. 2007). So, apart from GAD-mediated conversion, GABA is produced endogenously in plants by degradation of polyamines during salinity stress.

Exogenous application and endogenous production of GABA will lead to the differential expression of various stress-responsive genes. The differential expression of *Salt Overly Sensitive 1* (*ZmSOS1*) and intracellular Na^+/H^+ antiporters (*ZmNHX1*) genes, along with the high-affinity potassium transporter 1 (*ZmHKT1) gene*, helps plants survive under saline stress by excluding Na from the cytosol and protecting enzymes. Exogenous GABA application in maize plants leads to overexpression of *ZmHKT1* and downexpression of *ZmSOS1* and *ZmNHX1,* leading to a reduction in Na/K ratio selectivity (Aljuaid and Ashour 2022). Few other stress-responsive genes (*CIPKs*), such as *OsCIPK01*, *OsCIPK03*, *OsCIPK08,* and *OsCIPK15*, were upregulated in GABA-primed rice plants under salinity stress, which activates a number of downstream signaling pathways, including the accumulation of proline and soluble sugars (Sheteiwy et al. 2019). They also observed the overexpression of genes associated with secondary metabolites such as phenylalanine ammonia-lyase (*PAL1*), polyphenol oxidase (*PPO*), Shikimate dehydrogenase (*SKDH*), and Cinnamyl alcohol dehydrogenase (*1bCAD1*) which helps in ROS scavenging and inhibiting the radicle chain reaction (Sheteiwy et al. 2019). Li et al. (2020) revealed that enhanced salt stress tolerance by GABA has a great correlation with lipid, saccharide, and amino acid metabolism. And it is proved by the differential expression of genes such as magnesium-chelatase (*Mg-CHT*), aldehyde oxidase (*AO*), cytochrome P450 (*CYP450*), 3-ketoacyl-CoA synthase 6 (*KCS6*), acetyl-CoA carboxylase 1 (*ACC1*), zinc transporter 29 (*ZTP29*), and alpha-amylase 3 (*AMY3*), which are mainly involved in starch degradation, zinc homeostasis, and biosynthesis of fatty acid, abscisic acid, wax, and chlorophyll in creeping bentgrass subjected to salinity stress. From the above-mentioned reports, it was proven that GABA is an important biocompound for imparting salt stress tolerance in plants.

Conclusion

Salt stress limits crop development and yield, which poses a serious concern for agronomists. GABA is a nonprotein amino acid that plays an important role in plant stress tolerance. It acts in several plant functions, including metabolic processes and signaling under stressful circumstances. It acts as a signaling molecule synthesized in the chloroplasts of plant cells and is transported to the cytoplasm, where it interacts with GABA receptors present on the plasma membrane of plant cells, regulating the expression of stress-related genes. Under salinity stress, GABA acts as an osmoprotectant that helps maintain the osmotic balance of the cell. It can also reduce the growth restriction and oxidative damage caused by salt stress, and it supports the accretion of endogenous GABA and other osmolytes. In a dose-dependent way, the administration of GABA promotes plant growth, antioxidant metabolism, and the expression of antioxidant genes. The effects of salt stress on plants are lessened by GABA, which increases crop productivity. Overall, GABA plays an important role in plant stress tolerance, especially under salinity stress.

Declaration of Interest: None

Acknowledgments

RJ acknowledge Bayers fellowship program MEDHA. JMJ and NL gratefully acknowledge the University Grants Commission (UGC). AKS is indebted to the Council of Scientific and Industrial Research (CSIR), New Delhi, for the financial assistance through the award of a Junior Research Fellowship (JRF) and the Department of Botany, University of Calicut, Kerala. FP acknowledges the Department of Botany, KAHM Unity Women's College, Manjeri, Kerala.

References

Abbas, G., Saqib, M., Akhtar, J., and Basra, S.M.A. (2013). Salinity tolerance potential of two *Acacia* species at early seedling stage. *Pakistan Journal of Agricultural Sciences* 50 (4): 683–688.

AbdElgawad, H., Zinta, G., Hegab, M.M. et al. (2016). High salinity induces different oxidative stress and antioxidant responses in maize seedlings organs. *Frontiers in Plant Science* 7: 276.

Acharya, B.R., Sandhu, D., Dueñas, C. et al. (2022). Morphological, physiological, biochemical, and transcriptome studies reveal the importance of transporters and stress signaling pathways during salinity stress in *Prunus*. *Scientific Reports* 12 (1): 1274.

Aljuaid, B.S. and Ashour, H. (2022). Exogenous γ-aminobutyric acid (GABA) application mitigates salinity stress in maize plants. *Life* 12 (11): 1860.

Al-Quraan, N.A. and Al-Omari, H.A. (2017). GABA accumulation and oxidative damage responses to salt, osmotic and H_2O_2 treatments in two lentil (*Lens culinaris* Medik) accessions. *Plant Biosystems* 151 (1): 148–157.

Al-Quraan, N.A., Al-Ajlouni, Z.I., and Obedat, D.I. (2019). The GABA shunt pathway in germinating seeds of wheat (*Triticum aestivum L.*) and barley (*Hordeum vulgare L.*) under salt stress. *Seed Science Research* 29 (4): 250–260.

Ansari, M.I., Hasan, S., and Jalil, S.U. (2014). Leaf senescence and GABA shunt. *Bioinformation* 10 (12): 734.

Ansari, M.I., Jalil, S.U., Ansari, S.A. et al. (2021). GABA shunt: a key-player in mitigation of ROS during stress. *Plant Growth Regulation* 94: 131–149.

Arif, Y., Singh, P., Siddiqui, H. et al. (2020). Salinity induced physiological and biochemical changes in plants: an omic approach towards salt stress tolerance. *Plant Physiology and Biochemistry* 156: 64–77.

Balasubramaniam, T., Shen, G., Esmaeili, N., and Zhang, H. (2023). Plants' response mechanisms to salinity stress. *Plants 12* (12): 2253.

Benjamin, J.J., Lucini, L., Jothiramshekar, S., and Parida, A. (2019). Metabolomic insights into the mechanisms underlying tolerance to salinity in different halophytes. *Plant Physiology and Biochemistry* 135: 528–545.

Carillo, P. (2018). GABA shunt in durum wheat. *Frontiers in Plant Science* 9: 100.

Chen, H., Liu, T., Xiang, L. et al. (2018). GABA enhances muskmelon chloroplast antioxidants to defense salinity-alkalinity stress. *Russian Journal of Plant Physiology* 65: 674–679.

Cheng, B., Li, Z., Liang, L. et al. (2018). The γ-aminobutyric acid (GABA) alleviates salt stress damage during seeds germination of white clover associated with Na^+/K^+ transportation, dehydrins accumulation, and stress-related genes expression in white clover. *International Journal of Molecular Sciences* 19 (9): 2520.

Dabravolski, S.A. and Isayenkov, S.V. (2023). The role of the γ-aminobutyric acid (GABA) in plant salt stress tolerance. *Horticulturae* 9 (2): 230.

Deng, Y., Xu, L., Zeng, X. et al. (2010). New perspective of GABA as an inhibitor of formation of advanced lipoxidation end-products: it's interaction with malondiadehyde. *Journal of Biomedical Nanotechnology* 6 (4): 318–324.

Ghaleb, W.S., Sawwan, J.S., Akash, M.W., and Al-Abdallat, A.M. (2010). In vitro response of two *Citrus* rootstocks to salt stress. *International Journal of Fruit Science* 10 (1): 40–53.

Gill, S.S. and Tuteja, N. (2010). Reactive oxygen species and antioxidant machinery in abiotic stress tolerance in crop plants. *Plant Physiology and Biochemistry* 48 (12): 909–930.

Hasanuzzaman, M. and Fujita, M. (2022). Plant responses and tolerance to salt stress: physiological and molecular interventions. *International Journal of Molecular Sciences* 23 (9): 4810.

Isayenkov, S.V. and Maathuis, F.J. (2019). Plant salinity stress: many unanswered questions remain. *Frontiers in Plant Science* 10: 80.

Jalil, S.U. and Ansari, M.I. (2020). Physiological role of gamma-aminobutyric acid in salt stress tolerance. In: *Salt and Drought Stress Tolerance in Plants: Signaling Networks and Adaptive Mechanisms* (ed. M. Hasanuzzaman and M. Tanveer), 337–350. Cham, Switzerland: Springer International Publishing.

Jin, X., Liu, T., Xu, J. et al. (2019). Exogenous GABA enhances muskmelon tolerance to salinity-alkalinity stress by regulating redox balance and chlorophyll biosynthesis. *BMC Plant Biology* 19 (1): 1–15.

Kalhor, M.S., Aliniaeifard, S., Seif, M. et al. (2018). Enhanced salt tolerance and photosynthetic performance: Implication of γ-amino butyric acid application in salt-exposed lettuce (*Lactuca sativa* L.) plants. *Plant Physiology and Biochemistry* 130: 157–172.

Khan, M.I.R., Jalil, S.U., Chopra, P. et al. (2021). Role of GABA in plant growth, development and senescence. *Plant Gene* 2021 26: 100283.

Khanna, R.R., Jahan, B., Iqbal, N. et al. (2021). GABA reverses salt-inhibited photosynthetic and growth responses through its influence on NO-mediated nitrogen-sulfur assimilation and antioxidant system in wheat. *Journal of Biotechnology* 325: 73–82.

Kumar, A., Singh, S., Gaurav, A.K. et al. (2020). Plant growth-promoting bacteria: biological tools for the mitigation of salinity stress in plants. *Frontiers in Microbiology* 11: 1216.

Li, M.F., Guo, S.J., Yang, X.H. et al. (2016). Exogenous gamma-aminobutyric acid increases salt tolerance of wheat by improving photosynthesis and enhancing activities of antioxidant enzymes. *Biologia Plantarum* 60 (1): 123–131.

Li, Y., Fan, Y., Ma, Y. et al. (2017). Effects of exogenous γ-aminobutyric acid (GABA) on photosynthesis and antioxidant system in pepper (*Capsicum annuum* L.) seedlings under low light stress. *Journal of Plant Growth Regulation* 36: 436–449.

Li, Z., Cheng, B.Z., Peng, Y., and Zhang, Y. (2020). γ-Aminobutyric acid induces transcriptional changes contributing to salt tolerance in creeping bentgrass. *Biologia Plantarum* 64: 744–752.

Li, L., Dou, N., Zhang, H., and Wu, C. (2021). The versatile GABA in plants. *Plant Signaling & Behavior* 16 (3): 1862565.

Ma, Y., Wang, P., Chen, Z. et al. (2018). GABA enhances physio-biochemical metabolism and antioxidant capacity of germinated hulless barley under NaCl stress. *Journal of Plant Physiology* 231: 192–201.

Mekonnen, D.W., Flügge, U.I., and Ludewig, F. (2016). Gamma-aminobutyric acid depletion affects stomata closure and drought tolerance of *Arabidopsis thaliana*. *Plant Science* 245: 25–34.

Pérez-Tornero, O., Tallón, C.I., Porras, I., and Navarro, J.M. (2009). Physiological and growth changes in micropropagated *Citrus macrophylla* explants due to salinity. *Journal of Plant Physiology* 166 (17): 1923–1933.

Rahneshan, Z., Nasibi, F., and Moghadam, A.A. (2018). Effects of salinity stress on some growth, physiological, biochemical parameters and nutrients in two pistachio (*Pistacia vera* L.) rootstocks. *Journal of Plant Interactions* 13 (1): 73–82.

Renault, H., Roussel, V., El-Amrani, A. et al. (2010). The *Arabidopsis* pop2-1 mutant reveals the involvement of GABA transaminase in salt stress tolerance. *BMC Plant Biology* 10: 20.

Salah, A., Zhan, M., Cao, C. et al. (2019). γ-Aminobutyric acid promotes chloroplast ultrastructure, antioxidant capacity, and growth of waterlogged maize seedlings. *Scientific Reports* 9: 1–19.

Shabala, S.N. and Lew, R.R. (2002). Turgor regulation in osmotically stressed *Arabidopsis* epidermal root cells. Direct support for the role of inorganic ion uptake as revealed by concurrent flux and cell turgor measurements. *Plant Physiology* 129: 290–299.

Sharif, I., Aleem, S., Farooq, J. et al. (2019). Salinity stress in cotton: effects, mechanism of tolerance and its management strategies. *Physiology and Molecular Biology of Plants* 25: 807–820.

Shelp, B.J., Bown, A.W., and McLean, M.D. (1999). Metabolism and functions of gamma-aminobutyric acid. *Trends in Plant Science* 4 (11): 446–452.

Sheteiwy, M.S., Shao, H., Qi, W. et al. (2019). GABA-alleviated oxidative injury induced by salinity, osmotic stress and their combination by regulating cellular and molecular signals in rice. *International Journal of Molecular Sciences* 20 (22): 5709.

Shomali, A., Aliniaeifard, S., Didaran, F. et al. (2021). Synergistic effects of melatonin and gamma-aminobutyric acid on protection of photosynthesis system in response to multiple abiotic stressors. *Cells* 10 (7): 1631.

Sita, K. and Kumar, V. (2020). Role of gamma amino butyric acid (GABA) against abiotic stress tolerance in legumes: a review. *Plant Physiology Reports* 25: 654–663.

Soltabayeva, A., Ongaltay, A., Omondi, J.O., and Srivastava, S. (2021). Morphological, physiological and molecular markers for salt-stressed plants. *Plants* 10 (2): 243.

Tang, J., Li, M., Mao, P., and Jiang, Y. (2022). Effects of gamma-aminobutyric acid on seed germination, ion balance, and metabolic activity in perennial ryegrass under salinity stress. *Journal of Plant Growth Regulation* 41 (4): 1835–1844.

Ullah, A., Ali, I., Noor, J. et al. (2023a). Exogenous γ-aminobutyric acid (GABA) mitigated salinity-induced impairments in mungbean plants by regulating their nitrogen metabolism and antioxidant potential. *Frontiers in Plant Science* 13: 1081188.

Ullah, A., Tariq, A., Zeng, F. et al. (2023b). Application of GABA (γ-aminobutyric acid) to improve saline stress tolerance of chufa (*Cyperus esculentus* L. var. sativus Boeck) plants by regulating their antioxidant potential and nitrogen assimilation. *South African Journal of Botany* 157: 540–552.

Vijayakumari, K., Jisha, K.C., and Puthur, J.T. (2016). GABA/BABA priming: a means for enhancing abiotic stress tolerance potential of plants with less energy investments on defence cache. *Acta Physiologiae Plantarum* 38: 1–14.

Wang, Y., Gu, W., Meng, Y. et al. (2017). γ-Aminobutyric acid imparts partial protection from salt stress injury to maize seedlings by improving photosynthesis and upregulating osmoprotectants and antioxidants. *Scientific Reports* 7: 43609.

Wang, X., Dong, H., Hou, P. et al. (2019). Effects of exogenous gamma-aminobutyric acid on absorption and regulation of ion in wheat under salinity stress. In: *Computer and Computing Technologies in Agriculture XI. CCTA 2017. IFIP Advances in Information and Communication Technology*, vol. 546 (ed. D. Li and C. Zhao), 347–357. Cham, Switzerland: Springer International Publishing.

Wang, Y., Cao, H., Wang, S. et al. (2023). Exogenous γ-aminobutyric acid (GABA) improves salt-inhibited nitrogen metabolism and the anaplerotic reaction of the tricarboxylic acid cycle by regulating GABA-shunt metabolism in maize seedlings. *Ecotoxicology and Environmental Safety* 254: 114756.

Wu, X., Jia, Q., Ji, S. et al. (2020). Gamma-aminobutyric acid (GABA) alleviates salt damage in tomato by modulating Na^+ uptake, the GAD gene, amino acid synthesis and reactive oxygen species metabolism. *BMC Plant Biology* 20: 1–21.

Xiang, L., Hu, L., Xu, W. et al. (2016). Exogenous γ-aminobutyric acid improves the structure and function of photosystem II in muskmelon seedlings exposed to salinity-alkalinity stress. *PLoS One* 11 (10): e0164847.

Xing, S.G., Jun, Y.B., Hau, Z.W., and Liang, L.Y. (2007). Higher accumulation of γ-aminobutyric acid induced by salt stress through stimulating the activity of diamine oxidases in *Glycine max* (L.) Merr. roots. *Plant Physiology and Biochemistry* 45 (8): 560–566.

Xu, J., Liu, T., Yang, S. et al. (2019). Polyamines are involved in GABA-regulated salinity-alkalinity stress tolerance in muskmelon. *Environmental and Experimental Botany* 164: 181–189.

Yan, K., Cui, M., Zhao, S. et al. (2016). Salinity stress is beneficial to the accumulation of chlorogenic acids in honeysuckle (*Lonicera japonica* Thunb.). *Frontiers in Plant Science* 7: 1563.

Zhang, L., Ma, H., Chen, T. et al. (2014). Morphological and physiological responses of cotton (*Gossypium hirsutum* L.) plants to salinity. *PLoS One* 9: e112807.

Zhao, C., Zhang, H., Song, C. et al. (2020). Mechanisms of plant responses and adaptation to soil salinity. *The Innovation* 1 (1): 100017.

17

GABA and Nutrient Deficiency

Km Niharika[1], Shubhra Khare[2], Ajey Singh[1], Zeba Azim[3], Nimisha Amist[4], and Narsingh Bahadur Singh[3]

[1] *Department of Botany, University of Lucknow, Lucknow, Uttar Pradesh, 226007, India*
[2] *Department of Applied Sciences and Humanities, Invertis University, Bareilly, Uttar Pradesh, 243123, India*
[3] *Plant Physiology Laboratory, Department of Botany, University of Allahabad, Prayagraj, Uttar Pradesh, 211002, India*
[4] *Department of Botany, Ewing Christian College, Prayagraj, Uttar Pradesh, 211003, India*

Introduction

In the past couple of decades, various stress elicitors have raised an alarming concern invoking global loss in yield and productivity of major agricultural products. However, adequate mineral nutrients could boost plant endurance to better withstand extreme environmental conditions, thereby promoting growth. An adequate supply of essential nutrient elements is necessary to enhance the nutrient supplements of the human diet. Moreover, plants grown in nutrient-deprived soil could cause a severe deficiency of essential nutrients, ultimately inducing deficiency in humans after consumption. Therefore, the major population of the world suffers from challenges caused by nutrient deficiency.

To meet growing food demands, plant growers are compelled to use chemical fertilizers to maximize crop production by attaining higher yields with better quality. These fertilizers run off into the soil and could drastically deplete essential nutrients and soil biota (de Bang et al. 2021). Though, chemical fertilizers enhance plant productivity, their continued use can lead to various problems related to soil fertility, biogeochemical cycles, and environmental issues. In light of this, devising competent strategies to endorse the growth and development of crop plants as alternatives to chemical fertilizers could be put forward. Furthermore, exogenous Gamma-aminobutyric acid (GABA) supplementation has played a prominent role in contributing toward stress alleviation in several stresses, such as salinity, heavy metals, temperature, drought, mechanical injury, insect attack, diseases, etc., by strengthening the metabolic and adaptive pathways of plants (Guo et al. 2023). In this chapter, we document the literature from the past few years explaining the affirmative role of exogenous GABA treatment in mitigating nutrient deficiency.

GABA in Plants: Biosynthesis, Plant Development, and Food Security, First Edition. Edited by Samiksha Singh, Durgesh Kumar Tripathi, and Vijay Pratap Singh.

An Overview of GABA

GABA is a four-carbon non-proteinogenic amino acid that is present in almost all organisms, including both animals and plants (Li et al. 2020). In 1949, it was first isolated from the tissues of potato tubers, and later it was reported in brain extracts of mammals (Sita and Kumar 2020). Various studies support the presence of GABA in the tissues of the animal nervous system (Watanabe et al. 2002). Around the 1950s and 1960s, GABA was known for inhibiting impulses generated by crayfish stretch receptor neurons, which further led to its classification as an inhibitory neurotransmitter in animals (Guo et al. 2023). GABA acts as an endogenous signal molecule, regulating various developmental responses in plants. It is endogenously produced in plant cells and can alter metabolic processes involved both in normal and stressful conditions.

The metabolic pathway concerned with GABA biosynthesis is called the GABA shunt pathway. It works through a feedback loop mechanism to ensure the production and maintenance of an adequate supply of GABA in cells. The first step in the GABA shunt pathway is the decarboxylation of glutamate to GABA in an irreversible reaction catalyzed by the enzyme glutamate decarboxylase (GAD, E.C.4.1.1.15) in the cytosol. The next step is the reversible transamination, producing succinic semialdehyde (SSA) via GABA transaminase (GABA-T, E.C.2.6.1.19). Further, succinate is produced through SSA oxidation by NAD-dependent enzyme SSA dehydrogenase (SSADH, EC 1.2.1.16). Additionally, SSA is converted to γ-hydroxybutyric acid (GHB), catalyzed by succinic semialdehyde reductase (SSR), and enters the Krebs cycle (Simpson et al. 2008). Both GABA-T and SSADH are mitochondrial enzymes. The two GABA-Ts enzymes are α-ketoglutarate-dependent (GABA-TK) and pyruvate-dependent (GABATP) enzymes, which utilize either α-ketoglutarate or pyruvate to produce glutamate or alanine, respectively (Khan et al. 2021). The succinate produced via the GABA shunt is involved in important mechanisms by acting as an electron donor in the electron transport chain of mitochondria and the tricarboxylic acid cycle.

Moreover, another pathway for GABA synthesis is via the degradation of polyamines such as spermidine and putrescine. In this case, GABA is formed by gamma-aminobutyraldehyde, which is produced after polyamine degradation, catalyzed by enzymes diamine oxidase (DAO, E.C. 1.4.3.6) and 4-aminobutyraldehyde dehydrogenase (ABALDH) (Shelp et al. 2012). Alterations in the activity of components involved in the GABA shunt pathway can greatly influence diverse metabolic parameters of plant growth.

Role of GABA in Plant Development

GABA, as a potent signaling molecule, can modulate various responses related to plant developmental attributes. Several earlier reports suggest that it can alter many metabolic pathways through different mechanisms. A report by Xie et al. (2020) revealed that GABA could negatively inhibit adventitious root growth in *poplar*. They stated that exogenous application of GABA, GAD2 overexpression, and exogenous application of GABA-T inhibitor could enhance endogenous GABA levels, thereby decreasing adventitious root growth. Under modulated GABA levels, both pollen tube growth and its guidance into the ovary could be affected. It was noted that limited expression of the gene encoding the GABA-T enzyme could inhibit GABA-T activity, followed by increased GABA levels in ovule, resulting in abnormal pollen tube growth in *Arabidopsis* (Palanivelu et al. 2003).

Exogenously applied GABA induces endogenous levels of GABA, which influences dry weight and mineral nutrients in *Lemna* plants. A threefold rise in plant dry weight over the control was observed with GABA (10 mM) application in *Lemna* (Kinnersley and Lin 2000). Enhanced GABA accumulation in cellular tissues can manipulate levels of various endogenous amino acids, subsequently resulting in abnormal plant phenotypes. In this regard, Akama and Takaiwa (2007) showed that overexpressing GAD2 in transgenic lines of rice and tobacco induces higher GAD activity, resulting in increased GABA content. Furthermore, transgenic plants with accumulated GABA content exhibited reduced amino acids, which correlated with diverse phenotypic aberrations in plants.

Accumulated GABA levels caused cell elongation defects, which inhibited plant growth in GABA-T mutants (Renault et al. 2011). In addition, GABA-T mutants also exhibited early leaf senescence in *Arabidopsis* plants (Jalil et al. 2017). In another investigation, Koike et al. (2013) observed that transgenic lines of tomato plants with mutated GABA-T gene expression exhibited dwarf and infertile vegetative plants and delayed fruit ripening.

Role of GABA in Different Stress Tolerance

Various cellular metabolites show enhanced levels under stress conditions. GABA is one such metabolite that gets accumulated under unfavorable environmental extremes. The exogenous application of GABA promotes endogenous GABA levels, regulating multidimensional responses to sustain stress attributes.

The unfavorable environmental conditions could raise cytosolic Ca^{2+} levels, which further induce calmodulin-dependent GAD activity to promote GABA content (Bhattacharya et al. 2018). In muskmelon, exogenous GABA application can improve photosynthetic rate of plants under salt stress (Xiang et al. 2016). In an experiment by Kumar et al. (2017), GABA alleviates salt stress, promoting morphological characters and seed yield of *Vigna mungo* plants. Significant improvements were observed in plant height, dry weight, and pod number per plant of salt-stressed *Vigna* plants in GABA application.

GABA can efficiently ameliorate plant growth under heavy metal toxicity. Reports by Mahmud et al. (2017) indicated that GABA application improved various growth parameters of chromium-stressed *Brassica juncea* seedlings. Results showed that GABA might have reduced chromium uptake, thereby improving mineral content, relative water content, photosynthetic pigment, and antioxidant machinery, leading to decreased oxidative injury. Similarly, GABA can confer tolerance to cadmium toxicity in microalgae (*Monoraphidium* spp.) by inducing lipid metabolism and nutrient uptake correlated with detoxification of ROS (Zhao et al. 2020).

Exogenous GABA can efficiently alleviate growth of drought stressed white clover plants by maintaining parameters such as relative water content, lipid peroxidation, electrolyte leakage, proline metabolism, and polyamines synthesis, possibly by upregulating GABA shunt pathway (Yong et al. 2017). Similarly, Rezaei-Chiyaneh et al. (2018) revealed that GABA improved the antioxidant defense system, soluble sugar, and proline content of *Nigella sativa* plants under water deficit stress.

Both high and low temperatures could greatly alter the functional machinery of healthy plants. GABA treatment in heat-stressed mungbean plants improved membrane integrity, photosynthetic pigments, carbon assimilation, and exhibit higher osmolytes. Moreover, GABA also supported the reproductive health of flowers followed by higher seed yield (Priya et al. 2019). In *Agrostis stolonifera* plants, GABA mitigates heat-induced injuries by regulating the GABA synthesis shunt, photosynthesis, leaf senescence, and antioxidant machinery to confer stress tolerance (Li et al. 2016).

In a study conducted by Malekzadeh et al. (2017), chilling-stressed cucumber plants maintained oxidative damage by augmenting the synthesis of antioxidant enzyme activity, correlated with supporting cellular membrane permeability. Furthermore, tea plants exposed to low-temperature stress exhibited tolerance by maintaining carbon and nitrogen assimilation under exogenous GABA treatment (Zhu et al. 2019).

Exogenously applied GABA in maize seedlings exposed to waterlogging conditions improved photosynthetic content and antioxidant enzyme activity through reducing ROS generated due to lipid peroxidation and membrane damage (Salah et al. 2019). In a study on *Prunus* plants under waterlogging stress, GABA application can provide tolerance by increasing stomatal conductance and GABA shunt pathway, with lowered levels of hydrogen peroxide content (Salvatierra et al. 2016).

Likewise, in the case of biotic stress, GABA can provide tolerance and strengthen plant defensive systems. GABA can certainly accumulate in host cells to provide protection from pathogenic fungus or bacteria, infestation by pests or insects, and mechanical injury. Various reports evidence the role of GABA in providing tolerance against pathogenic fungi. Exogenous GABA can upregulate the transcriptional expression of genes related to chitinase, β-1,3-glucanase, phenylalnine ammonia-lyase, peroxidase, and polyphenol oxidase, protecting tomato plants from the fungus *Alternaria* (Yang et al. 2017). Similarly, exogenous GABA stimulated GABA shunt, which aided in forming chlorotic rings around the infected tissue and detoxified oxidative stress to minimize cell damage from the fungus *Botrytis cinerea* in tomato (Seif et al. 2013). Exogenous GABA can inhibit larvae of *Spodoptera littoralis* and *Choristoneura rosaceana* in *Arabidopsis* mutants. GABA can induce endogenous GABA accumulation, which in turn stimulated the activation of defense system in infected plants (Scholz et al. 2015). Enhanced GABA can downregulate quorum sensing against pathogenic bacteria. Moreover, they also utilize GABA as a sole source of carbon and nitrogen (Tarkowski et al. 2020).

These reports clearly depict the positive responses of exogenous GABA treatment in boosting plant endurance to cope with diverse abiotic and biotic stress factors prevailing in the surrounding.

Different Mineral Nutrients and Their Role in Plant Development

Nutrients play an essential role in governing various metabolic processes of plant growth and development. Plants need a balance of these nutrients for the growth and healthy functioning during their lifespan. The occurrence of various nutrient elements considerably depends on factors such as soil properties, seasonal and climatic fluctuations in different space and time.

Nutrients are classified into two categories: macronutrients, which are required in larger amounts, and micronutrients, which are required in lesser amounts. Among them, nitrogen (N), phosphorus (P), potassium (K), calcium (Ca), sulfur (S), and magnesium (Mg) are macronutrients, while chlorine (Cl), iron (Fe), boron (B), nickel (Ni), copper (Cu), manganese (Mn), zinc (Zn), and molybdenum (Mo) are called micronutrients.

Plants are equipped with approaches to induce sufficient uptake of different elements and to evade excess quantities that could cause toxic effects in the plant system. Moreover, higher levels of important essential mineral elements could impede cellular mechanisms through production of excess ROS and inactivation metalloproteins. However, plants also need to efficiently sustain under challenges caused due to nonessential nutrient elements. A stable equilibrium between the uptake and mobilization of necessary nutrients is needed in plant tissues to support normal health status of growth and metabolism.

Metal ions are absorbed from soil by roots, then gets loaded in xylem tissues and moved to all aerial plant parts through distribution in leaf tissues. This process of transporting and distributing metals ions from root to leaf tissues is governed via various metal transporters (Houmani and Corpas 2023). These metal transporters, along with chelating molecules, aid in the long-distance transport of nutrients and help maintain balance under fluctuating levels of available nutrients to plants. Metal transporter families in plants are effectively involved in influx of metal ions into the cytosol or their efflux from the cytosol (Colangelo and Guerinot 2006).

One of the essential macronutrients is N, it is an important component of nucleic acids, amino acids, proteins, and enzymes. Plants utilize nitrate and ammonium as inorganic and peptides and amino acids as organic N sources via different transporter proteins. N metabolism includes the conversion of nitrate to nitrite by the enzyme nitrate reductase followed by the conversion of nitrite to ammonium by the enzyme nitrite reductase, which forms different structural molecules.

P constitutes as a chief source of biomolecules such as ATP and NADPH known as the energy currency of the cellular system. Moreover, they are also required in nucleic acids and phospholipids in the plasma membrane. Various signal transduction enzymes catalyze phosphorylation, and dephosphorylation processes are regulated by P.

Plants utilize inorganic sulfate by roots as the main source, while atmospheric sulfur in the form of sulfur dioxide is taken by aerial plant parts. S assimilation involves the production of compounds namely, cysteine and methionine, thiamine, glucosinolates, glutathione, phytochelatins, and various prosthetic groups. Moreover, it is also involved in the Calvin cycle of light reaction as it is a component of the ferredoxin and thioredoxin system (Hasanuzzaman et al. 2018).

Ca is an immobile element, a structural unit of the membrane system, and a component of the middle lamella of the cell membrane that links phosphate and carboxylate groups of phospholipids and proteins to support cell rigidity. Additionally, the role of Ca with calmodulin in activating signal transduction pathways to facilitate varied metabolic processes is also known (Sharma et al. 2018).

Mg is a key component of the tetrapyrrole ring in chlorophyll molecules and also participates in light-harvesting complex of the Photosystem II reaction center. Besides this, they are also involved in the regulation of various enzymes. Similar to Mg, K also contributes toward the optimal functionality of certain essential enzymes. They take part in osmoregulation to further support plant water relations and maintain the membrane potential of cells (Shahid et al. 2019).

Zn is one of the essential micronutrients involved in the activation of numerous enzymes participating in the electron transport chain, antioxidant enzyme system, carbohydrate metabolism, synthesis of photosynthetic pigment, and protein synthesis. A variety of proteins contain Cu as a structural unit, or they behave as an enzyme cofactor (Kumar et al. 2021). They are involved in membrane integrity, hormone signaling, respiration, and assimilation of carbon and nitrogen in plants. As compared to other nutrients, plants utilize Mo in lesser amounts. Mo primarily constitutes an active site of enzymes such as nitrate reductase, xanthine dehydrogenase, sulfite oxidase, and aldehyde oxidase (Tripathi et al. 2015). Mo is required for nitrogen metabolism in root nodules of plants that fix nitrogen symbiotically.

Like Zn, Mn behaves as a cofactor for various enzymes namely, catalase, pyruvate carboxylase, malic enzyme, phosphoenol pyruvate, carboxykinase, and isocitrate lyase to trigger various important metabolic pathways. Likewise, Mn is essentially needed in oxygen-evolving complex during photophosphorylation and is involved in redox reactions to scavenge excess ROS. Ni acts as a cofactor and plays an important role in urease activity. During nitrogen metabolism, urea converts to ammonium through the enzyme urease (DalCorso et al. 2014).

Different Nutrient Deficiencies in Plants

A deficiency of macronutrients could decrease the growth attributes of plants through ROS accumulation, membrane injury, and reduced photosynthetic rate (da Cruz Ferreira et al. 2020). Anthocyanin and flavonoids were accumulated in plants undergoing N deficiency reflecting protection of stressed plants from oxidative damage by ROS (Tsay et al. 2011). In a study by Janpen et al. (2019), N-deficient mint plants were observed with hampered cellular permeability leading to severe electrolyte leakage.

In *Arabidopsis*, a decrease in primary root growth was observed under P deficiency. Due to reduced meristematic activity in the root tip, cells showed growth inhibition (Gutiérrez-Alanís et al. 2018). Nutrient deficiency could alter hormonal metabolism in plants affecting their overall growth and development. Regarding this, low P levels enhanced expression of genes involved in DELLA protein, whereas downregulation of gibberellic acid-dependent genes was observed (Jiang et al. 2007).

S deficiency could severely impact photosynthetic efficiency leading to chlorosis in leaves. Moreover, S-deficient plants showed affected carbon assimilation and electron transport chain resulted due to ROS overproduction in tissues (Courbet et al. 2019). Plants grown under low S levels exhibit reduced growth, chlorotic spots, enhanced anthocyanins, disrupted flowering, and seed setting (Jobe et al. 2019).

Plants experience a decreased rate of photosynthesis, relative water content, limited carbon supply, disrupted sucrose mobilization, disturbed sucrose movement in phloem cells, and inhibited K movement to aerial plant parts subsequently leading to loss of cell growth (Tsay et al. 2011). Adequate supply of K in *Arabidopsis*, triggered synthesis of enzymes related to ethylene and jasmonic acid pathway correlated with ROS accumulation in stressed plants showing growth inhibition (Li et al. 2017).

Ca is primarily involved as a constituent of cell membrane, under low Ca levels membrane permeability is increased leading to spillage of electrolytes and membrane damage with reduced growth (Hawkesford et al. 2012). However, Ca in the cell membrane could provide protection against pathogenic fungi and bacteria subsequently attaining improved crop quality after harvesting (Winkler and Knoche 2019).

Low Mg levels in plants could disturb phloem transport with accumulated sugar content and downregulated expression of enzymes and proteins involved in the Calvin cycle resulting in hindered photosynthesis (Yang et al. 2019). Higher activity of enzymes such as catalase, ascorbate peroxide, superoxide dismutase, and guaiacol peroxidase in plants undergoing Mg deficiency indicated their probable role in stimulating the ROS scavenging mechanism (Liu et al. 2018).

Zn deficiency could decline levels of hormones like gibberellins and auxins leading to a reduction in plant growth and elongation (Kumar et al. 2021). Moreover, they could induce sugar levels; show chlorotic and necrotic leaf spots, membrane damage owing to oxidative bursts in stressed plants. Insufficient Zn content possibly can impact the structure and functional catalytic activity of major enzymes of essential metabolic pathways, namely protein synthesis, carbohydrate metabolism, chlorophyll biosynthesis, and respiration.

In the case of Cu deficiency, the cell wall is disturbed, leaves become twisted and deformed, and altered xylem transport is probably due to affected lignin content (Gui et al. 2022). Additionally, Cu deficiency can also promote susceptibility toward pathogenic fungi, stunted plant growth, and delayed yield and quality of crops. Both polyphenol oxidase and diamine oxidase contain Cu and participate in lignin biosynthesis (Assunção et al. 2022).

Mo plays a crucial role in mediating nitrogen assimilation in leguminous crops representing nitrate as a main source for N status in these plants (Houmani and Corpas 2023). In addition, insufficient Mo content declines stem elongation, distorted leaves, premature tissue differentiation, reduced flower setting, and early senescence.

Phenylalanine ammonia-lyase is a precursor enzyme that catalyzes the synthesis of different phenolic compounds. As Mn behaves as a cofactor for phenylalanine ammonia-lyase, a decline in Mn levels could significantly limit the biosynthesis of phenolic compounds (Fan et al. 2021). Chloroplast is majorly affected by Mn deficiency impacting several processes such as impaired photosynthetic rate, carbon assimilation, pollen fertility, grain yield, and root growth inhibition.

Insufficient Ni levels could prominently inhibit urease enzymes followed by higher urea content (DalCorso et al. 2014). Leguminous plants symbiotically fixing N under Ni deficiency form urea containing necrotic lesions inhibits nodule formation which subsequently results in growth inhibition (Gui et al. 2022). Ni deficiency decreases seed germination and growth, grain yield, and induces early abscission.

Role of GABA in Nutrient Deficiency

GABA as a potent signaling molecule could modulate various developmental pathways encouraging the growth and productivity of plants under different mineral nutrient deficiencies (Table 17.1 and Figure 17.1). A study conducted by Guo et al. (2020) stated

Table 17.1 Response of exogenously applied GABA under different nutrient deficiencies in plants.

S. no.	Plant species	GABA concentration	Nutrient deficiency	Response under GABA	Reference
1.	*Arabidopsis thaliana*	10 mM	Phosphorus	AtGAD1-mediated deficiency attenuation was noticed. Enhanced anthocyanin content in both wild type and *atgad1* plants. Downregulated expression of genes involved in phosphorus deficiency. AtGAD1 promoted GABA accumulation by activating GABA shunt in both root and shoot.	Benidickson et al. (2023)
2.	Apple	0.1–5.0 mM	Phosphorus	Growth improvement due to better photosynthesis and reduced ROS accumulation. Triggered auxin signaling resulting in higher endogenous auxin levels. Increment in auxin levels aided in root growth, which simultaneously stimulated phosphorus uptake via root.	Chen et al. (2023)
3.	*Crocus sativus*	0, 0.5, 1, and 2 mg/l	Phosphorus	Improvement in flower number, proline, growth of daughter corm, and phosphorus uptake resulted in better yield.	Afshari et al. (2023)
4.	*Cucumis sativus*	0, 5, 10, 20, and 40 mmol/l	Iron	Significant improvement in photosynthetic capability, iron uptake, ferric-chelate reductase activity, with reduction in chlorosis and pH. Activation of genes relating to iron uptake. Higher transcript levels of genes involved in auxin metabolism.	Guo et al. (2020)
5.	*Andrographis paniculata*	0.1–5.0 mmol/l	Nitrogen	Improved genes and enzymes associated with carbon and nitrogen metabolism, thus maintaining protein content. Upregulated gene expression of nitrate transporters (*NRT2.4* and *NRT3.2*) via its receptors facilitated nitrate uptake to maintain nitrogen status in plants.	Huang et al. (2023)
6.	*Sorghum bicolor*	50 mg/l	Nitrogen	Promoted shoot length, plant biomass, photosynthetic pigments, nitrogen, and carbon metabolism. Essential enzymes related to nitrogen and carbon metabolism were increased by many folds. Higher nitrate uptake and its assimilation were noticed.	Kchikich et al. (2021)
7.	*Eriobotrya japonica*	0.5, 1.0, 1.5, and 2.0 g/l	Phosphorus, nitrogen, potassium	Promoted nutrient uptake by increasing root and shoot biomass, chlorophyll and carotenoid content, and thereby nutrient availability.	Yang et al. (2023)

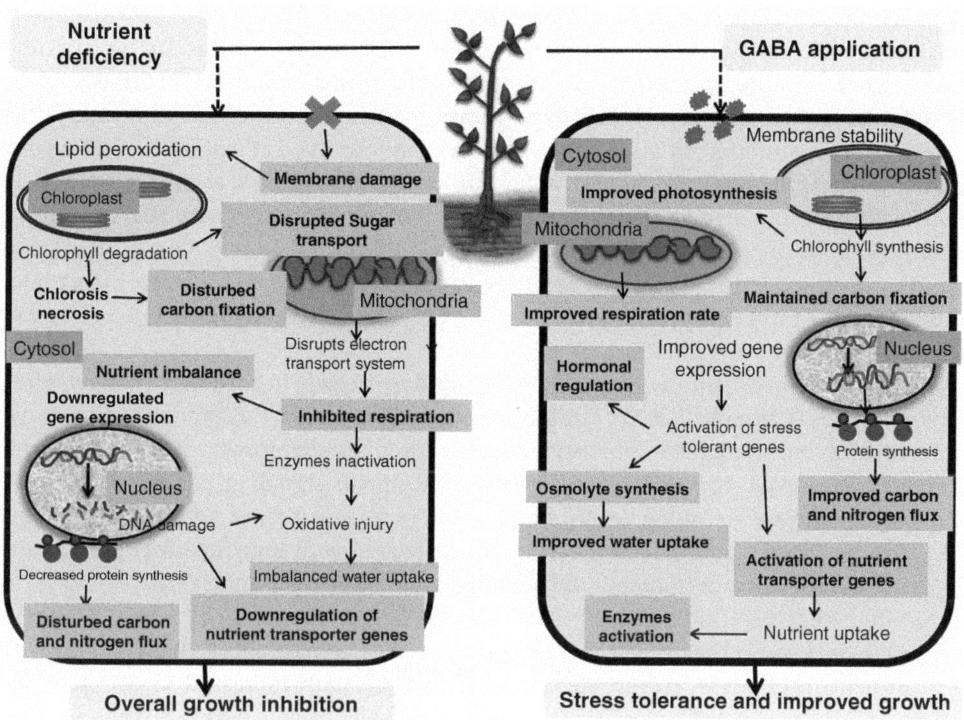

Figure 17.1 Representation of detrimental responses to nutrient deficiency and the role of exogenous GABA application in supporting stress tolerance in plants.

the affirmative responses of GABA in alleviating iron (Fe) deficiency in cucumber plants. However, treatment with 20 mM GABA showed better amelioration in comparison to other GABA concentrations. Authors reported enhanced GABA levels in roots and leaves of Fe-deficient plants. Exogenous application with GABA reduced chlorosis, improved photosynthetic efficiency, and Fe content in plant parts followed by increased growth. Moreover, GABA reduced the pH of nutrient solution with stimulated ferric-chelate reductase (FCR) activity and triggered the expression of genes such as *HA1*, *FRO2*, and *IRT1* which are related to Fe uptake in Fe-deficient roots. Furthermore, it was indicated that GABA-induced expression of genes related to auxin metabolism, namely, YUC4, IAA1, and PIN1. However, treatment of 1-naphthylphthalamic acid (inhibitor of auxin transport) disturbed GABA-mediated alleviation of Fe deficiency. This evidence proves the probable response of auxin in conferring tolerance to plants under Fe deficiency.

GAD, an enzyme mediating one of the important steps of the GABA shunt pathway, could facilitate acclimation to low P supply in *Arabidopsis thaliana* plants (Benidickson et al. 2023). The GAD isoforms studied in this experiment were AtGAD1, which occurred in roots, and the other is AtGAD2, which appeared mostly in several plant parts. P deficiency showed decreased 2-oxoglutarate dehydrogenase cofactor thiamine diphosphate (ThDP), which declines the endogenous P levels and biomass of plants.

Restriction in ThDP, as a consequence of upregulated GABA shunt, will ultimately impede the tricarboxylic acid cycle and respiration rate. Both *atgad1* and *atgad2* mutant lines showed a more prominent reduction in shoot growth and overall biomass under P deficiency in comparison to wild-type plants. Significant inhibition of primary root growth and stimulation in lateral root growth were observed in wild-type and *atgad2* plants under P deficiency. Furthermore, anthocyanin content was also enhanced in both wild-type and *atgad1* plants to ameliorate the detrimental effects of P deficiency. Quantification of amino acid by HPLC analysis indicated that GABA content was greatly increased in the wild type, whereas it was lowered in *atgad* mutant lines under P deficiency. Transcript profiling using qRT-PCR of P deficiency inducible genes was attenuated under exogenous GABA exposure.

Similarly, in apple plants, exogenous GABA could participate in acclimation under insufficient P levels (Chen et al. 2023). GABA participated in lowering accumulated ROS levels with further improvement in photosynthesis to stimulate the growth of stressed plants. Besides this, GABA contributed in activating the auxin signaling pathway leading to enhanced endogenous auxin levels in plants. Transgenic plants with overexpressing *MdGAD1* enhanced gene expression of auxin correlated with increased root growth and higher P uptake from roots, thereby conferring tolerance under P deficiency.

Another study depicts that exogenously applied GABA alleviates inadequate N supply efficiently to support the growth of *Andrographis paniculata* plants (Huang et al. 2023). The growth in plants is represented to be associated with increased levels of N and nitrate. GABA was noted to increase α-ketoglutarate and malate which further promote C and N metabolism and protein synthesis in plants. Moreover, it also triggered the expression of genes and enzymes involved in improving the N status of plants. GABA upregulates the expression of nitrate transporters such as *NRT2.4* and *NRT3.2* via GABA receptor which facilitates the uptake of nitrate, thus inducing N metabolism in N-deficient plants. However, under primidone (inhibitor of GABA receptor) application, obstruction in nitrate assimilation and growth was noticed. Hence, it was indicated that GABA promoted the growth of N-deficient plants by enhancing nitrate uptake mediated by its specific transporters.

Likewise, the combined application of GABA and root colonization by arbuscular mycorrhizal fungus in low N levels in *Sorghum bicolor* plants was assessed (Kchikich et al. 2021). Results showed that combined treatment with GABA and arbuscular mycorrhizal fungi displayed better growth characteristics such as shoot length, plant biomass, photosynthetic pigments, and N and C metabolism. Various enzymes, namely glutamine synthetase, glutamate synthase, glutamate dehydrogenase, nitrate reductase, phosphoenolpyruvate carboxylase, aspartate aminotransferase, and isocitrate dehydrogenase, which play essential roles in C and N skeleton, were greatly enhanced multifold in the roots and leaves of deficient plants. All these factors induced nitrate uptake and improved N assimilation in N-deficient plants, thus attenuating growth attributes.

Saffron plants growing in semiarid regions in calcareous soils often experience P deficiency, resulting in lower yield (Afshari et al. 2023). P deficiency is related to a reduction in P absorption from roots in the soil due to high pH and carbonate compounds reducing P solubility. However, foliar-sprayed GABA exhibited alleviation of P deficiency by improving flower number and yield, proline, growth of daughter corm, and P uptake.

Concluding Remarks

Nutrient deficiency, accompanied by unpredictable climatic conditions, has significantly exacerbated the loss of productivity and yield in major crops grown in agricultural field. This situation has raised the need for precise and targeted interventions to address nutrient deficiency efficiently in farmlands. There is a growing demand for approaches that are not only cost-effective but also environment friendly, reducing the detrimental effects of chemical fertilizers. Therefore, sustainable alternatives to chemical fertilizers are encouraged, which will promote research in the invention of effective methods for improving crop performance under nutrient supplementation.

GABA improves plant adaptability by contributing to morphological, physiological, biochemical, and molecular processes to restore growth in nutrient-deprived soils. Additionally, GABA can overcome the challenges of developing plants that can efficiently acclimate stress with higher nutritional value and better yield. Although there is substantial literature on GABA as a tool in stress alleviation, its response under nutrient deficiency needs to be better elucidated. Reports on how GABA mitigates the effects of inadequate nutrient supply are scarce. Future studies should focus on schematic and elaborative mechanisms for interpreting multiple nutrient interactions and their signaling pathway under both normal and stressful environments. While numerous researches have explored the role of GABA under abiotic stress; however, studies elucidating the crosstalk of GABA with other important metabolic pathways need to be updated. Moreover, a deeper understanding of the GABA shunt, GABA receptors, and GABA transporters needs special attention for sustainable agricultural productivity and stress management.

References

Afshari, R.T., Seyyedi, S.M., and Mirmiran, S.M. (2023). Clean application of pistachio residues-based vermicompost with γ-aminobutyric acid can alleviate the negative effects of high soil pH on P uptake in saffron (*Crocus sativus* L.). *Industrial Crops and Products* 195: 116443.

Akama, K. and Takaiwa, F. (2007). C-terminal extension of rice glutamate decarboxylase (OsGAD2) functions as an autoinhibitory domain and overexpression of a truncated mutant results in the accumulation of extremely high levels of GABA in plant cells. *Journal of Experimental Botany* 58 (10): 2699–2707.

Assunção, A.G., Cakmak, I., Clemens, S. et al. (2022). Micronutrient homeostasis in plants for more sustainable agriculture and healthier human nutrition. *Journal of Experimental Botany* 73 (6): 1789–1799.

Benidickson, K.H., Raytek, L.M., Hoover, G.J. et al. (2023). Glutamate decarboxylase-1 is essential for efficient acclimation of Arabidopsis thaliana to nutritional phosphorus deprivation. *New Phytologist* 240 (6): 2372–2385.

Bhattacharya, S., Khatri, A., Swanger, S.A. et al. (2018). Triheteromeric GluN1/GluN2A/GluN2C NMDARs with unique single-channel properties are the dominant receptor population in cerebellar granule cells. *Neuron* 99: 315–328.

Chen, X., Tian, Y., Xie, Y. et al. (2023). GABA responds to low phosphorus stress by interfering with endogenous auxin levels in apple. *Plant and Soil* 490: 435–450.

Colangelo, E.P. and Guerinot, M.L. (2006). Put the metal to the petal: metal uptake and transport throughout plants. *Current Opinion in Plant Biology* 9: 322–330.

Courbet, G., Gallardo, K., Vigani, G. et al. (2019). Disentangling the complexity and diversity of crosstalk between sulfur and other mineral nutrients in cultivated plants. *Journal of Experimental Botany* 70: 4183–4196.

da Cruz Ferreira, R.L., de Mello, P.R., de Souza Junior, J.P. et al. (2020). Oxidative stress, nutritional disorders, and gas exchange in lettuce plants subjected to two selenium sources. *Journal of Soil Science and Plant Nutrition* 120: 1215–1228. https://doi.org/10.1007/s42729-020-00206-0.

DalCorso, G., Manara, A., Piasentin, S., and Furini, A. (2014). Nutrient metal elements in plants. *Metallomics* 6 (10): 1770–1788.

de Bang, T.C., Husted, S., Laursen, K.H. et al. (2021). The molecular–physiological functions of mineral macronutrients and their consequences for deficiency symptoms in plants. *The New Phytologist* 229 (5): 2446–2469.

Fan, X., Zhou, X., Chen, H. et al. (2021). Cross-talks between macro-and micronutrient uptake and signaling in plants. *Frontiers in Plant Science* 12: 663477.

Gui, J.Y., Rao, S., Huang, X. et al. (2022). Interaction between selenium and essential micronutrient elements in plants: a systematic review. *Science of the Total Environment* 853: 158673.

Guo, Z., Du, N., Li, Y. et al. (2020). Gamma-aminobutyric acid enhances tolerance to iron deficiency by stimulating auxin signaling in cucumber (*Cucumis sativus* L.). *Ecotoxicology and Environmental Safety* 192: 110285.

Guo, Z., Gong, J., Luo, S. et al. (2023). Role of gamma-aminobutyric acid in plant defense response. *Metabolites* 13 (6): 741.

Gutiérrez-Alanís, D., Ojeda-Rivera, J.O., Yong-Villalobos, L. et al. (2018). Adaptation to phosphate scarcity: tips from *Arabidopsis* roots. *Trends in Plant Science* 23 (8): 721–730.

Hasanuzzaman, M., Hossain, M.S., Bhuyan, M.H.M.B. et al. (2018). The role of sulfur in plant abiotic stress tolerance: molecular interactions and defense mechanisms. In: *Plant Nutrients and Abiotic Stress Tolerance* (ed. M. Hasanuzzaman, M. Fujita, H. Oku, et al.), 221–252. Singapore: Springer Nature Ltd. https://doi.org/10.1007/978-981-10-9044-8_10.

Hawkesford, M., Horst, W., Kichey, T. et al. (2012). Chapter 6 – Functions of mMacronutrients. In: *Marschner's mineral nutrition of higher plants*, (Third Edition).e (ed. P. Marschner), 135–189. San Diego, CA: Academic Press https://doi.org/10.1016/B978-0-12-384905-2.00006-6.

Houmani, H. and Corpas, F.J. (2023). Can nutrients act as signals under abiotic stress? *Plant Physiology and Biochemistry* 206: 108313.

Huang, X.J., Jian, S.F., Wan, S. et al. (2023). Exogenous γ-aminobutyric acid (GABA) alleviates nitrogen deficiency by mediating nitrate uptake and assimilation in *Andrographis paniculata* seedlings. *Plant Physiology and Biochemistry* 198: 107700.

Jalil, S.U., Ahmad, I., and Ansari, M.I. (2017). Functional loss of GABA transaminase (GABA-T) expressed early leaf senescence under various stress conditions in *Arabidopsis thaliana*. *Current Plant Biology* 9–10: 11–22.

Janpen, C., Kanthawang, N., Inkham, C. et al. (2019). Physiological responses of hydroponically-grown Japanese mint under nutrient deficiency. *PeerJ* 7: e7751. https://doi.org/10.7717/peerj.7751.

Jiang, C., Gao, X., Liao, L. et al. (2007). Phosphate starvation root architecture and anthocyanin accumulation responses are modulated by the gibberellin-DELLA signaling pathway in *Arabidopsis*. *Plant Physiology* 145: 1460–1470.

Jobe, T.O., Zenzen, I., Karvansara, P.R., and Kopriva, S. (2019). Integration of sulfate assimilation with carbon and nitrogen metabolism in transition from C3 to C4 photosynthesis. *Journal of Experimental Botany* 70: 4211–4221.

Kchikich, A., El Omari, R., Kabach, I. et al. (2021). Effect of arbuscular mycorrhizal fungus and the γ-aminobutyric acid treatment in nitrate assimilation under nitrogen deficiency in sorghum plant. *Russian Journal of Plant Physiology* 68 (5): 901–908.

Khan, M.I.R., Jalil, S.U., Chopra, P. et al. (2021). Role of GABA in plant growth, development and senescence. *Plant Gene* 26: 100283.

Kinnersley, A.M. and Lin, F. (2000). Receptor modifiers indicate that 4-aminobutyric acid (GABA) is a potential modulator of ion transport in plants. *Plant Growth Regulation* 32 (1): 65–76.

Koike, S., Matsukura, C., Takayama, M. et al. (2013). Suppression of γ-aminobutyric acid (GABA) transaminases induces prominent GABA accumulation, dwarfism and infertility in the tomato (*Solanum lycopersicum* L.). *Plant & Cell Physiology* 54: 793–807.

Kumar, S., Jangde, S., Priya, T., and Yashu, B.R. (2017). Effect of GABA on morphology, yield and yield attributes in Black Gram (*Vigna mungo* (L.) Hepper) under salt stress condition. *International Journal of Pure & Applied Bioscience* 5: 1223–1228.

Kumar, S., Kumar, S., and Mohapatra, T. (2021). Interaction between macro-and micro-nutrients in plants. *Frontiers in Plant Science* 12: 665583.

Li, Z., Yu, J., Peng, Y., and Huang, B. (2016). Metabolic pathways regulated by γ-aminobutyric acid (GABA) contributing to heat tolerance in creeping bentgrass (*Agrostis stolonifera*). *Scientific Reports* 6: 30338.

Li, G.Z., Wu, Y.F., Liu, G.Y. et al. (2017). Large-scale proteomics combined with transgenic experiments demonstrates an important role of jasmonic acid in potassium deficiency response in wheat and rice. *Molecular & Cellular Proteomics* 16: 1889–1905.

Li, Y., Liu, B., Peng, Y. et al. (2020). Exogenous GABA alleviates alkaline stress in *Malus hupehensis* by regulating the accumulation of organic acids. *Scientia Horticulturae* 261: 108982.

Liu, M., Zhang, H., Fang, X. et al. (2018). Auxin acts downstream of ethylene and nitric oxide to regulate magnesium deficiency induced root hair development in *Arabidopsis thaliana*. *Plant & Cell Physiology* https://doi.org/10.1093/pcp/pcy078.

Mahmud, J.A., Hasanuzzaman, M., Nahar, K. et al. (2017). γ-aminobutyric acid (GABA) confers chromium stress tolerance in *Brassica juncea* L. by modulating the antioxidant defense and glyoxalase systems. *Ecotoxicology* 26: 675–690.

Malekzadeh, P., Khosravi-Nejad, F., Hatamnia, A.A., and Mehr, R.S. (2017). Impact of postharvest exogenous c-aminobutyric acid treatment on cucumber fruit in response to chilling tolerance. *Physiology and Molecular Biology of Plants* 23: 827–836.

Palanivelu, R., Brass, L., Edlund, A.F., and Preuss, D. (2003). Pollen tube growth and guidance is regulated by POP2, an *Arabidopsis* gene that controls GABA levels. *Cell* 114 (1): 47–59.

Priya, M., Sharma, L., Kaur, R. et al. (2019). GABA (c-aminobutyric acid), as a thermo-protectant, to improve the reproductive function of heat-stressed mungbean plants. *Scientific Reports* 9: 1–14.

Renault, H., El Amrani, A., Palanivelu, R. et al. (2011). GABA accumulation causes cell elongation defects and a decrease in expression of genes encoding secreted and cell wall-related proteins in *Arabidopsis thaliana*. *Plant & Cell Physiology* 52: 894–908.

Rezaei-Chiyaneh, E., Seyyedi, S.M., Ebrahimian, E. et al. (2018). Exogenous application of gamma-aminobutyric acid (GABA) alleviates the effect of water deficit stress in black cumin (*Nigella sativa* L.). *Industrial Crops and Products* 112: 741–748.

Salah, A., Zhan, M., Cao, C. et al. (2019). γ-aminobutyric acid promotes chloroplast ultrastructure, antioxidant capacity, and growth of waterlogged maize seedlings. *Scientific Reports* 9: 484. https://doi.org/10.1038/s41598-018-36334-y.

Salvatierra, A., Pimentel, P., Almada, R., and Hinrichsen, P. (2016). Exogenous GABA application transiently improves the tolerance to root hypoxia on a sensitive genotype of *Prunus* rootstock. *Environmental and Experimental Botany* 125: 52–66.

Scholz, S.S., Reichelt, M., Mekonnen, D.W. et al. (2015). Insect herbivory-elicited GABA accumulation in plants in wound-induced, direct, systemic and jasmonate-dependent defense response. *Frontiers in Plant Science* https://doi.org/10.3389/fpls.2015.01128.

Seif, H.S., Curvers, K., De Vleesschauwer, D. et al. (2013). Concurrent overactivation of the cytosolic glutamine synthetase and the GABA shunt in the ABA-defcient sitiens mutant tomato leads to resistance against *Botrytis cinerea*. *The New Phytologist* 199: 490–550.

Shahid, M., Saleem, M.F., Saleem, A. et al. (2019). Exogenous potassium–instigated biochemical regulations confer terminal heat tolerance in wheat. *Journal of Soil Science and Plant Nutrition* 19 (1): 137–147.

Sharma, A., Shankhdhar, D., and Shankhdhar, S.C. (2018). The role of calcium in plant signal transduction under macronutrient deficiency stress. In: *Plant Macronutrient Use Efficiency: Molecular and Genomic Perspectives in Crop Plants* (ed. M.A. Hossain, T. Kamiya, D. Burritt, et al.), 181–196. London: Elsevier Inc.

Shelp, B.J., Mullen, R.T., and Waller, J.C. (2012). Compartmentation of GABA metabolism raises intriguing questions. *Trends in Plant Science* 17: 57–59.

Simpson, J.P., Di Leo, R., Dhanoa, P.K. et al. (2008). Identifcation and characterization of a plastid localized *Arabidopsis* glyoxylate reductase isoform: comparison with a cytosolic isoform and implications for cellular redox homeostasis and aldehyde detoxifcation. *Journal of Experimental Botany* 59: 2545–2554.

Sita, K. and Kumar, V. (2020). Role of gamma amino butyric acid (GABA) against abiotic stress tolerance in legumes: a review. *Plant Physiology Reports* 25: 654–663.

Tarkowski, Ł.P., Signorelli, S., and Höfte, M. (2020). γ-Aminobutyric acid and related amino acids in plant immune responses: emerging mechanisms of action. *Plant, Cell & Environment* 43: 1103–1116.

Tripathi, D.K., Singh, S., Singh, S. et al. (2015). Micronutrients and their diverse role in agricultural crops: advances and future prospective. *Acta Physiologiae Plantarum* 37: 1–14.

Tsay, Y.F., Ho, C.H., Chen, H.Y., and Lin, S.H. (2011). Integration of nitrogen and potassium signaling. *Annual Review of Plant Biology* 62 (1): 207–226.

Watanabe, M., Maemura, K., Kanbara, K. et al. (2002). GABA and GABA receptors in the central nervous system and other organs. *International Review of Cytology* 213: 1–47.

Winkler, A. and Knoche, M. (2019). Calcium and the physiology of sweet cherries: a review. *Scientia Horticulturae* 245: 107–115.

Xiang, L., Hu, L., Xu, W. et al. (2016). Exogenous γ-aminobutyric acid improves the structure and function of photosystem II in muskmelon seedlings exposed to salinity-alkalinity stress. *PLoS One* 11: e0164847.

Xie, T., Ji, J., Chen, W. et al. (2020). GABA negatively regulates adventitious root development in *poplar*. *Journal of Experimental Botany 71* (4): 1459–1474.

Yang, J., Sun, C., Zhang, Y. et al. (2017). Induced resistance in tomato fruit by γ- aminobutyric acid for the control of Alternaria rot caused by *Alternaria alternata*. *Food Chemistry* 22: 1014–1020.

Yang, L.T., Zhou, Y.F., Wang, Y.Y. et al. (2019). Magnesium deficiency induced global transcriptome change in *Citrus sinensis* leaves revealed by RNA-seq. *International Journal of Molecular Sciences* 20 (13): 3129. https://doi.org/10.3390/ijms20133129.

Yang Z. Xu Y, Song P, Li X., Zhou, J., Lin L, Xia H, Liang D, Luo X, Zhang H, Deng Q (2023) Effects of Gamma Amino Butyric Acid (GABA) on Nutrient Uptake of Loquat [*Eriobotrya japonica* (Thunb.) Lindl.] Seedlings. *Horticulturae*, 9(2):196.

Yong, B., Xie, H., Li, Z. et al. (2017). Exogenous application of GABA improves PEG-induced drought tolerance positively associated with GABA-shunt, polyamines, and proline metabolism in white clover. *Frontiers in Physiology* 8: 1107. https://doi.org/10.3389/fphys.2017.01107.

Zhao, Y., Song, X., Zhong, D.B. et al. (2020). γ-aminobutyric acid (GABA) regulates lipid production and cadmium uptake by *Monoraphidium* sp. QLY-1 under cadmium stress. *Bioresource Technology* 297: 122500. https://doi.org/10.1016/j.biortech.2019.122500.

Zhu, X., Liao, J., Xia, X. et al. (2019). Physiological and iTRAQ-based proteomic analyses reveal the function of exogenous c-aminobutyric acid (GABA) in improving tea plant (*Camellia sinensis* L.) tolerance at cold temperature. *BMC Plant Biology* 19: 43.

18

GABA and Plant-Derived Therapeutics

Lakshmi Jayaram[1*], *Deepthi Puttegowda*[1*], *V. H. Pushpa*[2], *Shashank M. Patil*[1], *and Ramith Ramu*[1]

[1] *Department of Biotechnology and Bioinformatics, JSS Academy of Higher Education and Research, Mysuru, Karnataka, 570015, India*
[2] *Department of Pharmacology, JSS Medical College, JSS Academy of Higher Education & Research, Mysuru, Karnataka, 570015, India*

List of Abbreviations

BDS	Benzodiazepines
CNS	Central nervous system
DA	Dopamine
DSM	*Diagnostic and Statistical Manual of Mental Disorder*
ENS	Enteric nervous system
EPM	Elevated plus maze
GABA	Gamma-aminobutyric acid
GAD	Generalized anxiety disorder
GD	Glutamic acid decarboxylase
HAM-A	Hamilton Anxiety Rating Scale
HAM-D	Hamilton Depression Rating Scale
IPSP	Inhibitory postsynaptic potential
NMRI	Nuclear magnetic resonance image
NO	Nitric oxide
OCD	Obsessive-compulsive disorder
RTCs	Randomized controlled trials
SNRIs	Serotonin and norepinephrine reuptake inhibitors
SSRIs	Selective serotonin reuptake inhibitors

* Equal contribution.

GABA in Plants: Biosynthesis, Plant Development, and Food Security, First Edition. Edited by Samiksha Singh, Durgesh Kumar Tripathi, and Vijay Pratap Singh.

Introduction

Gamma-aminobutyric acid (GABA) is an essential molecule in plant metabolism. More than 70 years ago, scientists first identified the presence of GABA in potato tubers in 1949. GABA plays a significant role as a neurotransmitter in mammalian brains (Roberts and Eidelberg 1960), facilitating navigation and enabling adaptation to various growth environments. It is a crucial neurotransmitter in the central nervous system (CNS) of mammals, including humans. It plays a fundamental role in regulating neuronal activity and is primarily associated with inhibitory neurotransmission. The CNS's principal inhibitory neurotransmitter is the amino acid GABA (Steward et al. 1958). It inhibits nerve transmission, thereby lowering neural excitability. The thalamus, hippocampus, basal ganglia, hypothalamus, and brainstem are regions that contain GABAergic neurons. For appropriate cell membrane integrity and neurologic function, the balance between excitatory neuronal transmission with glutamate and inhibitory neuronal transmission with GABA is essential (D'Cunha et al. 1996).

The dual role of GABA, both as a metabolite and as a part of signaling pathways, provides plants with a versatile mechanism to adapt to various environmental conditions. When externally applied, GABA produces effects similar to the naturally occurring molecule (Hanrahan et al. 2011). As a result, this exogenously applied GABA holds the promise of enhancing the overall vitality of plants. Several "phytomedicines," which are compounds derived from whole plants or plant extracts, have demonstrated potential pharmacodynamic properties that could contribute to their anxiolytic effects. These properties are often observed in preclinical research both in vivo and in vitro. The anxiolytic results from phytomedicines may be attributed to various constituents, including alkaloids, polyphenols, and terpenoid substances having a number of potential uses in the CNS (Petty et al. 2014). These functions can involve binding to receptor sites for opiates, altering membrane structure, and voltage-gated channels to induce ionic channel transmission, or altering the function of enzymes such as glutamic acid decarboxylase (GD) and GABA-T (Fait et al. 2008), Particularly known for their ability to positively modulate GABA-A receptors are flavonoids (Hanrahan et al. 2011; Schmeling et al. (2008). Traditional medicine has long harnessed the power of plant-derived compounds, highlighting their effectiveness in treating anxiety and related disorders in humans as shown in Table 18.1. Unlike conventional synthetic anxiolytic drugs, natural compounds derived from plants, including GABA, exhibit therapeutic potential without significant adverse effects, first illuminating its significance in promising therapeutic applications, as shown in Figure 18.1.

The Mechanism of GABA in Action: Neurotransmission and Its Effect on Neurons

Neurotransmitters are chemical mediators that facilitate communication between nerve cells, or neurons, in the brain. The signal-emitting neuron releases these chemicals, which attach to certain proteins that are receptors on the signal-receiving neuron. Through its interaction with these cells' GABA-A receptor, the neurotransmitter GABA reduces the activity of signal-receiving neurons. GABA serves as the primary inhibitory neurotransmitter in the CNS.

Table 18.1 List of plants as therapeutic strategies.

Botanical name	Common name	Active element	GABA substrate mechanism	Preclinical animal anxiety model	Clinical model
Passiflora incarnata	Purple passion flower	Flavonoids: b-carboline alkaloid	Benzodiazepines (BDZ): positive allosteric regulation of GABA-A receptors; affinity for GABA-A and GABA-B receptor subtypes. Awad et al. (2007); Chatterjee et al. (1992); Hamid et al. (2016)	Hole board, EPM, and horizontal wire rodent models Ceremuga et al. (2015); Wanasuntronwong et al. (2012); Wijeweera et al. (2006)	Anxiety disorders Bradwejn et al. (2000); Jana et al. (2010); Sampath et al. (2011); Wolfman et al. (1994)
Piper methysticum	Kava kava	Kava lactones: Dihydrokavain and kavain	Positive modulators of GABA-A receptors enhance ligand binding across all subtypes, including calcium (Ca2+)-gated ion channels and sodium chloride (NaCl). Baxter et al. (2013); Boonen and Haberlein (1998); Chua et al. (2016); Davies et al. (1992); Gleitz et al. (1995); He et al. (1997); Jussofie et al. (1994); Magura et al. (1997); Martin et al. (2002); Mathews and MacLeod (2005).	EPM Garrett et al. (2003); Rex et al. (2002); Schulze et al. (2003)	Menopausal anxiety; nonpsychotic anxiety LaPorte et al. (2011); Pittler and Ernst (2003); Sarris. (2007); Sarris et al. (2011, 2013a, 2013b, 2013c); Sarris (2018); Singh and Singh (2002)
Withania somnifera	Ashwagandha	Glycosides alkaloids and steroidal triterpenoid lactones: anolides A, D, and G, glycowithanolides, anone, withaferin A, anoloids and anine, withasamnin	Agonists for ionotropic GABA-A receptors, specifically targeting the GABAρ1 receptor. Candelario et al. (2015); Yin et al. (2013)	EPM, social contact, delayed eating, and unfamiliar environments Bhattacharya et al. (2000); Kaur et al. (2016)	Bipolar disorder, chronic stress, generalized anxiety disorder (GAD), and general anxiety Andrade et al. (2012); Auddy et al. (2008); Cooley et al. (2009); Chengappa et al. (2013)
Valeriana officinalis	Valerian	Sesquiterpenoids: alkaloids, flavonones, valepotriates, and valverenic acid	Stimulation of glutamic acid decarboxylase (GD) is linked to GABA-A receptor binding, specifically in the GABA-A chloride channel receptor complex, particularly with the β2 or β3 subunit (N265 M), influencing GABA-A receptor allosteric modulation. Awad et al. (2007); Becker et al. (2014); Khom et al. (2007, 2016); Ortiz et al. (1999); Ramharter and Multzer (2009); Santos et al. (1994); Trauner et al. (2008)	Seizures mode, light/dark choice test, and EPM Benke et al. (2009); Khom et al. (2016); Hattesohl et al. (2008); Murphy et al. (2010)	GAD, OCD, HIV adults, healthy sample driving, and anxious children Ahmadi et al. (2017); Andreatini et al. (2002); Gutierrez et al. (2004); Pakseresht et al. (2011); Thomas et al. (2016)

(Continued)

Table 18.1 (Continued)

Botanical name	Common name	Active element	GABA substrate mechanism	Preclinical animal anxiety model	Clinical model
Scutellaria lateriflora	Blue skullcap	Flavonoids: wogoni, bacalin, baicalein, apigenin, scutelaterin A, and oroxylin A	GABA-A receptor positive-allosteric regulation. Awad et al. (2003); Wang et al. (2008); Xu et al. (2006)	EPM, open field Awad et al. (2003)	General anxiety Brock et al. (2014); Wolfson and Hoffmann (2003)
Melissa officinalis	Lemon balm	Flavonoids, volatile oils, triterpenoids: ursolic acid, rosmarinic acid, and oleanolic acid	Inhibiting GABA-transaminase (GABA-T) leads to increased affinity of GABA-A receptors with benzodiazepine (BDZ), resulting in the elevation of GABA levels in the brain. Awad et al. (2007, 2009); Abuhamdah et al. (2008); Huang et al. (2008); Mahita et al. (2014); Raines et al. (2009); Salah and Jager (2005); Yoo et al. (2011)	Open-field rodent models and EPM Ibarra et al. (2010); Taiwo et al. (2012)	Premenstrual dysphoria, general anxiety, sleep problems, and the stress-induction paradigm in healthy controls Akbarzadeh et al. (2015); Cases et al. (2011); Kennedy et al. (2002, 2003, 2004); Scholey et al. (2014)
Ginkgo biloba	Maidenhair	Terpenoids: gingkolides A, B, and C, and bilobalides	Antagonism of GABA-A and GABA-C receptors results in elevated GABA levels and increased activity of the GD enzyme. Huang et al. (2003); Ivic et al. (2003); Kiewert et al. (2007); Ng et al. (2017); Sasaki et al. (1999a); Sasaki et al. (1999b)	Water maze, EPM, open field Kuribara et al. (2003); Ma et al. (2012)	GAD Woelk et al. (2007)
Humulus lupulus	Hops	Terpenoids and flavonoids: xanthohumol, myrcene, and humulene	GD Inhibition Awad et al. (2007)	Avian model of nocturnal behaviors, EPM Ceremuga et al. (2013); Franco et al. (2012a); Schiller et al. (2006)	Menopausal women; a healthy sample under stress from job Franco et al. (2012b)
Matricaria recutita	Matricaria chamomilla	Flavonoids: apigenin, luteolin, rutin, herbiarin, umberlliferone, chlorogenic, and caffeic acids are examples of sesquiterpenoids, coumarins, and phenylpropanoids	GABA-BZD receptor–ligand binding Avallone et al. (1996, 2000); Johnston (2011); Viola et al. (1995); Willander et al (2009)	EPM, social interaction, open field Can et al. (2012); Shinomiya et al. (2005)	GAD, insomnia Amsterdam et al. (2009, 2011); Keefe et al. (2016); Mao et al. (2016); Zick et al. (2011)
Centella asiatica	Gotu kola	Asiatic acid triterpenoids	Activation of GD and selective agonist of GABA-B Awad et al. (2007); Chatterjee et al. (1992); Hamid et al. (2016)	Dark-light box, EPM, open-field test Jana et al. (2010); Wanasuntronwong et al. (2012); Wijeweera et al. (2006)	Acoustic startle response Bradwejn et al. (2000); Jana et al. (2010)

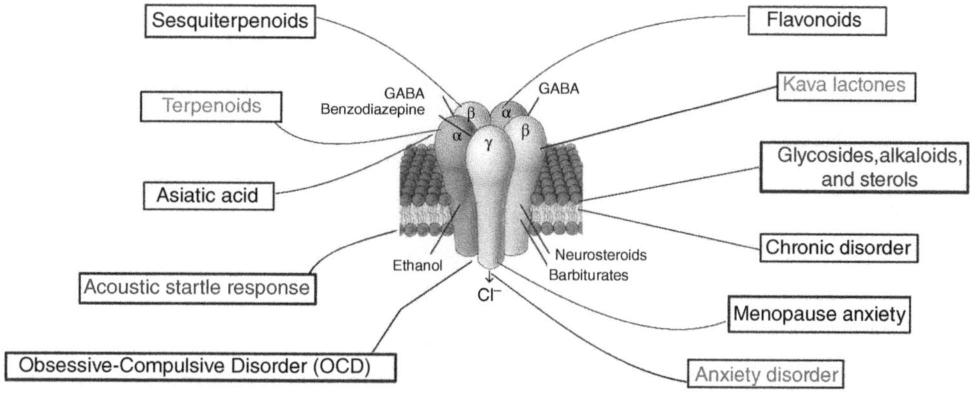

Figure 18.1 Plant-derived therapeutics.

The bidirectional signaling between the brain and the enteric nervous system (ENS) is vital in maintaining homeostasis (Cryan and O'Mahony 2011). Even though most research thus far has focused on the signaling from the brain to the gut, an increasing number of studies have explored the influence of gut microbiota on the brain. Research has shown that the bacteria in our gut could have a positive effect on our mental health, including reducing anxiety and improving mood, especially for those who suffer from chronic fatigue (Logan et al. 2009). GABA, an inhibitory neurotransmitter, is synthesized inside the presynaptic neuron's cytoplasm from the precursor glutamate. The enzyme glutamate decarboxylase is responsible for the synthesis process, which requires vitamin B6 (pyridoxine) as a cofactor. The next step involves the loading of GABA into synaptic vesicles by the vesicular inhibitory amino acid transporter. As an inhibitory neurotransmitter, GABA causes hyperpolarization of the postsynaptic neuron, generating an inhibitory postsynaptic potential (IPSP). Conversely, glutamate induces depolarization of the postsynaptic neuron, resulting in an excitatory postsynaptic potential (EPSP). The main method of communication between neurons and effector cells, such as other neurons, is through the generation of IPSPs resulting from the release of GABA (Meldrum 2000) (Figure 18.2).

Plants with Reported GABAergic Activity: A Novel Source of Therapeutics

Natural compounds from plants used in traditional medicine are effective in managing anxiety and associated complications (Wasowski and Marder 2011). Especially, the prolonged use of these natural compounds has not been associated with significant adverse effects in clinical settings. Conventional synthetic anxiolytic drugs used in clinical pharmacotherapy have been associated with various adverse effects.

Conventional medications that have been used to treat anxiety through a variety of mechanisms of action on neurotransmitter receptors (Bourin 2020) in recent years have both anxiolytic and antidepressant effects. There are several groups of medications used to treat anxiety, including benzodiazepines, selective serotonin reuptake inhibitors

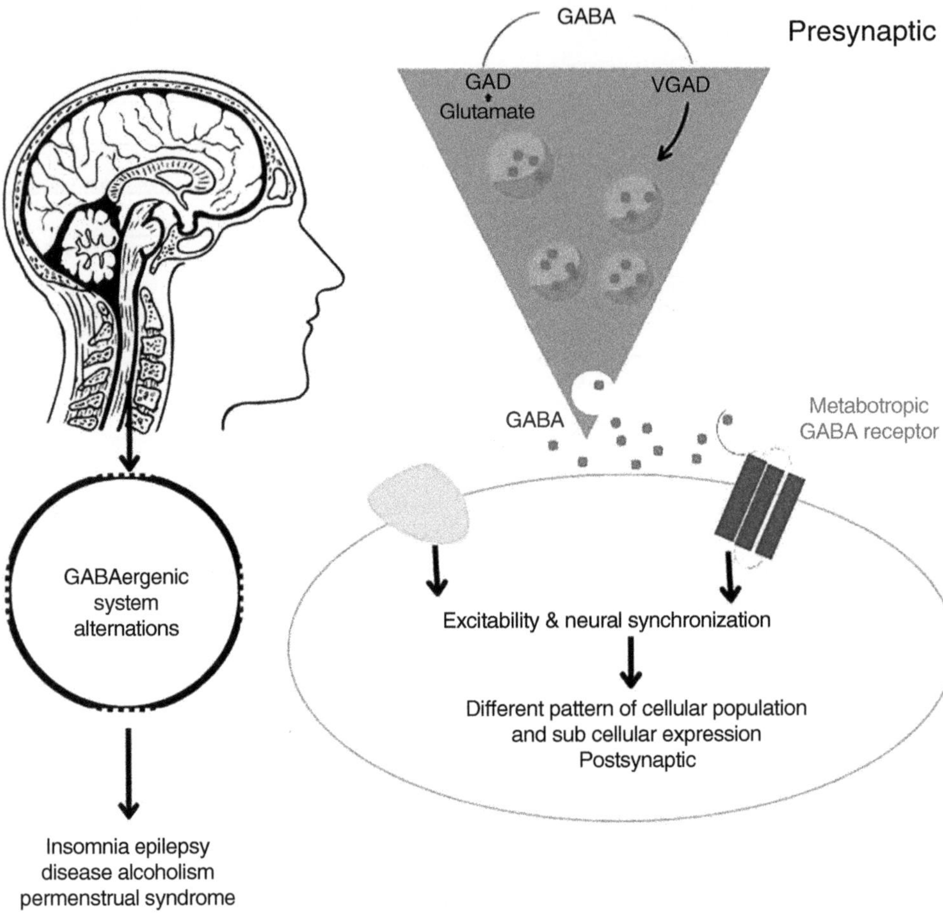

Figure 18.2 The GABAergic system in neurotransmitter.

(SSRIs), serotonin and norepinephrine reuptake inhibitors (SNRIs), pregabalin, buspirone, and moclobemide (Bandelow et al. 2017). Despite the therapeutic effectiveness of anxiolytic medications, documented side effects include memory loss, sleep disturbances, and pharmacological dependency. As a result, many patients encounter unfavorable side effects, especially from benzodiazepines, such as weariness, sleepiness, dizziness, and poor motor coordination. However, without therapy, it can progress to depression (Ernst 2006). It is important to evaluate the toxicity and efficacy of each individual plant or plant extract separately. A variety of natural chemicals found in plants have shown a potential to affect the CNS, including reducing anxiety. According to Sarris et al. (2013a), bioactive phytochemicals primarily affect the brain's neurotransmission systems, which include the gamma-aminobutyric aminergic (GABAergic), dopaminergic (DAergic), serotonergic (5-hydroxytryptamine, or 5-HT) system, and noradrenergic (NA) systems. These phytochemicals are members of several chemical families, including fatty acids, polyphenols, flavonoids, alkaloids, saponins, and triterpenoids (Fajemiroye et al. 2016). Herein, we discuss a few plants in which GABAergic activity has been reported.

Passiflora incarnata (Passion Flower)

Passiflora incarnata, commonly known as Passion Flower, is one such botanical remedy that has gained attention for its potential to modulate GABAergic activity (Appel et al. 2011). The climbing vine *Passiflora incarnata* is common in the Southeast region of the United States. Usage common among Native Americans for its calming properties, this plant has a rich history in folk medicine (Krenn 2002). The passion flower's unique blossoms are said to resemble the tools of the Crucifixion, leading to its association with Christian symbolism. In recent years, scientific research has delved into the compounds present in *P. incarnata* and their potential therapeutic effects (Meier 1995; Miroddi et al. 2013).

The GABAergic system regulates neuronal excitability and is linked to neurological and psychiatric disorders (Elsas et al. 2010). Passion Flower has been found to contain bioactive compounds that interact with the GABAergic system, influencing the release and uptake of GABA. *P. incarnata* is rich in flavonoids, including chrysin and apigenin, as shown in Figure 18.3. These compounds have demonstrated anxiolytic effects by modulating GABA receptors, particularly the GABA-A receptor. By binding to these receptors, flavonoids may enhance GABAergic neurotransmission, leading to a calming effect on the CNS (Aslanargun et al. 2012; Dantas et al. 2017). Harman and harmine alkaloids found in Passion Flower

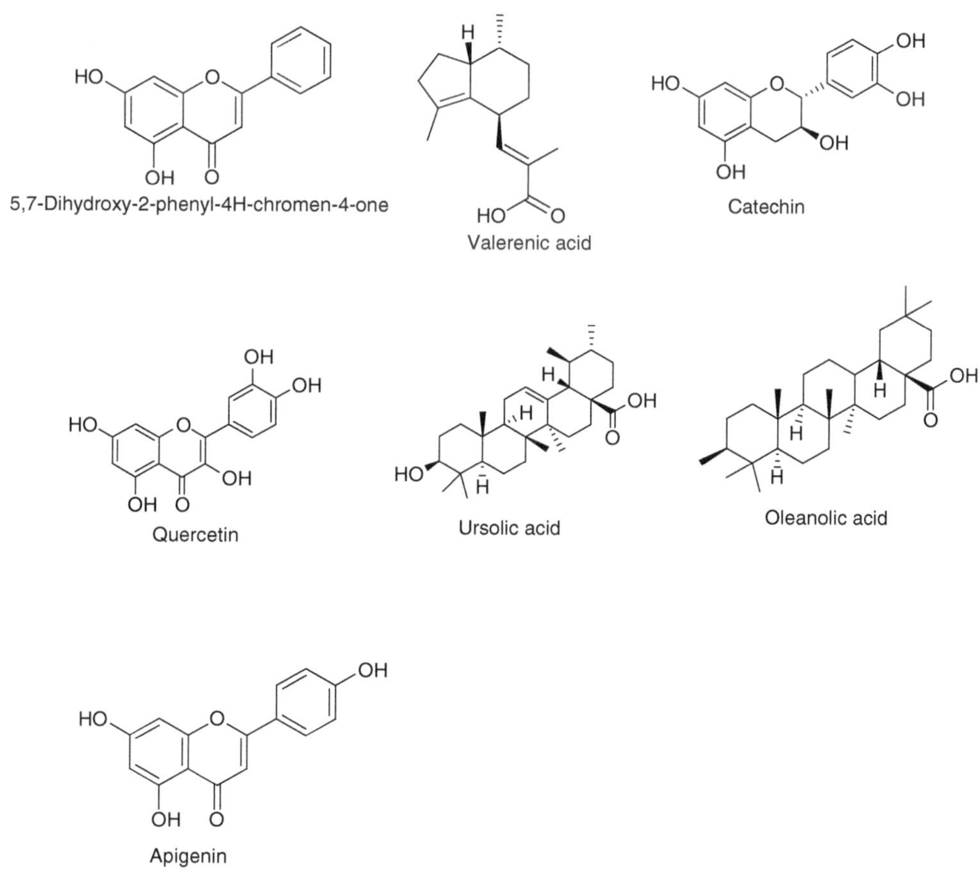

Figure 18.3 Phytochemical constituents with GABAergic properties.

have also been studied for their interaction with GABA receptors. Research suggests that these alkaloids may exert a positive allosteric modulation on GABA receptors, potentially enhancing the inhibitory effects of GABA (Wohlmuth et al. 2010).

The sedative properties of Passion Flower may make it a promising remedy for insomnia. By promoting GABAergic activity, it could contribute to improved sleep duration. Preliminary studies suggest that Passion Flower's interaction with GABA receptors may have anticonvulsant effects (Elsas et al. 2010).

Piper methysticum (Kava)

Piper methysticum (Kava) is widely used in the South Pacific, for its nootropic, anxiolytic, neuroprotective, nociceptive, and dysphoric effects. Kava lactones, specifically kavain and dihydrokavain, have anxiolytic effects (Tzeng and Lee 2015).

Preclinical models have provided insights into the potential mechanisms of kava's effects, primarily involving the modulation of GABA receptors. Several pathways have been proposed, including voltage-gated sodium ion channel blockade, enhanced ligand binding across subtypes of GABA-A receptors, rapid increase of GABA-A receptor function, and decreased release of excitatory neurotransmitters through calcium ion channel blockade (Mathews et al. 2005).

Numerous animal anxiety model investigations, including rodent Elevated Plus Maze (EPM) tests, have reported anxiolytic effects, characterized by reduced entry latencies into unfamiliar environments and increased time spent in these environments. Human clinical research has demonstrated that kava is superior to placebo in terms of reducing anxiety symptoms. Significant decreases in anxiety have been shown in favor of kava extract all over randomized controlled trials (RCTs), according to pooled effect size calculations based on Hamilton Anxiety Rating Scale (HAM-A) scores (Pittler and Ernst 2003). Preclinical studies in humans showed that the extract's various GABAergic activities play a major role in the kava's anxiolytic effects, at least in part. In expressed *Xenopus* oocytes, the kava phytoconstituent identified as "kavain" showed positive allosteric regulation across all GABA-A receptors (Pittler and Ernst 2003).

Withania somnifera (Ashwagandha, Indian Ginseng, Winter Cherry)

Withania somnifera, commonly known as Ashwagandha or Indian ginseng, is a revered herb in traditional Ayurvedic medicine known for its adaptable genic properties. Enriched with glycosides (withanolides), alkaloids (withanine, ashwagandhine, ashwaganidhine, and somniferine), and steroidal triterpenoid lactones, it is a botanical powerhouse belonging to the nightshade family (Mishra et al. 2000). Native to the Indian subcontinent, *Withania somnifera* is a small shrub with yellow flowers and red fruit, belonging to the *Solanaceae* family. It has been commonly used to reduce stress, boost energy levels, and improve overall life expectancy. In recent years, Ashwagandha has garnered attention in the scientific community for its diverse pharmacological effects, including its potential interaction with the GABAergic system. The GABAergic system centered around the neurotransmitter GABA, plays a crucial role in regulating neuronal excitability. *Withania somnifera* has been found to influence this system through various mechanisms. GABAergic actions of *Withania somnifera*, focusing on its constituents and clinical evidence support its role in anxiety and stress management (Choudhary et al. 2017). Studies indicate that certain bioactive compounds in Ashwagandha, including withanolides, may interact with GABA receptors. These interactions suggest a potential modulatory effect on GABAergic neurotransmission, influencing the inhibitory tone of the CNS. *Withania somnifera* may enhance the binding of GABA to GABA-A receptors. This augmentation of GABAergic activity could contribute to the anxiolytic and sedative effects associated with Ashwagandha (Bhattacharya et al. 2000). By modulating GABAergic neurotransmission, *Withania somnifera* exhibits neuroprotective properties. It may help regulate the balance between excitatory and inhibitory neurotransmitters, potentially mitigating neurodegenerative processes associated with excessive neuronal excitability. The GABAergic activity of *Withania somnifera* suggests a range of potential therapeutic applications. Ashwagandha's anxiolytic effects, mediated through GABAergic modulation, make it a promising candidate for managing stress and anxiety disorders (Candelario et al. 2015). One of its beneficial properties is its adaptogenic nature, which can aid in developing stress resilience and enhancing coping mechanisms. The sedative effects of *Withania somnifera*, possibly attributed to GABAergic interactions, position it as a natural remedy for sleep disorders. It may help regulate sleep patterns and improve sleep quality. The neuroprotective effects of Ashwagandha, linked to GABAergic modulation, suggest potential benefits for cognitive function (Choudhary et al. 2017).

Valeriana officinalis, (Valeriana)

Valeriana is a perennial plant with roots that are widely used as a sedative and anxiolytic. The plant is rich in flavanones, terpenes, and alkaloid compounds, among other bioactive substances. One specific compound that has been found to be a sesquiterpene that affects the GABAergic system is valerenic acid (Houghton 1988).

Preclinical research has demonstrated that valerian species have the capacity to increase GABA release while simultaneously blocking its breakdown or reuptake. The downregulation of GABA-transaminase (GABA-T) activity in brain tissue of animal models demonstrates its impact (Awad et al. 2007). The main component, valeric acid (Figure 18.3), acts as an allosteric modulator of GABA-A receptors, focusing on the β2/3 subunits. This demonstrates the special function of valerenic acid in regulating GABA-A receptor activity, especially in relation to these particular receptor subunits. In vivo studies on animals, including mice and rats, have shown anxiolytic and antidepressant properties of valerian extracts. Valerenic acid and valerenol influence GABA-A receptors, particularly β3 subtypes, leading to these outcomes. Research has confirmed that valerian extract can alleviate anxiety and has similar effectiveness to benzodiazepines in rat models (Murphy et al. 2010). A patient with HIV-positive revealed that daily supplementation of valerian extracts significantly reduced anxiety associated with antiretroviral therapy. In individuals with obsessive-compulsive disorder (OCD), valerian root extract was found to significantly reduce symptoms. A randomized controlled study found that valerian is more effective than diazepam in treating the symptoms of generalized anxiety disorder (GAD) (Andreatini et al. 2002).

Scutellaria lateriflora, (Scullcap, Blue Skullcap)

Scutellaria lateriflora has historically been appreciated for its antianxiety, calming, and antispasmodic effects in both Native American and Western medicine. The plant contains several flavonoids such as scutellarin, baicalin, baicalein, oroxylin A, apigenin, and wogonin. Significant elements including wogonin, apigenin, baicalin, and baicalein influence the GABA-A receptor's benzodiazepine (BDZ) site in a positive manner to produce their actions and thus, demonstrating GABAergic properties. Animal studies, particularly with male Sprague–Dawley rats, have shown behavioral changes reflecting GABA-mediated anxiolytic properties (Brock et al. 2014). Two human studies involving non-clinical samples reported improved mood after consuming *Scutellaria lateriflora* extract. In a double-blind crossover study, participants experienced enhanced mood after two weeks of consuming the extract, though there were no significant differences in anxiety outcomes due to a "floor effect" (Wolfson and Hoffmann 2003).

Melissa officinalis, (Lemon Balm)

Melissa officinalis, also known as *lemon balm*, was identified in fourth-century BC Eurasia. The plant boasts a diverse array of phytochemicals such as phenolic acids, terpenes, eugenol acetate, and tannins. In vitro studies highlight its potential to modulate GABA activity by inhibiting GABA-transaminase (GABA-T), while triterpenoids like ursolic enhance GABA-A benzodiazepine receptor affinity. It is native to Eurasia and has a rich history dating back to the fourth-century BC. It is composed of a number of phytochemicals, including tannins, phenolic acids, terpenes like caffeic and rosmarinic acids, and eugenol acetate. Studies in rodent models, including C57BL/6 Jico mice and Wistar rats, have reported dose-dependent anxiolytic responses to *lemon balm* extract, comparable to diazepam effects (Yoo et al. 2011). Human studies demonstrated decreases in mild-to-moderate anxiety and insomnia symptoms with *lemon balm* extract in stressed individuals. Another study in a healthy sample showed improvements in attention in a laboratory stress-induction mood. An RCT in school-age females reported the efficacy of *lemon balm* extract for tension and reducing stress associated with premenstrual dysphoric disorder over three months. *Lemon balm* appears to exert anxiolytic-like effects through various mechanisms, as demonstrated in both preclinical and human studies (Cases et al. 2011), highlighting its multifaceted impact on GABAergic modulation and stress-related conditions.

Ginkgo biloba, (Maiden Hair)

The ginkgo tree (*Ginkgo biloba*), one of the most ancient flowering plant species, has a rich history of therapeutic use, particularly in its nuts and leaves. Proanthocyanin, *ginkgo* and bilobalides, distinct terpenoids, and flavonoids including quercetin and catechin (Figure 18.3) constitute significant active compounds. Studies on human endothelial cell lines suggest neuroprotective, antiapoptotic properties, and vascular benefits through enhanced nitric oxide (NO). In female NMRI mice models and Wistar rats, research indicates modifications in cholinergic and monoaminergic pathways, suggesting potential cognitive and mood-enhancing effects (Fehske et al. 2009). Bilobalide and Ginkgolides A, B, and C act as antagonists at GABA-A receptors, with potential anticonvulsant effects by increasing GABA levels in striatum tissue and cortical hippocampal. Ex vivo assays in rat brain tissue showed selective inhibition of GABA-A receptors by Bilobalide and Ginkgolides (Sasaki et al. 1999a). In anxiety rodent models, Bilobalide and Ginkgolides demonstrated dose-dependent anxiolytic effects, particularly Ginkgolide A, which led to reduced anxiety in the EPM paradigm. However, flumazenil did not block these effects because, on the GABA and benzodiazepine receptor complex, it competitively inhibits the action of both benzodiazepine and non-benzodiazepine compounds that bind with benzodiazepine receptors. It has the ability to reverse how benzodiazepines attach to their receptors. Bilobalide in male Kunming mice found increased locomotor activity and reduced latency in the Morris water maze, proposing glucocorticoid receptor modulation as a potential mechanism. Patients diagnosed with GAD showed significant improvement after taking *Ginkgo biloba* extract, according to the HAM-A (Woelk et al. 2007). *Ginkgo biloba* exhibits a complex interplay of neuroprotective, anxiolytic, and potentially mood-enhancing effects through various mechanisms, including GABA modulation and glucocorticoid receptor involvement.

Humulus lupulus, (Hops)

Humulus lupulus plant belonging to the *Cannabaceae* family is known traditionally for its calming, sleep-inducing, antibacterial, and anti-inflammatory properties. It is primarily cultivated for the production of alcoholic beverages like beer. In modern medicine, *H. lupulus* is used for managing anxiety and insomnia due to its terpenoid and flavonoid compounds. Laboratory studies have shown that myrcenol, found in *H. lupulus*, enhances GABA-A receptor response in *Xenopus* oocytes (Franco et al. 2012). Hops produce sedative effects through GABA receptor activity. Preclinical research in male Sprague–Dawley rats indicated GABAergic effects of *H. lupulus* by inhibiting GAD activity. *H. lupulus* is limited, with the majority of studies concentrating on its sedative and hypnotic characteristics. The clinical efficacy of *H. lupulus* in reducing anxiety symptoms remains uncertain, despite preclinical indications of GABAergic mechanisms (Chamomile, German chamomile).

Matricaria recutita, (True Chamomile)

Matricaria recutita, a flowering plant from the *Asteraceae* family, has historically been used for its calming and sedative properties. It substance a range of bioactive constituents, including coumarins and herniarin sesquiterpenes, umbelliferone, flavonoids such as apigenin, luteolin, and rutin, as well as phenylpropanoids like caffeic acids and chlorogenic. Studies on tissue have indicated that specific constituents of *Matricaria recutita* modulate GABA receptors through ligand effects on the central benzodiazepine (BDZ) site, leading to decreases in GABA-activated chloride currents. Apigenin, a flavonoid in *M. recutita*, has shown anxiolytic effects as an inverse BDZ agonist, potentially acting on BDZ receptor subunits (Hanrahan et al. 2011). Animal studies support these findings, with *M. recutita* extract improving the time taken to sleep in male Wistar rats, suggesting GABAergic BDZ receptor modulation. Demonstrated significant drops in anxiety behaviors in Balb/c male mice through various testing models (Can et al. 2012).

In humans, patients with GAD, *M. recutita* treatment showed considerably higher decreases in anxiety compared to the control group after 8 weeks. These analyses suggested potential benefits for co-morbid depression. Additional short- and long-term studies with GAD patients revealed a "clinical response" in 58% of patients with significant decreases in anxiety scores. The long-term study indicated improvements in GAD-7 ratings for the continuation group compared to the placebo group. Another clinical study involving patients with *Diagnostic and Statistical Manual of Mental Disorder* (DSM-IV) primary insomnia showed that 28 days of *M. recutita* treatment led to significant improvements in fatigue, sleep quality, and latency, although it did not significantly affect anxiety symptoms (Zick et al. 2011).

Centella asiatica, (Gotu Kola)

An Asian native herb called *Centella asiatica*, commonly used in traditional Chinese medicine and Indian Ayurvedic medicine, has diverse applications, including antibacterial, antidepressant, anxiolytic, dermatotic, and cognitive benefits. Its main constituents include tannins, sterols, flavonoids, stearic acids, and triterpenoid glycosides (Gohil et al. 2010). GABAergic effects of *C. asiatica* and its constituents have shown increased GAD activity, with one study reporting a significant rise of more than 50% in vitro. Asiatic acid, a constituent of *Centella asiatica*, has been studied for its pharmacological effects. In vitro research has shown that *C. asiatica* acts as a negative modulator of specific GABA-A receptor subtypes, potentially influencing neurotransmission and offering insights into its effects on anxiety and other neurological conditions. Ex vivo studies in rats demonstrated that *C. asiatica* extracts increased whole-brain levels of GABA (Wijeweera et al. 2006). The use of rodent models and triterpenoid extracts derived from *Centella asiatica* demonstrated effectiveness comparable to diazepam in mitigating behavioral symptoms of anxiety across different stress scenarios (Kinnersley and Turano 2000),Anxiolytic effects were observed in EPM tests with different extracts and the isolated asiaticoside constituent. Animal studies also demonstrated that Asiatic acid had anxiolytic effects, and was a GABA receptor antagonist, indicating involvement in GABAergic benzodiazepine receptor activity (Wijeweera et al. 2006), attenuated these effects. Humans, those treated with *C. asiatica* extract showed a significant reduction in the acoustic startle response. In an open-label study involving individuals with GAD, notable enhancements in anxiety levels, as assessed by the HAM-A, were observed following a 60-day treatment period. *Centella asiatica* has potential anxiolytic effects through modulation of the GABAergic system (Jana et al. 2010).

Conclusion and Future Perspective

The study of GABA and plant-derived medicines is an exciting area where herbal medicine and neuroscience intersect. As an essential neurotransmitter in the CNS, GABA is critical for controlling brain activity and preserving overall mental health. The identification of its existence in several plant species has created new opportunities for using the capacity of organic substances to regulate GABAergic function. Treatments developed from plants that are enhanced with GABA or molecules that mimic GABA present a viable substitute for treating illnesses including anxiety, insomnia, and mood swings that are linked to neurotransmitter imbalances. These natural compounds have several advantages, such as the ability to improve sleep, lower stress levels, and foster serenity without having negative side effects that are sometimes associated to manufactured medications.

Moreover, the increasing number of studies on GABAergic substances originating from plants emphasizes how crucial it is to consider traditional herbal medicines in contemporary medical procedures. The incorporation of plant-based therapies into conventional medicine is becoming more and more important as the desire for natural and sustainable healthcare choices grows. However, using treatments derived from plants must be done carefully and with a strict scientific methodology. Even while the prospective advantages seem good, further study is required to completely comprehend the mechanisms of action, determine the ideal dose, and guarantee the safety and effectiveness of these botanical treatments. Additionally, individual differences in how these chemicals affect a person must also be taken into account, highlighting the importance of individualized and patient-centered treatment plans.

In conclusion, combining GABA and plant-based therapies offers a promising holistic approach to enhancing mental health and well-being. The secret to realizing these natural substances' full potential in enhancing mental resilience and general health is to proceed with scientific research while carefully incorporating traditional wisdom.

This book chapter could address challenges and knowledge gaps in the field of GABA and plant-derived medicines. It may also suggest future directions for research and development. Future endeavors may focus on developing tailored formulations to optimize the delivery and bioavailability of GABAergic compounds, exploring innovative extraction methods, encapsulation techniques, and synergistic combinations. A personalized medicine approach, tailoring treatments based on individual differences, could enhance outcomes, while exploring synergies with conventional therapies may offer a more comprehensive treatment strategy with enhanced efficacy and reduced side effects. Collaborative efforts in global ethnopharmacology, combining traditional knowledge with modern scientific methods, may lead to the identification of GABAergic-rich plants in various indigenous medicinal systems worldwide, offering a rich source for novel therapeutics.

References

Abuhamdah S, Matar A, Elashkar A. The influence of GABA-T inhibition on GABAergic neurotransmission. *Psychopharmacology (Berl)* 2008;196(3):463–472. https://doi.org/10.1007/s00213-007-0987-3.

Ahmadi, A., Amini, M., Pasha, S. et al. (2017). The efficacy of *Withania somnifera* in reducing anxiety symptoms in patients with generalized anxiety disorder: A randomized controlled trial. *Journal of Alternative and Complementary Medicine* 23 (4): 298–304.

Akbarzadeh, S., Dehghani, F., Ghaedi, K. et al. (2015). The effects of *Glycyrrhiza glabra* on stress-induced anxiety in humans. *Phytotherapy Research* 29 (6): 894–901.

Amsterdam, J., Li, Y. et al. (2009). Anxiolytic effects of herbal supplements in GAD: A systematic review. *Journal of Clinical Psychopharmacology* 29 (5): 447–451.

Amsterdam, J., Li, Y. et al. (2011). The relationship between anxiety disorders and insomnia: A review. *Sleep Medicine Reviews* 15 (2): 83–91.

Andrade, C., Aswath, A., Kanchan, P. et al. (2012). A randomized controlled trial of *Withania somnifera* for the treatment of generalized anxiety disorder. *Indian Journal of Psychiatry* 54 (3): 245–249.

Andreatini, R., Sartori, V.A., Seabra, M.L.V., and Leite, J.R. (2002). Effect of valepotriates (valerian extract) in generalized anxiety disorder: a randomized placebo-controlled pilot study. *Phytotherapy Research* 16 (7): 650–654.

Andreatini, R., De Lima, T., de Lima, F. et al. (2002). Anxiolytic-like effects of *Bacopa monnieri* in the elevated plus-maze model in rats. *Phytotherapy Research* 16 (5): 476–478.

Andreatini R, Sartori VA, Seabra ML, Leite JR. Effect of valepotriates (valerian extract) in generalized anxiety disorder: a randomized placebo-controlled pilot study. *Phytotherapy Research: An International Journal Devoted to Pharmacological and Toxicological Evaluation of Natural Product Derivatives.* 2002 Nov;16(7):650-4.

Aslanargun P, Cuvas O, Dikmen B, Aslan E, Yuksel MU. Passiflora incarnata Linneaus as an anxiolytic before spinal anesthesia. *Journal of anesthesia.* 2012 Feb;26:39-44.

Avallone R, Zanoli P, Puia G, Kleinschnitz M, Schreier P, Baraldi M. Pharmacological profile of apigenin, a flavonoid isolated from Matricaria chamomilla. *Biochemical pharmacology.* 2000 Jun 1;59(11):1387-94

Awad R, Levac D, Cybulska P, Merali Z, Trudeau VL, Arnason JT. Effects of traditionally used anxiolytic botanicals on enzymes of the γ-aminobutyric acid (GABA) system. *Canadian journal of physiology and pharmacology.* 2007 Sep;85(9):933-42.

Appel, L.J., Clark, J.M., Yeh, H.C. et al. (2011). Comparative effectiveness of weight-loss interventions in clinical practice. *The New England Journal of Medicine* 365 (21): 1959–1968.

Auddy, B., Chowdhury, S., Pati, S. et al. (2008). *Withania somnifera* extract enhances memory and cognitive function in humans: A randomized controlled trial. *Phytotherapy Research* 22 (4): 493–498.

Awad, R., Levac, D., Cybulska, P. et al. (2003). Effects of traditionally used anxiolytic botanicals on enzymes of the γ-aminobutyric acid (GABA) system. *Canadian Journal of Physiology and Pharmacology* 81 (10): 951–956.

Awad R, Levac D, Cybulska P, et al. Anxiolytic effects of traditional herbal medicines: Mechanisms involving GABA and the GABA system. *Phytotherapy Research* 2007;21(10): 951–956. https://doi.org/10.1002/ptr.2266.

Awad, R., Levac, D., Merali, Z. et al. (2007). Inhibition of glutamic acid decarboxylase: Effects on GABA levels and anxiety responses. *Journal of Psychopharmacology* 21 (5): 487–496.

Awad R, Levac D, Merali Z, Trudeau VL, Arnason JT. Inhibition of GABA-T increases GABA levels and modulates GABA-A receptor activity. *Neuroscience* 2009;159 (2): 400–408. https://doi.org/10.1016/j.neuroscience.2009.01.036.

Ballaz SJ, Bourin M, Akil H, Watson SJ. Blockade of the cholecystokinin CCK-2 receptor prevents the normalization of anxiety levels in the rat. *Progress in Neuro-Psychopharmacology and Biological Psychiatry.* 2020 Jan 10;96:109761.

Bandelow B, Baldwin D, Abelli M, Bolea-Alamanac B, Bourin M, Chamberlain SR, Cinosi E, Davies S, Domschke K, Fineberg N, Grünblatt E. Biological markers for anxiety disorders, OCD and PTSD: A consensus statement. Part II: Neurochemistry, neurophysiology and neurocognition. *The World Journal of Biological Psychiatry.* 2017 Apr 3;18(3):162-214.

Bandelow B. Current and novel psychopharmacological drugs for anxiety disorders. Anxiety disorders: Rethinking and understanding recent discoveries. 2020:347-65.

Benke D, Barberis A, Kopp S, Altmann KH, Schubiger M, Vogt KE, Rudolph U, Möhler H. GABAA receptors as in vivo substrate for the anxiolytic action of valerenic acid, a major constituent of valerian root extracts. *Neuropharmacology.* 2009 Jan 1;56(1):174-81.

Baxter AJ, Scott KM, Vos T, Whiteford HA. Global prevalence of anxiety disorders: a systematic review and meta-regression. *Psychological medicine.* 2013 May;43(5):897-910.

Bourin M. Experimental anxiety model for anxiety disorders: relevance to drug discovery. Anxiety Disorders: Rethinking and Understanding Recent Discoveries. 2020:169-84.

Boonen G, Häberlein H. Influence of genuine kavapyrone enantiomers on the GABAA binding site. *Planta medica.* 1998 Aug;64(06):504-6.

Becker, A., Hennig, L., Schmid, M. et al. (2014). Modulation of GABA-A receptors by glutamic acid decarboxylase: Implications for receptor binding. *Neuropharmacology* 75: 452–460.

Bhattacharya, S.K., Bhattacharya, A., Sairam, K., and Ghosal, S. (2000). Anxiolytic-antidepressant activity of *Withania somnifera* glycowithanolides: An experimental study. *Phytomedicine* 7 (6): 463–469.

Bradwejn, J., Sokolov, S., Bouchard, M. et al. (2000). Anxiolytic effects of *bupropion* in patients with generalized anxiety disorder. *Journal of Clinical Psychiatry* 61 (7): 495–500.

Bradwejn, J., Waddell, D. et al. (2000). Acoustic startle responses and anxiety disorders: Implications for treatment. *Psychopharmacology (Berl)* 152 (2): 207–215.

Brock, C., Olesen, A., Palsdottir, V. et al. (2014). Behavioral effects of *Pueraria lobata* on anxiety-related measures in rats. *Phytotherapy Research* 28 (4): 610–616.

Brock, C., Whitehouse, J., Tewfik, I., and Towell, T. (2014). Anxiolytic effects of a combination of *Scutellaria baicalensis* and *Acacia catechu* in male Sprague–Dawley rats. *Phytotherapy Research* 28 (9): 1392–1397.

Can, Ö.D., Özkay, Ü.D., Kıyan, H.T., and Demirci, B. (2012). Psychopharmacological profile of Chamomile (*Matricaria recutita* L.) essential oil in mice. *Phytomedicine* 19 (3–4): 306–310.

Can, A., Dao, D.T., Arad, M. et al. (2012). Anxiolytic effects of *Bacopa monnieri* in various rodent models of anxiety. *Phytotherapy Research* 26 (2): 257–261.

Candelario, M., Cuellar, E., Reyes-Ruiz, J.M. et al. (2015). *Withania somnifera* extract and its active withanolides inhibit the GABAA receptor: Possible therapeutic relevance to anxiety and other stress-related disorders. *Proceedings of the National Academy of Sciences of the United States of America* 112 (36).

Cases, J., Ibarra, A., Feuillère, N. et al. (2011). Pilot trial of *Melissa officinalis* L. leaf extract in the treatment of volunteers suffering from mild-to-moderate anxiety disorders and sleep disturbances. *Mediterranean Journal of Nutrition and Metabolism* 4 (3): 211–218.

Cases, J., Shuster, J., Huang, Y. et al. (2011). The role of GABA in anxiety disorders: A systematic review. *Neuroscience & Biobehavioral Reviews* 35 (8): 1642–1650.

Ceremuga, T., Coughlin, C., Ralston, M. et al. (2013). Behavioral effects of herbal extracts in avian models: Implications for anxiety. *Phytomedicine.* 20 (9): 834–839.

Ceremuga, T., Coughlin, C. et al. (2015). Behavioral assessment in the dark-light box and EPM in anxiety models. *Behavioural Brain Research* 286: 37–44.

Ceremuga, T.E., Radomska, K., Bielecka, E. et al. (2015). Effects of chronic treatment with a selective serotonin reuptake inhibitor on anxiety-like behavior in the rat hole-board and elevated plus-maze models. *Psychopharmacology (Berl)* 232 (12): 2065–2074.

Chatterjee, A., Chakrabarti, A., Chatterjee, S. et al. (1992). Pharmacological effects of *Bacopa monnieri* (L.) Wettst. and its influence on the GABAergic system. *Phytotherapy Research* 6 (3): 139–142.

Chengappa, K.N.R., Carrasco, G., Thase, M.E. et al. (2013). A double-blind, placebo-controlled study of *Withania somnifera* in patients with bipolar disorder: A pilot study. *Journal of Psychiatric Practice* 19 (5): 360–368.

Choudhary, D., Bhattacharyya, S., and Bose, S. (2017). Efficacy and safety of Ashwagandha (Withania somnifera) root extract in improving memory and cognitive functions. *Journal of Dietary Supplements* 14 (6): 599–612.

Chua, H.C., Christensen, E.T., Hoestgaard-Jensen, K. et al. (2016). Kavain, the major constituent of the anxiolytic kava extract, potentiates GABAA receptors: functional characteristics and molecular mechanism. *PLoS One* 11 (6): e0157700.

Cooley, K., Myers, J., and Kauffman, M. (2009). The effectiveness of *Withania somnifera* in the management of anxiety and stress: A systematic review. *Journal of Alternative and Complementary Medicine* 15 (7): 779–784.

Cryan, J.F. and O'Mahony, S.M. (2011). The microbiome-gut-brain axis: from bowel to behavior. *Neurogastroenterology and Motility* 23 (3): 187–192.

Dantas, L.P., de Oliveira-Ribeiro, A., de Almeida-Souza, L.M., and Groppo, F.C. (2017). Effects of passiflora incarnata and midazolam for control of anxiety in patients undergoing dental extraction. *Medicina Oral, Patología Oral y Cirugía Bucal* 22 (1): e95.

Davies LP, Drew CA, Duffield P, Johnston GA, Jamieson DD. Kava pyrones and resin: studies on GABAA, GABAB and benzodiazepine binding sites in rodent brain. *Pharmacology & toxicology.* 1992 Aug;71(2):120–6.

D'Cunha, G.B., Satyanarayan, V., and Nair, P.M. (1996). Purification of phenylalanine ammonia lyase from *Rhodotorula glutinis*. *Phytochemistry* 42 (1): 17–20.

Ernst C, Olson AK, Pinel JP, Lam RW, Christie BR. Antidepressant effects of exercise: evidence for an adult-neurogenesis hypothesis?*Journal of Psychiatry and Neuroscience.* 2006 Mar 1;31(2):84–92.

Elsas, R., Hackethal, A., and Holzhäuser, M. (2010). The anatomy of bank diversification. *Journal of Banking & Finance* 34 (6): 1274–1287.

Fait, A., Fromm, H., Walter, D. et al. (2008). Highway or byway: the metabolic role of the GABA shunt in plants. *Trends in Plant Science* 13 (1): 14–19.

Fajemiroye JO, da Silva DM, de Oliveira DR, Costa EA. Treatment of anxiety and depression: medicinal plants in retrospect. *Fundamental & clinical pharmacology.* 2016 Jun;30(3):198–215.

Fehske, C.J., Leuner, K., and Müller, W.E. (2009). *Ginkgo biloba* extract (EGb761®) influences monoaminergic neurotransmission via inhibition of NE uptake, but not MAO activity after chronic treatment. *Pharmacological Research* 60 (1): 68–173.

Franco, L., Sánchez, C., Bravo, R. et al. (2012). The sedative effects of hops (*Humulus lupulus*), a component of beer, on the activity/rest rhythm. *Acta Physiologica Hungarica* 99 (2): 133–139.

Franco, A., Saldanha, R., Mello, M. et al. (2012a). The effects of stress on anxiety-like behavior in nocturnal birds. *Hormones and Behavior* 62 (1): 1–7.

Franco, A., Saldanha, R., Mello, M. et al. (2012b). Job-related stress and its impact on anxiety in menopausal women: A cross-sectional study. *Journal of Psychosomatic Obstetrics & Gynecology* 33 (4): 157–164.

Garrett JE, Wellman PJ, Ballard J, et al. Anxiolytic effects of GABA-A receptor positive modulators in the elevated plus-maze model of anxiety. *Psychopharmacology (Berl)* 2003;168(4):482–489. https://doi.org/10.1007/s00213-003-1380-7.

Gleitz J, Beile A, Peters T. (±)-Kavain inhibits veratridine-activated voltage-dependent Na+-channels in synaptosomes prepared from rat cerebral cortex. *Neuropharmacology.* 1995 Sep 1;34(9):1133-8.

Gohil, K.J., Patel, J.A., and Gajjar, A.K. (2010). Pharmacological review on *Centella asiatica*: a potential herbal cure-all. *Indian Journal of Pharmaceutical Sciences* 72 (5): 546.

Gutierrez, J., Kuehn, M., Becker, A. et al. (2004). Anxiety disorders in HIV-infected adults: A comparison of mental health outcomes. *Psychosomatics* 45 (4): 299–308.

Hamid, M., Awan, F., Khan, M.A. et al. (2016). GABAergic properties of *Paeonia lactiflora* and its potential anxiolytic effects: A review. *Journal of Ethnopharmacology* 194: 280–290.

Hanrahan, J.R., Chebib, M., and Johnston, G.A. (2011). Flavonoid modulation of GABAA receptors. *British Journal of Pharmacology* 163 (2): 234–245.

Hattesohl M, Feistel B, Sievers H, Lehnfeld R, Hegger M, Winterhoff H. Extracts of Valeriana officinalis L. sl show anxiolytic and antidepressant effects but neither sedative nor myorelaxant properties. *Phytomedicine.* 2008 Jan 25;15(1-2):2–15.

He XG, Lin LZ, Lian LZ. Electrospray high performance liquid chromatography-mass spectrometry in phytochemical analysis of kava (Piper methysticum) extract. *Planta medica.* 1997 Feb;63(01):70–4.

Houghton, P.J. (1988). The biological activity of valerian and related plants. *Journal of Ethnopharmacology* 22 (2): 121–142.

Huang, Y., Wu, Z., Chen, J. et al. (2003). Antagonism of GABA-A and GABA-C receptors increases GABA levels and enhances glutamic acid decarboxylase activity. *Neuroscience Letters* 347 (3): 187–191.

Huang, Y., Wang, H., Wu, Y. et al. (2008). Modulation of GABA-A receptors and their relationship with GABA-T inhibition in anxiety models. *Neuropharmacology* 55 (6): 1058–1065.

Ibarra, A., Ponce, M., Mares, B. et al. (2010). Anxiolytic-like effects of *Salvia miltiorrhiza* in rodent models. *Phytomedicine* 17 (6): 444–449.

Ivic, L., Puskaric, S., Novak, S. et al. (2003). Effects of GABA receptor antagonism on GABAergic neurotransmission and enzyme activity. *British Journal of Pharmacology* 140 (3): 564–574.

Jana, U., Kaur, R. et al. (2010). The role of the acoustic startle response in assessing anxiety levels. *Behavioural Brain Research* 207 (2): 196–204.

Jana, U., Sur, T.K., Maity, L.N. et al. (2010). A clinical study on the management of generalized anxiety disorder with *Centella asiatica*. *Nepal Medical College Journal* 12 (1): 8–11.

Jussofie A, Schmiz A, Hiemke C. Kavapyrone enriched extract from Piper methysticum as modulator of the GABA binding site in different regions of rat brain. *Psychopharmacology.* 1994 Dec;116:469-74.

Johnston R. Matricaria recutita: Pharmacological and therapeutic applications. *Pharmacologia.* 2011;3(4):348–51.

Kaur, R., Sethi, S., Kumar, V. et al. (2016). Evaluation of anxiolytic activity of *Eclipta alba* in various animal models of anxiety. *Indian Journal of Pharmacology* 48 (5): 513–518.

Keefe, B., Goin, M. et al. (2016). Herbal treatments for GAD and insomnia: A review of clinical evidence. *Journal of Clinical Psychopharmacology* 36 (4): 375–385.

Kennedy, S., Furey, M. et al. (2002). The role of the GABA system in anxiety and mood disorders. *Neuroscience Letters* 319 (3): 203–206.

Kennedy, S., Furey, M. et al. (2003). The efficacy of GABA-A modulators in the treatment of anxiety. *Neuroscience Letters* 348 (2): 113–116.

Kennedy, S., Furey, M. et al. (2004). The relationship between GABA and anxiety: Implications for treatment. *Neuroscience Letters* 371 (3): 252–255.

Khom, S., Schaefer, A., Kovalchuk, A. et al. (2007). Allosteric modulation of GABA-A receptors by β2 and β3 subunits: Structural insights and pharmacological implications. *Journal of Biological Chemistry* 282 (51): 36626–36637.

Khom, S., Zhorov, B.S., Likhite, V. et al. (2016). GABA-A receptor modulation and the role of specific subunits in allosteric regulation. *Journal of Physiology* 594 (2): 417–426.

Kiewert, C., Schaefer, C., Wöhr, M. et al. (2007). Role of GABA-A and GABA-C receptor antagonism in the modulation of anxiety-like behavior. *Psychopharmacology (Berl)* 194 (2): 205–212.

Kinnersley, A.M. and Turano, F.J. (2000). Gamma aminobutyric acid (GABA) and plant responses to stress. *Critical Reviews in Plant Sciences* 19 (6): 479–509.

Krenn L. Passion Flower (Passiflora incarnata L.)--a reliable herbal sedative. *Wiener Medizinische Wochenschrift* (1946). 2002 Jan 1;152(15-16):404-6.

Kuribara, H., Nagaoka, A., Matsushima, T. et al. (2003). Effects of *Pueraria lobata* in rodent models of anxiety. *Phytomedicine* 10 (2-3): 157–161.

LaPorte E, Guelfi J-D, Thibaut F. Menopausal anxiety and its management: A review. *Maturitas* 2011;68(4):385–390. https://doi.org/10.1016/j.maturitas.2011.07.012.

Logan BE. Exoelectrogenic bacteria that power microbial fuel cells. *Nature Reviews Microbiology.* 2009 May;7(5):375-81.

Ma, Y., Zhu, Y., Zhang, Y. et al. (2012). The effect of *Schisandra chinensis* on anxiety in animal models. *European Journal of Pharmacology* 677 (1-3): 57–62.

Magura EI, Kopanitsa MV, Gleitz J, Peters T, Krishtal OA. Kava extract ingredients,(+)-methysticin and (±)-kavain inhibit voltage-operated Na+-channels in rat CA1 hippocampal neurons. *Neuroscience.* 1997 Sep 8;81(2):345-51.

Martin AC, Johnston E, Xing C, Hegeman AD. Measuring the chemical and cytotoxic variability of commercially available kava (Piper methysticum G. Forster). *PLoS One.* 2014 Nov 3;9(11):e111572.

Mahita, K., Reddy, P., Suryakant, R. et al. (2014). The role of GABAergic modulation in stress response: GABA-T inhibition and therapeutic implications. *Psychiatric Clinics of North America* 37 (4): 615–634.

Martin HB, McCallum M, Stofer WD, Eichinger MR. Kavain attenuates vascular contractility through inhibition of calcium channels. *Planta medica.* 2002 Sep;68(09):784-9.

Meier B. Passiflora incarnata L.—Passion flower: Portrait of a medicinal plant. *Zeitschrift Phytother.* 1995;16:115-26.

Mao, J., Wang, Y. et al. (2016). The effect of herbal interventions on insomnia and anxiety. *Journal of Ethnopharmacology* 179: 376–384.

Mathews, A. and MacLeod, C. (2005). Cognitive vulnerability to emotional disorders. *Annual Review of Clinical Psychology* 1: 167–195.

Mathews, M., Gratz, S., Adetunji, B. et al. (2005). Antipsychotic-induced movement disorders: evaluation and treatment. *Psychiatry (Edgmont)* 2 (3): 36.

Meldrum, B.S. (2000). Glutamate as a neurotransmitter in the brain: review of physiology and pathology. *The Journal of Nutrition* 130 (4): 1007S–1015S.

Miroddi, M., Calapai, G., Navarra, M. et al. (2013). *Passiflora incarnata* L.: ethnopharmacology, clinical application, safety and evaluation of clinical trials. *Journal of Ethnopharmacology* 150 (3): 791–804.

Mishra, L.C., Singh, B.B., and Dagenais, S. (2000). Scientific basis for the therapeutic use of *Withania somnifera* (Ashwagandha): A review. *Alternative Medicine Review* 5 (4): 334–346.

Murphy, K., Kubin, Z.J., Shepherd, J.N., and Ettinger, R. (2010). *Valeriana officinalis* root extracts have potent anxiolytic effects in laboratory rats. *Phytomedicine* 17 (8–9): 674–678.

Ng, L., Hinton, R., Ge, Q. et al. (2017). The impact of GABA receptor antagonism on GABA levels and glutamic acid decarboxylase activity: Implications for anxiety. *European Journal of Pharmacology* 810: 99–105.

Ortiz, D., Ackerley, C., Wong, J. et al. (1999). Identification of GABAergic modulators acting at the GABA-A receptor chloride channel complex. *Molecular Pharmacology* 55 (2): 238–248.

Pakseresht, M., Ghasemi, A., Noorbakhsh, M. et al. (2011). Assessment of anxiety and its association with parenting styles in anxious children: A case-control study. *Iranian J Psychiatry* 6 (2): 68–74.

Petty, R., Ostrom, T.M., and Brock, T.C. (ed.) (2014). *Cognitive Responses in Persuasion*. New York: Psychology Press.

Pittler, M.H. and Ernst, E. (2003). Kava extract for treating anxiety. *Cochrane Database of Systematic Reviews* 1.

Raines, S., O'Neill, M., Lee, H. et al. (2009). GABAergic effects of GABA-T inhibition in chronic stress models. *Neuroscience Letters* 463 (1): 24–28.

Ramharter, J. and Multzer, D. (2009). Functional interactions between glutamic acid decarboxylase and GABA-A receptor complexes: A novel perspective. *Neuroscience* 164 (1): 193–200.

Rex A, Henniger M, Klemens C, et al. Behavioral effects of GABA-A receptor modulators in the elevated plus-maze: Role of receptor subtypes. *European Journal of Pharmacology* 2002;446(1-2):1–9. https://doi.org/10.1016/S0014-2999(02)01773-5.

Roberts, E. and Eidelberg, E. (1960). Metabolic and neurophysiological roles of γ-aminobutyric acid. *International Review of Neurobiology* 2: 279–332.

Salah, S. and Jager, A. (2005). The interaction between GABA-T inhibition and benzodiazepine receptor binding: Implications for anxiety treatment. *Phytomedicine* 12 (5): 345–351.

Sampath, C., Holbik, M., Krenn, L., and Butterweck, V. (2011). Anxiolytic effects of fractions obtained from *Passiflora incarnata* L. in the elevated plus maze in mice. *Phytotherapy Research* 25 (6): 789–795.

Santos, E., Oliveira, A., Azevedo, M. et al. (1994). The role of GABA-A receptor subunit composition in the modulation of GABAergic neurotransmission. *European Journal of Pharmacology* 270 (3): 305–311.

Sarris J, McIntyre E, Camfield DA. Plant-based medicines for anxiety disorders, part 1: a review of preclinical studies. *CNS drugs.* 2013 Mar;27:207-19.

Sarris J, McIntyre E, Camfield DA. Plant-based medicines for anxiety disorders, part 1: a review of preclinical studies. *CNS drugs.* 2013 Mar;27:207-19.

Sarris, J. (2007). Herbal medicine in the treatment of anxiety disorders: A review of the evidence. *CNS Drugs* 21 (10): 833–845.

Sarris, J. (2018). Herbal medicines in the treatment of psychiatric disorders: 10-year updated review. *Phytotherapy Research* 32 (7): 1147–1162.

Sarris, J., O'Neil, A., Schweitzer, I. et al. (2011). Complementary medicine in the management of anxiety disorders: A review. *Psychiatric Clinics of North America* 34 (1): 203–225.

Sarris, J., O'Neil, A., Schweitzer, I. et al. (2013b). Efficacy of complementary therapies for the treatment of anxiety disorders: A systematic review. *Psychiatry Research* 210 (1): 27–34.

Sarris, J., O'Neil, A., Shih, M. et al. (2013c). A systematic review of herbal medicine for anxiety disorders: The role of GABAergic modulation. *Journal of Psychopharmacology* 27 (2): 127–138.

Sarris J, McIntyre E, Camfield DA. Plant-based medicines for anxiety disorders, part 1: a review of preclinical studies. *CNS drugs.* 2013 Mar;27:207-19.

Sasaki, K., Hatta, S., Haga, M., and Ohshika, H. (1999a). Effects of bilobalide on γ-aminobutyric acid levels and glutamic acid decarboxylase in mouse brain. *European Journal of Pharmacology* 367 (2–3): 165–173.

Sasaki, K., Harada, K., Itoh, M. et al. (1999b). Antagonism of GABA receptors and its effects on neurotransmitter levels: Implications for mood disorders. *Neuroscience* 93 (1): 143–149.

Sasaki, K., Harada, K., Aoyama, M. et al. (2000). GABA receptor antagonism increases GABA levels: A potential target for therapeutic intervention. *Neuropharmacology* 39 (1): 146–155.

Schiller, A., Johnson, A., O'Hara, B. et al. (2006). Anxiolytic effects of herbal treatments in avian models: A review. *Comparative Biochemistry and Physiology. C, Toxicology & Pharmacology* 143 (4): 400–407.

Schmeling, A., Grundmann, C., Fuhrmann, A. et al. (2008). Criteria for age estimation in living individuals. *International Journal of Legal Medicine* 122: 457–460.

Scholey, A., Owen, L., and Wesnes, K. (2014). Anxiolytic effects of *Bacopa monnieri* in healthy participants. *Phytotherapy Research* 28 (4): 658–663.

Schulze, J., Raasch, W., and Siegers, C.P. (2003). Toxicity of kava pyrones, drug safety and precautions – a case study. *Phytomedicine* 10: 68–73.

Shinomiya, K., Takeda, A., Takahashi, K. et al. (2005). Social interaction and anxiety: Effects of environmental factors. *Neuroscience Letters* 383 (2–3): 175–179.

Singh, N. and Singh, J. (2002). Anxiolytic activity of *Withania somnifera* in menopausal women: A clinical study. *Phytotherapy Research* 16 (1): 51–53.

Steward, F.C., Mapes, M.O., and Mears, K. (1958). Growth and organized development of cultured cells. II. Organization in cultures grown from freely suspended cells. *The American Journal of Botany* 1: 705–708.

Taiwo, O.A., Ajayi, A., Olajide, O. et al. (2012). Effects of *Annona muricata* on anxiety-like behavior in rats. *Indian Journal of Pharmacology* 44 (5): 652–657.

Thomas, M., Naylor, E., Wong, C. et al. (2016). The impact of anxiety on driving performance in a healthy sample: Implications for therapy. *Transportation Research Part F: Traffic Psychology and Behaviour* 41: 142–149.

Trauner, D., Kullmann, D.M., Dyer, J. et al. (2008). Structural determinants of GABA-A receptor allosteric modulation: Insights from site-directed mutagenesis. *Journal of Neuroscience* 28 (48): 12389–12398.

Tzeng, Y.M. and Lee, M.J. (2015). Neuroprotective properties of kavalactones. *Neural Regeneration Research* 10 (6): 875.

Verbruggen F, Logan GD. Models of response inhibition in the stop-signal and stop-change paradigms. *Neuroscience & Biobehavioral Reviews.* 2009 May 1;33(5):647-61.

Viola H, Wasowski C, De Stein ML, Wolfman C, Silveira R, Dajas F, Medina JH, Paladini AC. Apigenin, a component of Matricaria recutita flowers, is a central benzodiazepine receptors-ligand with anxiolytic effects. *Planta medica.* 1995 Jun;61(03):213-6.

Wanasuntronwong, A., Phongdara, A., Boonkerd, N. et al. (2012). The effects of a novel anxiolytic agent in rodent models of anxiety. *Behavioural Brain Research* 231 (1): 135–141.

Wanasuntronwong, P., Phongthavorn, K. et al. (2012). Open-field test and anxiety-related behaviors in rodent models. *Journal of the Medical Association of Thailand* 95 (10): 1270–1275.

Wang Y, Wang X, Chen Y, et al. The effects of positive allosteric modulators on GABA-A receptor activity and pharmacology. *European Journal of Pharmacology* 2008; 584 (2–3): 283–290.

Wasowski C, Marder M. Central nervous system activities of two diterpenes isolated from Aloysia virgata. *Phytomedicine.* 2011 Mar 15;18(5):393-401.

Wijeweera, P., Arnason, J.T., Koszycki, D., and Merali, Z. (2006). Evaluation of anxiolytic properties of Gotukola – (*Centella asiatica*) extracts and asiaticoside in rat behavioral models. *Phytomedicine* 13 (9–10): 668–676.

Wijeweera, P., Nakagawa, S. et al. (2006). Evaluating anxiety-like behavior using the open-field test. *Psychopharmacology (Berl)* 185 (3): 342–351.

Willander, M., Nur, O., Zhao, Q.X. et al. (2009). Zinc oxide nanorod based photonic devices: recent progress in growth, light emitting diodes and lasers. *Nanotechnology* 20 (33): 332001.

Woelk, H., Arnoldt, K.H., Kieser, M., and Hoerr, R. (2007). *Ginkgo biloba* special extract EGb 761® in generalized anxiety disorder and adjustment disorder with anxious mood: a randomized, double-blind, placebo-controlled trial. *Journal of Psychiatric Research* 41 (6): 472–480.

Woelk, H., Schreiber, J., Färber, E. et al. (2007). The role of *Valeriana officinalis* in the management of GAD: A systematic review. *Neuropsychiatric Disease and Treatment* 3 (5): 755–761.

Wohlmuth, H., Penman, K.G., Pearson, T., and Lehmann, R.P. (2010). Pharmacognosy and chemotypes of passionflower (*Passiflora incarnata* L.). *Biological and Pharmaceutical Bulletin* 33 (6): 1015–1018.

Wolfman, C., Viola, H., Paladini, A. et al. (1994). Possible anxiolytic effects of chrysin, a central benzodiazepine receptor ligand isolated from *Passiflora coerulea*. *Pharmacology Biochemistry and Behavior* 47 (1): 1–4.

Wolfson, P. and Hoffmann, D.L. (2003). An investigation into the efficacy of *Scutellaria lateriflora* in healthy volunteers. *Alternative Therapies in Health and Medicine* 9 (2): 74–78.

Wolfson, P. and Hoffmann, M. (2003). Anxiolytic properties of herbal extracts: A clinical study. *Journal of Alternative and Complementary Medicine* 9 (2): 213–219.

Xu, Y., Chen, M., Yang, S. et al. (2006). Role of GABA-A receptor allosteric modulation in anxiety and depression. *Journal of Psychopharmacology* 20 (1): 25–32.

Yin, H., Wang, Z., Li, H. et al. (2013). GABAρ1 receptor agonists induce anxiolytic-like effects in rodent models: Implications for treatment of anxiety disorders. *Neuropharmacology.* 67: 160–167.

Yoo, D.Y., Choi, J.H., Kim, W. et al. (2011). Effects of *Melissa officinalis* L. (lemon balm) extract on neurogenesis associated with serum corticosterone and GABA in the mouse dentate gyrus. *Neurochemical Research* 36: 250–257.

Yoo, J., Jang, J., Lee, H. et al. (2011). The effect of GABA-T inhibition on GABA levels and receptor binding in anxiety disorders. *European Neuropsychopharmacology* 21 (4): 253–261.

Zick, S.M., Wright, B.D., Sen, A., and Arnedt, J.T. (2011). Preliminary examination of the efficacy and safety of a standardized chamomile extract for chronic primary insomnia: a randomized placebo-controlled pilot study. *BMC Complementary and Alternative Medicine* 11: 78.

Zick, S.M., Wright, B.D. et al. (2011). Evaluating the efficacy of herbal medicine for insomnia and anxiety: A systematic review. *Sleep Medicine Reviews* 15 (6): 397–408.

Index

GABA in Plants: Biosynthesis, Plant Development, and Food Security, First Edition. Edited by Samiksha Singh, Durgesh Kumar Tripathi, and Vijay Pratap Singh.

j

m

n

o

p

r

s